A. D.	
150	Ptolemy recognizes spherical shape of earth and orbits of planets, and draws maps; Hipparchus constructs first table of trigonometric rates.
250	Diophantus writes *Arithmetic,* using symbols for the variables.
825	Al-Khwarizmi (Arab) writes first algebra book and introduces Hindu numerals with place value.
1150	Bhaskara (Hindu) proposes a system of multiplication.
1200	Fibonacci (Italian) writes *Liber Quadratorum,* summarizing arithmetic and solutions of equations.
1250	Contest in Norman court for solving $x^3 + 2x^2 + 10x = 0$.
1340	Planudes (Greek monk) invents long-division algorithm.
1400	First mathematics book printed: Euclid's *Elements* (1482); Columbus majors in mathematics and science at University of Pavia; first mathematics book using "+" and "−" appears (1489). Pacioli writes first book with modern method of multiplication (1494).
1500	Copernicus and Brahe extend knowledge of astronomy.
1550	Cardan (Italian) solves cubic equations; Robert Recorde (English) writes arithmetic texts; Gregorian calendar established (1582); Stevin (Belgian) introduces decimal fractions (1585).
1600	Science explodes as result of men such as Galileo and Kepler and measuring instruments such as thermometer (1597), telescope (1609), barometer (1643), microscope (1650); Napier (English) invents logarithms (1614); Descartes (French) invents analytical geometry.
1650	Pascal (French) invents calculating machine, extends projective geometry, and studies probability; Fermat (French) extends theory of numbers.
1700	Euler (Swiss), an expert in analysis of problems, Gauss (German), one of the three top mathematicians of all time, extends many fields—especially number theory.
1800	Lobachevski (Russian), Bolyai (Hungarian), and Reimann (German) develop non-euclidean geometry; Galois (French) establishes group theory; Abel (Norwegian) proves binominal expansion; Peano (Italian) establishes postulates for natural number system; Boole (English) invents the algebra of symbolic logic.
1900	Cantor (Dane) invents the theory of sets and ways to deal with infinities; Poincaré (French) creates ideas in differential equations and function theory; Hilbert (German) establishes a new structure for geometry; Einstein (German) creates the theory of relativity; Von Neumann (Hungarian) invents the theory of games and develops computer science.

Guidelines for Teaching Mathematics

Second Edition

Guidelines for Teaching Mathematics

Second Edition

Donovan A. Johnson
University of Minnesota

Gerald R. Rising
State University of New York at Buffalo

Wadsworth Publishing Company, Inc.
Belmont, California

Designer: Russell K. Leong

Editor: Ellen Seacat

ISBN-0-534-00189-0

L. C. Cat. Card No. 72-85027

Printed in the United States of America

3 4 5 6 7 8 9 10 . . . 76 75 74

Contents

PART FIVE: Use of Teaching Aids

PART SIX: The Future

Preface

This revised edition follows the general plan of the first edition, whose success has encouraged us to believe that we are describing effective ways to teach mathematics. We have, however, drawn on our own experiences over the last five years and have responded to the many suggestions of virtually hundreds of correspondents in making revisions. Three entirely new chapters (3, 5, and 7) have been added. Others have been reordered. For example, Part Two is placed early in the text because many teachers use the book as a guide when they enter schools as teachers for the first time. Chapter 5, in particular, is addressed to practical matters faced by the beginner. It is included—without apology—because beginners say it is helpful and *is not provided elsewhere*. Every chapter has been carefully reviewed; none appears unchanged.

Mathematics teachers face a multitude of decisions every day. We must decide what to teach, how to teach it, and how much emphasis to give certain ideas. We must decide what materials and activities are appropriate for students with different interests, abilities, and goals. In addition, we have to be able to evaluate the effectiveness of our own instruction. A successful mathematics teacher must have a broad background in mathematics and must be able to communicate his knowledge to students. He must use enlightened examples, appropriate anecdotes, and challenging activities. His students must be drawn into the cooperative exploration of new ideas. Importantly, his students must be instilled with some of his own love of teaching, as well as of mathematics itself.

When mathematics teachers are asked what their greatest difficulties are, they invariably name motivation, individual differences, and discipline. These problems cannot be solved by any textbook. They must be faced and dealt with by the teacher in the classroom. But when mathematics is properly and thoughtfully taught, many of these teacher problems are resolved. For exactly this reason we emphasize pedagogy designed to make teaching and learning a successful adventure.

In this book, we attempt to provide a framework on which the mathematics teacher can build his teaching activities. First, we set the stage by considering the foundations for mathematics instruction: the current status of the mathematics curriculum, some psychological–philosophical bases, and the goals and objectives of mathematics instruction. Next, we turn to the highly pragmatic problems of day-to-day classroom teaching, focusing first on the specific problems of beginners, and even experienced teachers, then on classroom management, lesson planning, and constructing evaluation tools, and finally on the master teacher. A third section explores in detail instructional activities related to specific aspects of the mathematics program as they are found at all levels of instruction. Part Four is devoted to ways to provide for individual learning differences. Instructional aids and how to use them effectively are discussed in the next section, and in the concluding chapter we predict the future of mathematics instruction.

In writing this text, we have been concerned with the balance between discussions of content and pedagogy. There are many fine books that discuss the specifics of mathematical content, but there are few that concern themselves with problems of presentation of that content. While both content and pedagogy are important, we emphasize the latter here in an attempt to fill the gap. A glance at the text will show that there is a great deal of mathematical content in these pages; here, however, the content provides the examples for pedagogical principles, rather than the basis for the presentation. Therefore, we strongly recommend that you supplement the study of this text with intensive study of several of the fine modern secondary school texts and teachers' manuals. Such activity will bring into sharper focus our more general discussion.

You, the reader, must adapt our suggestions to your own individual interests, personality, and methods. You must supplement and, at the same time, temper our point of view with your own. You must always utilize additional resources—the library, the audio-visual department, local teachers and supervisors—whenever appropriate and possible. One other new feature of this edition is the basic reference list immediately before Chapter 1. Chapter references and problem sets refer *only* to these resources. Colleges will wish to place these books on reserve shelves; schools may wish to draw from this list books for a mathematics teaching reference library.

In the first edition of *Guidelines for Teaching Mathematics* we wrote, "What has gone into these pages is the result of two people drawing on their experiences and those of their many friends and co-workers. In many ways, therefore, we think of this book as a continuing discussion." As we hoped, that discussion has flourished. We express here our thanks to those teachers and students who have made so many fine contributions to our thinking. We regret only that we cannot credit them individually. All readers are invited to enlarge this interchange.

Finally, we offer our best wishes for success in the very important task you have accepted—teaching mathematics.

D. A. J.
G. R. R.

Reference Tools

We recommend that these books form a minimum library closed reserve for reference use with this text. The books also represent a minimum secondary school professional library. References in this text are largely restricted to these books and journals.

Journals

The Mathematics Teacher
The Arithmetic Teacher
National Council of Teachers of Mathematics (1201 Sixteenth Street, N.W., Washington, D. C. 20036)
School Science and Mathematics
Association of Science and Mathematics Teachers (P. O. Box 246, Bloomington, Indiana 47401)

Yearbooks

National Council of Teachers of Mathematics
21 *Learning of Mathematics: Its Theory and Practice*, 1953.
24 *The Growth of Mathematical Ideas K-12*, 1959
26 *Evaluation in Mathematics*, 1961
27 *Enrichment Mathematics for the Grades*, 1963
28 *Enrichment Mathematics for High School*, 1963
32 *A History of Mathematics Education*, 1970
Readings in the History of Mathematics Education, 1970
33 *The Teaching of Secondary School Mathematics*, 1970
National Society for the Study of Education (University of Chicago Press)
69 (Part I) *Mathematics Education*, 1970

Other Books

Aichele, Douglas B. and Robert E. Reys, eds., *Readings in Secondary School Mathematics* (Boston: Prindle, Weber and Schmidt, 1971)

Bassler, Otto C. and John R. Kolb, *Learning to Teach Secondary School Mathematics* (Scranton, Pa.: Intext, 1971)

Bloom, Benjamin S., V. Thomas Hastings, and George F. Madaus, *Handbook on Formative and Summative Evaluation of Student Learning* (New York: McGraw-Hill, 1971)

Bruner, Jerome, *The Process of Education* (Cambridge, Harvard Univ. Press, 1960)

Butler, Charles H., F. Lynwood Wren, and J. Houston Banks, *The Teaching of Secondary Mathematics* (New York: McGraw-Hill, 1970)

Fawcett, Harold P. and Kenneth B. Cummins, *The Teaching of Mathematics from Counting to Calculus* (Columbus, Ohio: Charles B. Merrill, 1970)

Fremont, Herbert, *How to Teach Mathematics in Secondary Schools* (Philadelphia: W. B. Saunders, 1969)

Kidd, Kenneth P., Shirley S. Myers, and David M. Cilley, *The Laboratory Approach to Mathematics* (Chicago: Science Research Associates, 1970)

McIntosh, Jerry A., *Perspectives on Secondary Mathematics Education* (Englewood Cliffs, N. J.: Prentice-Hall, 1971)

Polya, Georg, *Mathematical Discovery,* Volumes I and II (New York: Wiley, 1962, 1963)

———, *How to Solve It* (New York: Doubleday Anchor, 1957)

Rising, Gerald R. and Richard Wiesen, *Mathematics in the Secondary School Classroom* (New York: Thomas Y. Crowell, 1972)

Sobel, Max A., *Teaching General Mathematics* (New York: Prentice-Hall, 1967)

Holt, John, *How Children Fail* (New York: Dell, 1964)

Röde, Lennart, ed., *The Teaching of Probability and Statistics* (New York: Wiley, 1970)

Cundy, A. M. and H. P. Rollet, *Mathematical Models* (New York: Oxford University Press, 1961)

Willoughby, Stephen S., *Contemporary Teaching of Secondary School Mathematics* (New York: Wiley, 1967)

Textbooks

As wide a selection as possible of contemporary secondary school mathematics textbooks—both commercial and experimental.

Foundations of Mathematics Education

ONE

Learning to Teach Mathematics 1

The teaching of mathematics is a challenging, exciting adventure. It has its dangers, successes, discouragements, and delights. Both its difficulties and its pleasures are derived from the subject matter, the student, and the classroom situation. The real satisfaction of teaching mathematics comes from the fact that we are teaching the subject we enjoy, to individuals who are important, in a way that we find effective.

As mathematics teachers we should be grateful for the opportunity to work in a professional field where the scope and pace of change are truly astounding. We should enjoy the opportunity to guide the learning of our future scientists and citizens. We should be excited about teaching a subject that is as remarkable as mathematics is. But, at the same time, we should be disappointed with the limited learning that takes place amid the daily classroom stresses.

The Teaching Process

Learning how to be an effective teacher is complex—even more difficult than learning mathematics, complicated as that sometimes is. The flow chart of figure 1.1 illustrates the many factors involved in the teaching of mathematical ideas. These aspects provide an outline for the text and will be discussed in depth in the later chapters of this book.

As the chart shows, the teaching program is dependent on previous instruction. The teacher's first step is to draw upon his knowledge of that previous instruction and upon the general goals of education (preparing for citizenship, preparing for college, advancing our society, establishing personal values) to develop *objectives* for the particular course unit or topic. These more specific objectives direct us to what the learners should do (skills), know (concepts), and feel (attitudes).

The next step is to determine appropriate *strategy* to implement instruction of the particular mathematical ideas. Both mathematical strategies (for example, what algorithm should be used) and pedagogical strategies (for example, what general teaching style best fits the topic) should be considered.

On the basis of this, both general and specific *plans* for carrying out the classroom instruction are developed. These detail the methods and materials for use with students.

These plans are then *implemented* in the classroom. This is of course the live action. You and your students work together to carry out the plans you have laid.

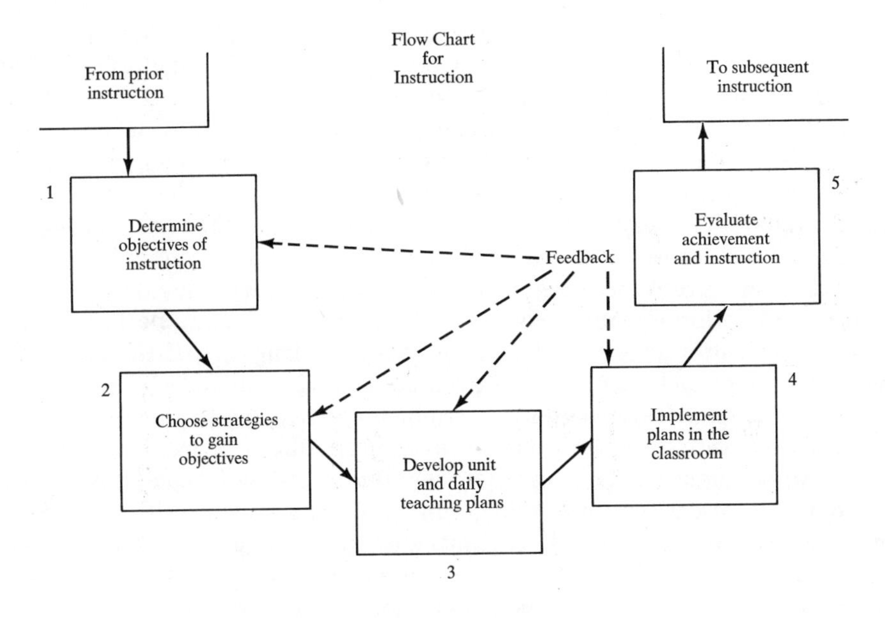

Figure 1.1

You must be ready to modify your organization on the spot to take care of unexpected opportunities or things you failed to foresee. You must modify not only the presentation itself but also the pace of your dialogue with your students.

Then you must *evaluate* your work. What content was appropriate, and what teaching techniques were successful? Your evaluation should not be based just on the evidence of tests, important as you may consider them. Your own subjective response should also be taken into account. This evaluation should provide you with feedback to revise the cycle just completed and to plan for the future.

The Power of Mathematics

Mathematics is a creation of the human mind, concerned primarily with ideas, processes, and reasoning. Thus, mathematics is much more than arithmetic, the science of numbers and computation; more than algebra, the language of symbols and relations; more than geometry, the study of shape, size, and space. It is greater than numerical trigonometry, which measures distances to stars and analyzes oscillations. It encompasses more than statistics, the science of interpreting data and graphs; more than calculus, the study of change, infinity, and limits.

Primarily, *mathematics is a way of thinking*, a way of organizing a logical proof. It can be used to determine whether or not an idea is true or at least whether it is probably true. As a way of thinking, it is used to solve all kinds of problems in the sciences, government, and industry. As a way of reasoning, it gives us insight into the power of the human mind and becomes a challenge to intellectual curiosity.

Mathematics is also a language, a language that uses carefully defined terms and concise symbolic representations, which add precision to communication. It is a language that works with ideograms, symbols for ideas, rather than phonograms, symbols for sounds. The equation $3 + 5 = 8$ means the same to a Swede, a Russian, or a Japanese, no matter how he reads it. Furthermore, because of their clarity and precision, ideograms serve as mental laborsaving devices. They enable us to perform computations, solve problems, and complete proofs that would be difficult if not impossible in any "natural" language. These ideograms make algorithms and manipulations accurate and efficient.

Mathematics is an organized structure of knowledge in which each proposition is deduced logically from previously proved propositions or assumptions. The structures of mathematics, somewhat like the structures of philosophy and theology, are logical structures that begin with certain undefined terms. These undefined primitive terms, such as point, line, and plane in geometry, are used to describe essential ideas. Certain assumptions—called axioms, postulates, properties, or laws—are made about these ideas or operations. In a sense, the primitive terms are "defined" by these postulates. Additional terms are then defined by means of the primitive terms and postulates. And, finally, these terms and assumptions are used to prove theorems. When one understands this basic organization of mathematical structures, he can more readily study new mathematical schema.

Mathematics is also the study of patterns—that is, of any kind of regularity in form or idea. Radio waves, molecular structures, sequences of numbers, orbits of celestial bodies, and the shape of a bee's cell—all have patterns that can be described and analyzed mathematically.

Mathematics is finally an art. As in any other art, beauty in mathematics consists in order and inner harmony. The mathematician tries to express a maximum number of ideas and relations with the greatest economy of means. The beauty of mathematics can be found in the process whereby a chaos of isolated facts is transformed into logical order. The exploration of new ideas and the invention of new mathematical structures are challenges to the creativeness, the imagination, and the intuition of the mathematician.

What is the relationship of mathematics to the other sciences? Not only its content but also its method is different. The method of the physical sciences, such as physics, chemistry, and biology, is that of induction or experimentation. Scientific induction is a method of reasoning from the particular to the general. A physicist, for example, may be interested in finding out the effect of tension upon a bar. To do this, he experiments with large numbers of metal bars and observes their behavior under stress. He draws conclusions based on what he sees.

On the other hand, a mathematician does not accept a generalization based only on observation. He must develop hypotheses based on his observations, but he further requires proof based upon a logical scheme of deduction. From properties that characterize his system, he sets out to prove that his conclusions are true or have some probability value. Once he has done this, he has confidence that his deduction is always true when the conditions fit his properties. In this fashion, mathematics frees itself of the restriction of the inductive method and becomes more universally applicable.

The Art of Teaching Mathematics

Teaching, as the great teacher William Lyon Phelps says in his autobiography, is "an art so great and so difficult to master that a man or a woman can spend a long life at it, without realizing much more than his limitations and mistakes, and his distance from the idea." Ideally, as Georg Polya points out, the teacher's "art" requires the acting of drama, the repetitions and variations of music, the elegance of poetry, and the originality of a painting. How can a mathematics teacher acquire this art? How does he go about this task of becoming a truly creative teacher?

To be creative in any art, one must have a broad background in technique, materials, and experience. The artist must also know the person or situation he is portraying or the subject he is painting. Similarly, the art of teaching mathematics requires a strong background in the subject, in techniques of presenting ideas, and in the materials available. The mathematics teacher needs a set of professional tools from which he chooses the one he and his students enjoy or the one which he thinks is most effective. Of course, one does not become a master of the art without practice and experience. Years alone do not provide this experience: some teachers have one year of experience repeated many times.

Most people are agreed that a first requirement for success in teaching mathematics is knowing mathematics. If we are to teach mathematics so that it is understood, so that it makes sense, so that it can be applied, we must have an adequate background in mathematical content. We must not be an articulate reservoir of misinformation. We need a reservoir of applications, historical side-light, unusual problems—yes, even tricks and puzzles—if we are to build appreciations, curiosity, and loyalty to mathematics. We must have a broad background in order to place the mathematics we teach into its wider context. We must also like mathematics. If we do not enjoy mathematics, if we do not enjoy learning more mathematics, we can never expect our students to have confidence in us or to be excited about learning mathematics themselves. Knowing mathematics then is necessary for success as a mathematics teacher, necessary but not sufficient.

A second quality of equal importance is an understanding and acceptance of students: their interests and needs, their difficulties and abilities, the ways they best learn mathematical concepts. Here we recognize that human beings are more complex, more delicate, more intricate than mathematics.

Today's adolescent is more sophisticated, more independent, and better informed than ever before. He has seen more things, done more things, and read more books than the teen-ager of a generation ago. His world is broader, his search is deeper, his frustrations greater than before. It is time to listen to what he is saying. He is likely to have something worth while to say. He may be saying to us:

"You tell us how to solve mathematical problems. We want to learn how to solve our personal problems."

"You tell us to solve our mathematical problems with logic, yet most of our classroom problems are met with emotion."

"You tell us to learn to compute accurately and quickly, but you make lots of mistakes."

"You tell us to enjoy learning mathematics, but I haven't ever seen you studying and enjoying a mathematics book."

"You tell us to be neat in our work, but I don't think your board work is very neat."

"You tell us that equations can be used to describe lines. Can you write the equation of your face?"

"You ask us questions but you don't wait for an answer."

"You tell us algebra is important, but my father hasn't used it since he was in school."

Teaching mathematics, then, involves more than knowing and enjoying the subject. The mathematics teacher must motivate his students, he must communicate his knowledge to them, and he must guide them to discover ideas. Methods and materials serve as links between the knowledge of subject matter, principles of psychology, and actual practice in the classroom. Creative teaching requires new materials and thoughtful activities that have often been neglected in the mathematics classroom.

Mathematics, with its abstract symbolism, its logical structures, its wide application, has unique learning problems. At one extreme, it involves learning simple skills, calculation, facts, and procedures where memory and practice are most essential. At the other extreme—the analysis of a problem, the proof of a theorem, the application of a generalization, the building of a mathematical structure—it requires a high level of creative thinking. Thus, the teacher of mathematics must know when and what concepts to teach, when and why students are having difficulty, how to make concepts meaningful, when and how to practice skills, and how to stimulate productive thinking. The current emphasis on discovery, problem solving, and attitudes poses problems of adaptability and flexibility in the classroom—problems that require far greater skill than that required for the usual day-to-day lecture-recitation presentation.

Today the spirit of innovation is perhaps the outstanding characteristic of mathematics education. Revolutionary changes in school mathematics are altering traditional content, practices, classroom organization, and even basic views of learning. Thus, the mathematics teacher of today needs a background in calculus, number theory, foundations, linear algebra, non-Euclidean geometry, set theory, probability, and statistics. More than this, he must be able to read mathematical literature so that he continues to learn mathematics independently and to evaluate and, when desirable, incorporate new developments. He must be up-to-date in his mathematical language, symbolism, and structures for his classroom presentations. He should find pleasure in reading mathematics books and in presenting newly discovered mathematical ideas to his students when the ideas are appropriate. In this connection, Professor Arnold Ross of Ohio State University suggests that the beginning teacher teaches all he knows and more; the experienced teacher teaches all he knows; but the master teacher selects, from what he knows, material appropriate to his students. Finally, the creative teacher must motivate his students, he must communicate his knowledge to them, and he must—partly through the use of instructional aids—guide them to discover ideas.

The instructional aids currently available for the mathematics teacher have multiplied tremendously in recent years. We now have programmed texts and mathematics laboratories, computers, supplementary books and pamphlets, charts, films, filmstrips, models, overhead projectuals, games, exhibits, chalkboard devices, and construction materials such as pegboard, cardboard, plastic, balsa strips, and modeling clay. The creative teacher needs to know what materials are available and where and how to use these materials to enrich the learning of mathematical ideas.

All these things must then be adapted to specific classes and even to specific students. The teacher must select appropriate goals for instruction of individual units and plan a variety of lessons and units to achieve these goals. He must stimulate the learning of mathematics—by developing desirable attitudes and appreciations of mathematics and by teaching the student how to study mathematics independently. He must guide the student to discover mathematical concepts; develop ability to solve mathematical problems; and build understanding, accuracy, and efficiency in computational skills. He must evaluate the student's achievement of concepts, skills, and problem solving. In addition, he must provide a program of enrichment and acceleration for gifted students and plan an effective program for the slow learner; evaluate new curriculum proposals; procure, evaluate, and use new instructional aids; find new applications, new ideas, and new materials; and—in all his activities—constantly evaluate student achievement and his own instruction.

Learning Exercises

1. It is often said that artistic talent is inborn: You either have it or you do not. Draw upon examples from the world of the arts: painting, sculpture, music, drama, writing, etc., to show that talent alone is not enough. What are some additional qualities common to many top-flight artists?

2. What aspects of mathematics do you most enjoy? Why?
3. What aspects of mathematics do you enjoy least? Why?
4. Here is a problem that could be assigned to a student at almost any grade level: *What is the smallest number of segments you must draw in an obtuse triangle to divide its interior into acute triangles*? Try this problem. What do problems like this—and there are many—suggest about the intellectual demands that *can* be made at any school level?
5. Explain by means of examples what you think Professor Ross means. See page 7.
6. What are the characteristics of an effective mathematics teacher based on your observations as a student?

Suggestions for Further Reading

Bell, E. T., "Buddha's Advice to Students and Teachers of Mathematics," *Mathematics Teacher*, 62, 5 (May 1969), 373–383.

Daniells, Roy, "A Space to Live In," *Mathematics Teacher*, 63, 8 (December 1970), 673–679.

Kinney, L. B., "Why Teach Mathematics?," *Mathematics Teacher*, 35, 4 (April 1942), 169–174.

Newsom, Carol V., "A Philosophy for the Mathematics Teacher," *Mathematics Teacher*, 62, 1 (January 1969), 19–23.

Polya, Georg, "On Learning, Teaching, and Learning Teaching," in Douglas B. Aichele and Robert E. Reys (eds.), *Readings in Secondary School Mathematics*, Prindle, Weber, and Schmidt, Boston, 1971, pp. 323–337.

Rosskopf, Myron F., "Mathematics Education: Historical Perspectives," in *The Teaching of Secondary School Mathematics*, Yearbook 33, National Council of Teachers of Mathematics, Washington D. C., 1970, pp. 3–29.

Steiner, Hans-Georg, "Some Aspects of a Modern Pedagogy of Mathematics," *Mathematics Teacher*, 63, 5 (May 1970), 441–445.

Wagenschein, Martin, "Mathematics Teaching—A Tragedy," in Aichele and Reys (eds.), *Readings in Secondary School Mathematics*, pp. 316–322.

Wilder, Raymond L., "The Beginning Teacher of College Mathematics," *CUPM Newsletter* 6 (December 1970), 1–10.

Young, J. W. A., "Teaching Mathematics," in Aichele and Reys (eds.) *Readings in Secondary School Mathematics*, pp. 303–315.

The New School Mathematics Curriculum 2

New Math. To a great many laymen this phrase has a negative connotation. The public thinks of new math as something exotic and impractical: Students know how to do problems but they never get the right answer. They know the difference between number and numeral, but they cannot calculate with accuracy —let alone speed.

This is a false and pernicious view of new mathematics curricula: false because it implies that the older programs did a better job, pernicious because in its focus on side issues it misrepresents the many important changes that have virtually universal support from mathematicians and teachers. Today as always there is disagreement among concerned mathematics educators about the details of content and method, but no critic is calling for a return to the mathematics teaching of the 1920s, teaching that John R. Clark has called "routinization with minimum insight." [1]

Factors Causing a Change in School Mathematics

New mathematics has been created. Mathematics is a dynamic, exploding field of knowledge, and its exponential growth is evident in the quantity as well as the quality of recent mathematics. As an example of the increased rate of development of mathematics, it is fascinating to note that during the past decade as much new mathematics was developed as was developed in all previous history!

New uses for mathematics have been discovered. As new mathematical ideas have been created, mathematics has proved of further use to the sciences, the humanities, and even to the arts. Psychologists are using mathematics to build learning models; social sciences are using probability and game theory to study politics, crime, and economics; even linguists are using mathematical analysis to study language and literature. One of the most dramatic examples of applied mathematics is in the creation and development of the electronic computer (see Chapter 25 for a discussion of its use in the school). Computers are now used to control space ships, plot political strategy, compose music, and instruct students.

[1] "Mathematical Education: Yesterday and Today," *Today's Education*, 59, 9 (December 1970), 50.

Our scientific society needs greater numbers of persons with high mathematical competence. More mathematics must be taught to more students, so that these students will be able to transfer their mathematical knowledge to a variety of situations. In addition, more than ever before, basic education must provide for the continuation of learning after the student has left school.

Many people misread the high unemployment levels among well-educated as well as poorly educated people as an indication that education makes no difference. They are wrong. What is needed is better education *for independence*, better general education that will help adults to continue to acquire new skills and to grow intellectually. Extreme specialization is dangerous whether it be specialized training as a comptometer operator, an aeronautical engineer, or even a surgeon. Change is the way of life today: Learning how to attack and solve original problems—an integral part of any good mathematics program—is training to meet the new problems of the future.

At the same time as our technological society escalates, the requirements of an informed citizenry—the basic rationale for public education—make ever greater demands on mathematical understanding.

New knowledge about how children learn mathematical ideas has been discovered in recent years. In order to attain society's new demands for mathematical competence, research has provided new information on the process of learning. We know that students can learn complex mathematical ideas quickly when they are given the opportunity to participate in appropriate learning activities, and that young children can comprehend complex ideas if these ideas are communicated in meaningful, original ways. Our new concept of the learning processes stresses the importance of understanding, with a special emphasis on building understanding through individual discovery. Numerous teaching aids have been prepared to facilitate and increase discovery activities. However, it is evident that each student has his own optimum learning style.

Increased financial support of experimental projects by private foundations and federal governments has stimulated innovations. Financial support has made it possible to organize writing teams, test new materials, and provide the profession with information concerning the new curricula. Government support through the National Science Foundation and the United States Office of Education in this country and from similar agencies in other countries has been the most powerful vector promoting change.

Probably one of the greatest factors in creating a need for new mathematics was the lack of success of traditional content and method. Students attained a low level of competence in skills and had little understanding of what they were doing. Not only that, but they usually came out of mathematics classes with a dislike for mathematics. When mathematics becomes a grind and a distasteful activity lacking intellectual flavor, then retention and application are minimal. Our new knowledge of principles of learning has shown us the importance of *understanding*. Since one of the ways of building understanding is through involvement of the learner, this is the type of learning emphasized today.

Generous support of school mathematics curriculum development has been provided by top-flight university mathematicians. Many internationally recognized mathematicians have contributed vast amounts of time and energy to curriculum projects. While their influence occasionally pulls the mathematics curricula in the direction of the highly specialized and highly abstract, their insights have proved valuable at all levels of instruction.

Twentieth-century Mathematics

The new topics in today's school mathematics include numeration systems, sets, mathematical structures, vectors, matrices, linear programming, probability, statistics, symbolic logic, non-Euclidean geometry, transformations, and computer programming. Most of these topics originated during the past hundred years. Others represent extensions of old fields with new ideas and new applications. For example, the binary numeration system was invented by Baron von Leibnitz in the seventeenth century. Little attention was given to this "impractical" idea until the invention of the electronic computer, which requires binary numerals for its operation. Galileo in about 1600 had used the idea of a one-to-one correspondence, but not until 1874 did Georg Cantor apply this tool in the development of set theory and the study of infinity.

Probably the most significant aspect of the past century has been the attention given to mathematical structures. Through the development of non-Euclidean geometries, by Nikolai Lobachevski and Farkas Bolyai about 1830, mathematicians could accept assumptions that were independent of physical representation. The mathematician was now free to use his imagination in creating new "arithmetics," "algebras," and "geometries." These new structures might be concerned with a finite or an infinite number of elements. George Boole in about 1850 established a symbolism for logical deductions, and this Boolean algebra provided the necessary tools for programming logic into circuits in the computer. In this century, David Hilbert established a new foundation for the structure of Euclidean geometry which had itself been used as a model structure for centuries. Giuseppe Peano later stated basic assumptions for arithmetic and built a new structure for the counting number system.

Group theory, originated by Évariste Galois in the early nineteenth century, has been called the supreme structure of mathematical abstraction. A mathematical group is the basic structure for many of the branches and topics of mathematics such as number systems, transformations, vectors, matrices, and sets. Today topology and category theory extend the hierarchies to even more abstract levels.

One new field of mathematics of this generation is the theory of games, developed by John von Neumann. This theory treats laws of strategy not only in games but in business, government, and international affairs. Another field of great activity is computer programming. Using numerical analysis, probability and game theory, symbolic logic and binary numerals, the computer is building a new world of automation, prosperity, and leisure. The most obvious mark of new mathematics programs is the inclusion of new topics such as these.

As an illustration of how a new topic unifies and clarifies mathematics, consider the topic of sets. A set is described as a collection of objects, ideas, or symbols. Sets are used to give meaning to the idea of number and operations such as addition. Sets are used to designate the roots of equations, to describe geometric figures, to define basic operations. An angle is defined as a set of points. Likewise, a graph is a set of points. The integers are a subset of the rational numbers. Probability is described in terms of sets of events, and functions are defined as sets of ordered pairs. Scientific classification is accomplished by set intersection. Even deductions are illustrated by the intersections of sets. And set-builder notation such as $\{(x, y) : x^2 + y^2 = r^2\}$ becomes a means of defining terms such as a circle.

The New Content of Secondary-School Mathematics

As a result of the introduction of new content, some secondary-school courses have changed drastically. The seventh- and eighth-grade course is no longer devoted merely to applications of arithmetic. These courses now include numeration systems, sets, plane and space geometry, equations, finite systems, statistics, and probability as well as new treatments of whole numbers, fractions, decimals, percent, measurement, and graphing. Ninth-grade algebra continues to include the classical topics of algebra (directed numbers, graphs of equations, systems of equations, factoring, quadratics, and radicals) but also introduces proof, emphasizes number systems, and includes inequalities.

Tenth-grade geometry, no longer merely plane geometry, incorporates coordinate geometry and solid geometry. However, geometry continues to be the most criticized course in secondary mathematics. Because all mathematics courses now emphasize proof and mathematical systems, geometry has lost its significance as a course on the logic and structure of mathematics. Critics are suggesting a complete revision of the content of geometry. They are suggesting that vectors, transformations, projective geometry, combinatorial topology, convex sets, non-Euclidean geometry, and coordinate geometry all be part of the new content.

The eleventh-grade mathematics course tends to be a fusion of intermediate algebra including logarithms, complex numbers, and analytic trigonometry. The emphasis is now on functions and structure.

The twelfth grade continues to be a dilemma. Some advocate that it be devoted to the study of calculus; others support analytical geometry. The Commission on Mathematics suggests a semester on probability and statistics, while others propose that it be largely concerned with linear algebra. Other recommendations include a topics course, providing brief exposures to number theory, history of mathematics, or limits with short monographs rather than a full text providing the basic content.

With all these additions to the secondary-school program, adjustments have to be made. To provide sufficient time for the new areas, some topics have been eliminated and others reduced drastically. The application topics (insurance, installment buying, stocks and bonds, taxation, banking, consumer problems) have been largely eliminated. These social applications are curtailed as mathematical topics for several reasons. Since the mathematics involved in these topics is minimal, they are better dealt with in connection with the social and commercial aspects treated in social science and business courses. Furthermore, in these changing times and with the advent of the electronic computer, the specific applications of today may be unimportant in the future. The number of drill problems of arithmetic and algebra has also been reduced, as has the number of proofs of geometry and the computational aspects of logarithms and trigonometry.

In addition to the elimination or reduction of topics, traditional topics are now covered in less time, and some topics are introduced earlier in the curriculum. For example, the number line, equations, and geometric figures are treated extensively in the elementary school; and many concepts traditionally taught in the ninth and tenth grades are now introduced informally in the seventh and eighth grades. Hence, modern senior high school courses already include most of the ideas of traditional college algebra and analytical geometry. Many colleges now offer calculus as a first-semester freshman course, and some high school graduates are given advanced placement as a result of high school studies. The Cambridge Conference goes so far as to suggest that in another generation what is now the first three years of college mathematics will be taught in high school, thus repeating the downward spiral that saw all of college mathematics of a hundred years ago moved to the high school.

Emphasis on Structure and Language

One hallmark of the new school mathematics is the careful and consistent use of language and symbols. The idea of an equation, for example, is clarified through the use of the following hierarchy of terms:

A *mathematical sentence* ($x + 5 > 3, \Delta A \sim \Delta B, 4x - 9 = 7, 5 \geq 7$)

A *mathematical statement*—a mathematical sentence that may be judged true or false ($5 \geq 7, 3 + 9 = 12, x + x = 2x$)

An *open sentence*—a mathematical sentence that becomes a mathematical statement when variables are specified ($3x + 6 = 12, 2x^2 - 9 < 15$)

An *equation*—a mathematical sentence containing an equal sign ($x^2 = 5, x/x = 1, 5 = 7$)

The intent of this careful use of language is to add precision and clarity, to avoid confusion, and to encourage insight. There are of course gains and losses in precision. Examine in this regard the difference between old and new notation for lines and angles, as shown in Table 2.1. Many mathematicians argue that "old" notation cuts down on vocabulary while at the same time being clear from the context in which it is written. For this reason some newer programs have retreated from extremes of formalism.

Table 2.1

Old	New	
AB	$\overline{AB}$	A segment
	AB	Measure of a segment
	$\overrightarrow{AB}$	A ray
	$\overleftrightarrow{AB}$	A line
$\angle ABC$	$\angle ABC$	An angle
	$m^\circ \angle ABC$	Measure of an angle in degrees
	$m_r \angle ABC$	Measure of an angle in radians
	int. $\angle ABC$	Convex region enclosed by an angle

Another important point of emphasis in the new programs is the structure of mathematics, the framework that supports all mathematical activity. This emphasis on structure is related to current learning theory. Psychologists today say that we learn most effectively when we see the structure of the topic, problem, or subject being studied because the structure of a subject helps us remember it and enables us to apply our knowledge to new situations. Hence, understanding the structure of mathematics and mathematical laws and procedures appears to be the most appropriate foundation for continued study and application of mathematics.

An example of structure may be seen in applications of the distributive law. By applying this one law a variety of problems formerly taught as distinct processes are seen to be related:

$$3 \cdot 32 = 3(30 + 2) \overset{\text{Ⓓ}}{=} 90 + 6 = 96$$

$$5 \cdot 3\tfrac{1}{7} = 5(3 + \tfrac{1}{7}) \overset{\text{Ⓓ}}{=} 15 + \tfrac{5}{7} = 15\tfrac{5}{7}$$

$$\tfrac{4}{7} + \tfrac{2}{7} = 4(\tfrac{1}{7}) + 2(\tfrac{1}{7}) \overset{\text{Ⓓ}}{=} (4 + 2)(\tfrac{1}{7}) = 6(\tfrac{1}{7}) = \tfrac{6}{7}$$

$$5ax + 15ax^2 = 5ax(1) + 5ax(3x) \overset{\text{Ⓓ}}{=} 5ax(1 + 3x)$$

The final point, and perhaps most important, is the new spirit of mathematics in these new programs. Mathematics is presented as an elegant invention of the human mind. New content emphasizes the power, the unity, and the uniqueness of mathematics. Mathematics then becomes a field with aesthetic qualities similar to those of art, music, or literature. This emphasis is reflected in the variety of mathematics books and pamphlets now available that are suitable for independent, recreational reading.

The rapid development and adoption of new school mathematics programs demonstrates the remarkable production that can result when competent people work together with adequate financial support. Most of the experimental projects in mathematics have involved the cooperative efforts of mathematicians, educators, psychologists, and school teachers supported by federal or foundation grants. Early participation in these new programs is an exciting experience for teachers and students. This enthusiasm is contagious, and other teachers and schools are encouraged to participate. In fact, some schools have adopted new programs without having a staff prepared to teach the new content; and some programs are being accepted without anyone asking where they lead or for what students they are designed. This is the most common reason the programs have been criticized.

New Mathematics: The Fifties and Sixties

Today the first round of modern mathematics revision is essentially complete. A brief review of that period is, however, in order because it sets the stage for the second round, the one currently under way in the seventies. In bare outline, here are the major activities of this time period arranged in order of their inception.

UICSM: University of Illinois Committee on School Mathematics

Originally set up in 1952 as a committee of mathematics, education, and engineering faculty members at the University of Illinois to recommend course content for secondary schools, the committee soon found that their recommendations would carry little meaning without textbooks that would reflect their thinking. Led by Max Beberman, UICSM expanded to include secondary-school classroom teachers, produced texts for grades 9 to 12, and developed intensive in-service training programs (originally required as a prerequisite for teaching the program). The texts went through several revisions and are now published commercially by D. C. Heath. They are well worth examining because they reflect both heavy use of technical language and much attention to detail and rigor on the one hand and thoughtful presentation on the other.

Commission on Mathematics of the College Entrance Examination Board

College Board examinations have an important influence on the secondary-school curriculum. Recognizing this and the possibility this gave them of exerting a leadership rather than a conservative role, CEEB appointed in 1955 a Commission on Mathematics under the direction of Albert Meder of Rutgers University. This group drafted proposals focusing on grades 9 to 12, sponsored a series of meetings with mathematicians and secondary-school teachers across the country to receive reactions, modified and finally published their recommendations in 1958. In addition to recommending new topics (sets, inequalities, fields, circular functions) and new approaches (an emphasis on structure and coordinate geometry), the report provided a valuable appendix that illustrated its ideas in sample teaching sequences. These appendixes proved to be of great value to both experimental projects and individual textbook authors.

An important follow-up to the commission report has been the reflection of new math in CEEB examinations. By moving ahead in this way, they were able to head off a problem that had an almost devastating effect on the PSSC (Physical Sciences Study Commission) physics course. Students in the physics course had to be provided a score increment to bring their scores on the traditional examination up to the national average. At no time has this been the case in mathematics.

The commission also sponsored a writing team headed by Frederick Mosteller of Harvard who wrote a probability and statistics text for secondary schools. This book is published in revised form by Addison-Wesley.

UMMaP: University of Maryland Mathematics Project

Under the direction of John Mayor, this group produced seventh- and eighth-grade texts in the fifties. Until these books were written, mathematics in grades 7 and 8 was essentially review of grades 1 through 6 with practical applications. This void was filled by material on sets, logic, statistics, probability, number and numeration systems, informal geometry, and algebra. Because of the overlap between the membership of UMMaP and SMSG (School Mathematics Study Group) writing teams, much UMMaP content and spirit is reflected in the better known SMSG program for these grades. UMMaP texts are reflected in some Holt, Rinehart & Winston texts.

SMSG: School Mathematics Study Group

Organized in 1958 by the American Mathematical Society, the National Council of Teachers of Mathematics, and the Mathematical Association of America, this group has been directed from its inception by Edward Begle of Stanford University. More than any of the other groups, SMSG benefited from the impetus of Sputnik, the Russian space vehicle that shocked the United States into the realization that its scientific preparation was suspect. It has been heavily financed by the National Science Foundation, a government agency that supports many of the other projects listed here and also sponsors an extensive program of teacher retraining. By 1960, SMSG had published through Yale University Press (later through Vroman's Bookstore and Random House) textbooks and supplementary publications for grades 1 through 12, developed by writing teams made up mainly of university mathematicians and secondary-school classroom teachers.

One feature of the texts that is often missed is the fact that different writing teams produced the texts for the various grades largely independently. Thus different texts reflect the idiosyncrasies of the major mathematicians who served on each team. In the case of geometry, differing approaches led to three different texts for the same course. Despite their differences, the texts all reflect precise language or formal presentation, unification through structure, a middle road on notation, and a movement away from both applications and mechanical manipulation.

For a time in the mid-sixties, the SMSG text out-sold commercial mathematics textbooks; but, as the sponsors hoped, the ideas are now represented in commercial textbooks, many of which are written by former members of the SMSG teams.

Other Groups

Many other projects were active during this period. These include the Ball State Mathematics Project (Merrill Shanks of Purdue University and Charles Brumfiel and Robert Eicholz of Ball State College) best known for the geometry text based on Hilbert postulates; the Greater Cleveland Mathematics Program (GCMP); the Minnesota School Mathematics and Science Teaching Center (Paul Rosenbloom of the University of Minnesota), which developed a number of ideas for correlating mathematics and science; and the Boston College Mathematics Project (Reverend Stanley Bezucka).

Criticisms of the New Programs

The new math provided a field day for cartoonists, monologists, and polemicists. Criticisms range from whimsical to deadly serious, from superficial to penetrating. Some are patently unfair (SMSG: Some Math, Some Garbage), but others are insightful and are proving useful today. The important point to keep in mind is that the criticisms themselves indicate some of the power of the new programs: This was no minor modification of curriculum; it was a major discontinuity in the history of mathematics education. No such drastic change involving so many writers, teachers, students, and parents should expect to be implemented smoothly.

There are several major criticisms of the programs:

The focus was on content almost exclusively, with little attention to the associated teaching problems. Even the expanded teachers' manuals of the SMSG program—thicker in most cases than the texts themselves—are largely explanations of the mathematical content and contain few teaching suggestions.

The texts were best suited to top students, especially those who will continue in mathematics beyond the high school. Little attention was devoted to the average or below-average student. There is a valid excuse for this: The experimental programs were largely funded to develop materials to train students of better ability to higher levels of mathematical competence. Some managed to "bootleg" materials for other students by broadly interpreting this change.

The applications of mathematics were largely ignored. Morris Kline of New York University, an applied mathematician, has been especially vehement in his criticisms of this failure. His message is that mathematics should derive from and apply to the real world or it loses its vitality. He has backed up his criticisms with textbooks that carry out his point of view.

Rigor, precision, and symbolism were often overdone and sometimes became an end in themselves. This, of course, varies from program to program. Two effects of this, however, have resulted: (1) a peculiar form of notational snobbery among some students (if you don't use symbols like $\exists$ and $\forall$ you cannot be a mathematician) and (2) a decline in interest on the part of students whose concerns are more practical.

The conceptual emphasis was so great that computation skills have suffered. Here again the curriculum developers have a response: Older programs were loaded with computation practice and this was what teachers were stressing. The new programs assume that teachers will continue to evaluate students and incorporate appropriate skills instruction. Many schools that have added very small doses of computation instruction (as little as 10 minutes a week) have found that new programs then provide positive rather than negative increments on standardized tests of computation.

The volume of content made good teaching difficult. ("I would like to use good techniques but I don't have time.")

Sometimes the mathematicians did not communicate to classroom teachers what they were trying to accomplish. Edwin Moise, the senior author of the SMSG *Geometry*, has suggested that experimental texts are too often like "the shot heard 'round the immediate vicinity": well taught in the hands of the author and his immediate disciples, but poorly taught beyond that inner circle. This was, of course the strength of the UICSM program when the early teacher-training requirements (teachers had to attend training programs before teaching the content) were in force, but such restrictions delay almost indefinitely the implementation of new programs.

It must be stressed again that much of the criticism of the so-called new math has been due to a lack of understanding on the part of the public and even the teachers themselves. The public still does not know what the new programs are or what they seek to do. Some people assume, for example, that lack of skill in computation is due to the new math rather than recognizing that students of traditional mathematics also lacked computations skill.[2] Similarly, teachers do not always use good judgment in their handling of new programs. They assume that the new topics will entirely replace important traditional topics. For example, some teachers spend a great amount of time on numeration systems—even insisting that students show competence in computing in a base other than ten; these teachers do not recognize the purpose of numeration systems—namely, to help students understand numerals and operations rather than to develop skill in computing with new numerals.

Continuing Change in the Seventies

By 1970, it was difficult to find elementary or secondary schools that had not adopted a "modern" mathematics program. Textbook titles with "contemporary" or "modern" abound today. Modern math had in effect won by default: Virtually all textbooks at least claimed to represent the contemporary trends, and the secondary-school curriculum had standardized along the lines of the SMSG texts. The stage was set for the second act.

To some extent, the current changes are responses to criticisms of the first round of change. But an examination of materials of the Comprehensive School Mathematics Project (CSMP) and Secondary School Mathematics Curriculum Improvement Study (SSMCIS) groups especially shows that the content of the new programs of the seventies goes well beyond that of the sixties.

It is possible to indicate some general directions that mathematics curriculum reform is taking today:

Pedagogical concerns are assuming increasing importance. Remarkable strides have been made by people like Robert Wirtz in translating concepts reserved for graduate schools a few years ago into teaching units suitable for secondary and even elementary schools. Consider in this regard a lesson Wirtz has developed for second-grade (!) students: A candy store in which students pick out the fewest boxes to fill orders for a certain kind of candy. The clerk has made out a table to tell what kind of boxes to give (see Figure 2.1). Young children in this informal setting are converting decimal to binary notation, a process reserved for graduate-school number-theory classes 20 years ago (and sometimes still today).

---

[2] In 1931, Schorling, studying more than 200,000 students in grades 5 through 12, found that only 20 percent of the twelfth-grade students could compute 2.1 percent of 60. In 1937, Taylor studied more than 2,000 freshmen in teachers' colleges and found that more than half could not divide 175 by .35. In 1942, Admiral Nimitz reported that 68 percent of 4,200 freshmen at 27 United States universities and colleges were unable to pass the arithmetical-reasoning portion of the examination for entering the Naval Reserve Officers' Training Corps. In 1943, Brueckner, conducting a national survey, found that the arithmetical competence throughout the country was even worse than the Nimitz report indicated.

Pressure to move mathematical content ever downward is continuing.

Changes in classroom organization and instructional procedures are an integral part of many of the new programs. Individualization, team teaching, small-group teaching, and mathematics laboratories are part of new programs.

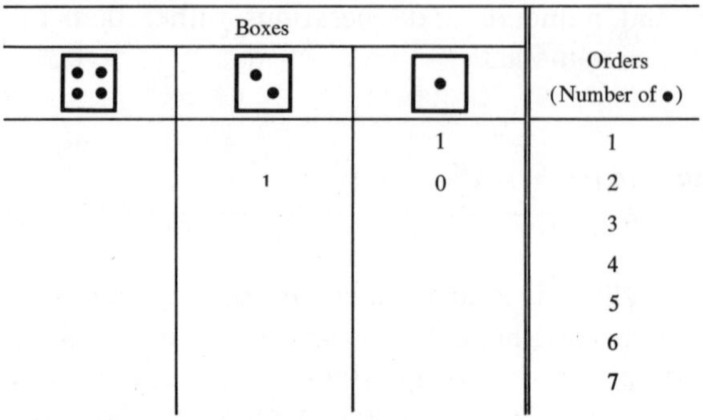

Boxes			Orders
••••	••	•	(Number of •)
		1	1
	1	0	2
			3
			4
			5
			6
			7

Figure 2.1

Traditional content boundaries are breaking down. In particular, geometry is changing to fit the modern stress on function. This leads naturally to coordinate and vector techniques but more particularly to transformation geometry. Algebra is encroaching on all areas of mathematics—geometry included—and finite processes are forcing new divisions of content.

To a much larger extent, commercial publishers are participating in curriculum revision by supporting writing projects rather than waiting for government-supported projects to carry out all experimentation.

Attention is being focused on average and below-average students and particularly on culturally deprived adolescents.

There is concern for skills, but that concern does not lead to mechanical drill. Instead instruction and practice on basic skills is embedded in conceptually oriented and interesting problems.

New instructional materials and equipment are being integrated into instructional programs.

Despite the extensive changes in mode of presentation and academic level of instruction, only a small fraction of classical, traditional mathematics (certainly less then 20 percent) has been pruned from the mathematics curriculum.

Alternate approaches to the same concept are being developed so that students who have difficulties with one learning program need not repeat exactly the same process. This development enlivens review by presenting it in new dress.

Innovation is now worldwide rather than restricted to the United States.

Major Innovative Programs of the Seventies

Like new car models that are issued well before their style date, the second round of modern innovation—the round of the seventies—started well before 1970. But teachers in the sixties were too busy incorporating the long round of changes to notice these early tremors. Also a number of factors are operating today to make the second revolution a quiet revolution: Pedagogical changes are not so exciting to the public as content changes, higher levels of mathematical sophistication among beginning teachers makes further modification less shocking, and the involvement of commercial publishers as well as government-supported projects takes away a source of irritation. These factors delayed general response to and realization of new new math until 1970.

Here are some of the important components of Round Two:

Activities in the United States

The Cambridge Conference on School Mathematics. Three separate conferences, which were held between 1963 and 1967 under the administration of the Educational Development Center with National Science Foundation support, were the starters. These conferences were concerned with (1) goals and curriculum of the K-12 mathematics of the future, (2) the training of teachers in order to implement innovative programs, and (3) the correlation of mathematics and science in school mathematics. Participants in all three conferences were almost exclusively university mathematicians and scientists.

Critics were quick to note some of the exotic aspects of the curriculum proposed: matrices, indirect proof, logarithms, and interpolation in intermediate grades; complex numbers, derivatives, and Chebychev's inequality in grades 7 and 8; vector spaces, tensors, multidimensional calculus, and Fourier series in senior high school. A careful reading of the report places into focus those goals, which the report calls "exploratory thinking with a view to a long-range future, . . . not concerned with the sort of practical considerations which govern the work of the next few years."[3] Despite these restrictions, the reports of these conferences, especially the first, deserve close attention. Here are some of its major points:

A three-year time saving is possible through the total abandonment of drill for drill's sake, that "technical practice can be woven into the acquisition of new concepts" (p. 8).

Support for the special curriculum.

---

[3] *Goals for School Mathematics: The Report of the Cambridge Conference on School Mathematics*, Houghton Mifflin, Co., Boston, 1963, p. 3.

Mathematics should play a role as confidence builder.

Vocabulary and symbolism should be developed to communicate ideas and must therefore be kept in balance.

A heavy recourse to intuitive development of concepts is stressed.

A variety of approaches are encouraged.

Independent and creative thinking should be fostered by the discovery approach, but this should not be overdone. Directed learning often saves time.

Motivation is an important component of teaching.

Applications have an important role but must be handled with care.

These recommendations carry much sense and justify a more complete and balanced analysis of the content recommendations. As will be seen, several groups are not waiting for the "long-range future" to attack these goals.

Comprehensive School Mathematics Project: CSMP. Organized and directed by Burt Kaufman originally at Nova High School in Fort Lauderdale, Florida, CSMP moved in 1967 to Carbondale, Illinois, where it operates under the sponsorship of the Central Midwestern Regional Educational Laboratory in cooperation with Southern Illinois University. This project set as its difficult goal, the "implementation of the Cambridge report today." It has two major components, the EM text series for secondary schools and a "package program" of individualized units for elementary school. The EM series is the most rigorous mathematics program yet designed for school use. In abstract algebra the books go beyond the recommendations of the Cambridge report. Use of the materials is limited to supervised trial centers and to teachers who have been carefully indoctrinated.

CSMP has also sponsored international conferences on the teaching of probability, algebra, and geometry.

Secondary School Mathematics Curriculum Improvement Study: SSMCIS. This project, supported by the United States Office of Education and Teachers College, Columbia University, is directed by Howard Fehr of Teachers College. This is another highly abstract mathematics program for the gifted students of six years of secondary school. The content is heavily influenced by European mathematicians and there is some overlap in direction with CSMP.

University of Illinois Committee on School Mathematics: UICSM. Despite the death of Max Beberman in 1970, UICSM continues to carry on curricular reform in mathematics. They provided a major breakthrough in the teaching of low achievers by developing two courses: *Stretchers and Shrinkers* (grade 7) and *Motion Geometry* (grade 8). *Stretchers and Shrinkers* develops operations with rational numbers through the use of stretching and shrinking machines (operators that multiply and divide). The *Motion Geometry* program develops geometric concepts through transformations. Each program includes four student workbooks and an activities handbook for teachers. The *Stretchers and Shrinkers* activities handbook especially is a rich source of teaching techniques. For once, the weaker students are provided with rich material first. Unfortunately, this advantage is only temporary because the concepts and techniques of the program are being incorporated in many texts for better students (see Chapter 20).

School Mathematics Study Group: SMSG. SMSG has developed for grades 7 through 9 a program of second-round texts designed to cover all mathematics necessary for functioning of an adult (nonscientist).

National Council of Teachers of Mathematics: NCTM. Two committees of NCTM developed activity-oriented materials for slow learners called *Experiences in Mathematical Discovery* and *Experiences with Mathematical Ideas.* A joint committee of American Statistical Association and NCTM members has prepared a sourcebook of statistics lessons for secondary schools. This committee has also published a book on statistical applications.

European Activities

The British School Mathematics Project (SMP) has produced secondary-school texts that are available in the United States. SMP provides a rich source of mathematical applications and activities for discovery learning. Continental authors like Jean Dieudonné and Gustave Choquet have pressed teachers to make school mathematics more rigorous and the Belgian couple, Georges and Frederique Papy, have developed some attractive pedagogical techniques for teaching abstract ideas. The Papy texts *Modern Mathematics* I and II are available in the United States. Of particular interest in these books is the extensive and dramatic use of color and graphs.

Non-curricular Factors

In this review of new math, we have focused on content and pedagogy changes as reflected in textual materials. But these are not the only changes. One major change is growing from the individualized programs like I.P.I. (Individually Prescribed Instruction) and PLAN (Programmed Learning According to Need). These programs are passing on to secondary-schools, students with an ever-wider range of backgrounds than in the past. A variety of patterns of secondary-school reorganization have been proposed to meet this problem: individualized instruction, team teaching, small-group instruction, mathematics laboratories, programmed instruction, and computer-assisted instruction. These pedagogical approaches will be discussed elsewhere in this book.

New Math and the Classroom Teacher

What should be the response of the classroom teacher to the new programs? Above all else, the continuing revolution in mathematics teaching makes it imperative that the classroom teacher keep abreast of current developments. To do this, he must examine new textbooks and attend local, state, and national meetings, paying particular attention to displays of new commercial and experimental materials. He should read reviews of new programs in professional journals and participate in in-service training related to new content and pedagogy.

Should he adopt a new program? Unless he has assured himself that he has adequate background in content and can translate the textual materials into viable classroom units, a teacher should be wary of "going it alone." With help or in concert with other staff members, trying new texts can be a worthwhile teaching and learning experience. But bandwagon-jumpers and those who will try nothing new are equally bad: Deliberate evaluation of new programs is extremely important. (See Chapter 22.)

If a new program is adopted, it is best to embark on it with a strong positive attitude. Be ready to question specific aspects of the program but try to see and follow the intent of the program. Ask yourself whether or not this is a program that attains the most important goals of mathematics instruction.

Learning Exercises

1. Compare the table of contents of a seventh-grade textbook copyrighted before 1960 with the table of contents of a seventh-grade textbook in use today. Choose a topic that appears in both books and compare the development.

2. Compare the table of contents of a contemporary geometry textbook with a geometry text copyrighted before 1960. Where do major differences appear?

3. Read sections 2 to 4 of the *Cambridge Report* (see references). List points with which you agree and points with which you disagree. On the basis of your results, would you describe yourself as a modern or a conservative mathematics educator?

4. Read the review by Marshall Stone of the Cambridge Report (see references). Do you feel that Stone's substitute recommendations are more traditional in content?

5. Read the debate between Morris Kline and Albert Meder (see references). Who is the better debater? What are some points on both sides with which you disagree?

6. Read one of the articles describing one of the new math programs. Indicate what the article says are program goals and special features. Does the program appear to you to be more attractive than the corresponding course you studied in high school? Why or why not?

7. Read the article "On the Mathematics Curriculum of the High School" and the replies to that article (see references). It has been said of this petition that many of the people being criticized signed it. Why do you suppose that happened?

8. Read the 1959 report of the NCTM Secondary School Curriculum Committee (see references). This report, prepared before Sputnik, predated all but early UICSM materials. List some points that are no longer valid.

9. PROJECT. Evaluate a new program in use in a local school. Do this by:
   (a) Examining the content of the text and the teacher edition in detail
   (b) Discussing instructional problems with teachers who are using the texts
   (c) Reading reviews of the materials in journals
   (d) Observing classes using the program

Suggestions for Further Reading

Adler, Irving, "The Cambridge Conference Report—Blueprint or Fantasy?," in Douglas B. Aichele and Robert E. Reys (eds.), *Readings in Secondary School Mathematics*, Prindle, Weber, and Schmidt, Boston, 1971, pp. 65–74.

Anderson, C. Arnold, "The International Comparative Study of Achievement in Mathematics," in Aichele and Reys (eds.), *Readings in Secondary School Mathematics*, pp. 94–109.

Begle, E. G., "SMSG: The First Decade," *Mathematics Teacher*, 61, 3 (March 1968), 239–245.

Bidwell, James K., and Robert G. Clason, (eds.), *Readings in the History of Mathematics Education*, National Council of Teachers of Mathematics, Washington, D.C. 1970, pp. 655–663 (UICSM), pp. 664–706 (Commission on Mathematics, CEEB).

Brumfiel, Charles *et al.*, "The Ball State Experimental Program," *Mathematics Teacher*, 53, 1 (January 1960), 75–84.

Braunfeld, Peter, Vincent Haag, and Burt Kaufman, *The CSMP Approach to Curriculum Improvement*, Comprehensive School Mathematics Program, Carbondale, Ill., 1972.

Cambridge Conference on School Mathematics, "Proposed Secondary School Mathematics Curriculum," in Aichele and Reys (eds.), *Readings in Secondary School Mathematics*, pp. 29–49.

Elements of Mathematics Program, Comprehensive School Mathematics Program, Carbondale, Ill., 1971.

Fehr, Howard F., "The Secondary School Mathematics Curriculum Improvement Study Goals—The Subject Matter—Accomplishments," *School Science and Mathematics*, 70, 4 (April 1970), 281–291.

Ferguson, W. Eugene, "The Junior High School Mathematics Program: Past, Present and Future," *Mathematics Teacher*, 63, 5 (May 1970), 383–390.

Garstens, Helen L., M. L. Keedy, and John R. Mayor, "University of Maryland Mathematics Project," *Arithmetic Teacher*, 7, 2 (February 1960), 61–65.

Hale, William T., "UICSM's Decade of Experimentation," *Mathematics Teacher*, 54, 8 (December 1961), 613–618.

History of Mathematics Education, A, Yearbook 32, National Council of Teachers of Mathematics, Washington, D. C., 1970.

Keller, M. Wiles, "Semantics and Mathematics," *School Science and Mathematics*, 68, 2 (February 1968), 103–113.

Kline, Morris, "The Ancients Versus the Moderns," and Albert E. Meder, Jr., "The Ancients Versus the Moderns—A Reply," in Aichele and Reys (eds.), *Readings in Secondary School Mathematics*, pp. 9–28.

Mosteller, Frederick, "Progress Report of the Joint Committee of the American Statistical Association and the National Council of Teachers of Mathematics," *Mathematics Teacher*, 63, 3 (March 1970), 199–208.

Mueller, Francis J., "The Public Image of 'New Mathematics,' " *Mathematics Teacher*, 60, 7 (November 1967), 696–706.

"On the Mathematics Curriculum of the High School" (and replies by Edward Begle and Phillip Jones), *Mathematics Teacher*, 55, 3 (March 1962), 191–198.

Secondary School Curriculum Committee of NCTM, "The Secondary Mathematics Curriculum," *Mathematics Teacher*, 52, 5 (May 1959), 389–417.

Stone, Marshall, H., "A Review of the Cambridge Conference on School Mathematics," in Aichele and Reys (eds.), *Readings in Secondary School Mathematics*, pp. 50–61.

Psychological Bases for Mathematics Instruction 3

The classroom teacher must seek a philosophical basis for his teaching so that his daily work does not degenerate to *ad hoc* response to each of the thousands of daily problems that confront him. Certainly common sense has a role to play here. But common sense and intuition tend to be very conservative: They quite often are disguises for "the way we did it when I went to school." To become a professional, you must strive to compare your intuitive beliefs against what others think. You may in the end retain your initial beliefs: If you do they will have met the test. On the other hand, you may accept the evidence for change or may at least be willing to give alternate ideas a fair trial. In either case, you will have weighed the evidence and will be acting on the basis of that evidence rather than limited individual experience and emotion.

Your next recourse is the literature of mathematics and psychology. Mathematics provides a way of thinking, a way of organizing ideas and attacking problems, and occasionally even insights into the way we learn. Aspects of this will be considered in Part Three of this text. In this chapter we briefly survey psychology to see what it has to offer. On the concepts you meet in this chapter will be built the more specific applications to teaching of all later chapters.

Psychology Provides Ideas, Not Final Answers

The classroom teacher should search the literature of psychology for a broad and firm basis on which to erect an instructional program. He should read research on the problems he identifies as most serious. The answers he does find will often be helpful. He may be surprised to find too that some psychologists, like some school administrators, are bandwagon riders, for there are temporary fads in psychology just as there are in curriculum and school organization.

Despite the almost complete absence of clear prescriptions in the psychological literature, the teacher can find there some interesting results and many more hypotheses about mathematics instruction and learning. While it is true that theories—especially contradictory theories—do not provide a very permanent foundation for an instructional program; still they give the teacher an opportunity to choose his building platform. He can in fact pick and choose, even constructing different aspects of his teaching philosophy on what may appear to be contradictory theories.

It is important for the teacher to know something about psychology as it applies to mathematics education for several reasons:

Psychology does afford some answers, and the immediate future may well bring significant breakthroughs. General awareness of current psychology will give the teacher the tools with which to interpret these new ideas.

Psychologists have developed some broad theories that provide the teacher with bases for serious thinking about teaching.

A knowledge of psychological study techniques encourages any teacher to observe his students systematically and to monitor his instructional program with more care.

All these confront the teacher with scientific evidence that bears on his own beliefs. While nothing in educational psychology contradicts thoughtfully organized and well-executed classroom teaching, the teacher will find that psychological insights will cause him to modify his ideas and even his procedures in significant ways.

Some Psychological Concepts

As is true of any technical field—even mathematics—psychology has its specialized jargon. In the case of psychology more than most others, however, the technical vocabulary is commonly used by laymen. This has led to a variety of problems related to misinterpretations of psychological concepts. For example, one common lay interpretation of the meaning of intelligence is "the mental capacity you were born with." Talk of modifying IQ may then take on the sound of gene manipulation in unborn children. It also gives teachers an excuse for giving up on students who are harder to teach. In the light of these problems, it is appropriate that we look more carefully at this and a few other terms before turning to psychological theories and studies related to them.

Intelligence

Intelligence is usually thought of as the ability to solve problems, to think or to learn; that is, to carry on purposeful activity. (Edward L. Thorndike modified this meaning to the solution of *novel* problems and thought of intelligence as the ability to transfer learning. The term "transfer" will be discussed in the next section.) This overall intelligence is often broken into components (verbal, perceptual, memory for example); but here we are concerned with the general intellectual ability, essentially the union of these intersecting subsets.

Tests cannot measure intelligence directly. No current instrument does unless we follow the lead of some test users who choose conveniently to redefine intelligence as the measure obtained on an intelligence test. Such a definition is circular and very severely limited. It and the tests themselves have been attacked because of the extreme difficulty of freeing the tests from cultural artifacts. At one extreme, for example, consider the results of such a test when the tested student does not speak the language. Less obvious cultural differences affect the scores in less striking but similar ways. Scores of individual children vary over time also.

Despite these reservations (separation of test scores from intelligence itself, cultural and other special effects, and score variations), many teachers categorize students by the number assigned to them by an IQ test: "John should do better work than Mary; he has a 116 IQ score and she has only 112." Score variations make such specificity meaningless; and at best a score, even in the absence of additional invalidating factors, may be thought to represent a range of 10 or more points. Thus the best we may say about John and Mary is that they appear to be about the same in measured IQ.

It is interesting to note that Russian psychologists reject the idea of IQ. Instead of considering a relatively stable intelligence around which instruction must operate, they consider instruction to be a direct attack on teachability. N. A. Menchinskaya indicates the direction of their research:

> The investigations, conducted by laboratory associates and teachers, revealed the psychological features of pupils with low teachability... and showed that by giving supplementary exercises to pupils in this category, it was not hard to attain the necessary level of mastery of school material. However, later on, during the mastery of new material the pupils' low teachability reappeared. Low teachability was changed only after prolonged, purposeful work that had been constructed on the basis of a regard for individual differences and that had influenced not only mental properties but also the personality as a whole.[1]

The Soviets relate their psychology to their political doctrine of dialectical materialism, some aspects of which are repugnant to Western observers. But their stress here on the inner worth of each individual and on education as the attempt to draw out this quality should have attraction for all teachers. The classroom teacher who identifies the goal of improving his students' general problem-solving ability as more important than imparting to them specific procedures and techniques is approaching instruction from this point of view.

These notes on intelligence are largely critical of the way the concept has been used by teachers. As a way of identifying in gross fashion students with unusual

[1] "Fifty Years of Soviet Instructional Psychology," translated and reprinted in Jeremy Kilpatrick and Izaak Wirszup (eds.). *Soviet Studies in the Psychology of Learning and Teaching Mathematics*, Vorman's, Stanford, Cal., 1969, vol. I, p. 16.

abilities or deficiencies and as a basis for developing an attack on the "teachability" of a group, the measures are surely of value. So many dangers are inherent in them, however, that today many school systems are rejecting IQ tests and measures entirely.

A good rule for the classroom teacher is to use the individual IQ score as a minimum estimate of a student's potential. Using this rule, a student who scores 116 on an intelligence test would be categorized as well above average, possibly gifted. Another student scoring 92 may be slightly below average, average, or above average. Considered in this way, the IQ score does not close doors for children, it helps teachers to open them. An interesting study by Robert Rosenthal and Lenore Jacobson, reported in their book *Pygmalion in the Classroom*,[2] shows how doors are opened for students who are not relegated by low test scores to a life in the slow group. In this study, false indications of learning potential were communicated to teachers, who responded by noting better achievement for these pupils. This kind of result, called the self-fulfilling prophecy, says again that we must not pigeonhole our weaker-appearing students.

One other aspect of intelligence should be noted because it may have a profound effect in the classroom. P. W. Jackson and J. W. Getzels, Milton Parnes, and E. Paul Torrence, among others, have led educators to consider what they call creativity: thinking that stresses novel rather than routine outcomes. Standard intelligence tests, most classroom tests and even teachers place a premium on the routine answer and a negative connotation on the creative, nonroutine response. A Brooklyn College mathematician, Banesh Hoffman, has written scathing attacks on testing in which he stresses this point.

Getzels and Jackson[3] identified by various measures of creativity and intelligence two groups: (1) highly creative with moderate intelligence and (2) highly intelligent with moderate creativity. They found that despite an average IQ difference of 23 points the groups achieved equally well. But consider some of their other findings: Teachers much preferred students in the high-IQ group. The high-IQ group ranked qualities they would like to have and those they believe teachers favor very similarly; the high-creativity group tended to disagree with what they identified as teacher-favored qualities. In particular, the high-creative student placed emphasis on "sense of humor" while his high-IQ neighbor ranked this quality low on his list of desirable qualities.

Guilford and Torrence have developed measures of creativity, but these are seldom part of a standard school test battery. Teachers then must themselves be alert to this nonconforming ability and should seek indications of this different aspect of thinking. Parnes has indicated that creativity may be fostered by appropriate training and encouragement. This too should be considered by the classroom teacher. We will refer to this again in Chapters 8 and 15.

---

[2] Holt, Rinehart and Winston, New York, 1968.

[3] J. W. Getzels and P. W. Jackson, "The Highly Intelligent and the Highly Creative Adolescent," in Calvin W. Taylor and Frank Bareon (eds.), *Scientific Creativity*, John Wiley & Sons, Inc., New York, 1963, pp. 161–172.

Transfer

Frandsen defines transfer of training as "the application to new learning tasks or problems of principles, concepts, or skills learned previously in different situations."[4] This concept has had an interesting history in education and particularly in mathematics education. The idea evolved from the early concept of education as mental training or discipline. The brain was thought of much like a muscle whose growth was stimulated by problems of ever-increasing difficulty. Latin was supported as a school subject, not for its historical and linguistic values, but because it trained the mind. And the trained mind would then apply itself to other more practical things. Similarly mathematics, and in particular geometry, was justified because it could be made difficult.

This extreme view of transfer was rejected by psychologists half a century ago, although it is evident that some mathematics teachers still use this criterion for evaluating a course. The reaction was so strong, in fact, that transfer was all but thrown out with mental discipline and Latin. (Geometry lost much of its stature for a time, too.) Thorndike in particular believed that transfer occurred only where identical elements were present; that is, the student had to be able to identify the new work as related to the earlier concepts. This implies that if the logic learned in geometry is to be applied to real situations, then the teacher must teach those applications. More recently, psychologists and others have stressed that transfer takes place on a much more substantial scale. If it did not, students could not solve original problems, and in fact all science would grind to a halt. The important point, of course, is contained in a remark by Hilgard: "Some transfer of training must occur or there would be no use in developing a foundation for later learning."[5] In other words, we might as well give up any idea of a sequential development of curriculum—central to mathematics instruction—if we give up the concept of transfer.

We will be especially concerned with problems of transfer in Part Three of this text.

Readiness

We do not concern ourselves with teaching infants the calculus. There are certain prerequisites—reading is not the least of these—that the child must learn first. But there are more than curricular prerequisites. We must also consider mental maturity and experience. The infant not only cannot read but also he has had virtually no contact with concrete embodiments of calculus concepts. In both senses, we say he is not *ready* for this learning.

---

[4] Arden N. Frandsen, *Educational Psychology*, McGraw-Hill Book Company, New York, 1961, p. 586.

[5] Ernest R. Hilgard and Gordon H. Bower, *Theories of Learning*, Appleton-Century-Crofts, New York, 1966, p. 7.

This concept of readiness was frozen into the curriculum in some unfortunate ways in this country during the first half of this century. Some educators identified grade levels at which certain concepts should be taught—this on the basis of virtually no research. A cursory examination of modern elementary and even secondary-school textbooks shows the continuing results of this conformity. Only very recently have many of these early judgments been set aside.

A quite different attack on the concept of readiness has been mounted by Jean Piaget, a Swiss psychologist. Piaget is essentially an observer and a recorder. He devised a series of individual conceptual tasks and then tested students of varying ages against these tasks. He found a sequence of levels of understanding through which children grow quite naturally. Many of Piaget's tasks are mathematical in nature and provide important insights into mathematical growth. One, for example, provides the subject with a table—much like one side of a pool table. On this table is a target ball. The subject is given practice with a spring-operated shooter aiming another ball at the target ball. Then he is asked to try to hit the target ball by shooting his ball against the wall (see Figure 3.1). Very

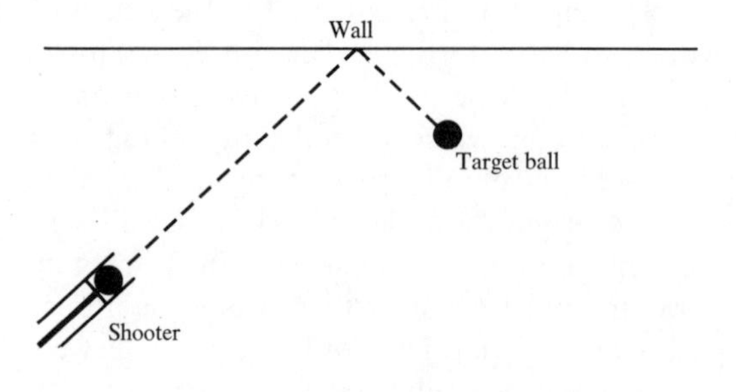

Figure 3.1

young children have no way of solving this problem. They often shoot over and over again at a point on the wall directly behind the target ball, thus passing well beyond the target. They never adjust. As older students try the test, they progress toward the solution until finally they develop themselves a good idea of the incidence-reflection relationship of Heron's formula.

Note that there is no teaching connected with this. This is not a standard school task (despite the tendency of curriculum developers to try to put such tasks into programs with the idea that they are attacking a concept rather than a specific task), so the growth in understanding may be thought of as a growth in readiness.

Piaget carried out his experiments in several geographic locations. He found varying ages at which his identified level of readiness are attained. Other evidence also bears on this phenomenon. Some tribes in interior African regions remote from civilization have little concept—even as adults—of a straight line. The nature of their environment precludes the concrete observation that supports this concept—there are no straight lines around them. Thus their environment has a strong effect on their readiness for certain kinds of learning.

Other Piagetian experiments demonstrate that conservation ideas develop in a chronological pattern similar to that of the reflection principle. Young students have little idea that number or volume is maintained even in physical experiments. Five things spread out are more than five things close together; tall narrow tumblers contain more than short fat tumblers even when the liquid in one exactly fills the other. Some learning disabilities even in secondary-school students may be traced to slower development of their readiness for such concepts. Interestingly, Piaget has also noted that geometric thinking progresses from topological to metric, the reverse of our usual school order of instruction.

Some mathematics teachers, notably Z. P. Dienes, are developing materials specifically designed to attack this problem of readiness. By providing their students with what they call "multiple embodiments" of very basic concepts of mathematics, they believe that they are providing earlier readiness for conceptual learning and thereby accelerating the process of education. Observers of Dienes's elementary-school students working on advanced group theoretic concepts at Sherbrook University near Montreal are struck with the success of his techniques. Dienes and many others stress "playing" with ideas at initial stages of learning. This offers strong support for models and mathematics laboratory activities discussed in later chapters.

Another psychologist, Jerome Bruner, has said that you can teach any subject in an intellectually satisfying way to a child of any age. This has often been taken to mean that Bruner opposes the idea of readiness. Instead Bruner is modifying the content to the child. Thus a seventh-grade student can understand some aspects of velocity and even acceleration without knowing the calculus. Instruction brings appropriate parts of the subject to him rather than bringing him to the subject. This is the basis for the spiral curriculum discussed in Chapter 11.

Some Theories of Learning

With the rejection by psychologists of mental discipline, the field was left wide open to new theories and many rushed to fill the gulf. (It should be noted here that many laymen and many classroom teachers as well do not reject the mental-discipline concept of teaching even today. It is not difficult to devise questions that elicit evidence of the continuing prevalence of this belief.) The newer theories fit more or less neatly into three categories: conditioning, gestalt, and mechanical. While these larger categories tend to blunt some important internal distinctions, they will serve here for this brief overview. Readers who are interested in such distinctions are directed to Hilgard and Bower's *Theories of Learning.*[6]

[6] *Op. cit.*

Conditioning

The earlier stimulus-response connectionism and conditioning theories were drawn together by an important and still-active Harvard psychologist, B. F. Skinner. Oversimplified, stimulus-response or S-R theory states that organisms—students included—tend to continue along rewarded paths and retreat from punished paths. In one sense, Skinner reversed the order to R-S by rewarding approximations to desired behavior.

This basic reorganization has had a striking effect on animal training, and Skinner himself has demonstrated his training procedures with animals. Within a very short time, he can teach a dog to turn around on command. To do this, he gives the command and rewards the—at first random—movements of the dog that are in the desired direction. At each stage, a little more movement is expected to get the reward: a food pellet or encouragement. (Failures may or may not be punished to extinguish them. In a sense, no response may be thought of as punishment so that all responses are rewarded or punished. Skinner himself plays down the role of punishment.)

Learning, according to this theory, is built up in the same way. The learning process may be thought of as a staircase of small steps. Larger concepts must be broken down into small discrete concepts; major objectives must be attacked in a series of short intermediate steps. An obvious result of this connectionist view of learning is programmed instruction, in which blocks of learning, even entire courses, are presented as a sequence of thousands of steps. Each step is a question, an easy question because the step is so small and relies so much on the previously mastered content. When he answers this question, the student immediately checks to see if his answer is correct (is thereby rewarded or punished) and then moves on to the next question.

Skinner has urged that this technique be applied in the classroom as well, that teachers be trained to break their teaching down into smaller, easier-to-master, discrete steps, that they manage their classroom along the lines of operant conditioning.

A related outcome of the conditioning theory of learning is the current stress on behavioral objectives: Set up the goals of instruction in terms of student behavior and then work backward through the program to develop a step-by-step approach to this behavior. In doing this, a hierarchy of prerequisites to the desired behavior is indicated and a "programmed" path for students outlined. This provides a basis for organized classroom activity. Robert Gagné is a contemporary learning theorist who has stressed this kind of specific task analysis of mathematics learning. He calls these learning hierarchies.

Much laboratory and classroom evidence supports various aspects of Skinnerian learning theory. Two areas in which it is often attacked are problem solving and transfer. Skinner and his followers have responses to these criticisms. Solving "larger" problems is merely a matter of manipulating variables at a level to which the student has been programmed; no originality is involved. An examination of a programmed text indicates what is meant by this. If you examine the first few exercises, they appear trivial, as they should according to the theory. But skip ahead. Without mastering the small intermediate steps, you will find that items a few pages later appear much more difficult. In other words, larger problems are merely the sum of smaller problem increments.

Transfer for Skinner and his followers comes when students respond to similar stimuli in different areas. This is quite similar to the identical-elements idea of Thorndike mentioned earlier.

David Ausubel is another contemporary psychologist whose work is usually associated with the Skinnerian group. He encourages analyses of learning hierarchies similar to those of Gagné and takes a somewhat more extreme position than the latter about content knowledge rather than learning to think as the central role of the school. But Ausubel adds to the concept of the learning hierarchy the principle of *advance organizer*. Given alternate task sequences to a given goal, Ausubel recommends the route that embodies the best overriding principle. James Bidwell (see reference) has described the application of Gagné's and Ausubel's ideas to teaching division of fractions. Examples of the three procedures for which hierarchies are established are given in Figure 3.2. Bidwell's preference is the inverse-operation process because he claims that the concept of inverse operation serves as a stronger advance organizer than the organizers imbedded in either the common-denominator or complex-fraction processes.

Common Denominator

$$\frac{2}{3} \div \frac{4}{5} = \frac{10}{15} \div \frac{12}{15} = \frac{10}{12}$$

Complex Fraction

$$\frac{2}{3} \div \frac{4}{5} = \frac{2/3}{4/5} \cdot \frac{15}{15} = \frac{10}{12}$$

Inverse Operation

$$\frac{2}{3} \div \frac{4}{5} = n \text{ means}$$

$$\frac{2}{3} = n \cdot \frac{4}{5}$$

$$\frac{2}{3} \cdot \frac{5}{4} = n \cdot \frac{4}{5} \cdot \frac{5}{4} = n$$

Figure 3.2

Benjamin Bloom[7] recommends a complete reorganization of the classroom to attack learning tasks. His students are allowed to vary the amount of time spent learning rather than the level of achievement, the reverse of common practice. Bloom calls for uniformly high levels of achievement, a constant, and individualized amounts of time to reach this achievement. Mathematics learning, which depends as much on previous attainment, should take into account Bloom's views.

[7] Benjamin S. Bloom, "Learning for Mastery," in Bloom *et al.* (eds.), *Handbook on Formative and Summative Evaluation of Student Learning*, McGraw-Hill Book Company, New York, 1971, pp. 43–57.

Gestalt

To many people the theories of the connectionists are dehumanizing. Is teaching the same as animal training? Is learning the accumulation of large numbers of discrete details? Gestalt psychologists attack the Skinnerians as focusing on memorization rather than understanding, on training rather than intellectual growth, on the end product rather than the organization of thinking. Meaning, the field psychologists claim, is neglected by connectionists; their own goal is improvement of the thinking process rather than the product.

The gestalt or field psychologists believe that there are larger learning increments and in particular that insight plays an important role in problem solving. A comparison of the two viewpoints is shown in Figure 3.3. Note how the quantum leap replaces the steady upward march.

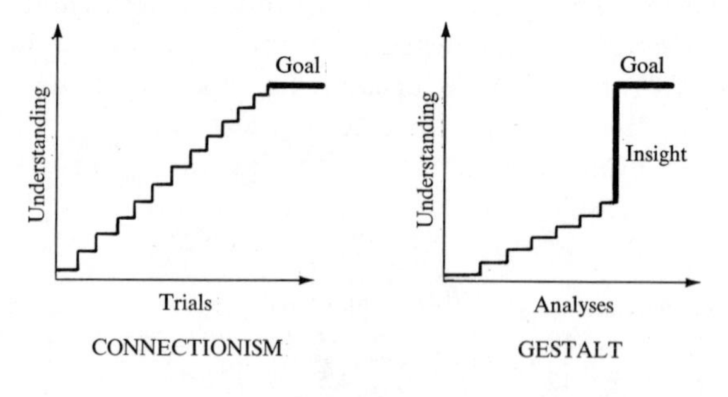

Figure 3.3

Jacques Hadamard[8] lists a series of steps in problem solving: (1) preparation, (2) incubation, (3) illumination, and (4) verification. These parallel an earlier sequence suggested by Dewey. The mathematical literature is, of course, full of anecdotal evidence in support of such a sequence: mathematicians to whom concepts crystallized in bed, on walks in the country, even as one, Poincaré, boarded a carriage.

Obviously, field theorists tend to turn their attention to "larger" or "more significant" problems. Howard Fehr, in his excellent analysis of gestalt psychology (see reference), suggests an example from geometry. In his problem the exterior angles of right triangle *ABC* are bisected, as shown in Figure 3.4. The bisectors meet at *D*, from which perpendiculars are drawn to $\overline{AC}$ and $\overline{BC}$. It is required to show that rectangle *CFDE* is a square. Direct attempts at congruency proofs are not productive. Insight is achieved when the properties of angle bisectors are

[8] Jacques Hadamard, *Psychology of Invention in the Mathematical Field*, Princeton University Press, Princeton, N. J., 1949.

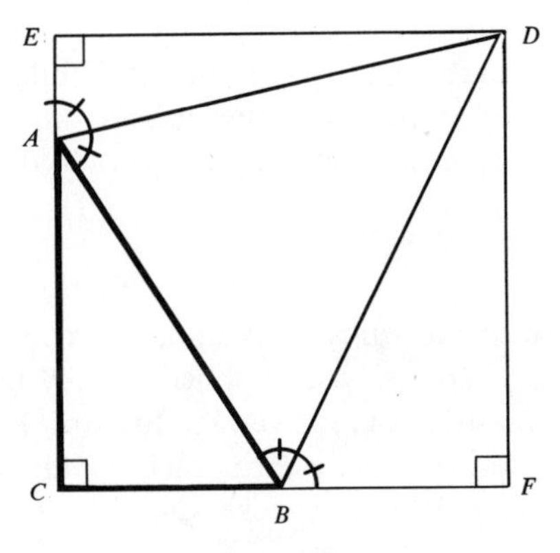

Figure 3.4

considered. A point (D) on an angle bisector is equidistant from each ray of the angle. This suggests drawing $\overline{DG} \perp \overline{AB}$ at G, resulting in $DE = DG = DF$. Note how the conditions had to be carefully set and considered with care to provide an *opportunity for* insight to occur. Looked at in this way, one important role for the teacher is to help the student to "set the stage" for his own insight.

One result of this theory is what is often described as teaching for discovery, a topic treated at length in Chapter 11. In this kind of teaching, the stage is set for students to make significant generalizations. One aspect of this teaching is the "Eureka" complex. The discoverer is given a strong charge of intrinsic motivation by success. This encourages him to make further, hopefully deeper explorations.

Peter Caws suggests that what is needed in problem solving is "a kind of single-minded concentration that keeps out irrelevant thoughts, and a facility for spotting wrong moves." He turns the tables by perhaps overstating: "As for everybody's not being a genius, the answer may be that everybody . . . *is*, until inhibiting factors supervene—*which almost always happens.*"[9] This would mean that the teacher should help the student to raise his concentration level so that he can make progress in learning and that he should help the student to avoid *losing* his in-born facility that helped him learn to walk and talk without a formal instructional program. Raymond Wilder,[10] in a discussion of the role of intuition in the process of learning and developing mathematics, suggests that we are in error teaching mathematics as a vast amount of knowledge. We should turn our energies to teaching students to participate in mathematical *activity* in order to develop the appropriate sets for insight to operate. Jerome Bruner, perhaps the major contemporary figure in this school, encourages keeping the learning

[9] "The Structure of Discovery," *Science*, 166 (December 12, 1969), 1377.

[10] "The Role of Intuition," *Science*, 156 (May 5, 1967), 609.

environment as fluid and individualized as possible in order to encourage generation of ideas.

Gestaltists then view learning in larger chunks. While they agree that much learning is accumulation of experience, they stress the reorganization of this past experience, the building of new configurations on old.

The Mind as Machine

Cybernetics is the scientific study of methods of control common to living organisms and machines, especially as applied to computer operations. One aspect of this activity is the study of artificial intelligence. Researchers in this area set out to make machines "think," that is, carry on humanlike activities. Computers have been programmed to play nim, checkers, and even chess. They have also been set up to prove geometric theorems.

Two general methods of machine "learning" have been explored. In application of one method, trial-and-error learning, to a simple game like tic-tac-toe, the computer plays a human opponent (or even another computer) a series of games. The computer "remembers" moves that lead to losses and does not repeat them in later play. IBM has made available for teachers a game of this type developed by Martin Gardner. Obviously, this procedure is limited to quite simple games. Even with its great speed, the computer cannot play all possible—or even all reasonable—complex games; and it cannot store the vast amount of information accumulated.

Consider in this regard the plays of chess. There are 20 opening moves, 20 possible responses, and so on, the number of possible plays on subsequent moves dependent on what specific earlier moves were made. It has been estimated that there are some 10^{120} possible games. Compare this with the number 10^{20}, which exceeds the number of seconds in a billion years.

A second approach to machine learning is by tree search. In this method, the machine is programmed to simulate human analysis by looking ahead. It adopts an "If I do this and he responds in this way . . ." approach at each stage of an activitity. As before, even in fairly simple activities the number of branches places a practical limit on how far ahead one can look. In checkers, for example, there are about seven possible moves per play at the beginning of the game. To look ahead five moves and responses then means considering 7^{10} or almost 300 million moves! Various pruning techniques, however, do allow machines to look ahead several moves and to evaluate the game situation at that point, returning down the best branch to select the appropriate play.

To date, game-playing machines have had limited success. A checker-playing machine has beaten a master player, but chess-playing computers are still at the level of passable amateurs at the time of this writing. Interesting articles on this and other more serious problems of artificial intelligence are found in Edward A. Feigenbaum and Julian Feldman, *Computers and Thought.*[11]

[11] McGraw-Hill Book Company, New York, 1963.

This analysis of machine approximation to human learning has led some psychologists to return the favor, to think of the mind as a computer and to consider its operation in terms of the way computers work. The brief description of artificial-intelligence activities gives some background for consideration of this analysis of learning and thinking.

The human central nervous system has some 10 billion nerves. This provides each person with computer resources that, it has been estimated, would require in terms of current equipment several buildings the size of the Empire State Building to house and several Niagara Falls to power. We each have working for us a very efficient machine.

The brain surely has procedures for *input:* accepting experiences obtained through the senses and possibly even some other ways from the environment. We know that it has *storage* both from the evidence of memory and also from physiological experiments that show accumulations in "storage areas"—the gray matter of the brain. *Retrieval* and *output* functions communicate the results of this magnificent instrument to our conscious state.

If we think of education in terms of the brain as computer, we may have to revise our values. For example, for a given computer, an increase of storage encroaches on other sections of the equipment. An unbalanced load of memorized factors may then actually inhibit conceptual learning by drawing energy and "room in the brain" from conceptual understanding.

We must also concern ourselves with how best to operate this computer and how to help students to operate their learning equipment most efficiently and most productively. Strong evidence indicates that we use our brains very inefficiently. Under hypnosis, for example, human subjects reveal "forgotten" or suppressed thoughts, perform more efficiently, and even control otherwise uncontrollable body functions like pulse rate.

How does this mechanical view of learning and instruction relate to other learning theories? At first observation it may appear to be merely a restatement of Skinnerian psychology in somewhat modified terms, but this is true only at a superficial level. The pathways along which input data travel must be extremely complex, and the ways in which these data are more or less efficiently organized are surely equally complex.

Some gestalt psychologists have noted *better than 100 percent* transfer of learning tasks. In other words, they (notably George Katona) have found that students sometimes do better on the new conceptually related task than on the original one. Is the brain providing some efficiency here that is not part of the input? Computer users know that this kind of serendipity is quite possible. Computers often produce unexpected bonuses of information. But is this merely a throwback to faculty psychology, the mind as a muscle? To some it seems so, but evidence points in other directions. Despite the power of the brain, its inherent inefficiency operates against direct attempts to train it as an athlete trains for a race. It is not sufficient to present harder and harder problems. Rather, we must devise educational sequences that provide an appropriate mass of information and explicitly stated patterns of use of this information. The brain then needs practice of a different type: for example, search patterns must be made as efficient as possible.

If the brain is comparable with a machine, conditions for learning become of increasing importance. Just as the computer center requires special air conditioning, so does the brain require a good environment. Insofar as possible, the clutter of inefficient programs must be cleared, continuing "noise" in input lines should be minimized, and the "electrical power" of motivation should be kept at peak loads. Again education takes on the appearance of protecting the student from himself, giving him the chance to utilize in the most efficient possible way the great gift with which he is born.

Translating Theory into Instruction in the Classroom

As we suggested at the beginning of this chapter, the value to the teacher of psychology in its current status lies more in the choice of fundamental constructs than in the availability of incontrovertible evidence to support specific modes of teaching. The teacher should spend some time thinking about a philosophical framework into which to fit his thoughts about teaching. Psychologists not only have provided a variety of such outlines, but also have brought to bear a volume of evidence in support of these frames.

Most teachers will not adopt one psychological theory to the exclusion of others. In fact, virtually no psychologist today lives in this kind of isolation. The advocate controversies of several decades ago are gone from this field. The teacher may then freely choose attractive aspects of varying theories and test them against the very real world of the classroom.

For example, two basic kinds of concerns face every classroom teacher: (1) having students learn a specified body of knowledge, and (2) having them develop intellectual independence. The connectionists have much to offer the teacher in his thinking about the first type of "guided learning," less about the second. The gestalt psychologists' contribution is exactly the reverse. The teacher may then draw upon both schools in his thinking. He should seek to promote discovery learning whenever possible, while carefully structuring his sequences to promote acquisition of knowledge.

The main point here is that philosophical issues should be a central concern of every classroom teacher. Too often we miss the forest for the trees. We trap ourselves in the day-to-day details of instructional programs and our total view is lost. Every teacher should challenge himself continually with questions about what he believes and why. In the remainder of this book, we draw heavily on the psychological ideas generated in this chapter. You, the reader, should continually test these constructs, not only against psychological "schools," but also against your growing personal philosophy.

We close this chapter with some less controversial precepts that derive from generally accepted psychological theory and that are directly applicable in the classroom. We draw heavily on summaries prepared by Ernest Hilgard and Gordon Bower[12] and by Goodwin Watson.[13] In the latter case especially, this list is offered defensively, to suggest that there is some common ground and that psychologists do have a positive contribution to make directly to the practice of teaching:

Students learn by doing. They should be active, not passive. Too much teacher direction leads to conformity, defiance, or nonparticipation.

Practice is still important. "Overlearning" guarantees retention. But sheer repetition alone is not good practice.

Rewards for desirable responses are important and are preferable to punishment. Immediate reward is stronger than delayed reward. Often fresh, novel experiences themselves are adequate rewards.

Criticism, failure, and discouragement destroy self-confidence, aspiration, and sense of worth.

Conflicts and frustrations arise inevitably in learning. They should be identified and resolved or accommodated.

The way a problem is presented to a student often determines its difficulty at least as much as the conceptual problem. Well stated is half solved.

Thinking takes place in response to accepted intellectual challenge.

Learning with understanding is more permanent and transferable than rote or formula learning.

Goals should be identified and their rationale provided to students.

Divergent or open-ended as well as convergent or specific-answer thinking should be stressed. Developing significant questions is as important an intellectual activity as answering those questions.

Tests have an inordinate influence on subject matter and teaching methods. If we seek to achieve process goals, we should see to it that our tests reflect those goals.

Anxiety levels should be monitored in the classroom. Anxious students may react negatively to comments on progress while low-anxiety students may benefit from such comments.

Motives and values strongly affect learning. What students perceive as relevant has an effect on achievement.

Group interactions are important in the classroom. Most students are more concerned about peer response than teacher response.

---

[12] *Theories of Learning*, Appleton-Century-Crofts, New York, 1966, pp. 562–564.

[13] *What Psychology Can We Trust?*, Bureau of Publication, Teachers College, Columbia University, New York, 1961.

These, then, are the contributions of the psychologists: general theories and a few specific prescriptions. You must weigh these ideas against your practical experience in the classroom in order to develop your own pragmatic philosophy of mathematics instruction.

Learning Exercises

1. Try to recall an incident when you experienced a "flash of insight" in solving a mathematics problem. Describe the incident.

2. Classify the following types of activities as most influenced by connectionist or gestalt psychological principles. Do not feel bound by obvious answers. Discuss reasons for your classifications:
   (a) Problem solving
   (b) Factoring quadratic trinomials
   (c) Proving geometric theorems (originals)
   (d) Learning definitions
   (e) Reviewing computation procedures
   (f) Applying mathematics to the real world
   (g) Developing structure and logic

3. It has been said that pressure to "cover" a great deal of mathematical content in a course obstructs the teaching of how to learn. Discuss this statement.

4. Describe your interpretation of the problem-solving steps of Hadamard (see page 36). Apply these steps to the way a bright student might respond to the geometry problem on pages 36–37.

5. In your experience as a student, have these problem-solving steps been taken into account by your teacher? If they were, describe some of the ways. If they were not, indicate some reasons you think your teachers did not foster their use.

6. Indicate some prerequisite topics to instruction in solving pairs of equations by graphing. Build a learning hierarchy for teaching this task. You may wish to refer to the articles by Bidwell and Gagné in the references to see examples of this process.

7. Comment on the statement: "Once the prerequisites are mastered, learning the next task is a snap." If this is true, why is school achievement not much greater?

8. Do you know your IQ? If you do not, estimate what you think it might be. Do you think that your tested intelligence would be higher, about right, or lower than your actual ability? Why?

9. Who is the most highly creative student you know? Describe some of his characteristics. How would you teach a student like this?

10. Most descriptions of the mind as a machine compare it with digital computers. What is an analog computer? Tell some human actions that fit this comparison.

Suggestions for Further Reading

Adler, Irving, "Piaget on the Learning of Mathematics," in Douglas B. Aichele and Robert E. Reys (eds.), *Readings in Secondary School Mathematics*, Prindle, Weber and Schmidt, Boston, 1971, pp. 210–221.

Ausubel, David P., "Limitations of Learning by Discovery," in Aichele and Reys (eds.), *Readings in Secondary School Mathematics*, pp. 193–209.

Bidwell, James K., "Some Consequences of Learning Theory Applied to Division of Fractions," *School Science and Mathematics*, 71, 5 (May 1971), 426–434.

Bruner, Jerome, "Bruner on the Learning of Mathematics—A 'Process' Orientation," in Aichele and Reys (eds.), *Readings in Secondary School Mathematics*, pp. 166–177.

———, *The Process of Education*, Harvard University Press, Cambridge, Mass., 1960.

Dienes, Zoltan P., "Dienes on the Learning of Mathematics," and William M. Bart, "More on Dienes," in Aichele and Reys (eds.), *Readings in Secondary School Mathematics*, pp. 222–251.

Gagné, Robert, "Gagné on the Learning of Mathematics—A 'Product' Orientation," in Aichele and Reys (eds.), *Readings in Secondary School Mathematics*, pp. 157–165.

Hartmann, George W., "Gestalt Psychology and Mathematical Insight," *Mathematics Teacher*, 59, 7 (November 1966), 656–661.

Rosskopf, Myron F., "Transfer of Training," in *The Learning of Mathematics: Its Theory and Practice*, Yearbook 21, National Council of Teachers of Mathematics, Washington, D. C., 1953, pp. 205–227.

Shulman, Lee S., "Psychological Controversies in the Teaching of Mathematics," in Aichele and Reys (eds.), *Readings in Secondary School Mathematics*, pp. 178–192.

———, "Psychology and Mathematics Education," in *Mathematics Education*, Yearbook 69, Part I, National Society for the Study of Education, University of Chicago Press, Chicago, 1970, pp. 23–71.

Skemp, Richard R., *The Psychology of Learning Mathematics*, Penguin Books, Baltimore, 1972.

Van Engen, Henry, "The Formation of Concepts," in *The Learning of Mathematics: Its Theory and Practice*, Yearbook 21, NCTM, pp. 69–98.

Goals and Objectives of Mathematics 4

In the construction of a new building, the architect does not design the building until he decides what effect he wishes to achieve; the builder does not select his materials until he knows what his blueprints specify; and the carpenter does not select a tool until he knows what operation he intends to perform. When the building is complete, it is considered acceptable only if it satisfies the specifications established before it was built. In a similar way, we must select the mathematics program for the effect we wish to attain. We must select materials that are appropriate for the program designed, and we must use the proper tools for the activity involved. Too often in the past, teachers have discussed the relative merits of a new mathematics text, an instructional aid, or a teaching method without specifying just what goals they hoped to achieve with the materials being used.

If we do not know what our objectives are, how can we evaluate a learner's progress? Only when we know where we want to go do we have a sound basis for selecting appropriate material, content, or instructional methods. Only when we know what progress has been made in attaining these goals can we evaluate the effectiveness of our instruction.

What, then, are appropriate goals for mathematics instruction? What mathematics is needed to attain these goals? What kind of student should strive to attain these goals? Finally, what behavior indicates that a student has attained these goals?

The words "goal" and "objective" are synonyms. In this chapter, we will use the word "goal" to represent larger, more general aims and "objective" to represent the more specific day-to-day aspects of the larger goals. It should be kept in mind, however, that the border between the two is hazy and ill defined.

The Selection of Goals

In selecting appropriate goals for mathematics instruction, we must take into account not only the needs of society but also the mathematical needs of our students. Almost every committee or commission that has worked on revising the mathematics curriculum has stated certain basic mathematical needs. The following list summarizes these needs:

The student needs to know how mathematics contributes to his understanding of natural phenomena.

He needs to understand how he can use mathematical methods to investigate, interpret, and make decisions in human affairs.

He needs to understand how mathematics, as a science and as an art, contributes to our cultural heritage.

He needs to prepare for vocations in which he utilizes mathematics as a producer and consumer of products, services, and art.

He needs to learn to communicate mathematical ideas correctly and clearly to others.

Mathematics instruction today must be broader and more inclusive than in the past if it is to meet the increasing demands being made on the mathematical competence of our students. The mathematics program must do more than develop the basic skills and techniques, although the broader goals of mathematics will include these skills and techniques. In other words, it must develop more than vocabulary, facts, and principles; more than the ability to analyze a problem situation; more than an understanding of the logical structure of mathematics. The mathematics program must, in addition, develop students who can use the logic of mathematics to distinguish fact from opinion, relevant from irrelevant material, and experimental results from proven theorems. This program must stimulate curiosity, so that the student will enjoy exploring new ideas and creating mathematics which is new for him even though it has been discovered by others. It must develop the reading skill, motivations, and study habits essential for independent learning of mathematics. In short, the mathematics program must produce students who know how to learn mathematics, enjoy learning mathematics, and are motivated to continue their learning.

These goals must in turn fit into the still broader social goals of the school: self-acceptance and self-reliance; respect for the rights and privileges of others, appreciation of the intrinsic beauty of the world in which we live, acceptance of the full responsibilities of society, and full recognition of the complex requirements of life in the days ahead. Such goals may seem far from the activities of the classroom, but much evidence points to the fact that citizenship is nurtured or injured in every classroom of every school.

Behavioral Objectives

In a sense, we make a contract with our students. They agree to pay a certain sum of effort in return for certain skills and knowledge. But most of the time we ask them to pay for something that is never carefully described. We fail to describe the given conditions under which we will evaluate their performance. If we fail to describe how we intend learners to be different after our instruction we are taking unfair advantage of them.

Describing our objectives in general hazy terms or—as is too often the case—not describing them at all leads to student confusion ("How do we know what you want?"), to inefficiency and disorganization in planning, and to unsatisfactory evaluation procedures. Yet, until very recently, few teachers recognized this basic failure in the design of their teaching. They thought in terms of teaching "quadratic equations" or "congruency proofs" or "factoring."

In response to this, educators developed the idea of much more specific behavioral objectives, objectives that identify in detail the terminal behavior expected of the student. Armed with these behavioral objectives, the teacher can then develop instructional sequences designed to accomplish his aims; he can communicate to his students what those aims are; and he can evaluate his students' performance and the success of his teaching sequence.

Preparing Behavioral Objectives

Preparing these specific objectives in behavioral terms is not an easy task. It requires attention to detail and careful judgement, but focusing these two ways of thinking on the teaching process is especially valuable for the classroom teacher. Many experienced teachers whose attitude is "That's just like what I've always done informally" soon find themselves studying actual classroom performance more clearly than they have ever done before.

Before we state our objectives we should recognize that an objective of instruction is a statement of what the learner is to be like when he has successfully completed a learning experience. It is a description of a pattern of behavior that we want the learner to be able to demonstrate.

When we state objectives, we often use terms such as "to know," "to understand," "to appreciate." These are general terms which may easily lead to a wide range of interpretation. More meaningful terms are "to write," "to identify," "to solve," "to state," "to compare," "to contrast," "to list," and "to differentiate." These words describe behavior that demonstrates the level of mastery attained by the learner. As we state objectives, then, we must know specifically what the learner does in a given situation and how he does it when he has achieved a given objective.

Robert F. Mager, in *Preparing Objectives for Programmed Instruction*,[1] states that objectives should satisfy three criteria. They should (1) describe *what the student will be doing* to demonstrate that he has attained the objective, (2) describe the *conditions* under which the student must demonstrate his competence, and (3) state the *standards of performance* expected of the student.

Suppose we state the objective "to know how to multiply positive and negative numbers." What are the given conditions? What will the learner do when he is

[1] Fearon Publishers, Belmont, Cal., 1962.

showing us that he has reached this objective? Will he recite the rules? Will he write a series of examples? Will he perform some computations correctly? Will he find the errors others make when they perform multiplications? The objective as stated does not tell us. It cries out for restatement in greater detail. Or consider the objective "to be able to solve equations." Although this statement does at least have a terminal act there are many shortcomings in the statement. What kind of equations will the learner be expected to solve? Is it important that the solution specify the domain and range? Do we just want a numerical answer? Will the learner be asked to justify each step in the solution? Will he be expected to draw a graph of the solution set?

These questions suggest the importance of specifying another aspect of the terminal behavior expected of the learner. We need to state the conditions under which the learner will demonstrate his mastery. Instead of simply saying, "to be able to solve an equation," we can specify our communication by saying: "Given a linear equation with one variable, the learner must be able to determine the domain and range, find the solution set, justify each step in his solution process, and draw a graph of the equation involved."

To guide you in identifying important aspects of the terminal behaviors, ask yourself these questions: What will be provided to the learner: what problems, what materials, what time? What will the learner be denied? May he use a book, slide rule, table of logarithms, formulas? Under what conditions will you expect the learner to perform: on a test, in the science class, on his homework? Such questions make the evaluation process specific so that there can be a one-to-one correspondence between objectives and test items.

After we have described what we want the learner to be able to do, we should indicate *how well* we want him to do it. Hence, we need to state the criterion of acceptable performance. If we can specify a minimum acceptable performance for each goal, we will have a basis for determining whether a program is successful. One criterion may be a time limit. We may specify that "The learner must be able to solve 30 of these equations in five minutes." Of course, a better criterion would be that "The learner must be able to justify orally each step in solving this type of equation." Another specification may be the minimum number of *correct responses* in a given situation. For example, given 20 linear equations, the learner can correctly identify the solution sets of at least 15 equations.

We have seen some examples of poorly stated objectives. Well-stated behavioral objectives are somewhat more difficult to illustrate. Here are two that relate to computation:

1. Given 20 equations of the form $ax + b = c$, a, b, and c in the set $\{m: {}^{-}20 \leq m \leq 20$ and m an integer$\}$, the student will produce solution sets with 90 percent accuracy in 20 minutes.

2. Given five factorable polynomials of the form $x^2 + bx + c$ with b and c natural numbers (or zero) and not more than 40, the student will factor over the integers at least four correctly in two minutes.

Often behavioral objectives are stated less formally with one or more of the required conditions unfulfilled. Especially with the current stress on mastery without reference to time, speed as an essential factor in computing answers has been played down. Also many teachers leave unstated some traditional restrictions on the parameters of a stated objective. Thus objective 2 above might appear more informally:

2′. Given five factorable polynomials of the form $x^2 + bx + c$, the student will factor at least four correctly.

Although the polynomial $x^2 + x(\sqrt{3} + \pi) + \pi\sqrt{3}$ now fits the strict interpretation of objective 2′, it would be apparent from the context of the content under study whether or not it would be suitable.

What is really important in these objectives is the way they indicate in detail to the teacher (and student too, if passed on to him) what must be accomplished. Now an instructional program to reach this objective can be mounted. This may involve pretesting to determine how close to the objective students already find themselves, development of a hierarchy of subskills—a task analysis—to pinpoint a route for instruction, or even establishment of a variety of patterns of self-instruction to reach the terminal behavior.

Objectives related to calculations and processes that are essentially algorithmic are amenable to statements in behavioral terms when care is exercised. Objectives related to higher conceptual activities are far more difficult to frame. Can you see why? Take, for example, the problem of stating a behavioral objective related to proving triangles congruent. Proofs of congruent triangles may be very simple and straightforward or extremely complex. Two possible ways around this difficulty (both often adopted) are illustrated:

3. Given three congruence proofs from the exercises on page 73 of the text, the student can write acceptable proofs of two in 10 minutes.

4. Given three congruence proofs like those in the exercises on page 73 of the text, the student can write acceptable proofs of two in 10 minutes.

Note that in the case of objective 3, the student can reduce the conceptual level of the task to the memorization level, just as many students do now when syllabi include "required proofs." In the case of objective 4, lattitude is left to the teacher (or test maker) to determine what kinds of exercises are "like" those on page 73. This does not, of course, solve the problem completely. In fact, both objective 3 and objective 4 illustrate how behavioral statements often orient teaching toward repeat-back-to-me processes when higher cognitive processes are intended.

These disadvantages do not significantly reduce the value of objectives 3 and 4, but they illustrate the exponentially increasing difficulty of expressing objectives for higher cognitive processes. Consider, for example, the problems associated with expressing in behavioral terms goals related to solving original challenging problems, demonstrating creative behavior, applying an abstract idea to a concrete problem, deriving enjoyment from the challenge of mathematics, interaction with other students in developing divergent mathematical ideas. These difficulties suggest that there is still room in mathematics teaching for goals stated in less concise terms.

Objectives and Evaluation

Because of this difficulty in stating the objectives in behavioral terms, some teachers prepare evaluation techniques at the time they are constructing their objectives. They then designate satisfactory performance on the test as their behavioral objective. These evaluation items become themselves the objectives of the learning sequence. For example, an objective might be stated:

2″. The student will be able to factor over the integers four of the following in two minutes:

$$x^2 + 5x \qquad x^2 + 8x + 16 \qquad x^2 + 10x + 24$$

$$x^2 + 16x + 15 \qquad x^2 + 16x + 39$$

This can, of course, degenerate into teaching for a specific test. More often it provides the teacher with a detailed set of specific targets at which to aim his instruction. Now he knows exactly what he is trying to accomplish. He can develop strategies, plan his classroom implementation methods, gather his materials, all with these specified objectives always in mind. No longer is he seeking vague and hazily defined goals.

Evaluation is discussed in greater detail in Chapter 9.

Types of Objectives

Two important publications have contributed significantly to the development of objectives. They are *Handbook 1: Cognitive Domain*[2] and *Handbook 2: Affective*

[2] B. S. Bloom (ed.), David McKay Co., Inc., New York, 1956.

Domain[3] of the *Taxonomy of Educational Objectives*. These two books provide a bench mark for a survey of objectives of classroom instruction. They also indicate a hierarchy of objectives running generally from simple to complex.

The general categories of the cognitive domain are:

Knowledge, ranging from knowledge of specific facts and terminology to knowledge of generalizations, theories, and structures

Comprehension, including translation, interpretation, and extrapolation

Application

Analysis of elements, relationships, and organization

Synthesis or organization of ideas into a report, a plan, or a system

Evaluation on the basis of internal or external evidence

The corresponding outline for the affective domain includes:

Receiving or attending

Responding or participating

Valuing or believing in the worth of something

Organizing values into a system

Characterization by a value or a value complex

A third domain, that of psychomotor or physical skills, is generally outside the province of mathematics education.

It is evident that most classroom instruction, most testing, most grading of students are associated with the lowest category, knowledge, of the cognitive domain. Too many teachers build their programs around content items that are fed to the students for regurgitation on examinations. Certainly attention must be given in our instruction and our evaluation programs to the higher categories of intellectual attainment. The mere fact that they are harder to measure and often require non-paper-and-pencil test evaluation procedures in no way dismisses these from the domain of responsibility of the classroom teacher.

Consider for example, some of the levels of conceptual understanding of a specific computational skill:

1. The student can perform the computation utilizing crutches.
2. He utilizes memorized facts to perform the computation.

[3] D. R. Krathwohl, B. S. Bloom, and B. B. Masia, David McKay Co., Inc., New York, 1964.

3. He employs shortcuts in the computation.
4. He relates the computation to another system like the number line.
5. He can estimate results and check accuracy.
6. He utilizes properties of the computational process such as commutativity.
7. He justifies the steps in the computational algorithm.
8. He can write a proof of the algorithm.
9. He can develop his own algorithm.

Advantages and Disadvantages of Behavioral Objectives

Behavioral objectives are making an important positive contribution to teaching. They provide a basis for tighter organization of an instructional program by specifying in detail what we want student behavior to be at the end of an instructional sequence. They also help teachers and curriculum developers to evaluate their product. They need only answer the question: Do students attain these objectives? In this regard, they provide a framework upon which accountability may be mounted. Once the specific objectives of our programs are spelled out, we can then better evaluate our teaching programs designed to reach those aims. If we can find better ways to invest our educational dollars, we should be seeking them out; and carefully stated objectives are a long overdue contribution to our ability to decide such questions.

Like so many other educational tools, the negative side of behavioral objectives is not inherent in the tool itself but rather in the widespread misuse of this tool. We have pointed out how the difficulty of stating behavioral objectives for higher cognitive levels increases exponentially. Parallel to this is the difficulty of establishing simple paper-and-pencil measures of achievement of these higher cognitive processes. What happens? They do not get measured! And in effect the more important goals of teaching—independence, critical thinking, creativity, skill in attacking truly original problems and solving them—are lost in the shuffle.

Spokesmen for behavioral objectives claim that all objectives may be stated in behavioral terms and their achievement measured! Challenged with the problem of stating goals like the ability to transfer conceptual understanding from one system to another, they respond that lack of current satisfactory measurement tools does not imply lack of techniques in the future. You cannot fight that.

But this is not really the problem. Thoughtful psychologists, educators, and test constructors (notably the test makers for the National Longitudinal Study of the School Mathematics Study Group) are stating higher cognitive goals in behavioral terms and are devising test items to provide measures of our success in reaching those goals. This tends to support the stand of the behavioral psychologists. What is wrong is that the easily stated objectives that are matched by repeat-back-to-me-what-I-have-just-told-you or do-ten-more-of-the-same testing items are too often the only items.

Standardized tests and commercial and even experimental textbooks are just as guilty as classroom teachers of focusing on the easy-to-test as opposed to the tough-to-identify and correspondingly difficult-to-teach. And here is where accountability goes astray. When our objectives focus on the trivial and the superficial and the easy, it turns out that tax dollars can often be better spent on machines than teachers. When instead our objectives focus on higher cognitive processes and we develop instructional programs that attack those objectives, we are on much more secure ground.

Learning Exercises

1. The hierarchy of objectives on pages 50–51 applies to teaching addition of integers. Give examples of behavioral objectives related to levels 1, 6, and 9 as they apply to this topic.
2. Develop a hierarchy like that on pages 50–51 for problem solving.
3. Develop a hierarchy like that on pages 50–51 for writing proofs.
4. Write a behavioral objective related to each of the following:
   (a) Teaching the definitions of regular polygons
   (b) Teaching how to solve a system of equations
   (c) Teaching how to prove $a \cdot 0 = 0$ using the distributive law
   (d) Teaching how to solve word problems in first-year algebra
5. Devise some activities you might use in an instructional sequence that relate *directly* to the behavioral objectives of your answers in exercise 4.
6. Go through several exercise sets in a contemporary high school mathematics textbook to determine the proportion of exercises that relate to the different categories of the Bloom taxonomy (see page 50).
7. What do you think should be the mathematical competences of the educated adult?
8. What do you think should be the minimum mathematical competences of the graduating high school senior?
9. Formulate an answer to the student who asks, "Why do I need to study geometry anyway?"

Suggestions for Further Reading

Barnard, James, and Jonathan Knaupp, "Using Behavioral Objectives to Teach Placeholders," *School Science and Mathematics*, 71, 6 (June 1971), 538–542.

Bloom, B. S., *et al.*, *Taxonomy of Educational Objectives*, David McKay Co., Inc., New York, 1956.

Hannon, Herbert, "All about Division with Rational Numbers—Variation on a Theme," *School Science and Mathematics*, 71, 6 (June 1971), 501–507.

Johnson, Donovan A., "Behavioral Objectives for Mathematics," *School Science and Mathematics*, 71, 2 (February 1971), 109–115.

Lick, Dale W., "Why Not Mathematics?," *Mathematics Teacher*, 64, 1 (January 1971), 85–90.

Ruchlis, Hy, "Putting Reality into Mathematics," *Mathematics Teacher*, 64, 4 (April 1971), 369–371.

Wood, Robert, "Constructing Teacher Objectives," in Douglas B. Aichele and Robert E. Reys (eds.), *Readings in Secondary School Mathematics*, Prindle, Weber, and Schmidt, Boston, 1971, pp. 272–294.

The Mathematics Teacher in the School

TWO

The Beginning Teacher 5

No one can be entirely ready for his first teaching assignment. You want to create a good first impression and wonder how to start. You wonder what your students will be like and how they will respond to your questions and react to your assignments. You wonder how you will face your first discipline problem. And you have thousands of questions that are basic restatements of the general question, "What do I do?" In this chapter, we set out to provide some answers; common sense will help with the rest.

No matter what your background and experience in a teacher education program a basic suggestion to you is in order: *Assume personal responsibility*! If your opportunity to visit classrooms, to familiarize yourself with texts and programs, or to meet students of varying ability has been limited, you will be at a disadvantage. Like so many of our students today, you will be *entering school as a remedial problem.* Now you must take charge of your preparation: You must look for ways to extend your preparation. Visit schools and classrooms. Talk to students, teachers, counselors, administrators. Attend teacher meetings and professional conferences. Read professional journals. Get started now! If you think that there are conflicting demands on your time before you teach, you will be shocked to find how they multiply when you enter the classroom.

In this chapter we ask you to learn about yourself, your new school, your coworkers, and your students.

Know Yourself—An Important First Step

Two serious problems faced by beginners are underconfidence and—less often—overconfidence. Before you embark on actual classroom practice, inventory the assets and liabilities you bring to this task. Be frank with yourself. You know your personality; you know how you best get along with others and what problems you face in social interaction; you know a good deal about your command of mathematics content. Our experience has been that in virtually every case your inventory will provide you with a positive total, an adequate base for entering that first classroom.

There are among us some born teachers, some good teachers, some weak teachers, some lazy teachers, and some who fail for one reason or another. What you must now realize is that each of us, because of the qualities he brings to teaching, will be limited in the range of what he can do. There are few among us comparable with Socrates or Emerson or Mark Hopkins. But though each of us is "assigned" a range, that range of quality is quite broad. On Figure 5.1 is denoted a sample range of teaching achievement for a specific person. *His responsibility is to move as rapidly as possible toward the upper limit of that range.* By inventorying your basic make-up, you are in effect identifying the range of your possibilities as a teacher. Use it to identify a course of action.

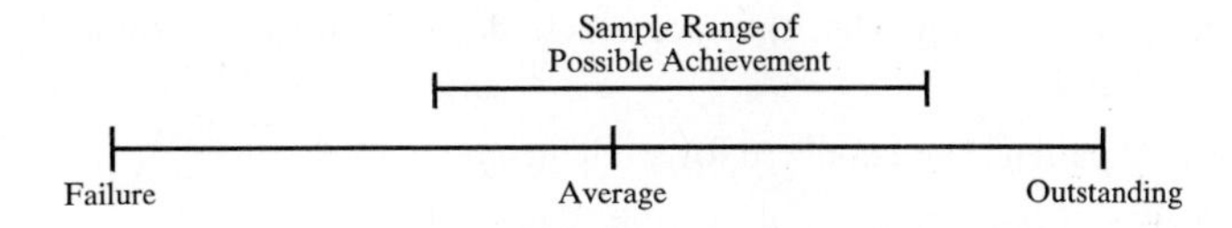

Figure 5.1 The Range of Teaching Achievement

Match your assets against those of many beginning teachers:

Physical energy. Your stamina is probably considerably greater than that of older teachers.

Youth. This carries with it an understanding of what is modern living—or "where it's at."

Subject knowledge. Your knowledge of mathematics is probably more current than will be that of many of your coworkers. Some of you will enter a school to find yourself the staff member with the most mathematical background. Even when you are not first in quantity you will have the valuable asset of recency—so important with today's developing mathematics and changing mathematics vocabulary.

A fresh, open, accepting attitude.

Enthusiasm and idealism, both basic ingredients for success.

Curiosity and willingness to try new things.

Add to this list the special aspects of your personality that supplement or replace these characteristics. Especially valuable is a sense of humor; so, too, are a friendly, outgoing personality and emotional stability. All these characteristics are not peculiar to either beginning or experienced teachers, but they are, of course, characteristics that promote good teaching.

Include, too, qualities that you can have only though effort on your part. Important among these are industry and responsibility. You can decide to short-change the first and dodge the second quality. More often, however, you will choose to apply both to your work in the schools.

Balancing these assets are a variety of liabilities often identified in beginners by other teachers:

Inexperience in working with young people.

Lack of judgment. Often beginners try to respond to problems on the basis of too little evidence. Emotion has a way of overwhelming reason at first.

Unfamiliarity with school regulations, personnel, students, and program.

A tendency to copy what they see others do without evaluation and modification to suit their own style.

Lack of patience and concern for students and coworkers.

Inability to separate themselves from their students.

Uncertainty about goals of instruction.

Inability to sense or diagnose learning difficulties.

You may bring other limitations to your work. It is especially important for you to recognize physical limitations. One of these is of particular concern: Do you have a small, weak, or high-pitched voice that will require extra effort to make it audible? There are highly successful teachers who teach from wheelchairs, who are blind, who have major deformities, who stutter. But these are people who do not merely recognize their limitations; rather, they identify them and work around or even through them. They face up in a variety of ways to these special challenges.

Once you have identified your assets and deficits, try to think of ways to add to the first and subtract from the latter. You are best equipped to do this for yourself, but there are many people in the school who will go out of their way to help you. Alexander Pope, a poet with gifted insights into human nature, said that the best way to win a friend is let him do you a favor. This twist on the usual form of avuncular advice should be remembered by all novice teachers. Within reasonable limits, people enjoy helping others.

Know the School Mathematics Curriculum

Obtain copies of the texts currently in use at all levels. If the school or mathematics department has curriculum guides, obtain copies of them too. Familiarize yourself with not only the courses you will be teaching but others as well. Remember that your courses fit into a total program.

Determine the special provisions of the curriculum for students of differing ability. Are there special classes for these students (homogeneous grouping) or are you expected to respond to their differences in the regular class?

Know School Routines

If it is at all possible you should visit the school in which you will teach before you take full responsibility as an instructor.

Many schools have brochures developed for their students and teachers. These are of special value to you. This is the kind of information you need:

A map of the school giving locations, not only of rooms, but also of administrative, departmental, and guidance offices; teachers' lounges; cafeteria; and other facilities. You will gain in confidence when you know your way around the building.

A listing of key personnel: principals, guidance counselors, department chairmen, faculty, and often student leaders. Give yourself a head start in learning some names.

Bell schedules. Overcrowded schools today have a wide variety of complicated schedules. Initial familiarity with modules, luncheon routines, and assembly modifications will be useful to you. See to it that you know and adhere to the time schedules set for teachers. To many teachers and administrators, tardiness is closely akin to deviltry. See to it that you avoid this highly visible error.

Student regulations and procedures. These provide information about student obligations that you are committed as a faculty member to support.

Miscellaneous information about the school—as much as you can find.

Some schools have brief codes of behavior for teachers; many do not. Teacher conduct codes are more often a matter of an accumulation of customs as are such conduct codes in many other positions. A few regulations regarding teachers are part of Board of Education policy and should be communicated to teachers. Usually these regulations include limitations on the disciplinary measures that may be taken by teachers. If a summary of such regulations is not available in a faculty handbook, ask about them. In addition to this, it is useful to read codes of ethical teacher behavior published by the major teachers associations and unions.

If you are a student teacher, you are a guest of the school. For very little return to them, the school staff is providing you and your sponsoring college this service. In almost every instance, you will find them very hospitable, and you should respond in kind. There may or may not be good reasons for all school regulations, but they *are* the regulations and you should help to support them. In particular you should go out of your way to conform to standards of dress and appearance, not because you are giving in on an issue of importance to you, but because you are a guest who would prefer to be less comfortable himself than to make others uncomfortable.

The new full-time staff member does not have this rationale to support a particular style of dress or grooming. Many young people ask others to judge them by their intellectual and moral standards and qualities rather than by their appearance. The point that they miss is that they will be judged on *all of these.* Since grooming is so immediately visible, it plays what may be a disproportionate role. Bad grooming, exotic dress, poor personal hygiene, these can make a difficult role far more difficult, sometimes impossible. You will be, as a teacher, one of the most important models of behavior presented to impressionable youngsters. This places on you special responsibilities that you should seriously consider. In the final analysis, the model you present in moral conduct, intellectual attitude, industry, and personal appearance should be determined with this special consideration in mind. Whatever you decide, do not take this lightly.

One final word about school regulations is in order. Many of you will never find yourselves in a position in which you must speak out against the standards of the school. Others will not be so fortunate. You may identify school problems that seem to call for opposition to school policy. We counsel you only to husband your resources. If you make yourself a constant gadfly to the school staff, you will find your influence weakened ("He's against everything"). Do not then waste your energies tilting with minor and unimportant issues. When the significant issues arise, you will then be in a far better position to speak out.

Know Your Fellow Staff Members

The Cooperating Teacher

For the student teacher the most important person in the school is his cooperating teacher, the secondary-school staff member who oversees his practice teaching or internship. For many beginners, this person will most influence their whole career in teaching. You will be in close contact with this person for a fairly extended period, so you will want to develop a positive relationship with him. Common sense should suggest to you many ways to make your working conditions together optimum. Some of these are:

Look for ways to help. Do not always wait for your cooperating teacher to assign your tasks. Ask to share in correcting homework even before you have taken class responsibility.

Take the initiative when it is appropriate. Helping individual students is a good way to start. Take responsibility for the class if the senior teacher is called from the room.

Support your cooperating teacher with his students. Even if you and he have differences, keep those differences private. More than this, you should exert a positive influence on students whenever you can. A remark to a student about something well done by the other teacher will always make everyone feel better. The student appreciates his instruction more because an outsider has approved; he will make it easier for the teacher to do his job; and you will know that you have improved classroom atmosphere.

Seek evaluation of your work. Do not wait until the end of your assignment to ask for criticism and suggestions. And do not let remarks like "You're doing okay" serve. You can help your sponsoring teacher to focus on particular points by asking more specific questions. Is my planning okay? Did you feel that I executed my plan well? Am I involving students enough? How can I make this part of the lesson go smoother? Is there any way that you can suggest to work with Bill? Do I speak clearly? How can I motivate Mary? How can I vary assignments?

Participate in school routines in addition to your assigned classes. Use your free time constructively. Encourage your cooperating teacher to help you to become a part of the school.

Good judgment on your part is demanded to make this relationship optimum. Quite naturally you will evaluate the teacher with whom you are working. You will wish to add many of his teaching techniques to your own ideas about teaching. You will probably want to adapt to his style, in some ways temporarily, in other ways permanently. No matter how much responsibility you are given, you should always remember that this is his class and only partly yours.

Slavish copying is not demanded, nor is it in order. From your first day in the classroom, you should try to be creative. Almost all critic teachers will respect your ideas; many will be delighted to hear them and see them in action. Expect them to reject some of your ideas, especially at first. They will help you to give your own thoughts the perspective of experience.

In your first teaching, you should start with a lesson format quite close to that of the supervising teacher. In this way, you will have a less disruptive effect on the class. But as you gain confidence and as the class and your coworkers gain confidence in you, you should begin to modify your initial teaching to be more *your* style. Try different techniques or different kinds of lessons.

Hopefully, you and your cooperating teacher should both learn during your stay. When he is able to say that he too has learned, you can be proud of your work.

The Mathematics Department Head

The department head is a primary reference person. It is to the chairman that the beginner must go for assistance on problems, for books and supplies, and often for personal guidance. The department head will usually wait to visit classes until they are well underway, but then he will occasionally attend in order to provide assistance and to evaluate progress. If you have an especially interesting lesson coming up, invite your chairman to sit in.

The Principal

The principal and his assistants are the most important administrators for an individual school. The principal is responsible for the school program—your teaching—and school personnel—and you. He is also responsible for thousands of other things, from budgets to public relations, but these two are the important ones for you. He usually sets the tone for the school, and identification of this tone is important for you. In particular, you should identify whether he runs a taut or loose ship.

Too often, the only contact of a beginner with the principal is related to details rather than total program: discipline cases, attendance and grade reports, parental problems, personal illness, and the like. You must recognize that the principal's life can often degenerate into a succession of these details. The details related to you in turn may sum to his evaluation of you. Do not let this happen. Early in your teaching career, *you* should initiate periodic appointments with your principal to discuss your progress. It is best to make these formal and brief. Tell him of your successes and tell him too of areas in which you need help. Ten minutes in a meeting like this can completely change your relation with your principal.

District Mathematics Supervisors

Few mathematics teachers realize the resources the district supervisor has available for the individual teacher. Too often they view him as a distant administrator who holds occasional meetings, perhaps makes up examinations, and works through committees to select texts. He does do these things; but, more important for you as an individual teacher, his office is usually a repository for most of the new things in mathematics education: sample texts and a wider variety of commercial models and visual aids. He is the person in the district who has most direct access to these materials. Salesmen go directly to him.

You should make a point of meeting this person and visiting his office. If you want to try a new program, ask him; he can often help you. For example, suppose you have a low-ability seventh grade for whom you would like to introduce a unit on desk calculators. Ask the district coordinator. He may be unable to be of help, but on the other hand he may be able to purchase or borrow thousands of dollars of equipment for your use.

Other Teachers

In most schools, the individual classroom is isolated from other staff members. When you close that door, you are essentially on your own. (There are partial exceptions to this in team-teaching programs.) You will know only by hearsay of the success or failure of even the teacher next door. But you will naturally come in contact with other teachers outside your classes. For your own well-being, it will be helpful to establish friendly relations with all of your coworkers.

A few cautions are in order:

Especially as a beginner, do not encourage confrontations with other teachers. You may sharply disagree with many points of view. Listen.

Concentrate on positive comments about students and teachers. If you have something good to say about someone, say it; if you have complaints, discuss them only with the person who can best respond.

Don't be a "lounge lizard," spending all your free school time relaxing in the teacher's lounge. As a mathematics teacher, you would find this difficult anyway, but you should be aware that teachers may evaluate you by what they see of you.

Other mathematics teachers can be a big help to you. Discuss with them your teaching. Experienced teachers will usually be delighted to share their ideas with you. Often a tightly knit staff can share some of the obnoxious jobs like test construction. Be sure that you contribute to as well as take from your coworkers. If an older teacher lets you use a review sheet or test of his, you should try to return the favor in some way.

Know Your Students

There is an old saying among teachers that schools are fun in summer when the kids are gone. Of course, schools were not built for vacations, and your role is bound up with those 100 to 150 (and in a few situations, more than 200) students that you will meet in your mathematics classes. Your job is to develop and supervise a program in your classroom that will help these students to gain mathematical maturity. This entire book is concerned with various aspects of this program; here we comment only on your social relations with your students.

It will help you to recall your own student experiences and attitudes. This will make you realistic. If you were like most students, you were more interested in peer approval than teacher approval, you often sought to get by with minimum effort, you tested regulations to determine just what were the limits on behavior, you enjoyed breaks in routine especially when they substituted for work. In fact, most of your strategies were designed to *protect you from learning*. (This is the thesis of John Holt's *How Children Fail*,[1] an excellent book on classroom interaction between students and teachers.)

Despite this, most students will progress. You must seek the best way to aid (or, in the case of some very bright students, not to impede) this progress. For some, this may best be done by prodding, for others by counseling, for still others by challenging them. If you had the opportunity to deal with one student at a time in a tutorial setting, you could probably determine the best approach to suit an individual personality. In a classroom, you cannot do this nearly so well. There are, however,

[1] Dell Publishing Co., New York, 1964.

some suggestions that may make your work with students somewhat easier:

Seek to gain your students' respect, not their friendship. Be businesslike and serious about your instruction, especially at first. You may be able to relax later.

Establish classroom routines and stay as close to them as possible. Avoid letting students disrupt these routines. For example, if a student is tardy, do not break into your lesson to discuss his problem. Merely ask him to see you at the end of class and go on with your work. (Do not forget to see him then!)

Be fair and provide equal treatment to all students.

Learn your students' names immediately. Record them and memorize them. Refer to them by name.

Remember that you are a new challenge to your students just as they are a new challenge to you. This gives you a few days in which to establish your routines, your discipline, your personality in the classroom. Use this time. It will be more difficult to do later.

Give all students your attention. You will find this difficult to reach beyond those extroverted attention-seeking students to serve the quiet retiring introverts or unattractive nonconformists who may need even more help.

Always stress the positive. Call attention to success. Try not to reward negative behavior. (Remember that any attention is a reward to some students, including the time you spend scolding them.)

Some educators would have you say, "I don't teach mathematics; I teach children." We believe that the stress here is equally unfortunate in either direction. Your job is to *teach children mathematics.* Certainly you should provide general guidance to the youngsters in your classroom. You want to help them to mature, to take responsibility, to make decisions. But you want them to attain these appropriate goals as part of their study of mathematics. The two are not distinct and should not be considered separate.

Sometimes you will need help in working with individual students. Do not try to go beyond your depth in dealing with severely disturbed youngsters. You can get informative records about your students from guidance counselors, who can also give you personal insights into their behavior. Use such information with discretion. Some teachers prefer not to utilize previous records on students or the impressions of others about these young people. They prefer to form their own independent evaluation, in this way to give the student a chance to start fresh. A balanced approach is best. Seek records and teacher comments when you feel they can help you to solve problems, but do not let them fix a student in one mold. It is, after all, exactly your job to modify behavior in positive ways. When outsiders can help, refer to them. Do not let them limit you.

Mathematics Teaching as a Job

Mathematics teaching is hard work. There are highly creative artistic aspects of this work. You will be center stage for several hours each day. There are also hours of planning necessary before you "perform" in the classroom. And finally there are hours of work correcting homework papers and tests.

To do justice to this task you must organize your time and budget your energy. As a beginning teacher, keep your other responsibilities and planned activities light. Teaching, like many other tasks, gives extra rewards to those who work harder at the beginning. A useful analogy here is learning itself. If your learn fundamentals well, later learning is easier.

As a beginning teacher you should plan with care your overall program as well as your daily work. Try to make each lesson the best you can develop; keep these plans. Make each test a valuable educational experience for your students; retain copies in your files. Construct models to illustrate concepts; save them. In this way, you build up a basis for an easier program in future years. You should certainly not plan to reuse the program in exactly your first format, but your first plans will form a sound basis for improvement. Failure to dig deeply into your work first time around not only weakens your teaching program now, but also merely postpones work that must be done if you are ever to develop into a good teacher.

In the chapter on classroom management are comments on homework administration. Here we add only that you should be sure to schedule your time in such a way that homework and tests are returned promptly. Their value to your students varies inversely with correction time.

Beginners must recognize that student teaching, as close as it approximates regular teaching, is still an artificial experience. When you enter a classroom as a beginning professional in early September, you have a major and most important task: to establish yourself and your program. In many modern classrooms, this is a battle that drains a teacher's energy for months. To many beginning teachers, the outcome is in doubt for more than a semester. Self-doubts even spill over into subsequent years; usually, however, a second year in the same school setting is much easier.

Since the earliest days of modern education, experienced teachers have told beginners to adapt their program to this extra problem of the new teacher. For almost as long, beginners have thought they were different and would not face these difficulties. In almost every instance, the beginners who do not recognize the difficulty add to their problems and extend the period of battle.

A few pointers that experienced teachers and supervisors often suggest are:

"Overplan." Have class activities highly organized.

Keep more distant from your students at this time.

Seek ways of cutting time in transition periods short. These are the times when students are changing from one activity to another. They are always a source of disruption.

Monitor student work more closely at this time. Homework patterns are established early and should be checked carefully.

Postpone striking deviations from standard classroom organization until you have established your basic routines.

Pay much more attention to discipline at this time than later in the school year. Try to be prepared ahead of time for disruptions. In this way, you can often minimize them. For example, you may wish to have prepared a special worksheet of problems related to your class discussion. When the expected disruption occurs, ask the student in a quiet voice to work by himself on the worksheet. If this is well done, you will have avoided a confrontation; you will have partially segregated the disruptive student; and, most important, you will not have broken seriously into the progress of the rest of the class. (This is, of course, just one of hundreds of tricks for turning off disruptions. Unfortunately such tricks are more important to the beginner who has not yet developed a number of them.)

A Final Suggestion

Lest some of the difficulties that have been detailed in this chapter overwhelm you, we leave you on a positive note: Be yourself! Let your personality come through to your students. Enjoy the many funny things that happen in school—even when the joke is on you. Enjoy your students. A close examination of even the most disruptive or antagonistic of your students will show him to have many good qualities. In fact, you will find that many times meeting these problem students outside the classroom in more relaxed surroundings provides the basis for friendship. Enjoy the respect that most people in the community feel for the difficult work you are doing. And enjoy too the feeling of renewal that young people give. Few people have as much a part of the future as do the classroom teachers in our schools.

Learning Exercises

1. Make an inventory of the qualities you bring to teaching. Include both positive and negative factors.
2. Why did you choose mathematics teaching as a career? What people influenced this choice?
3. Name the specific teacher who was the best you encountered in your education. This teacher may have taught you at any educational level: elementary, secondary, or college. Describe in detail the qualities of this teacher that made him good. Describe any negative qualities he had. Why did these not lower your estimate?
4. Describe a teacher of yours who was a failure. What did he do wrong?

5. Describe the classroom atmosphere you would like to achieve. Tell some ways that you would work to accomplish this atmosphere.

6. Describe a delinquent student you have known. (It may be you.) What are some things teachers could have done to help this student? Did teachers contribute to his delinquent behavior? How?

7. What values or attitudes do you have which you wish to transmit to your students?

Suggestions for Further Reading

Schorling, Raleigh, "An Evolving Bill of Rights for Teachers," *Mathematics Teacher*, 44, 2 (February 1951), 100–103.

Syer, Henry, "A Core Curriculum for the Training of Teachers of Mathematics," *Mathematics Teacher*, 41, 1 (January 1948), 8–21.

Classroom Management 6

As human beings, we all want to succeed in our endeavors. We want our colleagues, our administrators, our students, and their parents to say, "Well done." However, one of the major stumbling blocks in the way of success as a teacher is the failure to deal with the day-to-day problems of student activities. Consequently, this chapter will suggest ways to meet these problems, even though they are not all unique to the mathematics classroom. A major step toward success is taken when the teacher finds the key to efficient classroom management.

How does a teacher prevent discipline problems? How does he meet them when they occur? What can he do to establish (or even regain) his authority in the classroom? We believe that the basic answers to these questions lie not in a "law enforcement" program, but rather in careful organization of classroom activities and in proper motivation of students. There will always be some students who resist all efforts by adults to help them. Sometimes teachers themselves lack the qualities needed to control a class. The majority of discipline problems are, however, a result of poor classroom management and poor instruction on the part of a teacher who would otherwise succeed.

The adage "Teacher and student suffer together" applies best—or perhaps worst—to the poorly managed classroom. Students who do not know or understand what is expected of them become discipline problems. These students irritate the teacher, who in turn reacts harshly and unwisely; and the cyclic pattern, the self-serving spiral of difficulty, resentment, and antipathy is initiated.

The best time to control this ever-present teaching danger is at the beginning of the school year. At this time, classroom routines should be established and students should be made aware of what is expected of them and what they may expect of you. Unfortunately, it is the beginning teacher who most needs to set up careful ground rules for his teacher–student interaction and who at the same time most often fails to do so. Because of this early failure in their teaching assignments, some teachers leave the profession. Others learn this lesson, salvage their first teaching year as best they can, and start off better in subsequent years. The less fortunate never learn; they merely stagger along, usually hating their work and blaming their lack of success on students or school administrators. Meanwhile, those who start out on the right foot, who provide their students with the real *security* of knowing the rules of learning, who establish in a friendly atmosphere reasonable standards of conduct and work before problems arise, reap the benefits of a much more satisfying experience.

Although we focus on classroom rules and routines in this discussion, each teacher should set out to establish class regulations in a friendly, supportive atmosphere. The teacher should explain why these regulations are necessary, indicating his genuine and primary concern for the student and his progress. The teacher must be sincerely interested in students or such statements will sound foolish; students today are quick to react negatively to insincerity.

Beginning of the Term

Some class time at the beginning of any course should be devoted to giving students a clearer idea of the content they are to study and the techniques of instruction to be utilized. Specifically, you will wish to answer the following five questions:

1. What content will the course cover? Course titles like "Plane Geometry," "Trigonometry," or even "Intermediate Algebra" usually mean little to students. A student's ideas of what a specific course involves are often far from the truth. For example, many students enter a geometry course thinking that it will merely continue and extend the informal—even superficial—study of geometry of the upper primary grades: that is, computing areas and perimeters, identifying figures, and perhaps drawing more complicated diagrams. In this regard, there have been too many cases of weak students being counseled into geometry because they like to draw.

In discussing the course, the major point to make is why this course is important. What are the major goals of the course? What are your goals? Why is the course organized the way it is? Why are certain topics included? It might also be appropriate to refer to the way in which a course has developed or changed in recent years.

It is not enough merely to tell what the course is. Examples are needed in this introduction. They are especially useful when they relate to what is already part of the students' experience. For example, in geometry a careful proof, in the basic form you plan to use for the course, of an algebraic theorem familiar to the students will show the formal nature of what is to follow.

In many courses, you will be traversing ground already familiar to the student but doing so in greater depth than in the earlier courses. In college or advanced algebra, for example, logarithms are usually re-examined. An example or two comparing the level of attack you plan to employ with the earlier method of intermediate algebra will quickly demonstrate the approach of the new course:

In intermediate algebra we were mainly concerned with the use of logarithms to compute problems like the simplification of

$$\frac{3.24(23.7)^2}{.0674}$$

We will now be more concerned with deeper understanding of the logarithm as a function, the properties of its graph, and such problems as

$$\text{Prove: } \log_N b = 1/\log_b N$$

2. How is the course related to the rest of the academic program? Students should be told how the course will develop and to what future program it may lead. Consider the following statement:

Advanced algebra may extend the ideas of elementary and intermediate algebra programs to polynomial equations of degree higher than two, introduce concepts of probability, and provide the algebraic and some of the analytic foundations for a more careful study of the calculus.

While this statement may be understood by a person familiar with mathematics, it is virtually meaningless to the student. Examples, however, tend to clarify the relation:

You recall that in intermediate algebra we learned to solve equations of the form $ax^2 + bx + c = 0$, to recognize relationships between roots and coefficients such as the sum of the roots is b/a, and to identify properties of the graph of $f(x) = ax^2 + bx + c$. Now we will examine higher-degree equations, like $ax^5 + bx^4 + cx^3 + dx^2 + ex + f = 0$, to develop similar methods of attack.

In your introductory discussion it would be well to relate the course to science, business, or education—to mention but three areas in which mathematical applications can be found and in which students have some background and interest. For instance, in an introduction to intermediate algebra you might say:

In junior high school you learned the simple interest formula $i = prt$. In this course we will develop a formula that extends this idea to the calculation of the return, A, at compound interest: $A = p(1 + r)^t$.

Such formulas, solved by appropriate short cuts, using calculators or logarithms, provide the basis for the solution of many problems in banking.

Be sure to indicate to students how the course they are now taking fits into the pattern of prerequisites for future study:

Algebra provides basic tools for solving problems in many fields of study from physics to history, from botany to foreign languages and at the same time for further study in mathematics. Thus, it is one of a sequence of courses required for admission to most colleges. It lays the groundwork (with geometry) for the study of the calculus itself, a basic tool of the sciences, and for the study of finite mathematics, probability, and statistics—all important to study in the social sciences.

3. What will be the teaching techniques used? The serious student is interested in *how* you plan to teach so that he can adjust to your mode of presentation. The less serious student also needs to know what techniques you will use, if only to avoid later confusion. Will out-of-class assignments merely be applications of skills developed in class, or will they require the learning of new materials to supplement and extend ideas presented there? Do you expect students to take notes of ideas developed in class discussions? Will your course be formal—concerned with mathematics developed from a postulational approach—or will it be more concerned with problem solving, discovery, applications of principles, or even immediate utility? A teacher taking a middle-of-the-road approach might say:

In this class we will be interested in basic principles, of course, One of these is the distributive law which says that for all a, b, and c:

$$a(b + c) = ab + ac$$

We will learn how this law applies to many examples like $(x + 3)(x - 5)$ and $3\frac{1}{7} \times 5$, and even 32×2. Once we have noted how and when to apply the law, I will expect you to use any shortcuts in computation available to you. However, whenever you use a shortcut, you should know its basis so that you could defend it as being mathematically correct.

Your teaching may also rely strongly on student participation. Such participation must be the result of good interaction based on mutual trust, liking, and respect between students and teacher. But it is often well to explain when you want students to participate informally and when you do not, how you expect them to answer your questions, and what are your goals in doing this. Without such background, some students will fail to understand that a lesson that is fun for them is also developing good mathematics. (Some teachers actually have had complaints from parents or administrators who are upset by hearing students comment that their math class is "a blast.") Students should understand the serious purpose underlying your most informal classroom activities.

You might even involve your students in deciding on methods to use. Ask them how they would teach a given topic. Ask them what method is most helpful. Explain why you do things the way you do. Then, tell them that you are searching for ways to make learning mathematics pleasant and effective. Discuss how the class can work together in learning ideas. In other words, let the students feel involved, let them know that the course content and techniques are designed for them and not for the teacher. Let them know you want to help them and that you will listen to them.

Inexperienced teachers would, in fact, do well to use fewer informal techniques early in the year than they wish to use later. Informality in the classroom is very useful in relaxing students, in giving them an incentive to participate; but the teacher whose reputation is not already established has difficulty in controlling such a class. It is easier to move from much control to little control than to move in the reverse direction.

4. What will be the role of the student? Students should know from the outset what will be expected of them in class and out. Are they to participate freely in the development of new ideas? Are interruptions, even contradictions, welcomed?

Will homework be collected regularly? Corrected? Returned? Graded? When can students ask questions? When are they expected to get help with difficult homework—before, during, or after class sessions in which the homework is due? Is cooperative effort on homework between students encouraged or discouraged?

Here is one of many possible plans with regard to homework:

> Homework assignments will usually be given daily. These assignments will be written on the board and should be copied in an assignment notebook or in the upper left corner of the paper to be used for the work. The written part of an assignment should be on $8\frac{1}{2}''$ by $11''$ paper with your name, the date the assignment is due, and the class period recorded in the upper right corner of the first page.
>
> Although I expect reasonable standards of neatness and legibility, I will leave style of presentation flexible and will talk to you individually about it only when I am not satisfied with your work.
>
> Answers for some exercises will be provided so that you have a check on your work. If you have difficulties, make a record of them and call them to my attention in class. Do not forget them! Mark the number or numbers of problems about which you still have questions in a circle on the top of your paper. I will go over your work on those problems in detail and will try to write suggestions or corrections. If necessary, I will ask you to see me for more extensive assistance.
>
> Papers should be turned in not later than the second school day after the assignment is given. Although I will check some individual papers each day and all papers on some days, I will always carefully check the problems you indicate.
>
> One final word about homework. Homework is important. I do not assign it merely to keep you busy and out of trouble in study halls. It is assigned to help you to learn mathematics and to give direction to independent study—study that is a necessary part of learning the mathematics in this course. In fact, it might well be the most important part of the course for you. Don't shortchange yourself by failing to put real effort into this work.

Along with this information, the teacher should indicate where and when extra help is available and state a policy regarding absence. Many teachers require a student conference after any absence of more than a day or two, both to check on the student's understanding of material studied independently and to provide extra assistance if it is needed.

5. What methods of evaluation will be applied? In these days of the increasing pressure of college entrance and academic success, grading is a serious problem for both student and teacher. Next to problems related to misbehavior, the most common teacher–student difficulties are related to this aspect of the teaching assignment. Your grading policies should be made clear to students early in the term. A written statement directed to parents as well as students is often appropriate. This statement should outline your grading practice in clear and careful terms.

High school students expect and demand equal treatment. They rebel at anything that smacks—no matter how remotely—of favoritism. For that reason, a stated grading policy (obviously one that is not in conflict with school policies) to which the teacher closely adheres gives students real confidence. Here is part of one such statement:

> Semester grades will be based on four factors: (1) one-hour unit tests, (2) short announced and unannounced quizzes, (3) assignments, and (4) my classroom evaluation. The test average, (1) and (2), will determine a basic grade, unit tests counting four times as much as the shorter quizzes. The last two factors will then be used to modify this grade. Homework, because it is not completed under supervision, will not raise grades. Failure to complete assignments without any acceptable reason will reduce borderline grades to the lower grade. Finally, my evaluation will reflect my judgment of your ability to achieve and the effort you evidence in participation in classroom activities. Conduct, good or bad, is evaluated on a separate part of the semester report and will *not* affect the academic grade.

This statement is objective—although teachers who wish to record items in terms of points may find it too subjective. The teacher should modify his grading program, using his professional judgment when necessary and appropriate. However, the more subjective the grading system is, the more it is open to question by student and parent. More important, a subjective system that does not provide for adequate recording tends to make the student unsure about receiving a "fair" grade.

In discussing grading policy with students, the teacher should be frank to admit the inaccuracies of these measures and the bias in favor of good "test takers." He may even wish to discuss with students their own preference for evaluation procedures. Interestingly enough, students—aside from those clearly identifiable as goldbricks or classroom politicians—tend to favor objective over subjective ratings. Many teachers also carry this one step further, arranging individual conferences at or before the assignment of grades. At this time the teacher takes the student's self-evaluation into consideration and compares it with his own evaluation. Often misunderstandings can be averted in this way.

When the conference is held a few days before the end of the marking period, students are given the opportunity to make up missing work and are often given the (admittedly superficial and extrinsic) motivation to study hard for that last quiz.

Procedures such as these set the tone for a course, give the student an understanding not only of what he is to study and what is expected of him, but also of what he can expect in return from his teacher. They provide him with a good measure of security.

How to Cover Review

"Deadly" is perhaps the best adjective to describe the extensive review built into many mathematics textbooks and unfortunately translated directly into classroom practice by many teachers. After a summer vacation, all students need a review; but a long sequence of review lessons, uninterrupted by anything that could be called new material, impairs student morale.

Perhaps the most striking example of such misguided review is illustrated by an experience of one of the authors of this text. Introduced as a speaker by the president of a junior high school PTA, he was stunned by the final sentence of the introduction: "Perhaps our speaker this evening will be able to tell us why our sons and daughters in the seventh grade are still—in January—studying the same content they studied last year."

While the teacher should recognize the need for review, he should seek to introduce new ideas early in the course to motivate students and to build their enthusiasm. If this is done in the first few weeks, review can be made part of the overall program. This program of introducing new material at the outset has an additional advantage not readily apparent to many teachers. During the first week or two of a school year, students are rusty. Their past studies are not at their fingertips yet, and they often give the unthinking teacher the impression that they have been poorly taught. (This is one basis for an interesting phenomenon often noted: the tendency for grades to improve through the school year.) Tests administered early in September show weaknesses that very probably would be less striking a few weeks later. For this reason, it is better to start with fresh material before embarking on review activities in order to give students time to reaccustom themselves to school and to accommodate themselves to you.

But review is necessary and, in fact, is central to the idea of the spiral curriculum, the program in which at higher grade levels you study the same concepts but in more sophisticated form and in greater depth. For example, in intermediate algebra the solution of quadratic equations is studied, but the treatment there is in greater depth than the corresponding treatment given in elementary algebra. Here the emphasis is on structure and relationships such as those between roots and coefficients. If such a program is to function properly, topics will be restudied. This does not mean that they should be restudied in the same form. New strategies should be utilized that make review more palatable and that make it, in fact, as exciting as studying new content.

Too often teachers do not think about instruction of a review topic in the same way they think about the instructional program for a new topic. They too, are bored by repetition. This is a truly unfortunate situation because teaching a topic in a different way can be an exciting challenge, which demands real creativity. (Chapter 10 discusses some of these alternate approaches.) The teacher should seek to develop his own approach, and he should continually search the literature to find ways others have used. In any case, he should be sure that his students recognize the role of review.

All this does not mean that there is no room for straightforward review of content only partly learned or retained. Such review is sometimes necessary and should be part of the program of maintenance for your own course as well as those taught during previous years. For this kind of maintenance, a summary examination is useful to identify general difficulties and to pinpoint specific weaknesses. Usually, such a test will indicate which students need special help and which may be exempt from the review program. These latter students may devote their time more profitably to the study of some optional topic while their classmates seek to strengthen their knowledge of fundamentals.

Every mathematics teacher should accept the responsibility for reviewing concepts and skills of previous mathematics courses. Thus, the algebra teacher includes review of arithmetic skills, and the geometry teacher reintroduces algebraic concepts whenever possible. Similarly, previously taught concepts are reviewed by including problems from previous lessons in current assignments. Many teachers incorporate some exercises of the previous day's lesson in their daily assignments. They also include some exercises from lessons or units taught weeks or months previously. Others confine their homework assignments to review exercises so that students have the necessary background for learning new ideas in class.

Special Devices for Review

Another useful device, one that may supplement the test, is the frequent use of a series of short worksheets. The worksheets are constructed in such a way that the problems on them take a total of about five minutes to complete. Students work a single sheet and correct their work by reference to the answers provided. They note the number of errors they made and seek to determine the cause of their errors. When necessary, they seek help from the teacher or another student. In the next review session, they retake the same sheet, or, if they made no errors, move on to the next one in the sequence. Since the stress in such a program is on individual independent activity, the teacher can usually work with a small group of students or with an individual having a specific problem.

Teachers have tried a variety of other approaches to review. Some utilize a short arithmetic-oriented warm-up at the beginning of each class: "Start with 5, add 2, subtract $1\frac{1}{2}$, multiply by ..." Such a procedure not only provides a short review but it alerts the students, readying them for the classwork ahead.

Elliott Pierson, a mathematics supervisor in Weston, Connecticut, has used a game called mathematical baseball with great success. In this game the usual activities of baseball—strikes, hits, outs—are replaced by mathematical questions. The class is divided into two teams and the contest begins. In an activity of this sort, students are excited by the contest and are motivated to do well for their team, a form of motivation that forms the thesis of James S. Coleman's book *The Adolescent Society.*[1]

Review and maintenance activities should be justified to the students, their importance explained and stressed. One way to do this is to frame questions related to a topic in class in such a way that student responses indicate clearly any need to retrace steps. For example, once a factor of a polynomial is located by an application of the factor theorem, the students may be asked to carry out the division to find the other factor, the depressed polynomial. This latter activity may make it apparent that review of this topic is necessary (before introducing such shortcuts as synthetic division).

Another way to review previous work is to prepare a statement or list of the skills or basic concepts needed for the next unit. These could be stated as behavioral objectives or as a series of questions. These lists should also include references for the student so that he can get help on those skills or concepts on which he still has weaknesses.

Review, reinforcement, and maintenance may best be considered cooperative activities involving student and teacher together. Sometimes parents or fellow students may also play a role in helping students. It is always helpful to make clear to a student what his difficulties are and how he stands in comparison to his classmates in regard to background. It is also appropriate to indicate to him the relative importance of his difficulties so that he can attack the most important problems first. Of course, the sequential character of the mathematics program often makes this difficult. For example, a student may have difficulties in addition and multiplication. Multiplication may be a more important tool for the work you are doing, but since it depends on addition—in most developments—addition must be stressed at least equally.

Classroom Routines

Each teacher must seek a mean between a classroom program that becomes boring because of its predictable sequence and a program that makes students insecure because they do not know what to expect. In a reasonable program, this kind of imbalance does not usually become a problem, however. Instead, the

[1] The Free Press, New York, 1961.

routines of attendance taking, announcements, assigning and collecting homework, and record keeping are uniform. Students know that the routine activities of yesterday will be the essential pattern for routine activities today and tomorrow. On the other hand, the mode of presentation of new material, the length of time spent on homework, and the entire structure of the teaching program are modified from day to day to suit content, teacher and student preferences, and even such factors as the day of the week, proximity to a holiday, or time of day.

Routine school tasks in the main require no professional training. Consequently, many schools are turning over such duties to nonprofessionals (or, to use the pedagogical term in current favor, paraprofessionals). Whether or not this is the policy in his school, each teacher should seek to minimize or at least to put into balance the amount of time and energy he expends on these tasks. It is easy for a teacher to become so weighted down with these peripheral activities that they keep him from his central task of planning and executing his teaching program. This does not mean that these "housekeeping" duties should be taken lightly. They are necessary to a successful program. What it does mean is that these duties should not intrude on other classroom responsibilities.

Some teachers use an interesting device for reviewing homework. Students who have difficulty with specific exercises on an assignment put the number of the problem on the board before class. Each student checks the board as he arrives. If he sees the number of an exercise he has solved, he writes out the solution on that panel. This is all accomplished during the first few minutes of class while the teacher makes announcements, takes attendance, or confers with individual students. It not only saves class time and frees the teacher from another directive activity but it also fosters self-reliance and cooperation on the part of students. Once the problem solutions are written out, a few extra minutes for general discussion or further questions completes this part of the class activity.

Some teachers appoint class clerks or assistant teachers. Such an assignment may be rotated if it appears to be a burden for any one student. The clerk takes over the duties of taking attendance and reading announcements, if the latter are part of class activity. Clerks or aides may also help with the preparation of dittoed or mimeographed materials, or getting films or tapes ready for class use. Occasionally, with adequate supervision and control, a student may collect and record completed homework and correct objective tests. In fairness to students, such sharing of responsibilities should not only free the teacher but should result in greater return to the students in the form of additional individual attention and an improved learning program. Such student cooperation should be voluntary. Their main activity should be learning.

Some teachers appoint a student as a class host or hostess. Whenever a visitor arrives in the class, this host or hostess goes to the visitor's assistance, finds a seat for him, gives him a textbook or other material, and describes the lesson being taught. Sometimes a seating chart is provided so that the visitor can identify each student.

Students may also be involved more directly in the teaching program. One student's helping another may be justified on the basis of the contribution this makes to *both*. Any teacher knows he learns his subject better when he teaches it than when he studies it only for his own benefit. In the same way, the student who tutors a fellow student or even teaches a group of students is forced to organize his information better and in so doing to learn it better himself. On the other hand, some of the bars to free exchange are let down when a student works with his peers, and some students facing difficulty can better explain and resolve their problems in this setting. However, the teacher should not overwork this technique.

Many students enjoy using the classroom equipment. Some teachers give a special assignment in the use of this equipment to one or two students each day. Instead of preparing the entire assignment, the student prepares only one problem, writing it out in detail on either a ditto master or an overhead-projector transparency for presentation in class. Preparing dittoed copies of homework solutions is also a useful technique in helping a teacher save class time.

In today's secondary school classrooms, little grouping is attempted. In most classes the teacher works either with individual students or with the entire group. Secondary-school teachers could learn much from elementary-school classroom teachers about the values of smaller groups. There are, however, some bars to grouping at the higher level, the principal one being that students and teachers are unprepared to accept the technique. One useful procedure that does work well is utilization of a tape table and prerecorded lessons on tape.

Each student wears earphones that carry to him the voice of the teacher or of a fellow student prerecorded on tape. The taped voice tells him what to do on a worksheet, explaining procedures and teaching as the teacher would in a tutorial setting. The student's attention is concentrated on the voice and the worksheet. He is virtually oblivious to his neighbor at the table, to whom he cannot speak anyway. While the tape table has a variety of uses, from introducing new material to reviewing old, one of its main advantages lies in the fact that it "assumes complete responsibility for" a group of students. At the same time, the teacher can work with other students in a more intimate setting. The use of similar stations with videotape presentations may be expected in the near future.

Chapter 22 on Laboratory Lessons describes in detail some other procedures for grouping students and breaking the fixed classroom routines.

Discipline

Meeting the problems of discipline in the classroom is one of the most serious concerns of secondary-school teachers. It is also the major cause of teachers dropping out of the profession. A classroom in which the teacher has lost control of his students is emotionally disturbing for the students as well as the teacher. Learning mathematics does not happen in a chaotic classroom. What can be done to cope successfully with classroom behavior problems?

The first step is to try to prevent the occurrence of discipline problems. This requires much understanding, planning, and ingenuity. You need to understand why a student wants to disrupt the class. Does he want attention, recognition, identification with a given group? Does he resent failure, lack of freedom, previous treatment, or rejection by his peers? Is his home situation one which causes him to rebel, to cheat, or to be uninterested? Do other out-of-school factors such as jobs or personal problems encroach on his school attitudes? Are there physiological factors such as poor hearing, eyesight, health, or nervous tension? Are there personal factors such as low ability, emotional distress, egotism, low self-concept, self-consciousness, or cultural deprivation? Are there, on the other hand, teacher factors such as hostility, discourtesy, boring presentations, unacceptable values, or indifference? Are there classroom factors such as unattractive surroundings, uncomfortable seating, or insufficient materials?

After identifying the source of discipline problems, preventive measures should be taken. Of first importance is your response to students. You must set an example of concern, courtesy, industry, and fairness. You need to demonstrate dignity, common sense, and a sense of humor. Only when you have disciplined yourself are you ready to help your students to discipline themselves.

Recognition of the problem student as an important person is often a very difficult and complicated procedure for the teacher. The teacher can go just so far in accepting a rebellious, rude student; but the teacher should constantly attempt to think of such a student as at worst only a temporary combatant. Many teachers have had the pleasant experience of meeting a difficult student outside the school environment—perhaps as a clerk in a store or a service-station attendant—and finding him to be friendly and courteous in the different setting. For this reason, it is often best to deal with a student who is a discipline problem outside of the classroom where he may feel more relaxed and less antagonistic in a less formal atmosphere. Then too, he will be away from his peers and will not feel constrained to act "for them."

Students want attention, and they seek it in many ways. The best way is through effort and resulting achievement. The teacher should support such effort with strong positive reactions. Other students seek attention in less socially acceptable ways. If these ways do not gain the desired attention, they are not achieving their purpose. A busy classroom with continual activity works against such students, as does a soft word that doesn't give the sought-for class attention.

Common sense does many things for a teacher. It identifies student attitudes; it tells him if there is time to relax a little in the face of student high spirits; it spots difficulties before they start; and it helps to put the daily work into proper perspective. A sense of humor is only an offshoot of this. It buoys up the teacher and the class and lightens the burden of hard work. Students appreciate a sense of humor when it does not substitute for teaching. In the same way they enjoy fellowship with a teacher, but they soon recognize it as a facade if they discover it masks the style of a lazy teacher.

But most important is dignity. Dignity is not superficial. It is not familiarity—rather it is an acceptance of the basic role of the teacher as an adult acting *in loco parentis*. He is not and cannot be "one of the boys" even though his teaching methods may often bring him into cooperative activity with students. He enjoys his students but does not seek to join them; he tries to understand their motivations, their talents, their ways; but he does not adopt them as his own. He maintains his adult standards and in this way communicates them to the students. He is the authority in the classroom and must not be over-ruled by his students. If students are allowed to defy authority, classroom control becomes impossible.

One characteristic that a teacher might well add to his profile is willingness to accept an occasional defeat. Sometimes a classroom confrontation will turn out badly. A teacher will occasionally require help from an administrator in resolving a discipline problem. And sometimes a class period or even a school day will seem a shambles. The teacher to whom this does not happen is a rarity. The quality teacher accepts this for just what it is, a failure; but he rebounds and tries harder the next day, using new techniques but without hostility.

How to Avoid Trouble

The following suggestions are specific ways in which to prevent discipline problems:

Have lessons well prepared with all needed material at hand so that your class starts promptly and proceeds at a challenging rate. Have alternate plans and materials available in case your lesson does not work or emergencies arise.

Involve your students in all learning activities. Every student should have a definite responsibility for participating at all times whether it is a discussion, board work, laboratory lesson, independent study, or a game.

Create a spirit of success and optimism by assigning activities in which the student can successfully participate. Build confidence, initiative, and loyalty by praising achievement and demonstrating interest in each student's progress.

Hear all and see all that is happening in your classroom. Stop disorder as soon as it begins. Call on students whose attention is wavering. Arrange the seating so that problem students are separated and are readily visible.

Be sure students know what their responsibilities are in respect to learning as well as behavior. Have students participate in establishing attainable goals and standards of behavior. The students should know what the rewards or penalties will be for their performance and the reason why standards and rules are necessary. Never establish a rule or make a threat that you do not intend to enforce. Every effort should be toward self-control rather than control by penalties.

Even when every precaution has been taken, problems will arise. To meet these emergencies, you should decide in advance what you will do in dealing with students who defy authority, throw objects, fight, cheat, talk out of turn, sleep, eat, smoke, play games, do not participate, or walk out of class. Whatever the penalty is, it should be something unpleasant, should be related to the offense, and should be given only to the specific individuals involved in the misbehavior. If at all possible, the penalty should be given in private at a time when you are not angry and as a means of helping the person involved. Under no circumstances should learning assignments or tests be given as punishment. It should be a privilege, not a penalty, to learn mathematics. If you are fair and consistent, you will reduce the resentment and hostility that always accompany penalties.

Sometimes teachers get help from the students themselves. Be open with them about the difficulties of dealing with discipline situations. Ask them what they would do. Ask them what they think is a reasonable classroom situation. Share with them your concern for being fair and understanding. But also point out the necessity for order and regulations, punishment and protection, rules and authority.

Personality Conflicts

No teacher should expect to like every student. Cheats and bullies, lazy students, and openly contemptuous students will be part of the burden of every teacher. The teacher must seek to accommodate himself to these students and to teach them despite his reservations about their behavior. Above all, he must seek to be fair to these students.

Less often, there develops between teacher and student a personality conflict that appears to be unresolvable. Each class meeting then becomes an unnerving confrontation for both student and teacher. When the usual attempts to resolve such a situation fail, outside help should be sought. School guidance personnel and principals may be able to help. Occasionally transfers are in order; too often these are difficult to arrange. Sometimes a conference with a third person can help, especially when such a conference is *not* scheduled at a time when tempers are already boiling. In a calmer setting when the student recognizes an honest attempt on the part of the teacher to resolve his problems, some headway can be made.

An important point for all teachers to remember is that they are not practicing psychiatrists. Each teacher must recognize his professional limitations and seek appropriate help for disturbed students. Referral of students who appear to need help should be prompt. Do not exacerbate problems by delay. A special warning is in order here: Be alert for quiet, withdrawn students who may have problems just as serious as the extroverts. Special attention should be directed to these students in less extreme settings as well.

Of particular importance is the recognition that symptoms are based on underlying causes. The show-off may be displaying to hide his basic insecurity; the student who dogs your footsteps may find in his superficial relationship an adequate substitute for a loveless home life. When the teacher realizes that students' lives extend well beyond classroom and school, he soon understands that recalcitrant students are not solely concerned with making his life harder. Of course, it is difficult to look for motives in a situation fraught with tension, and most teachers lose their tempers occasionally. Later reflection always shows a loss of temper to be counterproductive. The biblical recommendation to turn the other cheek is as valid today as it was when first recorded.

Housekeeping

The classroom is a good reflection of the teacher. The well-organized teacher has a neat, orderly, attractive classroom. Bulletin boards have interesting and instructive displays; the teacher's desk is orderly; and work tables, student desks, book shelves and display cabinets are equally well kept.

This is not a matter of the grade level of students or the sex of the teacher. No matter who the students are, they will respond better in an orderly classroom. A few minutes of extra time can turn a sloppy classroom into a much more cheerful place in which to teach. Students will help with this, and some will show non-academic talents that will raise your estimates of them significantly.

This should extend as well to equipment and especially files. If you can quickly lay your hands on a dittomaster, an extra copy of an old test, a worksheet for an absent student, you will lighten your own work. Organize your files with future use in mind. Do not save everything, and the things you do save keep systematically so that they may be quickly retrieved.

Housekeeping is simplified when routines and responsibility are established. Passing in assignments, going to the chalkboard, returning books and equipment, dismissing the class should be orderly processes. Establish a routine that saves time and confusion. Decide on a fixed time when class begins, when a student is recorded as tardy. Establish who dismisses the class: you or the bell.

Questions to Stimulate Thinking

Asking questions has always been a major activity of mathematics teaching. Traditionally, questions have been asked to find out what students have learned. Better uses of questions are to stimulate thinking, to guide problem solving, to arouse curiosity, and to direct discovery activities. To do this, you must plan, so that when the student responds to the question he will make applications, analysis, and evaluations. Then answering questions becomes a real learning experience. You should think of your "level of questioning" in the same way you should consider the "levels of objectives" of Chapter 3.

The following examples indicate the kind of questions that promote learning by requiring thoughtful answers:

> How do you know that this figure is a parallelogram?
>
> What different ways can be used to find out if two segments are congruent?

What is wrong with John's method of finding the solution set?

Can you think of another example of where we need to find the area of a circle?

How can you prove that the probability of this event is $\frac{1}{2}$?

When is it better to use the median than the mean?

How can you use one-to-one correspondence to show that there are as many millions as there are ones?

What happens to the intercept of the graph of $y = mx + b$ if b is increased?

Do you agree with Mary's definition of a trapezoid?

What is the difference between a ratio and a fraction?

How would you tell a sixth grader how to add a positive and a negative integer?

How can you justify the cancellation of these terms?

What kind of a sample would you need to have confidence in the result?

How would you go about measuring the diameter of a basketball?

What is the relationship between the weight and the length of the elastic?

Why does this trick with numbers always work?

What number is closer to zero than any other number?

What are the hidden assumptions inferred by the statement, "Four out of five dentists interviewed use this toothpaste"?

Where would you find data about the track records for the Olympics?

What would you do to find the properties of a finite number system?

If you were asked to design a standard unit of time based on a decimal system, what problems would you face?

How can you prove that the product of an odd number and an even number is always even?

Characteristics of Good Questions

Such questions can be a means of arousing curiosity, of directing thinking, of checking understanding, of providing practice. To develop them, it is important that they be well planned. Good questions like these should have the following characteristics:

The wording of the question is brief, clear, and definite. A great deal of the difficulty involving questions is due to the fact that the student does not comprehend the question.

The question is adapted to the purpose for which it is used. Thus, discovery questions are open-ended, while drill questions are highly specific.

The question is adapted to the ability and background of the students. The language, difficulty, and content will vary from one class to another. Easy questions are directed to the low-ability student and challenging questions to the more talented. However, the question is normally directed to the entire class before designating the specific student for reply. This means that the student's name is affixed at the end rather than at the beginning of any question. There is no better way of building hostility to mathematics than asking questions that the student cannot answer. At the same time, do not follow this admonition so closely that you exclude the slow student from all but the trivial.

The question is stated loudly enough for all students to hear. It is not repeated for the student who is inattentive. Similarly, the student response should be loud enough for the entire class to hear. Expect the class to hear all questions and answers. Do not repeat answers. Require individual responses rather than class responses.

Seek ways to encourage students to ask questions. They should feel secure to ask, but their questions should also be stated clearly. Allow ample time for the student to think about his answer. This time should vary depending on the nature of the question and the student involved.

Motivation

Motivation is the key to learning as well as to an understanding of the dynamics of behavior. We must develop motives in the students that are psychologically sound and socially acceptable, for, when our students are properly motivated, learning becomes a pleasure and teaching becomes an exciting adventure.

As teachers, we usually rationalize our lack of success by saying that our students are uninterested, uncooperative, or unable to learn. Instead we might ask whether the trouble lies with the learner or with the teacher. If we could properly motivate our students, we would eliminate or at least reduce most of our behavior problems and problems of individual differences.

Motives largely determine what students learn. At the basic level, motives satisfy biological or physiological needs like hunger and sex. While these needs are not acceptable as motivation instruments, they are frequently factors in the classroom environment. Motives must also satisfy social and psychological needs that are the direct responsibility of the teacher. These include:

The need for emotional security, affection, and acceptance

The need for intellectual security, recognition, and status

The need for acceptance by and identification with peers

The need for philosophical security: understanding of values, ideals, goals, and interests

These basic needs should result in student and teacher concern for reducing emotional tensions such as anxiety, insecurity, anger, frustration, and doubt. In fact, concern for these problems should often override concern for questions of content and presentation.

It is sometimes proposed that motivation should always be intrinsic. We want our students to learn mathematics because it is important to them and it meets their needs. But we also want them to learn mathematics because it is a satisfying pleasant experience. The success of commercial enterprises in contracting for remedial instruction indicates how effective extrinsic motivation is. Awards of green stamps, candy, money, radios, or free time motivate students to exert the necessary effort to learn mathematical content. We do not advocate the use of these rewards by the mathematics teacher, but the success of these rewards must remind us that we can give students special privileges such as free time to read mathematical material of their choice or to play a mathematical game. And we must recognize the importance of success itself in tests, activities, and even out-of-class achievements, no matter what the motivation.

Motivating Techniques

Some approaches to motivation include activities such as the following:

1. Success breeds satisfaction and further success. Work for it in your classroom. Try to get each student in the success-enjoyment-success cycle and out of the failure-discomfort-failure cycle (see Figure 6.1). Adapt your activities to individual student abilities so that success is attainable. Provide the additional security of methods of checking work and using concrete models for abstractions so that the students can help themselves.

Figure 6.1

2. Promote a realistic understanding of his strengths and weaknesses on the part of the student. Encourage him to use his talents well but help him to avoid unreasonable expectations.

3. Seek always to establish a favorable emotional climate in the classroom. Bring the introvert into group activities. Help the aggressive student outside the classroom. Avoid sarcasm like the plague. Do not be patronizing.

4. Make your students feel part of your program, not just its target. Active cooperation rather than passive acceptance should be your goal.

5. As a check on student understanding of your goals and expectancies, have students evaluate themselves. Compare their grades and their rationale for those grades with your own. Where major discrepancies turn up, individual conferences are useful.

Student involvement does not make things easier for you. It often makes them harder. But we must encourage adult responsibility if we are to expect adult responsibility.

Learning Exercises

1. Select a teacher you had in high school who you feel handled discipline well. Indicate his methods.
2. Select a teacher you had in high school who you feel was a poor disciplinarian. Indicate what you think he did wrong. What were some things he might have done to improve?
3. How would you respond to the following classroom problems:
   (a) On the first school day in September classes meet for only 20 minutes.
   (b) Tardiness.
   (c) A student who sleeps in your class.
   (d) Obscene language.
   (e) Students tease a withdrawn boy about his grooming.
   (f) A third of your class misses two weeks of school illegally. They are now far behind the other students.
   (g) A student refuses to sit down and demands to know what you are "going to do about it."
   (h) Only about half of your students are preparing their homework before class. Others do some problems while you are answering questions in class.
   (i) You think that one of your students is under the influence of drugs.
4. What is your attitude toward grades? How would you prefer to determine grades for your students?
5. What would you do to motivate the following students?
   (a) A freshman with IQ 140 is taking general mathematics. He finds mathematics very easy but he is too lazy to do more than the required minimum. Due to carelessness his test scores are usually only average. He irritates the teacher by asking irrelevant questions at inopportune times. He is an excellent chess player and athlete.
   (b) A freshman with IQ 95 is taking general mathematics. He is having a difficult time. He blames the teacher and sees no reason for taking the course, except that it is required for graduation. He is interested in automobiles and hopes to work in a garage someday or to be a pilot. He comes from a culturally deprived home.
   (c) A sophomore with IQ 105 is taking geometry and doing D work. He is a transfer student from a school where he did C+ work in mathematics. His father is a wealthy and successful executive who expects his son to go to college. The student has been in for much extra help, mostly to no avail. His desire to do college-acceptable work has been reduced to an inner fear that he will fail the course.

(d) A junior with IQ 115 is taking eleventh-grade algebra. His parents have selected his course of study, expect him to do well in mathematics and science so that he can enter the university. He is not interested in mathematics and has done near-failing work. He feels that he would like to join the army or get a job when he graduates rather than go to college.

(e) A junior with IQ 125 is taking eleventh-grade algebra. He works many hours a week at a supermarket to help his widowed mother. He likes algebra, but is depressed by his situation and the possibility of not going to college. Recently he attended a high school event intoxicated. He has been put on probation by the principal.

(f) A junior with IQ 110 is taking first-year algebra. He owns a car and works in a garage. He says that he does not understand algebra, but refuses to complete his homework. He always has an excuse for a late assignment or a poor test score. He has been found to cheat on a test. He brags to his friends about his low grades and cheating on tests. He asks what else can life offer besides a car, a job, money, and friends.

(g) An eighth-grade student with IQ 130 dislikes mathematics intensely because her mark is always D. In most of her classes she gets A's and B's. Her computational skill is very poor (PR = 5). She seems to be content just to get by in mathematics. Her file shows that she missed a great deal of school during the second and third grades. A note from the counselor interview states that the girl enjoys English and is considering becoming either an English teacher or an elementary teacher.

6. Choose a teaching topic of interest to you (for example, a lesson on subtraction of integers). What are some thought-provoking questions that can be asked to guide students in developing this topic?

7. Which of these assignments seem most appropriate? Explain the reason for your choices.

(a) Work as many exercises on page 72 as are necessary so that you can subtract integers correctly and quickly.

(b) Work the odd-numbered exercises on page 72. Note which ones give you most trouble so that we can discuss them tomorrow. Ask yourself why they are difficult.

(c) For tomorrow learn how to subtract integers correctly and quickly. Read the instructions and study the examples on page 71. Work 10 exercises on page 72. Compare your answers with those given below. Make a drawing to show how one of the subtractions is related to distance on a number line.

8. Which of these classroom questions are the best? Explain the reason for your choices.

(a) How can you check to see if the answer to a subtraction problem is correct?

(b) What is an example of an everyday situation where we perform a subtraction with positive and negative integers?

(c) Is the difference $({}^{-}12 - {}^{+}12)$ a positive or negative integer or zero?

(d) How would you tell your ten-year-old brother that $({}^{-}5 - {}^{+}5)$ is ${}^{-}10$?

9. Rank these disorders in your classroom in order of seriousness of the offense; justify your choice of the most serious and the least serious: smoking, sleeping in class, throwing objects, refusing to obey the teacher, talking to neighbors, cheating on a test, walking out of class without permission, fighting.

Suggestions for Further Reading

Dodes, Irving Allen, "Planned Instruction," in *The Learning of Mathematics: Its Theory and Practice*, Yearbook 21, National Council of Teachers of Mathematics, Washington, D. C., 1953, pp. 303–334.

Gragney, William C., *Controlling Classroom Behavior*, National Education Association, Washington, D. C., 1965.

———, *The Psychology of Discipline in the Classroom*, The Macmillan Company, New York, 1969.

Lowry, William C., "Course Requirements and Grading," in Douglas B. Aichele and Robert E. Reys (eds.), *Readings in Secondary School Mathematics*, Prindle, Weber, and Schmidt, Boston, 1971, pp. 419–422.

Swineford, Edwin J., "Ninety Suggestions on the Teaching of Mathematics in the Junior High School," *Mathematics Teacher*, 54, 3 (March 1961), 145–148.

Wernick, William, "A List of Standard Corrections," *Mathematics Teacher*, 57, 2 (February 1964), 107.

7 Lesson Planning

A basic requirement for success in most of our activities is planning. The successful lawyer, engineer, politician, or salesman probably spends as much time planning his activities as he does executing his plans. In a similar way, the successful mathematics teacher needs to take time for preparation; time to plan daily lessons and in addition time to plan units, courses, and examinations. Even experienced teachers spend time replanning lessons that they have taught several times. This is the way in which they keep from going stale or getting into a rut. The lesson plan translates the goals and content of a curriculum into an operational plan.

A colorful teacher may seem to respond to his class in such a spontaneous manner that no lesson planning is apparent. However, the chances are that this teacher has—formally or informally—carefully planned his approach to the class. He has probably mastered the content so well that he is poised and secure in his response to the students. His objectives are so ingrained that his reactions are almost automatic.

The planning levels for a school program include courses of from a half year to a year and occasionally two years in duration, units that may be taught in a period of a week to a month, and daily lessons. Planning quite naturally runs from the more general development of courses for an instructional program to the specific daily lesson.

Course Planning

Many patterns of course development for individual schools are possible. Some states provide detailed syllabi for a variety of courses. Many local staffs choose a textbook with care, letting the book provide the basic outline for the course. Other schools design their own courses to meet the needs of specific groups of students. After determining these needs, interests, and abilities of the group, objectives are developed and a program mounted to accomplish those objectives. This program may include as textual materials: a laboratory manual, enrichment pamphlets and books, a computer workbook, and teacher-prepared materials or a more conventional text.

Most staffs use a combination of these approaches. The more traditional courses of the college-preparatory sequence are often outlined by a standard commercial text. A course for ill-prepared or low-achieving students must often be individually tailored. Other courses fall at intermediate points along the spectrum.

Every teacher has the responsibility to think carefully about the overall design of the courses he teaches. Even when the textbook or syllabus provides the basic organization, decisions about its use are in order. Should some sections be skipped or covered as optional topics for better students? What will be the approximate time allowance for the various topics? Do some topics call for supplementary texts? Such preliminary course planning can head off the empty feeling that comes to a few beginners who discover in February that they are only a quarter of the way through a mandated syllabus.

Unit Plans

Many teachers do not prepare unit plans; instead, they prefer to focus their thinking on the day-to-day activities and let the text structure the course. This is unfortunate because the unit plan helps put the daily lessons into broader perspective and at the same time forces a deeper analysis of the content to be taught. It also encourages the teacher to plan for special materials such as films. Such supplementary aids usually require advance requests or at least prior planning for classroom use. The day-by-day planner often thinks of such things too late.

Any experienced teacher who has not tried unit planning should try one unit in one course as a first attempt. If real justice is done to this initial effort, the results will well warrant further work. Here are some of the things that should be considered in organizing the plans for a substantial segment of classwork:

Why is this unit important? What are the objectives of the unit? How can I motivate my students? What concepts and skills are directly applicable? Which are the keys to future progress?

What are the central ideas and the unifying concepts of the unit, around which activities may be organized? What should be stressed most? How should the class time be divided? How much time should the entire unit take?

What teaching strategies are appropriate? Is this the first time students have met these ideas? How can they be tied to past work? How can students develop the ideas themselves? What materials are available to provide a varied attack on the unit? What alternatives are available?

What concepts, skills, and experiences are needed as background for this unit? How can the content be modified for students of varying ability? What extra practice can be provided for weak students? What special teaching techniques may be used with them? What can other students do when weaker students receive special attention? What enrichment topics should be included for all students? What topics should be assigned to bright students only?

What teaching techniques will best suit this class? What are the tough spots that require special attention or a different approach? How did I teach this material previously?

Should I change my approach or techniques? What lessons are appropriate as laboratory lessons?

What materials will this unit require? What supplementary books or pamphlets would be helpful for students? What models, films, or projectuals are appropriate? What should the bulletin board display? Are any field trips or excursions suitable? Who might be a suitable outside speaker or class participant?

What kind of evaluation should I use? What ways of evaluation are best suited to this content and this class?

What kinds of assignments should the students prepare? Are long-term assignments appropriate? Can the students learn part of the material independently?

These questions help to pin down amorphous thinking about teaching problems and help the teacher escape from the textbook. They provide the basis for developing a careful plan of attack that may be put into operation via the daily lesson plans and classroom procedures.

A unit plan should contain these elements:

1. A statement of objectives.

2. A pretest to determine the skills and background concepts of the students and their prior knowledge of the topic of the unit.

3. A selection of possible learning activities, including alternatives and a schedule.

4. The teaching procedures and techniques, including motivation and provision for individual differences.

5. A list of materials.

6. Lesson plans, including assignments and alternative work for reteaching and review.

7. An outline of written tests and other evaluation procedures.

Some Excerpts from a Unit Plan

Unit: Graphing

Objectives

1. Locate points, segments, and intervals on a one-dimensional graph.
2. Locate points, lines, and regions on a two-dimensional graph.
3. Translate in both directions

algebraic statement $\leftrightarrow$ coordinate representation

Pretest:

Instructions: This test will not be counted toward course grade. It is designed to supply information for planning purposes only. Some of the content has not yet been taught in this course. Try your best on each item but do not guess.

****

Learning Activities:

* * * * * * *

3. Locating points represented by ordered pairs.
   (a) Using the rows and columns of seats in a classroom
   (b) Playing the game battleship
   (c) Using a state highway map
   (d) Using a rectangular coordinate grid

* * * * * * *

5. Drawing graphs of sets of ordered pairs.
   (a) Have students at seats identified by ordered pairs stand to show conditions that give lines of students.
   (b) Graph the ordered pairs collected by measurement and experimentation.
   (c) Graph the ordered pairs produced by a guessing game.
   (d) Graph the ordered pairs produced by a function machine.
   (e) Graph the ordered pairs of the truth set of a linear equation.

* * * * * * *

Special Teaching Procedures and Techniques:

* * * * * * *

2. Stress working in pairs, especially on checking homework, to develop student interaction and to improve preparation.

* * * * * * *

Materials:

* * * * * * *

4. Overhead projector transparencies
   (a) Lattice points—half-inch grid
   (b) Number lines (make two to show development of coordinate systems with crossed number lines)
   (c) Half-shaded transparencies (to use for half-planes to develop inequalities)

* * * * * * *

Daily Lesson Plans: [see next section]

Testing: [see Chapter 9]

The Daily Lesson Plan

Now you have the unit as the framework for your daily lessons. You know where you are going and about how much time is needed to get there. You must translate this general plan into specific daily activities. These plans will give you a sense of security and help you to make efficient use of time and gain the confidence of the class.

Lesson plans vary widely: from fat and detailed to thin and merely a few notes. As in public speaking, beginners use detailed plans; but top-flight teachers, just like top-flight speakers, continue to plan, often giving greater attention to detail! Just as there are different learning styles for our students, there are different planning styles for different teachers.

However, no lesson plan in itself will ensure success. It is the way the lesson is carried out that is important. The response to unexpected questions, the sensitivity to student reaction, the enthusiasm and resourcefulness of the teacher are keys to success. There are some teachers who contend that lesson plans are not necessary, that they reduce flexibility and spontaneity and that they do not allow enough freedom to explore student interest. This need not be true. Lesson plans should not be followed slavishly no matter what the student reaction is. Instead, lesson plans should give you a feeling of security so that you have greater freedom to explore student questions. One way to attain this flexibility is to focus the plan on student activity rather than on your own behavior. Try to anticipate student responses and use them to give fresh, non-text approaches to a topic.

All of this book applies to planning lessons. Your thinking about goals and objectives and your philosophy of teaching apply. Likewise the later concerns of this text must be taken into consideration: testing techniques, the approaches to different kinds of teaching, the special provisions for students of varying ability, and the specific kinds of materials. All these will expand the initial framework that we discuss here.

In this chapter, our concern is to provide you with a few alternate ways for planning different kinds of daily lessons. Your ability to write lessons will improve as you use them in the classroom.

Regardless of format, the following items are usually included in a lesson:

Objectives—What am I trying to do today?

Motivation—How can I get the class started?

Techniques and activities—What is the best strategy for this topic?

Materials—What materials are available to lend variety?

Assignments—What and how will an assignment be made?

Evaluation—What activities of this lesson were most successful? What were unsuccessful? How can I improve it next time?

Although objectives seem too theoretical for daily lessons, they should provide a setting for the entire lesson. Share your objectives openly with your students so that they recognize their responsibility and your basis of evaluation. Teachers who give students the objectives of a lesson, sample test items, and performance specifications usually get from students a positive response and also improved achievement.

Beginning and Ending the Class

As you plan your lesson, be sure to have a strong beginning. The first five minutes are crucial because they establish the tone for the whole period. Begin promptly and enthusiastically with an activity that needs the immediate attention of every student. A crisp announcement, a problem stated, a comment about an event of the day or a local personality, an anecdote, or a mental exercise suggest some possibilities. Do not begin by asking, "Are there any questions about your assignment?" If you do this, you will probably devote the best part of your lesson to the answers. The first 20 minutes of your class is the time to have your class exploring new ideas.

Your lesson should also have a definite ending. This should not be the bell or the statement, "Your assignment for tomorrow is . . .." Rather your lesson could end with a summary, a recapitulation of what has been done, an open-ended question for further exploration, an application, or a projection of where this lesson leads in subsequent days (all before the closing bell rings).

Motivation

Every lesson needs some activities to energize learning and to capture the interest of students, as described in Chapter 15. The key for lesson planning is to have a variety of lessons. Try to bring to the classroom surprising results, unusual problems, humor, and a change in pace. Try an unstructured or nondirective approach in which the students decide how they want to learn a certain topic. Try an independent work contract in which they select the level of achievement to attain. Use a contest or game situation as a new way to learn an idea. Perhaps they can collect or make up problems to exchange rather than using the text problems. Pose a problem and ask the students to collect information about it or explore it with experiments.

Method

If possible, try different methods in your classes to learn which one works for you with your students with respect to a specific objective. One maxim that seems to apply to all lessons is "Students must be participating." They must respond and react whether the lesson is a classroom discussion, a laboratory lesson, or independent study. Hence, the lesson plan should give emphasis to student activity rather than teacher activity.

One of the important parts of the lesson is the questions and problems that are to be used. List the key questions that you think will guide the student to discover the generalization. Select the examples you want to use to illustrate a process or concept. Use different examples than those given in the text so that the student can use the text example to supplement the class example. Before the class meets, work all problems or complete proofs to be presented to be sure that you will not be embarrassed by not being able to do it for the class. However, if you have difficulty, admit it and ask the class to work with you in completing the exercise.

Many of the most significant events of a lesson are those that are unplanned. The successful teacher is likely to be one who can respond spontaneously to classroom events. Sometimes student questions and reactions can be anticipated. However, if you are open to your students, if students feel free to ask questions, or if students demonstrate interest in an idea that extends a lesson in a new direction, then you will need to respond spontaneously. You can prepare for these extemporaneous responses by feeling secure about your mastery of the content and by having control of the class based on trust rather than threat. You will also be prepared for these unpredictable events if you are sensitive to students and have well-established values that subconsciously direct your spontaneous response. You will then remain poised, be able to think on your feet and give maximum attention to your students and their interactions.

Materials

With the variety of instructional materials described in Part Five of this text, careful choice should be made for each lesson. Explore the use of colored chalk on the chalkboard. Find commercial visuals or prepare your own for use with the overhead projector. Duplicate the assignments occasionally so that students recognize that you have prepared material especially for them.

Search for ways to illustrate or make physical representation of subject matter. Try to find ways in which students can collect original data or local applications to supplement the text exercises. In any event, be sure to list the materials needed for your lesson so that you will have them at hand when needed.

Questioning

Asking questions is a two-way street to learning. The teacher asks questions to stimulate the student's thoughts. The student, however, asks questions to get answers to his own thoughts. Both types of questions play a significant role in the mathematics classroom, but of the two the latter is the more important. If this is to happen, the student must feel free to ask questions without fear of embarrassment or punishment. He should never be ridiculed for asking a "foolish" question. Keep the classroom climate open by listening to any question which a student poses. It does not matter whether it is insignificant to you, whether you have just given information which answers the question, or whether it is irrelevant to the topic at hand. However, this does not mean that you give a complete answer to every question asked. Sometimes your response is a question that may clarify the question asked; sometimes you may delay the answer until the idea questioned will be studied; sometimes you give information on where to get the answer; and sometimes you agree to give the answer in a conference with the student.

In a similar fashion, students should be free to make mistakes. We learn by making mistakes. You should point out to students that their mistakes even help you know what to teach them. Then we use student errors to find the cause of the error and how to correct it. Another aspect of learning from mistakes is to make guesses or estimates. These are almost invariably incorrect but with experience from these errors, students grow in their ability to make better estimates.

Assignments

As mathematics teachers, we almost always make daily assignments but are rarely successful in planning appropriate assignments. We are very confident that homework is necessary but the little evidence we have is that the contribution of homework toward mathematics achievement is minimal. We continue to assign the odd-numbered problems on page 57, and the student probably works the exercises in as short a time as possible with a minimum of thought. Hence little learning takes place; and, at the same time, hostility of the student toward mathematics is increased.

Assignments may contribute to learning if they are carefully planned. The first consideration should be to differentiate the assignment according to need. The low achiever needs to build an understanding of basic ideas rather than to practice making errors in his computation. The high-ability student needs to explore open-ended problems that are interesting and challenging. In any event all students should know why they are doing the assignment, how to do it, and how to get help if they have difficulties.

Evaluation

After teaching a lesson, take some time to reflect on its successes or failures. One of the advantages of a written plan is that it reminds one of the activities that took place and furnishes a basis for thinking about ways to approach the lesson differently. Sometimes one can even ask the students why the lesson did not work. Consult with colleagues if a lesson has presented considerable difficulty. Make notes on the lesson and file it so that next year you will be able to profit by your success or failure.

As you plan lessons, try to be creative. Try a new format, a different sequence, and unique activities. For example, a laboratory session may have the entire lesson in the materials provided. You need only say "Go!" and students report to stations and perform activities outlined. However, whatever format is used, be sure that it is focused on the learner so that he attains the objectives. Also plan your lessons so that the student gains understanding and mastery rather than superficiality or frustration.

Varieties of Mathematics Lessons

Too often, instruction in mathematics classrooms is barren and uninspiring. The typical lesson is as follows: First of all, homework is discussed; then the teacher demonstrates a new procedure or theorem; then he assigns the next exercises—and class is dismissed. Such a procedure does little justice to the exciting content, varied instructional aids, and emphasis on student participation of mathematics today. Besides, there are many more interesting and rewarding procedures. Some of these procedures (which will be discussed in detail in subsequent chapters) are listed and discussed briefly here. These lesson types, however, are not mutually exclusive; most good lessons, in fact contain several types.

Laboratory lessons. In these lessons, pupils make measurements (often using simple equipment, such as rules and compasses); play games, use calculators, program at computer terminals, collect data by experimentation or surveys, make drawings and models, make computation devices, and perform experiments with kits such as the Probability Kit. These can be true discovery lessons, and teachers who have used them are most enthusiastic about the results.

Audio-visual presentations. The overhead projector is a versatile device for which the teacher can now order a great number of commercial projectuals as well as develop equally useful homemade projectuals. There are new films such as *Sets, Crows, and Infinity; Possibly So, Pythagoras;* and *Donald in Mathemagic Land* that provide original and unique learning aids. Tape recorders and film loops provide opportunities to individualize instruction. Even the use of colored chalk can enliven presentations.

Games. Many of the commercial games available or those prepared in the classroom by the students themselves can promote the learning of mathematics in an interesting and entertaining way.

Student-directed class discussions. Student demonstrations, reports on enrichment topics, or even the teaching of a regular lesson can be a wholesome experience for the student demonstrator as well as the class.

Discovery lessons. The teacher can promote student discovery by asking good questions, posing problems, suggesting an experiment or an investigation. Processes, properties, and assumptions are developed rather than crammed into the students.

Small-group instruction. Although materials are not so plentiful for teaching mathematics as they are for the teaching of reading, it is just as appropriate to have mathematics groups as reading groups. Possibly some of the better students can help to conduct these small groups; as they do so, they will learn mathematics, they will learn to communicate, and they will get experience in a leadership role.

Enrichment lessons. There are many wonderful and exciting mathematical ideas for discussion that are not in the regular text—topics such as historical incidents, applications, space travel, game theory. We need to present these ideas to our young people, for the mathematics classroom might be the only place where the students will encounter such topics.

Communication lessons. These are special lessons on communication and learning how to learn mathematics. They teach listening, reading, writing, and speaking correctly about mathematical ideas.

Individualized instruction. Here each student investigates an idea independently. Remedial teaching—using programmed units, remedial kits, tapes, or guide sheets—is helpful. A laboratory or conference room adds to the effectiveness of this type of lesson.

Creative learning lessons. Writing original problems, solving problems, establishing theorems with original proofs, discovering and stating relationships in one's own language, drawing an original design—these are all beginning experiences in creative thinking. Communicating mathematical ideas in an original fashion—through demonstrations, proofs, exhibits, poems, or research projects—gives further opportunity for originality. The development of a new numeration system, the building of an original model, or the discovery of a new idea or a new application of mathematics—all illustrate creative work at a high level.

Team teaching. The introduction of flexible scheduling has made it possible for teams of teachers to teach a given course. Several classes may be combined for a large-group presentation. This group may view a film, take a test, or listen to a teacher's presentation of a topic. Other lessons may involve small-group instruction for remedial work or enrichment lessons. Some lessons may have individual instruction. Thus the team is involved in a variety of activities—planning the course, planning lessons, preparing tests, keeping records, preparing materials, and teaching different groups. Usually a teacher concentrates on those aspects of the team's responsibilities for which he has the greatest aptitude and interest.

Use of the computer. Although computers are not generally available in schools today, in the near future they will be involved in mathematics instruction in many ways. They can be used as aids in problem solving (with students asked to prepare flow charts or write programs). They may be used as an instructional aid in learning mathematical ideas and as mediators in individualized instruction (with a computer terminal providing a lesson and a computer recording pupil responses).

Out-of-doors lessons. There is a need to capitalize on the great variety of mathematics outside the textbook and outside the classroom. The school grounds are available for measurement activities. Parks and playgrounds have examples of mathematics in nature. Streets and stores are full of products to survey and problems to solve. Lessons about the local community are one way to add relevance to mathematics.

Banded lessons. A banded lesson is one that has several bands of activities. These activities may include recreational activities such as games or puzzles, practice activities such as short worksheets or oral drill, discovery discussions, a laboratory activity, or a film.

It is similar to any good lesson except that it specifies that the lesson include a variety of diversified activities with a time allotment for each.

All of these types of lessons have advantages and disadvantages. Particular lessons and even particular subjects lend themselves best to certain styles of presentation. For example, the need to "take apart" a diagram makes the overhead projector a useful tool for teaching congruency; stick models are especially appropriate to instruction in solid geometry; films can display trigonometry's continuous functions, which are so difficult to develop by other means. But the right choice still remains a major problem for the teacher. Industry refers to the organization of such choices as *systems development*. Systems development is the coordination of materials, personnel, and procedures to meet a given goal in the most effective way. For mathematics instruction, this means that each unit, each lesson, must be programmed to use the right resource and the right teaching technique at the appropriate time.

Sample Lesson Plans

The lesson plan may have any convenient format and may vary in its thoroughness. Ordinarily, it is arranged in the order in which activities are to take place. For the novice, a time schedule is helpful. In any event, the plan should be flexible enough to allow for student response. Also, the teacher should know it well enough so that he need not refer to it throughout the lesson.

Lesson: Integers (a lesson in eighth-grade algebra)
Objectives:

Students will relate subtraction of integers to problems in the real world.
Students will begin to develop subtraction algorithms.

Materials needed:

Drawing of thermometer scale on board
Number line, overhead projector, transparency
Worksheet 37: temperature, date, and integer scales
Red and white dice

Activities:

1. Report (from assignment) on highest and lowest U.S. temperatures (from almanac)
2. Key questions:
   (a) What is the difference between these two temperatures?
   (b) Can you suggest a way of showing
      (i) Going from lowest to highest?
      (ii) Going from highest to lowest?

3. Record problems from last two questions as subtraction exercises on board in two forms and discuss.

   (a) $\begin{array}{r} {}^{+}134 \\ -\,{}^{-}82 \\ \hline \end{array}$ $\quad {}^{+}134 - ({}^{-}82)$

   (b) $\begin{array}{r} {}^{-}82 \\ -\,{}^{+}134 \\ \hline \end{array}$ $\quad {}^{-}82 - ({}^{+}134)$

4. Discuss why by comparison with subtraction of natural numbers

$$\begin{array}{r} 8 \\ -5 \\ \hline \end{array} \Leftrightarrow \text{From 5 to 8}$$

5. Distribute worksheet 37.
   Assign pairs to work together on questions 1 to 8.
   Note: These require students to plot integers on number lines, draw arrows from subtrahend to minuend, and by the procedure of the discussion in activity 4 to find difference. They include ${}^{+}12 - {}^{+}5$, $0 - {}^{-}5$, ${}^{-}7 - {}^{-}2$, etc.
6. Oral discussion of answers.
7. Dice game for subtraction:
   Demonstrate several plays.
   Partners from activity 5 play as time allows.
   Note: Game is played with one red (−) die and one white (+) die and number line in integers from ${}^{-}20$ to ${}^{+}20$. A marker is placed at zero and one player is assigned negative and the other positive. Dice are thrown and difference between left die and right die is added to the score until a winner is determined.

Assignment:

Complete worksheet 37. *Note:* This leads to next lesson on discovery of shortcut for subtraction of integers.

Lesson: Ratio

Objectives:

Given two numbers, the student is able to write a ratio.

Given two ratios, the student can compare them by writing the ratios as decimals.

Review:

Place a flow chart on the board to show how a fraction is changed to a decimal.

Illustrate with an example.

Introduction:

Here is a car ad. It says the gear ratio is 3.5 to 1.

What does this mean?

What are some other examples of gear ratios? (Bicycle, watch.)

Activities:

1. Use two gears in mesh to show what happens.

   Compare the turns. Count the number of teeth for each gear.

   Write the ratios for the gears.

2. Illustrate ratios in team standings from the sports page.

   Show the need for a comparison base. Convert sports ratios to decimals.

Assignment:

Work as many exercises on page 72 as necessary in order to be able to change a ratio to a decimal.

Optional: Find the gear ratio for a bicycle. Find how this ratio changes for different shifts.

Recreation:

Play a wastebasket ball game with a friend. Use a large paper wad for a ball. Keep a record of how many baskets you can make from a given distance. Compare your records by using the ratio:

$$\frac{\text{Number of tosses in the basket}}{\text{Total number of tosses}}$$

Lesson: Introduction to Reasoning

Objectives:

Given sources of information, the student is able to select those that have correct information.

Given a situation, the student is able to identify the authority that makes decisions.

Material:

Newspaper for each student.

Guide sheet for surveying information in the newspaper.

Introduction: Who decides the truth of these situations? (1) A foul in a basketball game, (2) smoking is harmful, (3) a check is worthless, (4) a person is guilty of a crime.

Sometimes an authority, person, book, law decides the truth of a statement.

Examples for discussion:

Which of these publications are usually accepted as correct? (1) Encyclopedia, (2) newspaper, (3) athletic rule book, (4) advertisement, (5) textbook, (6) cartoon, (7) magazine, (8) atlas, (9) dictionary.

Which of the publications above would you use to get the following information? (1) Population of a city, (2) method for computing probability, (3) size of a tennis court.

What governmental agency establishes the truth of these statements?

1. A 15-year-old can drive a motorcycle.
2. An insecticide can be used to spray an orchard.
3. A hunter can shoot a deer.
4. A drug can be sold by a drugstore.

What are other ways in which a true statement is established?

Assignment:

Each student has a newspaper and a guide sheet.

Use the guide sheet to find information in your newspaper such as:

1. What are some number facts reported in sports?
2. What are some facts determined by a judge?

3. What are some number facts from government agencies?
4. What are some products for which different prices are reported?
5. What are some statements made in advertisements which are probably not true?

Evaluation:

Lesson too long.
Concentrate only on authorities.
Relate this to authorities in the school situation.

Lesson: Parallelogram

Objective:

Given a parallelogram, the student is able to identify its properties.

Review:

The definition of a quadrilateral.
The angles, sides, and diagonals of a quadrilateral.
The types of quadrilaterals: general, rectangle, square, parallelogram.

Material:

Flexible quadrilateral with elastic thread connecting midpoints of adjacent sides.

Introduction:

Name some shapes which are parallelograms.
From what you know about parallel lines cut by a transversal, what angle relationships do you know?

Learning Activities:

Measure the angles and sides of several parallelograms.
How do these measures compare?
What generalizations can you state based on these measurements?
(Illustrate these relationships with the flexible quadrilateral.)

Assignment:

Write the proof of this theorem: A diagonal of a parallelogram divides the parallelogram into two congruent triangles.

Experiment:

Draw several different quadrilaterals. Draw the lines connecting the midpoints of adjacent sides. (Illustrate with the flexible quadrilateral.) Compare the measures of these lines. What kind of a quadrilateral seems to be formed? Prove that your generalization is correct.

Planning is a Means, Not an End

Lesson plans should always be viewed as a tool, not as a product. If your plans do not help you to improve your teaching, there is something wrong with them, and you should seek then to improve them. Perhaps you focus too much on content, not enough on practical ways to meet classroom problems. Perhaps you need more details for opening, closing, and transition activities, less for the more set daily activities. Most likely you need alternate activities that will fill in the blank spaces in your program—unplanned time periods that invite discipline problems.

One teacher met this last problem by posting a series of diagrams on his bulletin board like those of Figure 7.1. He was always ready to fill in transition periods in his geometry classes with questions like: "On the third diagram you are given that $\overline{AD} \cong \overline{BD}$ and that E and C are midpoints of their respective sides. What conclusions can you draw?"

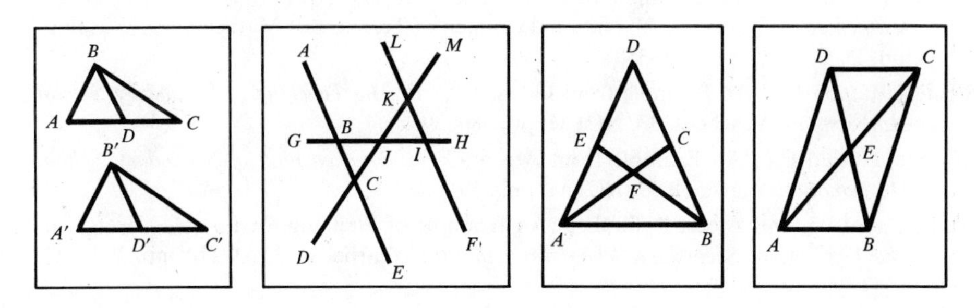

Figure 7.1

Finally the lesson plan must focus on the students, not on you the teacher. It is a practical here-is-what-we-are-going-to-do-today road map, not a philosophical discussion. As such, plans can be a most important aspect of improvement of instruction.

Learning Exercises

1. Read Irvin Allen Dodes, "Planned Instruction," in NCTM 21st Yearbook (see reference). Rephase in your own words Dodes' three laws of motivation. Give two additional examples of each type of motivation.
2. Write a detailed plan for an enrichment lesson to be given to a mathematics club.
3. Abstract from your plan in exercise 2 brief notes that you would use for your presentation.
4. Choose a topic in a text, state three specific objectives related to that topic, and plan a lesson or a series of lessons to accomplish those objectives. Use judgment in alloting time to achieve your objectives.
5. Prepare homework worksheets that are as original and creative as you can make them for two of the following topics:
   (a) Solving quadratic equations
   (b) Angle measure on the circle
   (c) Slope-intercept representation of linear equations
   (d) Scale drawing (in junior high school)

These worksheets are for use after classroom instruction on these topics.

Suggestions for Further Reading

Dodes, Irving Allen, "Planned Instruction," in *The Learning of Mathematics: Its Theory and Practice*, Yearbook 21, National Council of Teachers of Mathematics, Washington, D. C., 1953, pp. 303–334.

Heinke, Clarence H., "An Example from Arithmetic," in *The Teaching of Secondary School Mathematics*, Yearbook 33, National Council of Teachers of Mathematics, Washington, D. C., 1970, pp. 337–358.

Steinen, Ramon F., "An Example from Geometry," in *The Teaching of Secondary School Mathematics*, Yearbook 33, NCTM, pp. 380–396.

Trimble, Harold C., "An Example from Algebra," in *The Teaching of Secondary School Mathematics*, Yearbook 33, NCTM, pp. 359–379.

Wells, David W., and Albert P. Shulte, "An Example of Planning for Low Achievers," in *The Teaching of Secondary School Mathematics*, Yearbook 33, NCTM, pp. 397–422.

Evaluation of Achievement 8

Evaluation activities chart the progress of students toward the objectives outlined for them by the teacher and by themselves. Therefore evaluation is an essential aspect of instruction at all levels. It is a means whereby the quality of our mathematics programs can be constantly maintained and improved.

Evaluating a student's achievement is a teacher's constant duty. It is a time-consuming, frequently tedious activity, because of the clerical work involved. At the same time, it requires a highly technical proficiency and involves the teacher's professional value judgments. It can be discouraging to find out in evaluating students how little they have learned from what we thought was well-planned instruction. But evaluation is an indispensable task, which becomes increasingly important if we want students to achieve their optimum potential.

Evaluation involves activities such as the following:

Constructing and administering, marking, and evaluating tests, examinations, checklists, and questionnaires

Observing, recording, and evaluating student activities

Assigning, directing, and evaluating student projects, reports, written work

Recording evaluations, interpreting the records, and assigning grades

Conferring with parents, students, counselors, and employers

Writing recommendations for colleges and employers

No wonder teachers need time and clerical help for evaluation activities. And, for the following reasons, it is likely that future evaluation activities will increase:

Individualization of the instructional program demands constant monitoring.

More accurate information is needed for the assignment of students to classes according to ability, achievement, and motivation.

The plethora of published tests available require the teacher to expend a great deal of time evaluating these materials.

New machines are now available for administering and correcting tests, recording scores, and keeping records.

The new emphasis on research concerning the learning process and curriculum problems requires further measures for analysis.

Thus, every mathematics teacher needs to be competent and well informed in evaluation procedures.

Purposes for Evaluation

Bloom[1] has characterized evaluation under two designations, formative and summative. Summative evaluation, the general assessment of student progress often associated with grades, is the kind of evaluation that follows units or entire courses of instruction. Formative evaluation, on the other hand, is directed at the instructional process itself—planning, teaching, and learning—for the purpose of improvement of the process. Formative tests include pretests to help with development of instructional units, diagnostic tests to point out specific teaching-learning problems, and self-administered mastery tests that forecast summative evaluation results and help the student to attack his own learning deficiencies. Most stress in schools has been on summative evaluation, but every classroom teacher should give equal attention to formative testing.

A student's preparation for a test and his participation in evaluating the test may be a *worthwhile learning experience in itself*. The preparation involves learning activities such as summarizing, organizing, and outlining. Completing the test itself is an intense learning experience for him, which may involve problem solving, reflective thinking, deductions, and computations. Giving the student an opportunity to discuss a completed test and correct errors helps him locate his areas of misunderstanding and correct wrong methods and thereby reinforces a student's correct learning procedures.

However, a word of caution is needed here. Cramming for a test often involves memorizing facts which are quickly forgotten after the test. Furthermore, intense preparation for a test makes the completion of the test, rather than learning the material, the goal of learning.

Self-evaluation activities tend to *motivate the student to learn ideas and skills.* Competition with one's own record, the class record, or national norms can be a stimulating experience. However, learning mathematics should be accepted by the student as worthy of effort regardless of the achievement mark involved.

Communicating with others is a major aspect of evaluation. It involves collecting information for reporting a student's status to him, to his parents, to counselors, to admission officers, and to employers. Evaluations that are correct and meaningful are needed by all individuals who make decisions regarding a given student.

---

[1] Benjamin S. Bloom, J. Thomas Hastings, and George F. Madaus, *Handbook on Formative and Summative Evaluation of Student Learning*, McGraw-Hill Book Company, New York, 1971.

The measurements made in evaluating may be *useful data for research projects.* Research studies analyze data obtained by measurement to study learning and to evaluate the effectiveness of methods, materials, or curricula.

Some Basic Principles of Evaluation

The purpose of teaching mathematics is to provide experiences that help each student make progress toward the attainment of worthwhile objectives. We measure the amount of progress each student has made in attaining a given objective by the use of instruments such as examinations. Then we use a value judgment to evaluate this measure. Thus, evaluation involves construction, administration, and interpretation of measures.

The meaning of achievement can be clarified if we consider achievement as a vector. This vector is described by coordinates which are measures of achievement. The measures may be scores on tests of specific objectives. A simplified model will illustrate this concept and at the same time point out some of the real difficulties of achievement measurement. Consider the achievement of two students on two tests before and after exposure to a specific instructional program (Table 8.1).

Table 8.1

	Pretest		**Post-test**	
Student	**Concepts**	**Skills**	**Concepts**	**Skills**
Paul	7	32	13	50
Mary	19	23	43	52

In Figure 8.1, these results are recorded on a coordinate system. The progress of each student is indicated by vectors representing the difference between the initial and final achievement vectors. Some basic questions of measurement are indicated by this model: (1) Does a scale unit have meaning? (2) Can we assume that units on the two scales are the same? (3) Would a zero vector really represent the absence of achievement? (4) How can goals be indicated on such a model? (5) Do the measurements really indicate achievement of the concepts? In fact, are they even approximations to such achievement? (6) There are so many facets to be measured that the vectors would be multidimensional, perhaps including even more than ten components. Would metric-space hypotheses have any application to such vectors? (For example, would the length of these n-dimensional vectors provide a basis for comparing achievement increments?)

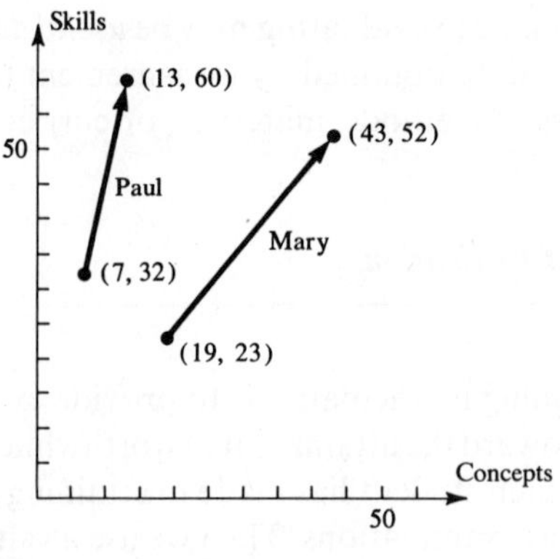

Figure 8.1

Thus, achievement may be seen to be extremely difficult to pin down. The model does, however, provide a framework on which to focus thinking about achievement testing along the following lines:

Achievement should be measured in terms of all the objectives of instruction. Our goal is located by coordinates that represent measures of all objectives. Hence, we can locate our students in this coordinate space only if we have all the coordinates for each student. To arrive at these coordinates, we need devices that will measure the level of attainment of each objective. Therefore, we must know what our objectives are, so that we can measure progress we have made toward attaining them. The evaluation of achievement in mathematics must be more than just measuring skills and knowledge, for we must also measure our students' status relative to such goals as attitudes and appreciations. We should test our students' progress in learning how to study mathematics or how to read mathematics. If we are attempting to teach how to apply mathematical learning to new situations, we need to devise tests of this ability. If we are building skill in thinking logically and building mathematical structures, we will need to build test situations or items in which the student can exhibit his ability to do these things.

Achievement should be measured in terms of growth, change, and progress in the attainment of our goals. This implies that we have a pretest or achievement record that locates the student's standing at the beginning of the term. It also means that we have measuring instruments that can determine the different levels of attainment of the concepts or skills involved. It also means that we must take into account the differences between students.

Measurement should emphasize the retention of learning over a long period of time. This suggests the need for comprehensive examinations.

Measurement should emphasize ability to use the learning involved. The coordinates of achievement should be in terms of ability to transfer knowledge to problem situations and to new applications. We teach mathematics for use, not as an end in itself.

Measurement should emphasize understanding of structure and concepts. The coordinates of achievement should be measures of comprehension, rather than measures of the memory of isolated facts. Only when higher cognitive demands are made on tests will students feel responsible for these levels of understanding.

Steps in an Evaluation Program

If measurement is to give coordinates in terms of objectives, several steps are necessary (see Figure 8.2):

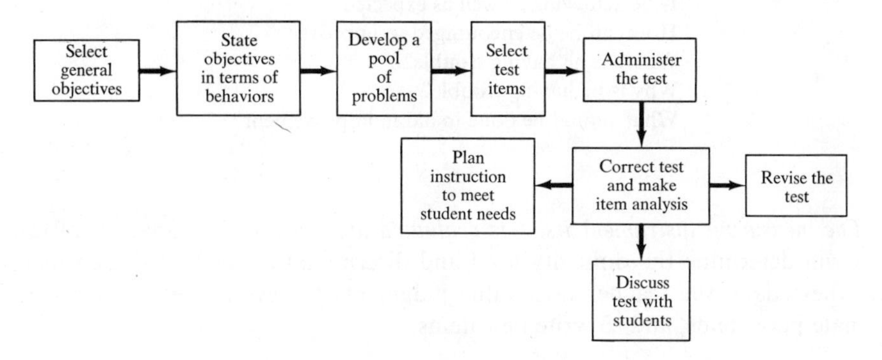

Figure 8.2 Evaluation Flow Chart

The objectives of instruction must be carefully selected. Usually these objectives are stated in terms of specific behaviors of students as they deal with the facts, formulas, principles, or theorems to be taught. Specifications of this type are necessary but not sufficient.

To function as coordinates, objectives must be stated in terms of behavior patterns on the part of the learner which indicates attainment of a particular objective. What does the student who has mastered a concept do, think, and feel? The behavioral description uses terms such as recall, state, draw, solve, analyze, describe, derive, apply, or prove. This is the format of the objectives stated in Chapter 4.

Sample situations, problems, activities must be selected that will demonstrate the student behavior outlined in the objectives. In mathematics this usually consists of a sequence of test items. However, the measurement of attitudes or creativity may involve the observation of behavior in an unstructured classroom situation or even outside the classroom. At other times the testing situation may involve observation of student use of the text, reference books, or laboratory equipment.

The situations, problems, and test items are presented to the student under optimum performance conditions. The student is given the time, tools, and materials he needs to perform to the best of his ability. The student knows what the test is about and why it is administered.

The student's responses are analyzed. An item analysis of a test will provide information for the next step in instruction. This analysis may suggest changes in objectives, learning activities, instructional methods, or materials. It may indicate what content needs to be revised, repeated, or eliminated. Through class discussion, correct answers are reinforced and errors eliminated; and individual needs are diagnosed and remedial instruction given. Questions to ask yourself regarding individual students include:

Is he achieving as well as expected?
How can he be encouraged to improve?
Where is he having trouble?
Why is he having trouble?
What should be done to aid in improvement?

The measuring instrument itself is evaluated and revised. The analysis of test items will determine the difficulty level and discriminatory power of each item. With these data, the teacher uses value judgments to review the test items, to eliminate poor items, and to write new items.

Types of Test for Measuring Achievement

The type of test to select depends largely on the objective to be measured and the purpose of the test.

Open-book tests. These tests emphasize understanding, application, and use of the text. They also have the advantage of stressing the transfer of knowledge rather than the memorization of facts. A sample item is as follows: "Identify the undefined terms, definitions, assumptions, and theorems that indicate the mathematical structure of the topic Area."

Reading tests. We need to build reading skills, study skills, and the desire to learn mathematics independently. A reading test presents a paragraph of unfamiliar mathematical material and then proceeds to ask questions about the content.

Performance tests. These tests require the student to discover a relationship through measurement, manipulation, experimentation, drawing, paper folding, investigation of patterns. This is a way of measuring ability to discover a new idea, apply facts, and do productive thinking. For a performance test, each item may be at a separate "station" or table in the classroom. Here is a sample item:

Station No. 10
Materials: Piece of board, 5 hooks, 18-inch ruler

At this station you are given a piece of board which is to be nailed to the wall in the laundry room. Various items are going to hang from the hooks, which will be screwed to the board.

______1. To the nearest $\frac{1}{4}$ inch, how long is the board?
______2. To the nearest $\frac{1}{8}$ inch, how far is the black center line from the sides of the board?
(a) $1\frac{1}{8}''$ (b) $1\frac{3}{4}''$ (c) $1\frac{5}{8}''$ (d) $1\frac{1}{2}''$
______3. The five hooks are to be screwed into the board along the center line such that the distance from the edge of the board to the outside hook is equal to the distance between any two adjacent hooks along the board. What should this distance be?

Essay or free-response tests. These tests include items that emphasize integration of ideas and communication skills and show the level of concept mastery. Here are two sample items:

Show that the (radian) measure of an acute angle is less than the arithmetic mean of its tangent and sine.

Compare the techniques of proof in coordinate geometry with those of synthetic geometry.

Attitude appraisal. Attitude tests are still in primitive form. However, rating scales, questionnaires, and anecdotes shed some light on attitudes. Here are some typical items:

I would like to take another mathematics course like this one.

My favorite subject in school is ________.

I enjoy mathematics problems as long as I get the right answer.

Tests of productive thinking. In problem solving, we should test the method of solution and the elegance of the proof rather than the answer. The planning, the organization, and the insight are significant aspects of productive thinking. A sample item is as follows:

Jim runs around a track in 40 seconds. Mark, running in the opposite direction, meets Jim every 15 seconds. How many seconds does it take Mark to run once around the track? Show your method of solution. If possible, find more than one method.

Achievement tests. These are tests that measure the extent to which the pupil has attained the specific objectives of a course or unit of instruction. A typical item would be:

If $(x - r)(x - s) = 0$, then
(a) $(x - r)$ must equal zero.
(b) $(x - s)$ must equal zero.
(c) $(x - s)$ and $(x - r)$ must both equal zero.
(d) $(x - s)$ or $(x - r)$ must equal zero.
(e) None of the above.

Diagnostic tests. This type of test is designed to discover the specific process, skill, type, or level of problem that creates difficulty for the pupil. This is useful in planning remedial instruction. Figure 8.3 shows the beginning section of a diagnostic test. Note how the columns and rows have common difficulties.

Type	Solve for a	Solve for b	Solve for c	Solve for d	Answers
I	1. $6a = 30$	2. $32 = 4b$	3. $18c = 6$	4. $15 = 8d$	1. $a =$ 2. $b =$ 3. $c =$ 4. $d =$
II	5. $\frac{a}{4} = 8$	6. $12 = \frac{2b}{3}$	7. $\frac{c}{3} = \frac{1}{5}$	8. $3\frac{1}{4} = \frac{3d}{4}$	5. $a =$ 6. $b =$ 7. $c =$ 8. $d =$

Figure 8.3 Portion of Diagnostic Test

Mastery tests. These self-administered and often self-corrected tests are designed for the pupil to check his own progress and needs. As a result, correct responses are reinforced, errors are corrected, and areas for further study are suggested. Alternate forms of the same test are provided so that students can review and retest themselves.

Inventory or survey or pretest or readiness tests. These are tests designed to determine a student's readiness for new work by measuring his background of previous experience and achievement. Such tests often include questions on the content to be studied in order to provide the teacher with an indication of the pace at which the new material may be presented.

Prognostic or aptitude tests. These tests predict a student's likely success in a given course. An item from an algebra prognostic test is the following:

Given: $x = y/n$. If y and n are always equal, how will x change in value if y and n increase?
(a) Remain the same. (b) Increase. (c) Decrease. (d) Cannot tell.

Contest tests. Tests sometimes provide the basis for a contest between two mathematics teams (sometimes called *mathletes*). State or national contests also use tests that probe high-level insight and mastery. The following are two sample items:

A circular piece of metal of maximum size is cut out of a square piece, and then a square piece of maximum size is cut out of the circular piece. The total amount of metal wasted is what fraction of the original square?

Given: the distinct point $P(a, b)$, $Q(c, d)$, $R(a + c, b + d)$, and $S(0, 0)$. Line segments $\overline{PQ}$, $\overline{PR}$, $\overline{QR}$, $\overline{PS}$, $\overline{QS}$, and $\overline{RS}$ are drawn. Depending on the location of P, Q, and R, is one of figure $PQRS$ or $PRQS$ I. a parallelogram? II. a trapezoid? III. a straight line?
(a) I only. (b) II only. (c) III only. (d) I and II only. (e) All three.

Besides teacher-made appraisal instruments, we need to capitalize on the wealth of available published evaluation materials. These tests are constructed by experts, are based on extensive experimentation, and have established norms. These published tests (see list of publishers in Appendix B) include prognostic, diagnostic, unit, and long-range achievement tests. Some of them emphasize the manipulative aspects of mathematics, while others emphasize structure and logic. All these tests can be scored objectively. Some of these tests can be used to measure year-to-year progress in mathematics while others measure achievement in specific topics or subjects. Published tests usually furnish norms such as grade equivalents or percentile ranks or standard scores. Norms make possible the comparison of class or individual performance with national or state achievement averages. However, the norms may be unsatisfactory as a standard for your class because your students may be very different in aptitude from the norm sample.

On the other hand, teacher-made tests have great advantages also:

Teacher-made tests may be adapted to the local situation—the students, the teacher, the community, and the school.

Teacher-made tests can be constructed to keep pace with curriculum changes. Published tests should not be the only basis for curriculum decisions, nor should they alone establish objectives—as they sometimes do.

Teacher-made tests are inexpensive as compared to published tests. However, the saving of time and the increased accuracy of measurement, more then justifies the costs of published tests.

Writing a test is a wholesome learning experience for the teacher. Constructing a good test forces the writer to consider the objectives of instruction and the individual differences of his class.

Whether the test used is a published one or a teacher-made one, it should meet the following criteria:

The test should be *valid*. It should be a true measure of the objectives it is supposed to measure.

The test should be *reliable*. It should measure consistently that which it is designed to measure.

The test should be *fair* to the student. Each statement should be clear and the answer determinable and definite. The language should be readable and correct.

The test should *discriminate* between the good and the poor achiever.

The test should be *comprehensive* so that it measures completely, not merely skimming ιe surface.

It should be as easy as possible to *administer* and *score*.

asuring Different Cognitive Levels

Chapters 4 and 6 stressed development of objectives in setting up the instruc- l program. It is best policy to construct a pool of test items at the time of planning. These test items form the criterion of successful learning of the They help you to focus your teaching on exactly what you want to accomplish. t just as objectives should be developed to encourage development of cognitive skills, just so should items be included in your inventory test items that are directed to these levels. This requires careful attention because the pressures are great to develop lowest common denominator tests that measure only the simplest skills and knowledge.

A number of classification schemes have been developed that suggest hierarchies of concepts. It is not necessary to follow the details of any one. In fact, Marion Epstein of Educational Testing Service has suggested a simplified practical set of categories for development of classroom tests: (1) computation; (2) routine problems; and (3) non-routine problems. Still, it is important for teachers occasionally to examine their test items in greater detail to see that their examinations are not skewed in specific directions.

Following the categories of Bloom's taxonomy,[2] discussed in greater detail in Chapter 4, here are some sample test items:

1.00 Knowledge

1.10 Knowledge of specifics

1.11 Knowledge of terminology

A meter is a unit of
(1) Energy
(2) Force
(3) Distance
(4) Area
(5) Volume

1.12 Knowledge of specific facts

The author of the *Elements*, a basic source of geometric information, is __________.

1.20 Knowledge of ways and means of dealing with specifics

1.21 Knowledge of conventions

The number of significant digits in 0.0340 is __________.

1.22 Knowledge of trends and sequences

In what sequence are the following area formulas postulated or proved in geometry:
(1) Triangle $A = \frac{1}{2}bh$
(2) Rectangle $A = lw$
(3) Parallelogram $A = bh$
(4) Trapezoid $A = \frac{1}{2}h(b + b')$

1.23 Knowledge of classifications and categories

Which one of the following is *not* a rational number:
(1) π
(2) $-3\frac{1}{5}$

[2] Benjamin S. Bloom (ed.), *Taxonomy of Educational Objectives: Handbook 1, Cognitive Domain*, David McKay Co., Inc., New York, 1956.

(3) 0
(4) 2.76
(5) 112_{three}
(6) .333

1.24 Knowledge of criteria

If we set out to prove two polygons similar, which one of the following is it usual to seek?
(1) Congruent sides
(2) Congruent sides and angles
(3) Proportional sides and congruent angles
(4) Congruent sides and proportional angles

1.25 Knowledge of methodology

In indirect proof, alternate conclusions, one of which must be ________ (true or false), are listed and then all but one are proved ________ (true or false).

1.30 Knowledge of the universals and abstractions in a field

1.31 Knowledge of principles and generalizations

If the product of two natural numbers greater than one is odd, their sum is:[3]
(1) Odd and less than their product
(2) Even and less than their product
(3) Odd and greater than their product
(4) Even and greater than their product
(5) Either even or odd

1.32 Knowledge of theories and structures

Which of the following is (are) subsystems of the real number system:
(1) Rational numbers
(2) Natural numbers
(3) Complex numbers
(4) Integers
(5) Algebraic numbers

2.00 Comprehension

2.10 Translation

Write an equation to represent the graph shown in Figure 8.4.

2.20 Interpretation

The solution set of the equation $x^2 + px + q = 0$
(1) Contains two elements if $p^2/4 - q < 0$
(2) Is the empty set if $p^2/4 + q = 0$
(3) Is $\{p/2\}$ if $p^2/4 - q = 0$
(4) Contains at least one element if $p^2/4 - q \geq 0$
(5) Is a real number only if $p^2/4 - q > 0$

---

[3] Adapted from Cooperative Mathematics Tests, Form A, Arithmetic, p. 8.

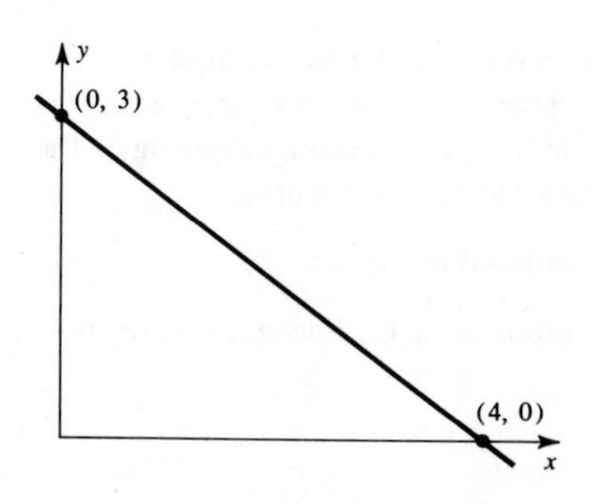

Figure 8.4

2.30 Extrapolation

Which of the following numerals represents the largest number?

(1) 1011_{two}
(2) 102_{three}
(3) 23_{four}
(4) 21_{five}
(5) All the numerals above represent the same number.

3.00 Ability to apply knowledge

If a and b are two prime numbers each greater than 10, which of the following is true?

(1) $a \times b$ is a prime number.
(2) $a - b$ is a prime number.
(3) $a \div b$ is a whole number.
(4) $a + b$ is an odd number.
(5) $a \times b$ is an odd number.

4.00 Ability to analyze relationships

4.10 Analysis of elements

If x is an integer and $\frac{(x + 5)}{2}$ is also an integer, then x could be

(1) Any negative integer
(2) Any positive integer
(3) Any even integer
(4) Any odd integer
(5) Any multiple of 5

4.20 Analysis of relationships

In computation of 54×23, what is the mathematical explanation of the reason we write the 8 under the 6?

$$\begin{array}{r} 54 \\ 23 \\ \hline 162 \\ 108 \\ \hline 1242 \end{array}$$

(1) We write the product under the multiplier.
(2) We write the product in the ten's place because the multiplier is a ten.
(3) We move over one place when multiplying by the second figure.
(4) We are using a shortcut that works.

4.30 Analysis of organizational principles

Compare the assumptions of Euclidean geometry with those of a geometry on a sphere.

5.00 Synthesis

5.10 Production of a unique communication

Given the symbols *a*, *b*, *c*, *d*, *ba*, *bb*, . . . as numerals for 0, 1, 2, 3, 4, 5, . . .

Use this new numeration system as answers for these questions:
(1) Write the numeral for forty-five.
(2) What number is represented by the numeral $b\,a\,d$?
(3) Write the numeral for the sum $b\,c + c\,d$.
(4) Write the numeral for the product $(d\,c \times c)$.
(5) What is an equivalent "dot" fraction (decimal) for c/d?

5.20 Production of a plan or proposed set of operations

Develop a procedure for laying out a baseball diamond by use of plane table and angle mirror.

5.30 Derivation of a set of abstract relations

Develop a finite mathematical system of four elements with a well-defined operation. Determine the properties and relationships of this system. What mathematical system does it illustrate?

6.00 Evaluation

6.10 Judgments in terms of internal evidence

Given H_1, and H_2, is C a valid conclusion?
H_1: All grasps are harpies.
H_2: X is not a grasp.
C: X is not a harpie.

6.20 Judgments in terms of external criteria

Occasionally some people state that Euclidean geometry is no longer acceptable, sometimes that it is not correct. Are these statements true? Defend your position. Consider in your answer what you mean by a statement and what you mean when you say a statement is true or false.

The foregoing test items indicate some of the wide variety of areas to be measured. They also indicate to some extent the variety of forms of questioning: completion, multiple choice, true–false, essay, proof, problem solution. Excellent books are available that consider the special values of the various forms of objective and subjective questions.[4]

Preparing Tests

When writing test items to measure mathematical achievement, the teacher should keep in mind these crucial factors:

Each item should be related to specific objectives.

Each test item should be mathematically correct.

Each test item should be technically correct. It should be correctly stated and have a specific answer.

The items should then be identified according to the specific objectives they have been meant to test so that the student's response will indicate a specific level of achievement.

Question construction is a difficult art. Even national examinations occasionally include errors, for it is difficult to avoid clumsy construction of items or a poor choice of test format. Great care should be taken in test construction, but the teacher must recognize that he will sometimes make errors and have to adjust grades.

The following is an example of the kind of unexpected difficulty that sometimes negates the value of an item:

Considering that $\sqrt{2} = 1.4$, the absolute value of $1 - \sqrt{2}$ is:
(a) $1 - \sqrt{2}$. (b) $\sqrt{2} - 1$. (c) 1.4. (d) -1.4. (e) -0.4.

Any answer may be justified on the basis that a false hypothesis may lead by correct logic to any conclusion. In this case we cannot "consider that" $\sqrt{2} = 1.4$. Omitting that introductory phrase and appending the word "approximately" after items (c), (d), and (e) would make the item satisfactory.

Careful thought should be given to each test item in order to aim at higher-level objectives. This does not mean that factual questions are to be eliminated.

[4] See William D. Hedges, *Testing and Evaluation for the Sciences*, Wadsworth Publishing Co., Inc., Belmont, Cal., 1966; *Evaluation in Mathematics*, National Council of Teachers of Mathematics, Washington, D. C., 1961; and publications of the School Mathematics Study Group related to the National Longitudinal Study of Mathematics Learning.

They should continue to enjoy a place on examinations. What should be sought, however, is a proper balance between such items and items requiring higher-level thinking processes. Consider in this regard the following two items. The first requires only a straightforward application of a basic concept, the second a more thoughtful application of that concept in a broader context:

1. If $N < 0$, which of the following is negative?

   I. N^2 II. N^3 III. N^4

   (a) Only I
   (b) Only II
   (c) I and II only
   (d) I and III only
   (e) None of these answers

2. If a, b, and c represent real nonzero numbers, which of the following expressions can equal zero?

   I. $a + b + c$ III. $a^3 + b^3 + c^3$
   II. $a^2 + b^2 + c^2$ IV. $a^4 + b^4 + c^4$

   (a) Only I
   (b) Only II
   (c) II and IV only
   (d) I and III only
   (e) None of these answers

Specific Points to Consider in Writing Test Items

The writing of good mathematics test items is a complex process. It requires background and skill in mathematics, writing, test construction, learning theory, and curriculum. This section gives only a few specific suggestions or "tricks of the trade" for test construction. For further study, the reader is referred to the test-construction references listed at the end of the chapter.

Plan the test so that major ideas, structures, principles, and skills are included. To make the test comprehensive, allocate a balanced number of items to each objective. Exclude insignificant and trivial items.

Devise items that test the student's ability to deal with the implementation of facts rather than with the mere recognition of them. Rather than "What is the formula for the area of a circle?" ask "How does the area of a circle with a 3-inch radius compare with the area of a circle with a 6-inch radius?"

Write a variety of test items including some in objective form, such as multiple-choice, true–false, matching, and some in subjective form requiring such things as the solution of problems and proofs. The type of question used should depend on what is being measured.

Write the items in simple, concise, correct language that every student taking the test will understand.

Have other mathematics instructors take the test so that they can evaluate it.

Write test items on index cards and file them according to topics or objectives for future use.

Use some open-ended or interpretive items that measure originality or creativeness. For example: "Select a set of four elements, invent a binary operation for combining these elements, and investigate the properties of your system."

Write test items that measure aspects of problem solving, such as estimating, selecting a search model, reasoning, and generalizing.

Use some essay items that measure communication skill, organization, reflective thinking, and ability to build a mathematical structure. Here is an example: "What is wrong with this definition? *A triangle is a polynomial with three sides.* Write a correct definition of a triangle."

Make each item independent of the others. Make sure that no item gives a clue to the answer to another item on the list.

Arrange the space for answers in a convenient place. Provide adequate space for writing answers and performing computations.

Avoid the use of trick questions. These questions produce hostility and resentment among the students and probably measure aptitude rather than the achievement of the objective to which the item is related.

Arrange the test items in each section of the test in ascending order of difficulty. Easy items build confidence, reduce tension, and encourage the low achiever to do his best. This encouragement should occur early in the test.

Discuss test construction, test taking, and test scoring with the students. Describe your method of selecting test items and how you score the test. Discuss time allotment, scoring for guesses, and the problem of cheating. Illustrate how test scores are only approximations of achievement and indicate your concern for giving correct marks.

Achievement Evaluated by Observation

Evaluation by observation of learning activities has decided advantages over paper-and-pencil tests. Some of these advantages are:

Observations can be made of the performance of the pupil in a natural, practical situation.

Observations permit the pupil to respond without restrictions or tensions that frequently are concomitant with testing.

Observations permit the evaluation of certain outcomes that cannot be obtained in any other way.

Observations can be continuous and part of the instructional activities.

Observations make possible immediate guidance and remedial teaching before undesirable or incorrect habits become established.

If these observations are to be used effectively, the teacher must know what behavior indicates the attainment of the objectives being evaluated. This will involve factors such as:

Understanding the scope of problems at hand

Planning the assembling of information or materials

Locating information and materials

Using instruments and materials skillfully

Organizing activities and recording information

Using ingenuity and resourcefulness in attacking the problem

Exhibiting enthusiasm, energy, interest in the activities

Working cooperatively with others

Completing projects promptly and independently

Mathematics Progress Chart

Problem, Unit, Activity__________ **Grade**______ **Date**_______

Performance Code: Excellent (+) Average (0) Unsatisfactory (−)	Paul	Mary	Bill
1. Planning activities			
2. Locating information			
3. Using measuring instruments			
4. Organizing information			
5. Recording data			
6. Computing accurately			
7. Cooperating with others			
8. Completing projects promptly			

Figure 8.5

Since the recording of these observations is time consuming, it is essential that the teacher use a checklist. The validity and reliability of these observations will also be enhanced if the results of observations are recorded on a progress chart or rating scale.

These rating scales may be constructed as shown in Figure 8.5.

Assigning Marks in Mathematics

After measuring instruments have been used to provide data, how should marks be determined for each student? A first requirement is that all measures be recorded correctly and be labeled completely so that they can be properly identified. Whenever an error in marking has been made, it should be admitted and corrected. Then these records should always be available for examination by the student, parents, counselors, or substitute teachers. Each student should know what items are taken into account in determining his final course mark. He should know how his teacher will go about determining his mark. And he should know his standing in relation to the comparison group involved. When a student inquires about his standing, the teacher should give complete information and use this opportunity to encourage and inspire better achievement.

Students of all ability levels are sensitive about their marks. They need to be assured that they will be given fair treatment and that their marks will as accurately as possible reflect their achievement.

Teachers use a variety of schemes for combining measures to determine a final mark. As each teacher does this he (as well as his students) should recognize that all measures are approximations. No matter how carefully he goes about his measurement, some students will be assigned marks that are not correct. Some A's should be B's or C's, and some C's should be A's or maybe F's.

Mathematics teachers like to use a quantitative system for combining measures to determine grades. To do this properly, they should use some system of standard scores to control the weights of different measures. There are teachers who keep a folder of material for each student, so that samples of the student's work are available for conferences and for help in rendering subjective evaluations. The more information available, the more confident the teacher can be in his marking. Hopefully, teachers of the future will have the necessary clerical help and computer facilities so that marking can be less tedious and more accurate.

With computer facilities now available for many schools, it would seem appropriate to keep significant information on a mathematics record card. This card should be a cumulative record of standardized test scores, marks in mathematics, and any other pertinent information. It should be kept in the mathematics department file so that the teachers have at hand significant information about each student.

If feasible, it is recommended that marks for achievement in mathematics be recorded on a special mathematics report form for parents and counselors. Only then can achievement be reported in sufficient detail to be meaningful. This report should indicate achievement in terms of specific objectives and it should also report this achievement with reference to specific comparison groups. Additional information such as standardized test scores or comments should be included. These comments should record special talents, certain difficulties, and unusual achievement. A suggested mathematics report card is illustrated in Figure 8.6.

Figure 8.6

Student Report Form for Mathematics

Name:________
Reporting Period: __First, __Second, __Third, __Fourth, Year_____
Course:__________
Section:__________

Mathematic objectives based on measurements	Comparison base				Comments
	Potential ability	Other class members	National norm	College entrants	
1. Understanding of concepts. 2. Skill in computation. 3. Ability to solve problems.					

General objectives based on subjective rating	Poor	Acceptable	Excellent	Comments
1. Attitudes such as appreciation. 2. Values such as respect for others. 3. Study habits.				

Mark: ____ Based on achievement in this class as compared with __________
Comments: ________
Instructor: ________

Individualized programs dictate a need for comprehensive records of progress and achievement. For mastery learning programs in which the student paces himself taking one or more self-checking mastery tests on each unit of work, the records are often kept cooperatively by teacher and student. Occasionally the two hold a conference to review problems, to discuss progress, and to plan ahead. In such programs, only the summative examinations that determine course grades are scored by the instructor. A report form for this kind of program might take the form:

Figure 8.7

Mastery Progress Reports

Student: ______________________________________
Course: ______________________ Section: __________
Topic: ______________________________________

	Mastery Tests					
	Form 1*		**Form 2**		**Form 3**	
Units	**Date**	**Level**	**Date**	**Level**	**Date**	**Level**
1						
2						
3						
4						

* Achievement level of 90% required before moving to next unit.

Marks for Students in Special Classes

The assignment of marks for slow learners in special classes and for the accelerated student poses certain questions. Should the top student in the general mathematics class get an A? What mark should the bottom student in the accelerated class get? Some schools prescribe that all grades be below B in general mathematics or that all grades be above C in the accelerated class. This prescription is based on two assumptions: (1) that the mark in a mathematics class is an indication of the absolute level of achievement in mathematics and (2) that the mark in a mathematics class should have its proper weight in determining class rank at graduation or for college entrance. Neither of these arguments is completely acceptable.

Every mark is a comparison, for the mark indicates relative status with respect to a group. An A indicates that this student ranks with the top students of the comparison group (the class, the local school, the city system, a state or national norm group, etc.), and an F indicates a very low rank. However, the crucial question here is of *which* comparison group is the student a member? This group may be the specific class involved, it may be all students of the same course, or all students of the same grade. Only when the A or F is related to the comparison group does it convey much meaning.

It would seem reasonable then that grades for every course in which the students of that course are the comparison group should have an entire range of marks. Thus, the general math student should be able to achieve an A; a calculus student should be able to fail. Only then is maximum information given about a student's relative achievement. Just as a typing class gives grades from A to F, so the general mathematics or accelerated class should allow a range of marks from A to F. It is not likely that an A in general mathematics will be equated to an A in accelerated mathematics as far as mathematical competence is concerned.

However, this may pose problems with respect to college entrance possibilities for the talented student in the accelerated class. Rank in a high school graduating class is based on the student's grades in all courses, and this rank is a significant factor in college entrance. If students are to accept assignment to an accelerated class they should not have their college entrance opportunities limited by this assignment. Some system such as the use of an "S" mark or a system of weighting marks in the accelerated class is needed for ranks based on absolute achievement. Another possibility is the use of an appropriate standardized test or common examination to determine the absolute level of mathematics achievement. Where there are multiple sections, common examinations make it possible to use the larger group as a comparison group. In the case of general mathematics, where motivation is already low, it would seem that successful students should be marked accordingly. Ideally, then, the report form should be such that the mark in any course can be properly interpreted in terms of course content and the comparison group.

Learning Exercises

1. There are several techniques used to assign conceptual levels to problems. One is to assign the highest level to which the question applies. Assign the following questions about quadratic equation to levels of Bloom's taxonomy (see pages 115 to 118) by this means:
   (a) Write the quadratic formula.
   (b) Give a formula that solves the following equation for y:

$$py^2 + qy + r = 0$$

   (c) Solve for x: $x^4 - 13x^2 + 36 = 0$.
   (d) Derive the quadratic formula.
   (e) Solve by completing the square $x^2 - 4x - 4 = 0$.
   (f) Write a flow chart for solving quadratic equations.
   (g) Find the y-range for the equation $y = x^2 - 4x + 5$.
2. Make up additional questions about quadratics that apply to different levels of the taxonomy.
3. Locate a chapter or unit test in a textbook of interest to you. Analyse the conceptual levels of the test questions.
4. Read pages 685 to 690 in Bloom, Hastings, and Madaus (see references) on questions for interests and attitudes. Explain how you could use some of these questions in your teaching.
5. Choose a topic (perhaps a chapter) in a textbook of interest to you. Indicate what would be your objectives in teaching this topic. Translate these objectives into test items. Be sure to include objectives at higher cognitive levels.

6. What are the pros and cons for each of the following:
   (a) Take-home tests (c) Mastery tests (e) Contract grading
   (b) Open-book tests (d) Formative evaluation (f) Eliminating marks
7. PROJECT. (a) Analyze and evaluate a published test. (b) Collect report cards from several schools. Analyze the pros and cons of each card.

Suggestions for Further Reading

Anderson, C. Harold, "The International Comparative Study of Achievement in Mathematics," in Douglas B. Aichele and Robert E. Reys (eds.), *Readings in Secondary School Mathematics*, Prindle, Weber, and Schmidt, Boston, 1971, pp. 94–109.

Bloom, Benjamin S., J. Thomas Hastings, and George F. Madaus, *Handbook on Formative and Summative Evaluation of Student Learning*, McGraw-Hill Book Company, New York, 1971, Chapters 3 to 10 and especially 19.

Burlow, Elsa H., "Tips on Testing," *School Science and Mathematics*, 64, 8 (November 1964), 709–714.

Cliffe, Marian C., "The Place of Evaluation in a Secondary School Program," in Aichele and Reys (eds.), *Readings in Secondary School Mathematics*, pp. 423–427.

Elder, Florence L., " 'Take-Home' Tests," in Aichele and Reys (eds.), *Readings in Secondary School Mathematics*, pp. 407–410.

Epstein, Marion G., "Testing in Mathematics: Why? What? How?," in Aichele and Reys (eds.), *Readings in Secondary School Mathematics*, pp. 394–406.

Evaluation in Mathematics, Yearbook 26, National Council of Teachers of Mathematics, Washington, D. C., 1961.

Guaru, Peter K., "A Time for Testing," *Mathematics Teacher*, 60, 2 (February 1967), 133–136.

Hedges, William D., *Testing and Evaluation for the Sciences*, Wadsworth Publishing Co., Inc., Belmont, Calif., 1966.

Jeffrey, Jay M., "Constructing Mathematics Tests—A Psychological Set," in Aichele and Reys (eds.), *Readings in Secondary School Mathematics*, pp. 390–393.

Merwin, Jack C., and Martin J. Higgins, "Assessing the Progress of Education in Mathematics," *Mathematics Teacher*, 61, 2 (February 1968), 130–135.

Myers, Sheldon S., *Mathematics Tests Available in the United States*, National Council of Teachers of Mathematics, Washington, D. C., 1970.

Payne, Joseph N., "Student Participation in Evaluation," in Aichele and Reys (eds.), *Readings in Secondary School Mathematics*, pp. 416–418.

Romberg, Thomas A., and James W. Wilson, "The Development of Mathematics Achievement Tests for the National Longitudinal Study of Mathematical Abilities," *Mathematics Teacher*, 61, 5 (May 1968), 489–495.

Salkind, C. T., *The Contest Problem Book: Annual High School Contests of the MAA, 1950–1960* (SMSG New Mathematics Library No. 5), Random House, Inc., New York, 1961.

———, *The MAA Problem Book II: Annual High School Contests of the MAA, 1961–1965* (SMSG New Mathematics Library No. 17), Random House, Inc., New York, 1967.

Weaver, J. F., "Evaluation and the Classroom Teacher," in *Mathematics Education*, Yearbook 69, Part I, National Society for the Study of Education, University of Chicago Press, Chicago, 1970, pp. 335–366.

Wolff, Harry, "Oral Testing," in Aichele and Reys (eds.), *Readings in Secondary School Mathematics*, pp. 411–415.

The Master Teacher 9

Every teacher owes it to himself to set a professional goal of becoming a master teacher. Such a long-term goal will help you to organize a specific program of improvement and to evaluate your progress. It will also help you to focus on long-term teaching improvement and to avoid overemphasis on the ephemeral aspects of daily chores. Finally, the goal of becoming a master teacher provides more satisfaction in progress toward fulfillment than in final accomplishment.

Years of teaching service are usually repaid by school systems with salary increments, tenured appointments, better teaching loads, and improved faculty status. But for some, those years of service are really the same year repeated many times. Satisfactory service cannot be equated with development. Just as achievement of competitive goals in sports is dependent on a rigorous conditioning program, so is growth as a teacher dependent on a planned program of improvement.

The time to start a self-improvement schedule is now. Like dieting, it is too easy to postpone. The beginning teacher has his problems with new assignments, new students, and new associates; but, in subsequent years, additional responsibilities both in and outside the school more than take up the slack created by familiarity with surroundings. Delay in starting means delay in improvement and failure to fulfill your teaching capacity as early as you might. On the other hand, there is no apparent "over 21" factor that relates to teacher improvement. Many teachers who drifted through years of teaching without extra effort have found the far greater satisfaction that extra attention to improved teaching finally brings them. Their only regret: their late start.

Characteristics of a Master Teacher

Every teacher has his own personality, his own teaching style, his unique strengths and weaknesses, and his inventory of basic qualities provides him with special improvement advantages or problems. However, some characteristics are common to many master teachers.

The master teacher knows mathematics thoroughly, not only at the level of the courses he teaches, but far beyond this level. Moreover, he is able to draw from this depth to illustrate, to amplify, to clarify, and to challenge. His knowledge is current, and he can counsel his students well. At the same time, he recognizes his limitations and is ready to admit lack of complete knowledge, to search for new answers, and to refer students for further assistance when necessary.

His classes are well organized and effective. Students learn, they participate, they extend their resources. There is plenty to do for everyone, and each student feels fully involved in each class. Students are successful in achieving the objectives of the master teacher; and this success breeds satisfaction, security, and self-confidence. They know what they are doing and why, yet each class has the spark of originality and reflects both the developed creativity of the teacher and the growing creativity of the students. Measured by any standard—student grades, student scores on standardized examinations, success of students in more advanced mathematics classes, evaluation by students—these classes rank high.

He recognizes his responsibility to his profession and assumes that responsibility. He participates in department and school activities and is active in professional organizations related to his teaching. He is, however, selective in his involvement and sees to it that out-of-class responsibilities do not reduce the quality of his instructional program. In particular, he is ready and willing to work with other staff members in program improvement; he contributes to and gains from such exchanges of ideas. This open-minded approach reflects his continued readiness to try new things, to experiment with both content and teaching technique—a readiness that is evident in the continuing freshness of his classes.

Above all, the master teacher enjoys respect—the respect of his colleagues, his school administration, and often others outside the school, but most important the respect of his students. This respect is his return for the hard work that has gone into his personal development. It has little or nothing to do with his personality, which may be outgoing or retiring, or his physical image, which may be robust or diminutive, or his social posture, which may be contemporary or old-fashioned. It does, however, have much to do with his value system: What he is and what he has done speak clearly to his responsibility to his students, to his chosen teaching field, and finally and most important to himself.

Most of us have come in contact with one or more of these master teachers. A remarkable proportion of beginning teachers cite their reason for choosing teaching and for choosing their teaching subject as their wish to emulate one of these teachers. Thus the established tradition of fine teaching has its opportunity to live beyond the teaching years of the master teacher.

How do you become a master teacher? In the section that follows we suggest some of the things that may be done.

Self-evaluation

To start an improvement program, it is well to know from where you start. An inventory of the advantages you enjoy as well as the difficulties you must overcome is in order. You may then build or repair on the basis of results of this examination. Such an analysis need not be entirely a self-examination. It is extremely useful to have a supervisor review with you your strengths and weaknesses. Some teachers invite students to tell them what they like and what they dislike about their classes. But in the end the teacher must look to himself.

One especially useful, often surprising, and occasionally even painful way to examine yourself is to make or have made audiotape or videotape recordings of several classes. Such a record often makes almost too apparent such things as irritating overuse of certain words, poor use of class time especially at transition points between topics, overemphasis on telling, poor questioning techniques, and the other common classroom errors. Retaining a few tapes for later comparison is appropriate when possible.

Here are some kinds of questions that you might include in a personal inventory:

Do my students know what I am trying to accomplish and why?

Do I do something each day to stimulate student interest in mathematics?

Do I ask questions that require reflective thinking?

Do I ask students to give reasons for their answers?

Do my students feel secure enough to ask questions?

Is my classroom attractive and conducive to learning?

Are my assignments reasonable and purposeful?

Do I vary the requirements and assignments for different individuals?

Do I use materials outside of the text frequently?

Do my students know the basis for their grades?

Do I give instruction in how to learn mathematics?

Are my tests based on my objectives?

Do I provide ample time and instruction for the mastery of key concepts and skills?

Do I listen to students so that I know their problems and needs?

Have I used library or department facilities adequately?

Do I relate new ideas to past experiences?

How well do I know each of my students?

Do I encourage students to respect each other's opinions?

Do I ever attack new problems with my students?

Do I point out applications of and ways to transfer mathematical knowledge?

Do I encourage students to pursue original ideas, solutions, or proofs?

Do I use illustrations, examples, devices that add meaning to concepts?

Do I present a personality and appearance that my students accept as desirable?

Do I use a variety of tests to improve my instruction as well as to measure the learning of my students.

Are routine activities administered efficiently?

Do I exercise care to present mathematical ideas correctly?

When did I last do anything new or different?

Have I learned any new ideas recently?

Have I shared with another mathematics teacher to improve our instruction?

Do my students appreciate mathematics more now than they did before they began my course?

Do I show my students the same courtesy I expect from them?

Do I have answers for my students and myself to justify what I am teaching?

Another way to evaluate your professional performance is to utilize standard rating scales like those often used by administrators and supervisors. While these lose the detail that questions like those just posed can pinpoint, they may better delineate major strengths and weaknesses. Figure 9.1 shows one such scale. The major danger of using scales like this for evaluations of any kind is the tendency to consider total scores as meaningful. Such a procedure fails to give credit to the many highly creative teachers who are not as good at some of the housekeeping aspects of teaching as they are at inspiring students.

Developing Quality Teaching

Once an analysis is made of the starting point, it is necessary to plan specific activities designed to improve the quality of your teaching. These activities may be generally considered in three areas: background, instructional program, and professional participation.

Broadening Your Background

Two of the most exciting characteristics of mathematics are growth and change. Whole new areas of mathematics (e.g., category theory and nonstandard analysis) have been developed in the past decade, and the accretion of new developments in traditional areas is tremendous. The college algebra course presents one of the most striking examples of rapid change. Compare the Birkhoff and McLane *A Survey of Modern Algebra* published in 1953 with the authors' newer text McLane and Birkhoff *Algebra* published in 1967 to note this extreme change.[1]

The secondary-school mathematics teacher often fails to understand the heavy pressure that this rapid development puts on him. He may tend to see his algebra or geometry class as isolated and self-contained rather than as an integral part of the broad stream of mathematics. To place specific courses in appropriate context requires more than superficial knowledge of many fields of mathematics. The horizon is virtually unbroken; pushing back that horizon is the task at hand.

Figure 9.1

Professional Competence Rating Scale

Directions: Rate each item according to the teacher's performance. A zero indicates very unsatisfactory performance and a 9 indicates very superior work. Comments can be used to identify the major basis for the rating.

	0	1 2	3 4	5 6	7 8	9
1. Preparation, planning, purpose Comments						
2. Selection of appropriate content Comments						
3. Methods of presentation Comments						
4. Materials of instruction used Comments						
5. Strategies for processes Comments						
6. Questioning and student response Comments						
7. Class control and direction Comments						
8. Expression, speech, communication Comments						
9. Appearance, manner, poise Comments						
10. Managing classroom routine Comments						
11. Measurements and evaluation Comments						
12. Provision for individual differences Comments						
13. Assignment of appropriate tasks Comments						
14. Provision for a healthful, attractive, comfortable classroom Comments						

Virtually all mathematics courses have at least indirect influence on the secondary-school program, but certain courses are more immediately relevant to the work at this level. When these kinds of courses are available, they deserve special attention by teachers:

History of mathematics

Mathematical foundations

[1] Macmillan, New York.

Probability and statistics

A problem-solving course (along lines laid out by George Polya[2])

Geometry, including transformation geometry in particular

Finite mathematics

Numerical analysis utilizing computers

Mathematical applications to physical, biological, or social sciences

Computer science or its equivalent

As a classroom teacher, you have specialized objectives in electing advanced mathematics courses. You should want to accomplish much more than passing another course or adding to an ever-increasing accumulation of credit hours. You should look for applications to your own teaching. You should place your teaching in the broadened context of the new work. You should question how your program relates to the concerns of your students, how best you can prepare your students for the changing programs of the colleges. You should involve yourself in problem-solving aspects of the courses in order not only to continue to grow as a mathematician yourself but also to gain insight into the parallel difficulties of your own students as they try to develop heuristic skills.

Parallel with your mathematical course work you should consider courses that add to your understanding of pedagogy and learning theory. Such courses as the following should be among those considered:

Courses or seminars related directly to courses that you are teaching

Courses that describe teaching programs of interest to you, such as laboratory teaching or computer-assisted instruction

Courses that relate to experimental mathematics programs, like Comprehensive School Mathematics Project, Secondary School Mathematics Curriculum Improvement Study, Programmed Learning According to Need, or British Primary School

Theories of instruction and learning

Adolescent psychology

Courses that relate to modern programs in the sciences, like Biological Sciences Curriculum Study (BSCS), Chem Study, Physical Sciences Study Committee (PSSC), Earth Science Study Committee (ESSC), Science, A Process Approach, or Harvard Project Physics

Courses that deal with new developments in mathematics education, such as computer-assisted instruction, flexible scheduling, performance contracting

Here again you will wish to focus your attention on those aspects of the courses that you can relate directly to your teaching.

[2] *Mathematical Discovery*, vol. I, John Wiley & Sons, Inc., New York, 1962.

All the foregoing are found in formal educational programs. Informal self-education can be even more important than participation in formal courses. Every classroom teacher should regularly read articles of interest to him in *The Mathematics Teacher*, *School Science and Mathematics*, *Mathematics Teaching*, *Arithmetic Teacher*, and the many other fine mathematics and mathematics education journals. Reading books on mathematics and teaching can be equally effective. A number of publications that fit these categories are listed in the appendixes.

Often school systems provide opportunities for in-service education programs in which teachers with common concerns can work together. Participation in such programs may be supplemented by attendance at workshops, meetings, and institutes.

A record system should be developed so that the gains of your study—formal or informal—will not be lost before they can be put to use. Such a record-keeping system is described in greater detail in Chapter 18.

Improving Your Instructional Program

This entire text is devoted to the improvement of instruction in mathematics. An improvement program based on a careful selection of ideas from this book would serve any teacher. However, framing these ideas into a program is not an easy task. In great numbers, ideas become overwhelming; and it is difficult to decide what to do when.

Here your study of objectives and the implementation of those objectives may be utilized. As a start, list some specific short-term objectives that you would like to reach as a teacher. These objectives should represent limited but real gains that you feel you can accomplish within a reasonable time. At first you may wish to focus on one or two such objectives. For example, you may be dissatisfied with the responses to your questioning; therefore, you could set as your first objective the improvement of your questioning skills.

Once you have stated your objectives, you can particularize them (more like behavioral objectives) by stating as precisely as you can the kinds of behaviors in yourself and your students that you want to produce. Continuing with the example, if you want to improve questioning skills, you might indicate that you want to develop techniques that (1) solicit more than one- or two-word replies, (2) encourage interaction between students (rather than always between you and one student), (3) develop responses from many students, and (4) call forth creative responses.

Now you should develop a systematic attack on this objective. Decide exactly what you can do and when. Your program would probably include reading appropriate sections in this text, reading some of the references on this topic in journals, observing teachers whose questioning techniques are good, making notes on types of questions that you think will work for you, and reworking your lesson plans to see that such questions are incorporated in your teaching program. Finally, you should evaluate your progress in the light of the specific goals you have set. This evaluation may help you to redirect your energies in your attack on this specific problem.

While development of communication skills is particularly important in improvement of the instructional program, there are many other areas of concern that also deserve your attention. Identifying priority problems is an important step in the improvement process. Here are some examples of actions you can take in several problem areas:

Encourage student initiative and responsibility.

Improve homework assignments.

Utilize audio-visual equipment more effectively.

Optimize personal relationships with students (or even an individual recalcitrant student).

Try an experimental text.

Develop values.

Improve problem-solving skills.

Implement laboratory techniques.

Develop problems for solution by a computer.

Note from this list the wide variety of problems. Some are directly related to content, some to pedagogy. Some, on the other hand, relate to the mental health aspects of instruction, to the sociological structure of the classroom, and to your own personality development. Every classroom teacher can identify his own specific areas calling for improvement or increased breadth.

Participating in Professional Activities

The many professional organizations for teachers today include local, state, national, and even international associations. In this section, we shall discuss only those devoted to mathematics and mathematics teaching. Other professional organizations may well be important to a teacher of mathematics but they are outside the scope of this text. Every teacher must keep in mind the importance of priorities. As you become an experienced teacher, there will be increasing demands on your energies, and you will have to set priorities and manage your resources. Certainly, you must direct some of your energy to those activities that relate to salary and teaching conditions. However, do not permit such involvements to divert you from professional activities that relate even more directly to your teaching program.

Professional activity in mathematics and mathematics education is intimately related to the other two aspects of developing quality teaching: broadening background and improving classroom instruction. Often the answers to teaching problems are found in professional journals or at professional meetings.

Unfortunately, many teachers look upon participation in professional organizations as a rather noxious duty; however, thousands of teachers have derived tremendous enjoyment and satisfaction from these activities. These organizations have much to offer the classroom teacher:

Local organizations (city or county) sponsor meetings and may publish journals. (Examples: Association of Mathematics Teachers of New York City, Nassau County Mathematics Teachers Association.)

State and regional organizations sponsor annual meetings, often conduct summer workshops of longer duration, and publish journals. (Examples: Association of Teachers of Mathematics in New England, California Mathematics Teachers—Northern Section.)

National and international organizations sponsor national and regional meetings; publish journals and yearbooks; and conduct many other activities, such as film and textbook production, usually through a well-developed committee structure. (Examples: National Council of Teachers of Mathematics, Association of Science and Mathematics Teachers.)

Many teachers belong to one or more organizations at each level, and multiple memberships are not as much a financial drain as it might appear. Local organizations often have nominal dues or none at all—an expression of interest is all it takes to become involved. State and regional organization dues are usually five dollars or less, and national organizations still cost less than ten dollars.

As is too often the case, a bare listing of activities fails to tell the whole story. In particular, attendance at a state or regional meeting can be a wholesome and exciting experience for a teacher. Among the stimulating offerings available to him at such meetings are:

A wide variety of exhibits including displays of enrichment materials, student projects, and (particularly important) virtually all available text books and other commercial materials

Speakers on a great range of topics

Formal and informal social affairs

Committee meetings and other business activities open to visitors

Even more important than these are the opportunities to meet other classroom teachers—to talk with them about mutual problems, to share solutions, and sometimes even to commiserate with each other when solutions do not seem to be accessible—in these ways to refresh each other and to send each back to the classroom recharged with energy. This fellowship of teachers with common problems and common interests can be one of the most rewarding experiences of a teacher's professional life.

Responsibilities of the Master Teacher

It may seem odd to speak of responsibilities for a teacher who has shown through his actions that he is a person with a strong personal commitment to his profession. Beyond his duty to his students and himself, the quality teacher should sponsor improvement of teaching quality in others as well. Next to self-improvement, this is the best way to help improve the quality and the image—and the status as well—of the teaching profession.

Especially important is assistance to student teachers and young staff members. Assuming responsibility for a student teacher calls for more than a technical introduction to teaching procedures. It calls for an introduction to the demands upon and the opportunities for teachers. Inspiration is an accurate description of a major responsibility of the supervisor to the teaching intern. And the teacher who not only demonstrates quality teaching to the beginner but encourages him to emulate the model makes a special contribution to teaching. He can take an avuncular pride in the radiating lines of quality for which he deserves much credit.

The master teacher can also contribute to the quality of his department by working on and supporting activities that reinforce good teaching. Here judgment is in order: invited advice and assistance are much better accepted than those forced on others.

Summary

A master teacher is the sum of many things. He is a person with a well-developed warm and accepting personality. He is industrious, interested even to the point of excitement about mathematics and mathematics teaching. He likes students and is challenged by each one to provide him with the best possible educational program. He is a leader: a leader of students, a leader among coworkers. He is active as a professional in local and regional activities. His concern for teaching and learning pervades his professional life.

This all seems like a large order. Are there such people? The answer is an unqualified "yes." They are found in large schools and small, urban and rural, public, private, and parochial. They teach A. P. Calculus courses; they also teach general mathematics classes. They are department chairmen and regular classroom teachers. They are men and women.

But many many more are needed!

Learning Activities

1. Describe a teacher you know whom you consider a master teacher or who closest approaches this status. What are his strengths and weaknesses?

2. Outline an improvement program for yourself.
3. Describe a difficult student. Indicate a program that you might develop to help this student academically and socially.
4. Analyze your current daily and weekly schedule. Tell how you might restructure this schedule to provide more time for self-improvement activities.
5. Find out and list the mathematics and mathematics education organizations on your local and state levels.
6. PROJECT. Find a rating scale used to evaluate teaching. Use it to evaluate a teacher whom you can observe teach.

Suggestions for Further Reading

Eagle, J. Edwin, "Helping Students to See the Patterns," *Mathematics Teacher*, 64, 4 (April 1971), 315–322.

Fawcett, Harold P., "Reflections of a Retiring Mathematics Teacher," *Arithmetic Teacher*, 11, 7 (November 1964), 450–456.

Hamachek, Don E., "What Makes a Good Teacher?," in Douglas B. Aichele and Robert E. Reys (eds.), *Readings in Secondary School Mathematics*, Prindle, Weber, and Schmidt, Boston, 1971, pp. 338–346.

Minnesota Council of Teachers of Mathematics Professional Standards Committee (L. Hatfield, chairman), "Patterns for Professional Progress," *Mathematics Teacher*, 62, 6 (October 1969), 497–503.

Attaining the Objectives of Mathematics Instruction

THREE

Strategies for Presenting Mathematics Content 10

In teaching a new concept, the teacher must decide what content will attain the objective of the study, then select the proper strategy for teaching the concept. "Strategy" here refers to the mathematical process, not the teaching process. This selection of strategy must precede selection of the method of instruction because it determines the method; also, strategy follows the selection of objective and content because they help to determine the strategy.

The strategy for teaching a given mathematical concept is the procedure, the algorithm, used to deal with the concept. The strategy selected by the teacher may begin from a number of directions; for example, in teaching numeration systems, we are faced with a number of questions:

Should we use a historical approach and begin with a discussion of ancient numeration systems?

Should we use a specific base such as 5 for a new numeration system or should several systems be considered simultaneously?

If we do use a single base, which one should logically be considered first?

Should we use the arabic digits for the new system or should a new set of symbols be created?

Should we use subscripts to identify the base or is some other means, such as color, more appropriate?

Should we treat an operation such as addition with one number base or compare results for several bases?

Should we develop complicated algorithms, such as division and square root, in the new base system?

Should we change nondecimal numerals directly to another base without recourse to base 10?

Should we include rational numbers in our study?

Should we consider problems of negative bases?

Should a finite modular system be related to the corresponding numeration system?

After answering these questions, the teacher selects the procedure that is the most efficient for his teaching style and most pleasant for his specific group of students to attain the objectives of the study.

The strategy selected depends on the topic, the class, the objectives, and the procedures known to the teacher. It is often appropriate to use a different strategy, a new approach, when the topic is taught a second time or when it is reviewed. When alternate strategies are utilized, they add life to a topic and generate the same kind of interest that study of an entirely new topic produces.

New strategies have a therapeutic effect on the teacher. They break the routine aspects of teaching—especially when the same subject has been taught several times or when several sections of the same subject are taught in the same year. The new approach makes the teacher's role a little more exciting and offers him a bit more challenge. All teachers should develop a repertoire of alternate strategies.

Strategies should be reviewed and a particular one selected at the time unit plans are developed, because the strategy for teaching determines the overall classroom presentation and the materials to be used.

Criteria for Selecting Strategies

There are several strategies available for teaching most mathematical topics. The following sections will suggest many strategies that can be used. Once the possible strategies are determined, the teacher must select the one most suitable to his particular situation.

Here are some guidelines for selecting a strategy:

The strategy should be mathematically correct. If alternative proofs of a theorem are possible, the one selected must be based on definitions, axioms, and proved theorems previously developed in the classroom.

The strategy should have meaning for the class. The mathematics should not be too sophisticated, that is, "over the heads" of students. Usually it should be possible to illustrate the concepts of an algorithm in terms of what they have learned previously. In developing the structure of the rational numbers, for example, the teacher will want to be sure that the class knows what an identity is, what closure means, and what an inverse does.

The strategy should meet the demands for a proper teaching procedure. It should be possible to formulate the process by means of concrete visual representations, to lead to the abstract representations, and to end with generalizations. The strategy should rely on a minimum of new concepts, specified conditions, and new procedures. Thus, a secondary-school teacher would hesitate to develop the real numbers on the basis of sequence limits because it would involve him in additional complex problems.

The strategy should provide a satisfying experience, so that students will be willing to exert the energy required to master the new technique. If factoring a set of algebraic phrases becomes a meaningless manipulation, it would be better not to have given the assignment.

That strategy is best which has the greatest application to future use of the concept. Thus, mathematicians recommend that dealing with the logarithm as a function is more valuable in advanced mathematics than dealing with logs as computational tools, a use essentially displaced by calculating equipment. The teacher should recognize that while a particular strategy may appear to be better on the basis of one or more points, it might be poor on others. For example, in the treatment of logs, the definition of the logarithm as the area under a curve, while more satisfactory mathematically, may be too sophisticated for some students (see following discussion).

Strategies for Teaching Some Basic Concepts

The following are examples of alternate strategies for teaching some basic concepts. The first example, taken from elementary school mathematics, is offered to give some insights into a problem at a more basic level. Awareness of the alternatives and the choices made by elementary teachers will be increasingly important to the secondary-school teacher as new programs with varied approaches develop in the elementary school.

The Definition of Addition

One of the first strategies the elementary-school teacher must select is one for the treatment of addition. What does addition mean? How is the transfer best made from concrete objects to the symbolic representation? How is the meaning of carrying (regrouping) made clear? How must the concept be modified as rational numbers, negative integers, and real numbers are introduced?

The usual introduction to addition begins with *combinations* of groups of objects:

$$A, B, C \text{ and } D, E \text{ combined are } A, B, C, D, E.$$

Recording the number of objects in each group or set, we have

$$3 + 2 = 5$$

Some teachers now deal with this situation in the symbolism and operations of sets:

$$\begin{aligned} K &= \{A, B, C\} \\ M &= \{D, E\} \\ N(K) &= 3 \\ N(M) &= 2 \\ \{A, B, C\} \cup \{D, E\} &= \{A, B, C, D, E\} \\ N(K) + N(M) &= N(K \cup M) \end{aligned}$$

A sum is the number of elements in the union of two disjoint sets. Note the high level of sophistication of this treatment, both in terms of symbols and vocabulary.

Another strategy involves the developing of number concepts in terms of the number line—addition is illustrated by jumps on the number line.

For example, $3 + 4 = 7$ is visualized in Figure 10.1.

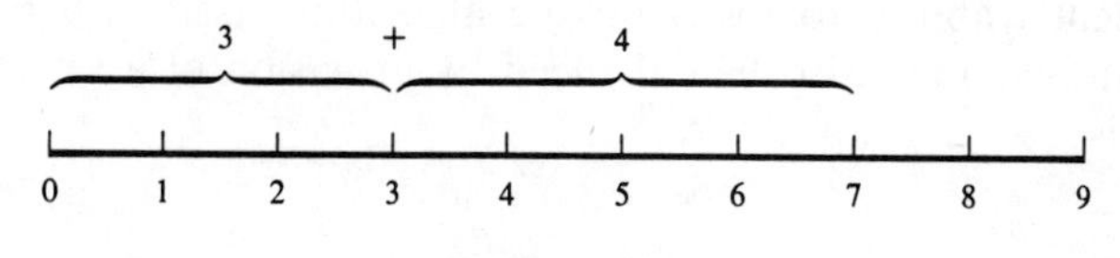

Figure 10.1

At a higher level, the addition operation is considered a mapping or a two-to-one correspondence.

The number-line representation has the advantage of giving the relationships of numbers, the order, the "distance" apart; therefore, it is useful in developing the properties of addition. It also lends itself without essential modification to the introduction of rational and real numbers, negative numbers, and graphing. Especially on the basis of graphing, which makes it useful at higher grade levels, the number-line representation seems the most appropriate strategy for discussing number concepts and operations.

Long Division

The algorithm for long division is typically "estimate a quotient, multiply, subtract, bring down, divide," and so on. It works but has little meaning to the student. Compare the ease and meaning of the process on the right with that on the left:

```
     25
23)584        23)584
   46             230| 10
   ---            ---
   124            354|
   115            230| 10
   ---            ---
     9            124|
                  115|  5
                  ------
                    9  25
```

The strategy on the left could have more meaning if it were introduced as follows:

$$
\begin{array}{rl}
20+5 & = 25 \\
23\overline{)584} & \\
460 & = 20 \times 23 \\
\hline
124 & \\
115 & = 5 \times 23 \\
\hline
9 &
\end{array}
$$

The student's ability to focus on the algorithm itself rather than isolated multiplication facts may also be enhanced by provision of a set of facts for the specific exercise:

23	23	23	23	23	23	23	23	23
×1	×2	×3	×4	×5	×6	×7	×8	×9
23	46	69	92	115	138	161	184	207

Most mathematics teachers agree that the end result should be the ability to use the first algorithm presented. But this may follow or be supplemented by the development on the right, which focuses on division as a process of repeated subtraction.

Division by a Rational Number

Division by a so-called fraction has rarely been made meaningful. The usual treatment is to "invert the divisor and multiply." What strategy will add meaning to this operation?

One easy strategy is to change the fractions to fractions with common denominators. Thus, $3/4 \div 2/3$ becomes $9/12 \div 8/12$. If the denominators are considered to be names for types of units, the problem becomes one of dividing nine of those units by eight of them, giving the quotient 9/8. This example may be justified by reference to the number line, numbers involving feet and inches, or pie cutouts.

Another more complex strategy is to relate the division problem to the definition of division:

$$\text{If } a \div b = c, \text{ then } a = c \times b$$

In terms of rational numbers:

$$\text{If } \frac{m}{n} \div \frac{r}{s} = c, \quad \text{then } \frac{m}{n} = c \times \frac{r}{s}$$

$$\text{Then} \quad \left(\frac{m}{n}\right) \times \frac{s}{r} = \left(c \times \frac{r}{s}\right) \times \frac{s}{r} \quad \text{or} \quad \frac{m}{n} \times \frac{s}{r} = c$$

$$\text{Thus} \quad \frac{m}{n} \div \frac{r}{s} = \frac{m}{n} \times \frac{s}{r}$$

Still another strategy utilizes a property of 1, the identity element for multiplication: $a \cdot 1 = a$ for any a. The division problem

$$\frac{m}{n} \div \frac{r}{s} \quad \text{is rewritten} \quad \frac{\frac{m}{n}}{\frac{r}{s}}$$

Using the identity element we have (since $x/x = 1$ for $x \neq 0$):

$$\frac{\frac{m}{n}}{\frac{r}{s}} \cdot 1 = \frac{\frac{m}{n}}{\frac{r}{s}} \cdot \frac{\frac{s}{r}}{\frac{s}{r}} = \frac{\frac{m}{n} \cdot \frac{s}{r}}{1} = \frac{m}{n} \cdot \frac{s}{r}$$

This technique leads to alternate approaches for specific examples:

$$\frac{5}{6} \div \frac{2}{3} = \frac{\frac{5}{6}}{\frac{2}{3}} \cdot \frac{\frac{3}{2}}{\frac{3}{2}} = \frac{15}{12} = \frac{5}{4}, \quad \text{or}$$

$$\frac{5}{6} \div \frac{2}{3} = \frac{\frac{5}{6}}{\frac{2}{3}} \cdot \frac{6}{6} = \frac{5}{4}$$

In the second case the choice of 6/6 for 1 instead of (3/2)/(3/2) simplifies the computation.

Percent and Percentage

The computation of percents and percentage has always been troublesome—partly because of the confusion of the similar terms "percent" and "percentage," partly because different numerals (e.g., 75%, .75, and 3/4) are used for the same percent with no distinction made between the rate and the numeral. But the main source of difficulty is the failure to recognize the difference between a percent and a rational number.

A percent is a rate. It is expressed as a ratio. But it is *not* a numeral for a rational number. Here is an example showing the kind of paradox that arises when a ratio and fraction are interchanged. Given two rational numbers 3/4 and 5/8, the sum $3/4 + 5/8 = 6/8 + 5/8 = 11/8$. But if 3/4 and 5/8 are ratios, then we may say $3/4 + 5/8 = 8/12$. To show why this is true, consider 3/4 as representing the rate "Our team won 3 out of 4 games in football." Then 5/8 may represent the rate "Our team won 5 out of 8 games in basketball." For both sports this statement follows: "Our teams won 8 out of 12 games this year." In other words, $3/4 + 5/8 = 8/12$ when 3/4 and 5/8 represent rates or ratios.

Of course, there are many different numerals to represent rates just as there are different numerals for rational numbers. Thus 3 out of 4 is equivalent to the rates 6 out of 8 or 15 out of 20. The rate 3/4 is one member of a set of equivalent rates or

$$\frac{3}{4} \in \left\{ \frac{3}{4}, \frac{6}{8}, \frac{9}{12}, \frac{12}{16}, \frac{15}{20}, \ldots, \ldots, \frac{75}{100}, \ldots \right\}$$

Consequently, 3/4, 75/100, .75, and 75% are four numerals for the same rate.

Various means are used to distinguish the rate 3/4 from the rational number 3/4. One program uses $\frac{3}{4}$ and 3/4. Another refers to a rate as a rate pair, and thus rate $\frac{3}{4}$ is represented by (3,4). This may be confusing to some students since (3,4) is also the notation for the coordinates of a point on the Cartesian plane.

Percents are usually used to solve three types of exercises such as the following:

1. What is 6% of 84?
2. What percent is 12 out of 60?
3. Seven is 9% of what number?

One past strategy was to classify a problem into one of these three categories. Then, the following rules were used to complete the exercises.

Case I. Change the percent to a decimal and multiply.

Case II. Divide the whole by the part and change the decimal to a percent.

Case III. Change the percent to a decimal and divide the part of the whole by this decimal.

These rate rules for Cases I, II, and III were meaningless. Percents above 100 were especially difficult in using these rules.

Another strategy has been to solve all percent problems by the use of the formula $br = p$ (base $\times$ rate $=$ percentage). Then by the substitution of the values given for two variables, the resulting equation can always be solved for the third variable. The main difficulty here is that the student too often cannot distinguish between b and p, the base and the percentage.

A third strategy uses a proportion for all rate problems. Since rates are ratios, it seems reasonable to use a proportion that is an expression of equality for two ratios. Then a percent is always a ratio in which the comparison base is 100. Then 75% is the ratio 75 to 100 or 75/100. All percentage problems now consist in finding two numerals for the same rate. Usually the problem is simplified because one of the numerals concerned has 100 as its comparison base.

This is how this strategy solves the three exercises above:

1. What is 6% of 84? 6 of 100 equals what of 84? Or

$$\frac{6}{100} = \frac{x}{84}$$

2. What percent is 12 out of 60? What of 100 is 12 of 60? Or

$$\frac{x}{100} = \frac{12}{60}$$

3. Seven is 9% of what number? 7 of what number equals 9 of 100? Or

$$\frac{7}{x} = \frac{9}{100}$$

To find the truth sets for each proportion, the student uses the equality of cross-products and equivalent equations. Thus, in exercise 3:

$$700 = 9x \quad \text{or} \quad x = \frac{700}{9}$$

Of these three strategies, the last seems the most appropriate. It solves all situations by the same method, namely, a proportion. It avoids the confusion of percent and percentage. It is mathematically correct in finding two equivalent rates. But, above all, it is a strategy that makes sense because it is based on the meaning of a percent.

Square Root

The square root of a number has several bases for exploration. At the intuitive level, it can be illustrated as the length of the side of a square whose area is given.

Thus, if 25 is the measure of the area of a square, the measure of each side is 5. Then the square root of 25 is 5.

At a higher level, the square root of a number is illustrated as one of the two equal factors of a number. This is satisfactory for perfect squares such as 36 but is not appropriate for numbers such as 17. Since factors are usually restricted to counting numbers, a difficulty might arise, for 17 does not have two equal factors that are counting numbers. The student might then conclude erroneously that there is no square root of 17.

This matter of equal factors is related to the use of exponents. For example, 7×7 is represented as 7^2 and called "seven squared" or "the second power of seven." Then, finding the square root of a non-negative number is the inverse of squaring a non-negative number. Then $\sqrt{49} = \sqrt{7^2}$ or $\sqrt{49} = 7$.

The approximate value of the square root of numbers that are not perfect squares can be found by a measurement approach. This approach is based on the Pythagorean formula for the right triangle, $a^2 + b^2 = c^2$ or $c = \sqrt{a^2 + b^2}$.

The strategy is based on drawings on graph paper, as shown in Figure 10.2.

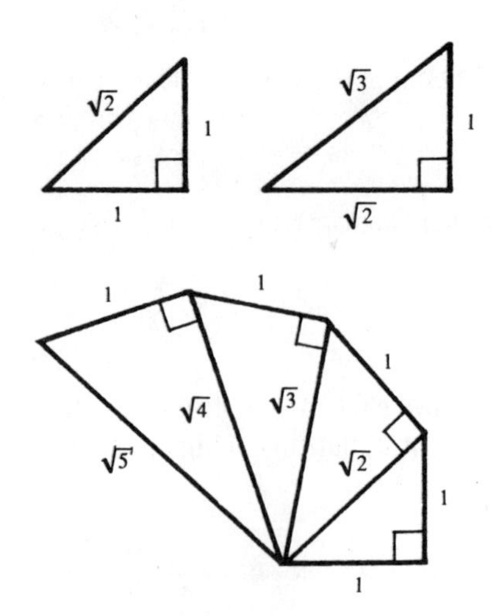

Figure 10.2

1. Draw an isosceles right triangle with each leg one unit long. Then the hypotenuse represents $c = \sqrt{1^2 + 1^2}$ or $c = \sqrt{2}$. The actual value of $\sqrt{2}$ can be determined by measuring the length of c with a separate strip of the graph paper.

2. Using the $\sqrt{2}$ as a length of one side and 1 as the length of the second side, draw another right triangle. The hypotenuse of this triangle represents $\sqrt{3}$. Continuing in the same way, the square roots of the integers may be determined. When constructed together, they form a spiral array.

Since the square roots of so many numbers are irrational numbers, the algorithm for the computation of a square root is troublesome for students. The usual strategy is to use the algorithm based on the square of a polynomial. It is widely accepted only because it "works" in finding a square root correct to any desired degree of accuracy. Here is how is applies to $\sqrt{746}$:

$$
\begin{array}{r|l}
 & \;\;2\;\,7.3 \\
 & \sqrt{746.00} \\
 & \;\;4 \\
\hline
40 & 346 \\
47 & 329 \\
\hline
540 & 1700 \\
543 & 1629 \\
\hline
 & \quad\;\; 71
\end{array}
$$

When students ask why this method works, it is helpful to show them a geometric and algebraic justification based on the square. It can be visualized somewhat as follows:

Given: A square whose area is 746 square units. What is the length of each side?

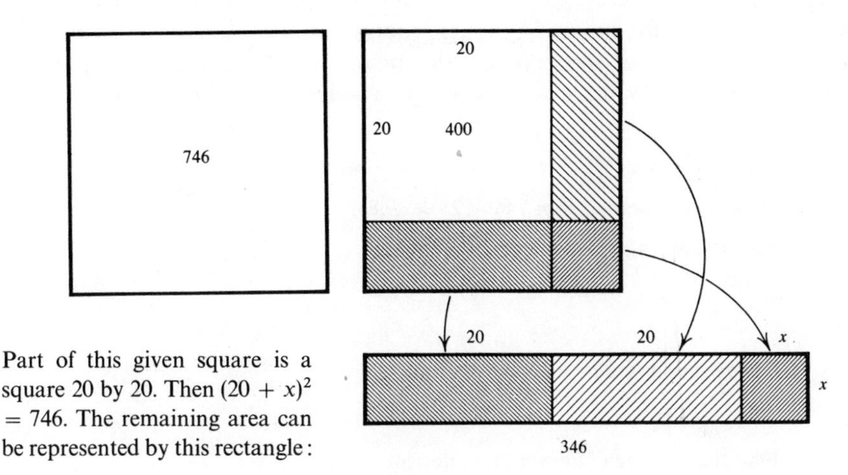

Part of this given square is a square 20 by 20. Then $(20 + x)^2 = 746$. The remaining area can be represented by this rectangle:

Figure 10.3

Since $(20 + x)^2 = 746$, $400 + 40x + x^2 = 746$, or $x(40 + x) = 346$, then x can be estimated by dividing 346 (the area of the rectangle) by 40, an approximation to the length of the rectangle ("double the first digit and add zero"). This suggests 8, but when 8 is added to 40 to get the total length of the rectangle, the product 48×8 is more than 346. Hence,

x is approximately equal to 7. The remaining area can be represented by y in the equation $(2 \cdot 27 + y)y = 17$ or by the rectangle.

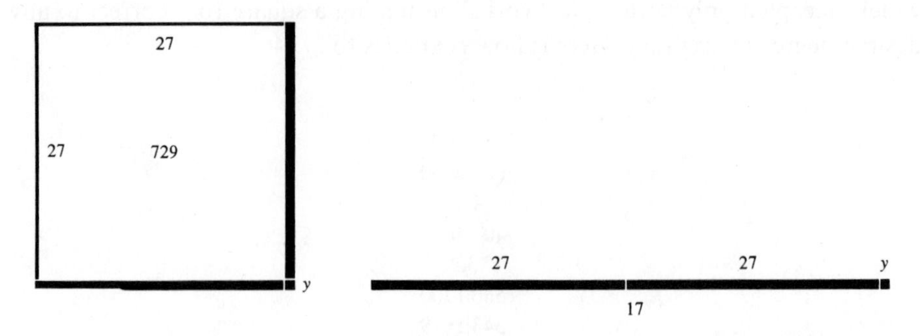

Figure 10.4

Consequently, dividing 17 by 54 gives an approximation for y, namely .3. This process can be repeated as often as necessary.

To be consistent with measurement theory, the square root should have the same accuracy as the given number. In this instance, it means the square root should have three significant figures.

The algorithm above is, of course, based on the square of a binomial $(a + b)^2 = a^2 + 2ab + b^2$. As such, it is a good extension of products of binomials. To be meaningful and remembered, this basis should be carefully established.

A second strategy for computing square root is the estimation-division method. To get the square root of 746, estimate the answer and then divide. Suppose we begin by estimating the square root as being 25 (since $20^2 = 400$ and $30^2 = 900$).

1. Divide by estimated root $746 \div 25 = 29.84$.
2. Find the average of divisor and quotient

$$\frac{25 + 29.84}{2} = 27.42$$

3. Divide the square by this new estimate

$$746 \div 27.42 = 27.206$$

4. Find the average of divisor and quotient

$$\frac{27.42 + 27.206}{2} = 27.31$$

In this strategy, note that no totally new algorithm needs to be learned. Here we use division and averaging. It is based on the idea that the arithmetic mean of two numbers is an approximation to the geometric mean, that is, $\frac{x + y}{2} \approx \sqrt{xy}$. Where $xy = N$, N is the number whose square root is to be determined. This last method, the one in most common use in today's texts, is as mathematically correct as the first and at the same time seems easier to remember. Answering some of the many questions left open by the algorithm is, however, not easy. To see this, examine some of the algebra involved.

Choose the first estimate

$$x_1 > \sqrt{n} > 0$$

Let the error in this first estimate be e_1, so that

$$x_1 = \sqrt{n} + e_1, \qquad e_1 \geq 0$$

The second estimate is found by averaging as described above:

$$x_2 = \frac{1}{2}\left(\frac{n}{x_1} + x_1\right) = \frac{n}{2x_1} + \frac{x_1}{2}$$

Let e_2 be the error in this second approximation

$$x_2 = \sqrt{n} + e_2$$

Then

$$e_2 = x_2 - \sqrt{n} = \left(\frac{n}{2x_1} + \frac{x_1}{2}\right) - \sqrt{n} = \frac{n - 2x_1\sqrt{n} + x_1^2}{2x_1} = \frac{(\sqrt{n} - x_1)^2}{2x_1} > 0$$

Thus:

$$(1) \qquad x_2 = \sqrt{n} + e_2 > \sqrt{n}$$

At the same time $x_1 > \sqrt{n}$ implies the following chain of reasoning:

$$x_1^2 > n; x_1 > \frac{n}{x_1}; 2x_1 > \frac{n}{x_1} + x_1; x_1 > \frac{1}{2}\left(\frac{n}{x_1} + x_1\right) = x_2$$

So:

$$(2) \qquad x_1 > x_2$$

Statements (1) and (2) place x_2 between x_1 and $\sqrt{n}$, thus making it a better approximation.

This complicated justification still leaves unresolved two more difficult questions: (1) How much better is the approximation? and (2) Does the sequence $x_1, x_2, x_3, \ldots$ converge to $\sqrt{n}$?

Note that the teacher is left in a real quandary by these alternate strategies. Each poses real problems if understanding is a basic objective. Because of these difficulties, some teachers do not teach any algorithm and direct their students to tables, logarithmic solutions, nomographs, slide rules, or graphs. Others, especially those with calculators available, use direct trial (by squaring) to approximate roots.

Multiplication of Directed Numbers

One of the most frequent, and unanswered, questions in an algebra class is "Why does a negative times a negative give a positive product?" To add meaning to this process, a variety of strategies such as the following are used:

1. Relate the multiplication process to trips on the number line. This will relate the multiplication process to the visualization of multiplication used in elementary school arithmetic. The new element introduced here is the idea of positive and negative directions, positive and negative time, and positive and negative distance.

Suppose that John lives on an east-west highway. Consider the highway as a number line with John's home at the origin. Using positive coordinates to locate points èast of John's home and negative coordinates for points west, consider distances to the east of John's home positive and distances west negative. Therefore, the direction of travel is positive when toward the east and negative toward the west. Future time is called positive and past time is negative, with the present being zero. (Note how many conditions must be established for this "intuitive" development of products of integers.)

Then $(^{+}50)\ (^{+}3) = {}^{+}150$ since travel toward the east at 50 mph for 3 hours in the future results in arriving at 150 miles east of John's house.

How then would the following be interpreted?

$$(^{-}50)(^{+}3) = {}^{-}150$$
$$(^{+}50)(^{-}3) = {}^{-}150$$

Finally $(^{-}50)(^{-}3) = {}^{+}150$ means that a truck traveling 50 mph west was 150 miles east of John's home three hours ago.

2. Gains and losses of the football field provide a similar travel situation.

3. Income and expenses resulting in being "in the red" or "in the black" give another application.

4. The UICSM uses the ingenious method of a motion picture projector which runs forward (+) or backward (−). Then a motion picture of a pump filling (+) or draining (−) a tank will result in an increase in water in the tank (+) or a decrease (−). The reversed projection (−) of a pump draining the tank (−) results in the tank being filled (+).

5. A different strategy uses patterns for predicting the product:

(a)
$$^{+}3\cdot{}^{+}2 = {}^{+}6$$
$$^{+}3\cdot{}^{+}1 = {}^{+}3$$
$$^{+}3\cdot 0 = 0$$
$$^{+}3\cdot{}^{-}1 = ?$$
$$^{+}3\cdot{}^{-}2 = ?$$

(b)
$$^{-}3\cdot{}^{+}2 = {}^{-}6$$
$$^{-}3\cdot{}^{+}1 = {}^{-}3$$
$$^{-}3\cdot 0 = 0$$
$$^{-}3\cdot{}^{-}1 = ?$$
$$^{-}3\cdot{}^{-}2 = ?$$

6. At a higher level, a deductive approach is possible

(a) $^{+}3 \cdot 0 = 0$	(a) property of zero
(b) $^{+}3(^{+}3 + {}^{-}3) = 0$	(b) substitution of $(^{+}3 + {}^{-}3)$ for 0
(c) $(^{+}3)(^{+}3) + (^{+}3)(^{-}3) = 0$	(c) distributive property
(d) $^{+}9 + (^{+}3)(^{-}3) = 0$	(d) substitution of $^{+}9$ for $(^{+}3)(^{+}3)$
(e) $(^{+}3)(^{-}3) = {}^{-}9$	(e) since $(^{+}3)(^{-}3)$ must be the additive inverse of $^{+}9$ in d.

The general forms are similar.

(a) $^{-}a \cdot 0 = 0$
(b) $^{-}a[b + (^{-}b)] = 0$
(c) $^{-}ab + {}^{-}a(^{-}b) = 0$
(d) $(^{-}a)(^{-}b) = -(^{-}ab)$
(e) $(^{-}a)(^{-}b) = ab$

It is likely that the total strategy for teaching the product of two negative numbers will start with a number line example, include the pattern of results and end with the deductive presentation.

Space Geometry

Space or solid geometry is now an integrated part of tenth grade geometry. But there are different strategies for integrating plane and space geometry.

Begin with a point and next consider lines as sets of points. Then a plane is a set of points or a set of lines. Finally space becomes a set of points, a set of lines, or a set of planes. This strategy is usually based on a development using coordinate geometry. The number line locates points on a line, two perpendicular lines provide coordinates on a plane, and three mutually perpendicular lines provide the coordinates for space geometry.

The second strategy consists of extending plane geometry concepts to the analogous concepts in space geometry.

Plane geometry	*Space geometry*
lines	planes
parallel lines	parallel planes
plane angles	polyhedral angles
polygons	polyhedra
circles	spheres

The third strategy treats plane geometry first and space geometry last. Often, inadequate time for the entire course means very little space geometry is included.

A basic requirement for success in space geometry, no matter what strategy is employed, is to allow ample time for intuitive development. One of the greatest difficulties is visualizing the relationships in three-dimensional space by two-dimensional drawings. Learning principles of orthographic projection, making models, and using blackboard stencils to get proper perspective help develop the perception of space.

Conic Sections

The conic sections may be taught from three approaches: locus, algebra, and space geometry. The first avenue stresses the construction of the conics in the plane by recourse to the locus-defining statements:

> *Ellipse.* A point on the plane the sum of whose distances from two fixed points on the plane is constant. (*Circle* is a special case of this, the two points coincident.)
>
> *Hyperbola.* A point on the plane the absolute value of the difference of whose distances from two fixed points on the plane is constant.
>
> *Parabola.* A point on the plane whose distances from a fixed point on the plane and a fixed line on the plane are equal.

The conic sections may also be approached by folding wax paper. Fold a line to represent the directrix and locate a point to represent the focus of a parabola. Make multiple folds so that the directrix is superimposed on the focus. The creases will form the envelope of a parabola. Relating the locus definition to the creases will give the mathematical analysis of the result.

In a similar way, an ellipse can be formed by using a circle and a point within the circle on wax paper. Folding many creases so that the circle is superimposed on the point will give the envelope of an ellipse. Again the locus definition is used for the mathematical explanation. If a point is selected outside the circle, creases formed by superimposing this point on the circle will give the envelope of an hyperbola.

Another avenue defines the conic figures in terms of algebraic statements from which graphs are constructed. (Rotations and translations are considered separately.)

Ellipse

$$\frac{x^2}{a^2} + \frac{y^2}{b^2} = 1$$

Circle is a special case of this, $a = \pm b$.

Hyperbola

$$\frac{x^2}{a^2} - \frac{y^2}{b^2} = 1 \text{ (or } -1)$$

Parabola

$$y = ax^2 \text{ or } x = ay^2$$

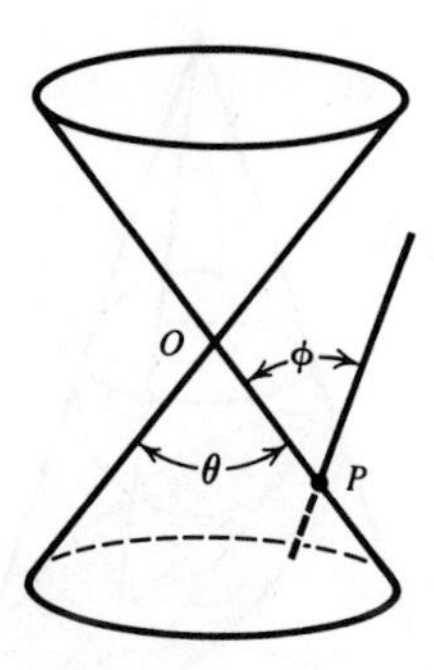

Figure 10.5

The third avenue defines the conics on the cone itself. Through a point P on an element of the right circular cone a plane is passed as on Figure 10.5. The figures that are the intersection of the plane and the cone depend on the relation between θ and ϕ.

Ellipse

$$m(\angle\theta) < m(\angle\phi) < \pi$$

Circle is a special case of this:

$$m(\angle\phi) = \frac{\pi + m(\angle\theta)}{2}$$

Hyperbola

$$0 < m(\angle\phi) < m(\angle\theta)$$

Parabola

$$m(\angle\phi) = m(\angle\theta)$$

Still another approach to conics is through mechanical construction of the curves.

Here is an example of a topic which may be approached from many directions, and often is in the classroom. Unfortunately, the real value of the multiple-strategy approach is lost here because students are seldom shown the interrelation between the approaches. The interrelationship is not hard to establish and in fact offers some insight into the unity of mathematics.

To show the locus-space geometry connection, reference is made to Dandelin's cone. To prove that the elliptical section of the cone through P in Figure 10.6 satisfies the locus definition, two spheres (like scoops of ice cream) are inserted in the cone in such a way that they are tangent to the plane of the ellipse at F and F' respectively. An element of the cone is drawn through P tangent to the spheres at

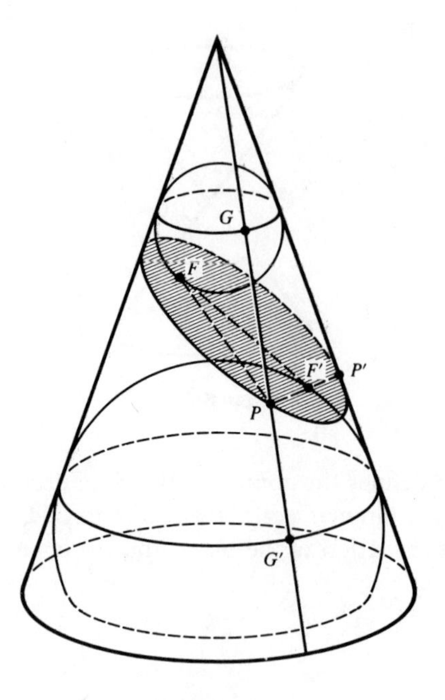

Figure 10.6

G and G'. For any position of P length GG' is constant. But $PG = PF$ and $PG' = PF'$ since two tangents to a sphere from an external point have equal measures. Since $GG' = GP + PG' = FP + PF'$, this latter sum is also constant and the connection is made.

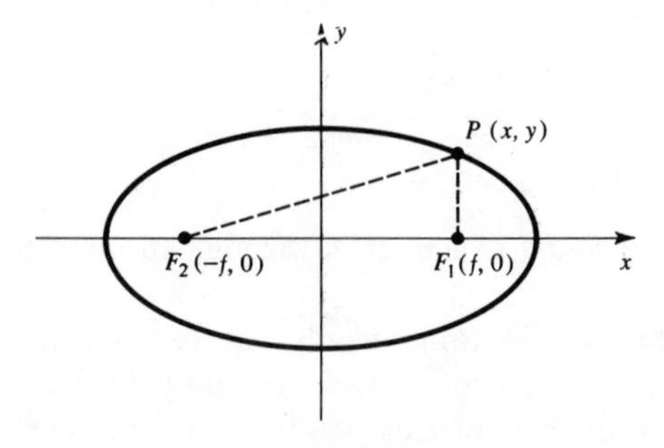

Figure 10.7

To establish the locus-algebra connection, place the ellipse symmetric to the origin, as in Figure 10.7. Let the fixed points (foci) be at $F_1(f, 0)$ and $F_2(-f, 0)$ and let the constant distance F_2PF_1 be $2a$. Assign the coordinates (x, y) to the locus point P. The formula for distance between two points provides the following equation:

$$\sqrt{(x+f)^2+y^2}+\sqrt{(x-f)^2+y^2}=2a$$

Eliminating the radicals in this equation and simplifying gives

$$\frac{x^2}{a^2}+\frac{y^2}{a^2-f^2}=1$$

Figure 10.8*a* shows why the constant distance $2a$ was chosen. When P is on the y-axis, $F_2P + PF_1 = pp' = 2a$. By symmetry, then, P is the point $(a, 0)$; $P'(-a, 0)$.

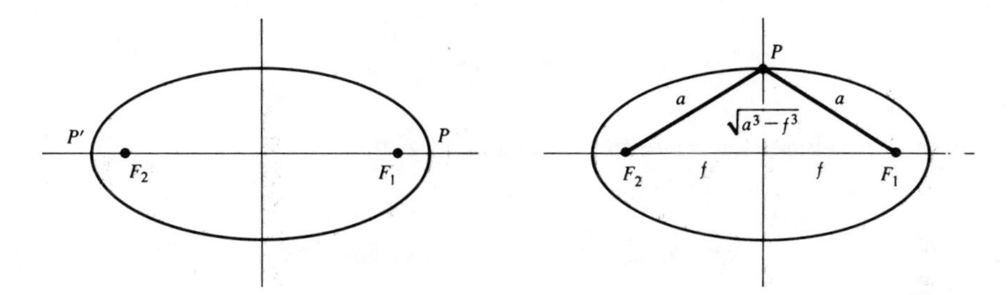

Figure 10.8

Figure 10.8*b* suggests the substitution $\sqrt{a^2 - f^2} = b$. This substitution completes the derivation of the algebraic equation

$$\frac{x^2}{a^2}+\frac{y^2}{b^2}=1$$

Since the arguments are transitive and reversible, the connection between the three definitions is established. Similar arguments may be made for hyperbolas and parabolas and are suggested as exercises for strong mathematics students.

Sum and Product of Roots of a Quadratic

The standard derivation of the formulas for sums and products of roots of a quadratic equation is by reference to the quadratic formula:

$$x_1 = \frac{-b + \sqrt{b^2 - 4ac}}{2a}$$

$$x_2 = \frac{-b - \sqrt{b^2 - 4ac}}{2a}$$

$$x_1 + x_2 = \frac{-2b}{2a} = -\frac{b}{a}$$

$$x_1 x_2 = \frac{b^2 - (b^2 - 4ac)}{4a^2} = \frac{4ac}{4a^2} = \frac{c}{a}$$

Alternate derivations may be used when other topics are being studied. The method of equating coefficients may be used. This theorem states that in the equation:

$$a_n x^n + a_{n-1} x^{n-1} + \cdots + a_1 x + a_0 = b_n x^n + b_{n-1} x^{n-1} + \cdots + b_1 x + b_0$$

when $m > n$ values of x make the statement true, then

$$a_n = b_n, a_{n-1} = b_{n-1}, \ldots a_1 = b_1, a_0 = b_0$$

To form a quadratic with roots x_1 and x_2 we set $(x - x_1)(x - x_2) = 0$. We may also set $ax^2 + bx + c = 0$ and write this in the form

$$x^2 + \frac{b}{a}x + \frac{c}{a} = 0$$

Equating the two we have an equation true for all substitutions of x (not just x_1 and x_2):

$$(x - x_1)(x - x_2) = x^2 + \frac{b}{a}x + \frac{c}{a}$$

This may be written:

$$x^2 - (x_1 + x_2)x + x_1 x_2 = x^2 + \frac{b}{a}x + \frac{c}{a}$$

Applying the theorem, we have

$$-(x_1 + x_2) = \frac{b}{a} \quad \text{or} \quad x_1 + x_2 = -\frac{b}{a}$$

$$x_1 x_2 = \frac{c}{a}$$

Another derivation (of the same formula) depends on symmetry properties of the parabola and the formula for the axis of symmetry. In Figure 10.9, the equation of the axis of symmetry of the graph $ax^2 + bx + c = y$ is $x = -b/2a$. Symmetry indicates that this is midway between (or the average of) the roots x_1 and x_2.

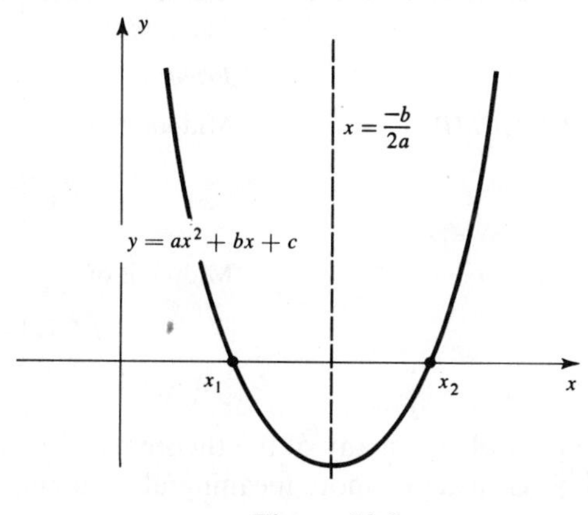

Figure 10.9

Apply the formula for finding the mean:

$$\frac{x_1 + x_2}{2} = \frac{-b}{2a} \quad \text{or} \quad x_1 + x_2 = \frac{-b}{a}$$

Analytic and Synthetic Proofs

Integrating coordinate geometry into the tenth-grade geometry course is now widely accepted. One of the reasons for this acceptance is that analytic proofs based on a coordinate system are considered earlier and more useful than synthetic proofs. Here is a simple example, based on Figure 10.10.

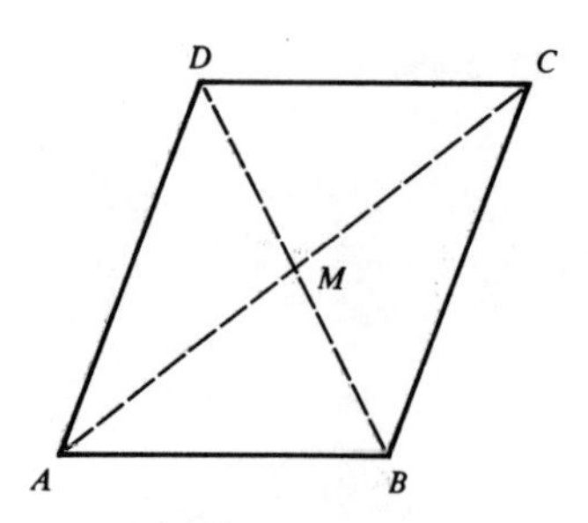

Figure 10.10

Theorem: The diagonals of a parallelogram bisect each other.

Synthetic Proof

Given: Parallelogram $ABCD$ with diagonals $\overline{BD}$ and $\overline{AC}$ intersecting at M.

Proof:

$$\Delta AMB \cong \Delta CMD$$

Analytic Proof

Given: Parallelogram $ABCD$ with vertices as follows: $A(c, 0)$, $B(b, 0)$, $D(a, c)$, $C(a + b, c)$

Proof:

Midpoint of

$$\overline{AC} \text{ is } \left(\frac{a + b}{2}, \frac{c}{2}\right)$$

Midpoint of

$$\overline{BD} \text{ is } \left(\frac{a + b}{2}, \frac{c}{2}\right)$$

Similar proofs are possible for many of the theorems of geometry.

Whether one method is easier or more meaningful than another has not been established.

Strategies for Specific Topics

For almost every topic taught in school mathematics, the teacher faces the problem of selecting the best or selecting several presentation strategies. Here are some further examples in addition to those discussed above:

Equations

Perform actual operations on each member of the equation.

Write equivalent equations based on properties of equalities.

Write a two-column proof justifying each step.

Use shortcuts such as "transposing" terms or "canceling" terms (not recommended by the authors unless careful groundwork is laid and the processes completely justified!).

Use flow charts. One chart "reads" the equation. Its reverse with inverse operations "solves" the equation.

Factoring

Find products by means of the distributive property. Reverse the process to find factors.

Relate the terms of a product to the areas of a rectangle. Factoring then becomes a process of finding dimensions of rectangles with given areas.

Classify products into given categories and then apply the appropriate rule for writing the factors.

Guess one factor, test it, and find another by division. Use parity (odd-even) relations for sums and products to lead to possible factors.

Congruence

Assume that triangles are congruent according to three given conditions.

Assume one basic congruence theorem and then prove the others on the basis of this congruence assumption.

Construct triangles according to given conditions. Compare them by superposition. Prove the congruence theorems by the traditional superposition proofs.

Consider congruence as a special case of similarity.

Follow Euclid's treatment of congruence.

Utilize geometric or algebraic transformations.

Logarithms

Define logarithms as exponents and operate as with exponents.

Change numerals to scientific notation. Use the results of computation with scientific notation to identify characteristic and mantissa.

Define logarithms as the area under a curve, specifically

$$\log n = k \int_1^n \frac{dx}{x}$$

where k determines the base ($\sqrt[k]{e}$) of the particular log system. Develop the properties of logs by recourse to the geometric analog.[1]

[1] School Mathematics Study Group, *Intermediate Mathematics*, Stanford, California, Vroman's, 1960, pp. 453–455.

Measurement

Estimate measures. Use English or metric units but do not convert from one to another. Rather, develop a mastery of the relative size of all units in each system.

Establish new, arbitrary units of measure. Use these units to make measurements and convert from one unit to another.

Learn the metric system with emphasis on the meaning of prefixes. Memorize key conversion factors for metric to English and learn to convert measures from one system to another.

Study measurement with emphasis on the approximate nature of measures and the necessary computational rules. Know the difference between precision and accuracy and learn to apply these ideas to computations with measures.

Trigonometric Functions

Define them in terms of right triangles. Build a table of functions by drawing right triangles, measuring the sides, computing ratios.

Define the functions in terms of a rectangular coordinate system.

Define the functions in terms of rational numbers and infinite sequences.

Develop the functions as line values on the circle (Figure 10.11).

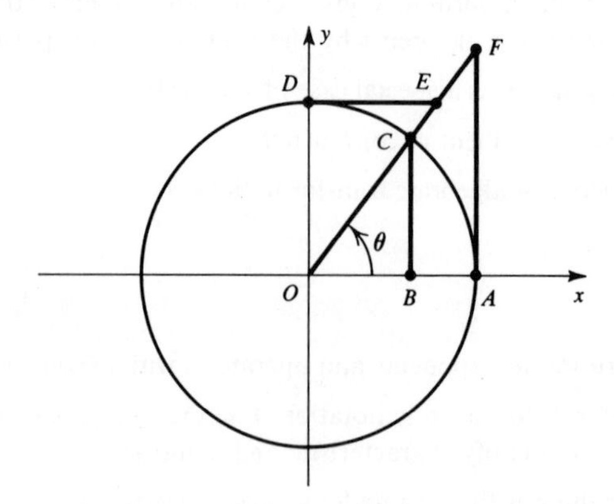

Figure 10.11

$BC = \sin \phi$	$OE = \csc \phi$
$OB = \cos \phi$	$OF = \sec \phi$
$AF = \tan \phi$	$DE = \text{ctn } \phi$

Define the functions in terms of a wrapping function.

On Figure 10.12, t is the distance measured counterclockwise on the unit circle from the point $(1, 0)$. (Tangent is defined to be $\sin t/\cos t$.)

Develop the functions from a vector approach.

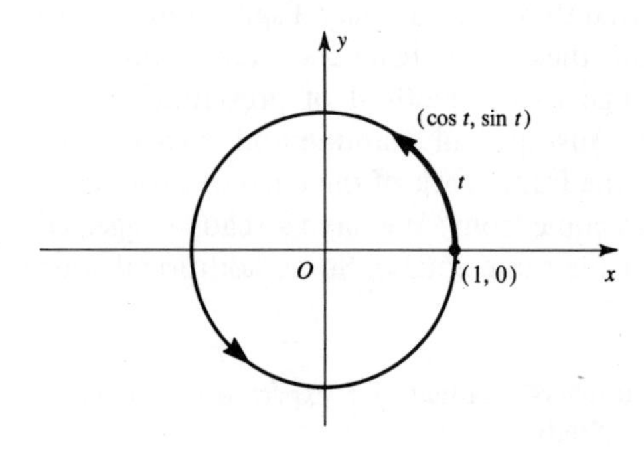

Figure 10.12

Quadratic Functions

A strictly algebraic treatment designed to provide techniques for solving quadratic equations by factoring, completing the square, and applying the quadratic formula.

A graphical study based on point-by-point construction of the graph $y = ax^2 + bx + c$.

A strategy based on the idea of a function machine $x \to ax^2 + bx + c$.

A study of properties of the quadratic equation such as sum and product of roots and properties of the discriminant.

A structural study of the graph of quadratics $y = x^2$, $y = ax^2$, $y = x^2 + p$, $y = (x - q)^2$, and finally $y = a(x - q)^2 + p$.

A study based on an application such as a freely falling body: $s = \frac{1}{2}gt^2 + v_0t + s_0$, where s is the height, t time, v_0 initial velocity, s_0 initial height, and $g = 32$ (ft/sec/sec), the constant of acceleration due to gravity.

Program the solution for a computer.

Systems of Linear Equations

Find the common solution as the intersection set of two truth sets.

Find the common solution by graphing.

Find the intersection of truth sets by substitution.

Find the common solution by determinants.

Find the solutions by a computer program.

Sources of Teaching Strategies

Many teachers are limited by their experience and background, and feel that they are not creative enough to use alternate routes to content development. Such teachers often feel that they should "play it safe" and follow the textbook presentation completely. But these same teachers often change textbooks and teach new courses. They change to the method of presentation of the new book readily enough. They could just as well examine and occasionally try new (or even old) approaches within the framework of the current program.

Where do ideas come from? We have already suggested one excellent source: other textbooks for the same course. Some additional sources include:

Texts and teachers' manuals for experimental programs, like those for School Mathematics Study Group.

Professional journals, including European publications, that stress classroom presentations, like *Mathematics Teaching* (English), and *Praxis der Mathematik* (German).

Reference books like Kasner and Newman's *Mathematics and the Imagination* (see Appendix C).

Other teachers.

Professional meetings.

Mathematics is a versatile science. Possible procedures, proofs, and methods of solving problems are legion. In teaching mathematics we select the most elegant algorithm or proof that has real meaning in terms of both content and pedagogical aims for our students.

Learning Exercises

1. Select a given mathematical topic from high school algebra or geometry. Find different strategies for teaching it in different texts. What strategy is most suitable for your class?
2. Illustrate different proofs of the Pythagorean theorem. Which proof is the most suitable as the first proof for a tenth-grade geometry class? A seventh-grade class? Why? (A useful reference for this exercise is E. S. Loomis, *The Pythagorean Proposition*, National Council of Teachers of Mathematics, Washington, D. C., 1968.)
3. Examine a modern mathematics textbook and a textbook of a generation ago. How do these texts differ in the strategies used for a specific topic?
4. Write the synthetic and analytic proofs of a theorem in geometry. Discuss the relative advantages or disadvantages of the different proofs.
5. Write a lesson plan for teaching division by a rational number. Use a strategy (different from that discussed in this text) that has meaning and is mathematically correct.

Suggestions for Further Reading

Baker, Betty L., "Developing a Meaningful Algorithm for Factoring Quadratic Trinomials," *Mathematics Teacher*, 62, 8 (December 1969), 629–631.

Braunfeld, Peter, Clyde Dilley, and Walter Rucker, "A New UICSM Approach to Fractions for the Junior High School," *Mathematics Teacher*, 60, 3 (March 1967), 215–221.

Coxford, Arthur, "Classroom Inquiry into the Conic Sections," *Mathematics Teacher*, 60, 4 (April 1967), 315–322.

Edmonds, George F., "An Intuitive Approach to Square Numbers," *Mathematics Teacher*, 63, 2 (February 1970), 113–117.

Forbes, Jack E., "The Most Difficult Step in Teaching of School Mathematics: From Rational Numbers to Real Numbers—With Meaning," *School Science and Mathematics*, 67, 9 (December 1967), 799–813.

Frandsen, Henry, "The Last Word on Solving Inequalities," *Mathematics Teacher*, 62, 6 (October 1969), 439–441.

Hardesty, James, "On Similarity Transformations," *Mathematics Teacher*, 61, 3 (March 1968), 278–283.

Mallory, Curtiss, "Intuitive Approach to $x^0 \equiv 1$," *Mathematics Teacher*, 60, 1 (January 1967), 41.

Ranucci, Ernest R., "Aspects of Combinatorial Geometry," *School Science and Mathematics*, 70, 4 (April 1970), 338–344.

Sanders, Walter J., and J. Richard Dennis, "Congruence Geometry for Junior High School," *Mathematics Teacher*, 61, 4 (April 1968), 354–369.

Scott, C. H., and Terry Rude, "Plane Geometry by Vector Methods," *School Science and Mathematics*, 70, 3 (March 1970), 230–238.

Spitznagel, Edward L., Jr., "An Experimental Approach in Teaching of Probability," *Mathematics Teacher*, 61, 6 (October 1968), 565–568.

Wiscomb, Margaret, "A Geometric Introduction to Mathematical Induction," *Mathematics Teacher*, 63, 5 (May 1970), 402–404.

11 Teaching Mathematical Concepts

How are mathematical concepts learned? What goes on in the mind of the student as he attempts to unravel new ideas? How should ideas be presented to the student so that a concept is learned quickly and correctly? If we understand the process whereby mathematical concepts are learned, then we can plan proper learning experiences.

Concepts are defined in many different ways. We will define a mathematical concept to be a mental construct. A concept is a mental abstraction of common properties of a set of experiences or phenomena. The elements of these sets may involve objects (set concepts), actions (operational concepts), comparisons (relational concepts), or organizations (structural concepts). The basic types of mathematical concepts are illustrated by these examples:

Set concepts. A number is the common property of equivalent sets.

Operational concepts. Addition is the common property of the union of disjoint sets.

Relational concepts. Equality is the common property of the number of elements of equivalent sets.

Structural concepts. Closure is the common property of a mathematical group.

In the case of the set concept, for example, the concept "three" is abstracted from the common properties of many sets. The mental construct is similar to the idea of color, also the common property of many sets. Note that in each case the concept is essentially abstracted from the object or objects themselves.

Necessary Conditions for Learning Mathematical Concepts

If concepts are to be learned, certain necessary conditions must be present. The learner must be *ready*, *willing*, and *able* to learn. Then, he needs *guidance*, *resources*, and *time* for the learning. Thus, the following conditions are essential in building new mathematical concepts:

The learner must have the necessary information, skills, and experiences so that he is ready to learn a new concept. Only when he has the background necessary to perceive common properties, relationships, patterns, and structure of ideas will he be able to generalize. For example, algebraic fractions cannot be considered if the learner has a meager understanding of rational numbers, the meaning of common denominator, the impossibility of division by zero, and the identity element for multiplication. Listing prerequisite concepts and skills and pretesting students is an important aspect of teaching.

The learner must have been motivated to the extent that he is willing to participate in learning activities. The learner learns what he is doing, seeing, feeling, or thinking. Hence, learning is possible only if the learner himself responds to the learning situation. And he will respond only when he thinks a response is desirable. Instead of asking a student to learn the associative law for addition for a future test, we would do better to ask him to look for a shortcut in the solution of the following exercises:

$37 + 52 + 48 \qquad 234 + 987 + 13 \qquad 7\frac{1}{3} + 3\frac{3}{7} + 5\frac{4}{7}$

The learner must have the necessary capability so that he is able to participate in the learning activities. Learning mathematical concepts is an intellectual process that involves activities such as manipulating, visualizing, listening, reading, computing, writing, thinking, verbalizing, abstracting, generalizing, and symbolizing. This means that concepts to be learned must be selected within the range of the learner's ability to do these things if he is to make progress. We should not expect a slow learner to compute cube root when he has not been able to learn how to compute square roots. We should not expect the student to learn to solve quadratic equations if he cannot solve linear equations.

The learner must be given some guidance so that motivation is preserved and learning is efficient. Learning by trial and error or by haphazard reflection may discourage him so that he never reaches the goal. Ideas should be presented to him so that he can perceive common elements. Thus, learning how to subtract a directed number is facilitated if the student is helped to compare the directed distances between points on the number line and the results of adding opposites:

Subtraction as the directed distance between two points on the number line

Subtraction as the sum of opposites

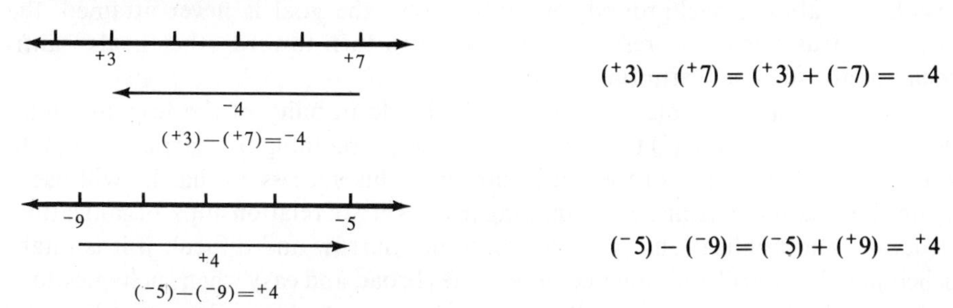

$$(^{+}3) - (^{+}7) = (^{+}3) + (^{-}7) = -4$$

$$(^{-}5) - (^{-}9) = (^{-}5) + (^{+}9) = {}^{+}4$$

The learner must be provided with appropriate materials (e.g., a text, a model, a film, or a tape) with which to work. For example, to learn how to add in a new number base, the student could make an addition slide rule to help him build an addition table.

The learner must be given adequate time to participate in learning activities. To discover a concept independently is time consuming. Learning is a growth process leading gradually to responses at an increasingly mature level. To master a concept, then, requires varied experiences, applications, and uses—all time-consuming activity. Too many teachers move on to new ideas before this absorption can take place. When teaching a basic concept such as ratio and proportion, a teacher should take time to develop mastery and use the concept in a variety of situations so that the student will apply it as needed in the chemistry class, social science class, the shop, or the home.

The Classroom Learning Situation

The way in which the student is responding to a learning situation can be illustrated by the chart in Figure 11.1.

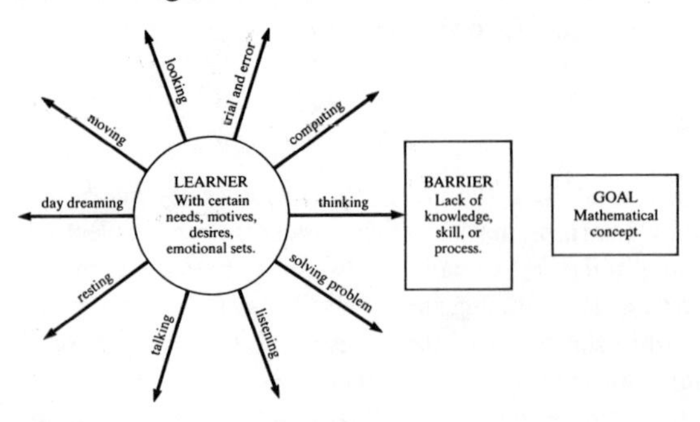

Figure 11.1

The goal of classroom activities is to learn mathematics. But the learner is reacting to a variety of disturbing stimuli in the classroom. He responds by doing what seems at the moment most desirable or most necessary for him. Most of these activities are not directed toward the learning goal. The barrier that separates him from the goal is ignorance or lack of skill. He cannot answer the question or solve the problem at hand. If the barrier is insurmountable—that is, if the learner has insufficient ability, background, or motivation—the goal is never attained, the learner is frustrated, and responses are likely to turn toward other goals—anti-social goals, such as disrupting the class.

It is the teacher's role (1) to establish the desirability of the learning goals described in Chapter 4, (2) to provide means of overcoming the barrier, and (3) to make the learner aware of the path that led to his success so that he will use it again. This path is established by making him aware of relationships, organization, structure, and applications. The path becomes narrow and difficult if it is established merely by drill and practice; it becomes broad and easy when such questions as how and why are answered. With this latter approach supplemented by time, experience, and reflection, concepts take on greater meaning and become useful in attaining other concepts. Thus, equations take on more meaning when related to inequalities and when used to solve problems based on formulas. At the same time, they make the attainment of goals related to inequalities and problem solving easier to achieve. Intuition and constructive thinking lead to deduction and analytic thinking.

How Mathematical Concepts are Formed

The typical sequence for learning a mathematical concept progresses from perception to abstraction to integration to deduction, is outlined in the flow chart of Figure 11.2. Concepts of the simplest order, such as addition, are formed as the

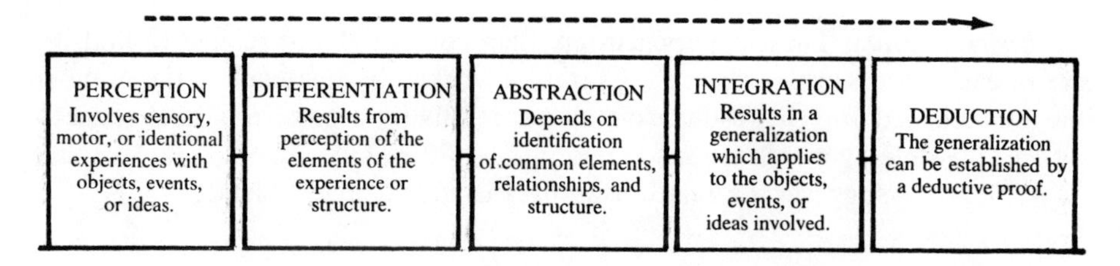

PERCEPTION
Involves sensory, motor, or identional experiences with objects, events, or ideas.
DIFFERENTIATION
Results from perception of the elements of the experience or structure.
ABSTRACTION
Depends on identification of common elements, relationships, and structure.
INTEGRATION
Results in a generalization which applies to the objects, events, or ideas involved.
DEDUCTION
The generalization can be established by a deductive proof.

Figure 11.2 ***Concept Formation Flow Chart***

result of repeated sensory or motor experiences. Concepts of higher order, such as functions, are formed when experiences and previously learned concepts are related through reflective thought. In short, we learn mathematical concepts in the following way: (1) We sort objects, events, or ideas into classes or categories. (2) We become aware of relationships within the classes or categories involved. (3) We find a pattern which suggests relationships or structure. (4) We formulate a conclusion which seems to describe the pattern of events or ideas involved. (5) We establish the generalization by a deductive proof.

Let us apply the flow chart of concept formation by seeing how a specific concept is formed: addition of a positive and a negative number (see Figure 11.3).

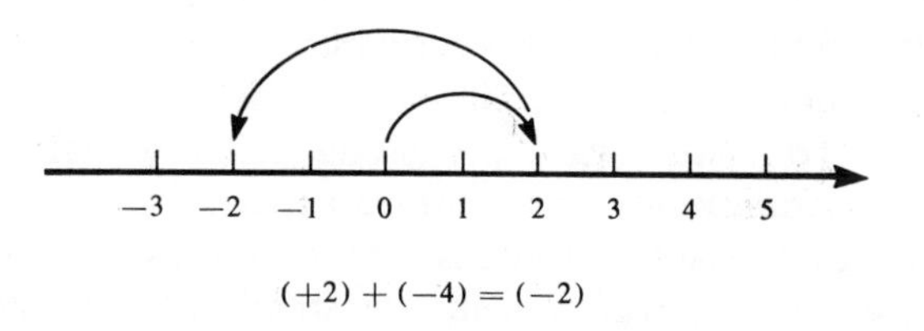

Figure 11.3

We first assure ourselves that the learner has the necessary background information about positive and negative numbers as coordinates of points on a number line and an understanding of absolute value. If he has mastered these ideas he is *ready* to learn the addition concept.

Experience. The first stage in learning a concept is the sensory experience with concrete objects or visual representations. The manipulative sensory stage for the addition of integers consists of illustrations of these sums in terms of "jumps" on a number line. The learner readily relates this process to his experience with the number line in adding two counting numbers.

Differentiation. These experiences are then compared and related to find the role of each process or each result. Examples worked by reference to the number line are recorded and the results examined carefully for a pattern of data suggesting a possible generalization. Thus, $^{+}5 + {}^{-}2 = {}^{+}3$, $^{-}2 + {}^{+}5 = {}^{+}3$, and $^{-}5 + {}^{+}2 = {}^{-}3$ suggests the relationship of the sum to the addends.

Abstraction. The first idea abstracted is usually that there is some kind of a subtraction involved in performing the operation. However, the learner is often able to find the sum of a positive and a negative number before he is able to verbalize the generalization. (Before he generalizes, the student often makes applications such as finding the truth set for $x + 7 = 2$.)

Integration. The learner finally relates the results of his additions to the use of absolute values and states a generalization: the sum of a positive and a negative number is the difference of the absolute values directed toward the larger addend. After considerable practice and guidance he is able to establish the generalization by a deductive proof. Then he determines how what he has learned fits with what has gone before and what questions it opens for future study. And he explores all the properties of the set under the operation of addition to establish the mathematical structure involved, in this case, a group.

Notice the number of steps and varied responses made before one reaches the deduction stage. Too often, when presenting mathematical concepts, teachers begin with the generalization or even the deductive proof of the concept. Instead they should learn and use the relationships of the media, the response, and the thought processes in learning a mathematical structure as illustrated by the model of Figure 11.4.

The Role and Sequence of Different Media for Thinking

Each of the different media used for learning concepts plays a particular role, the sequence usually running from concrete to visual to symbol—as indicated in Figure 11.4.

Concrete objects and models provide sensory experiences from which discoveries can be made. An odometer removed from an old car provides a concrete basis for an understanding of decimal relationships.

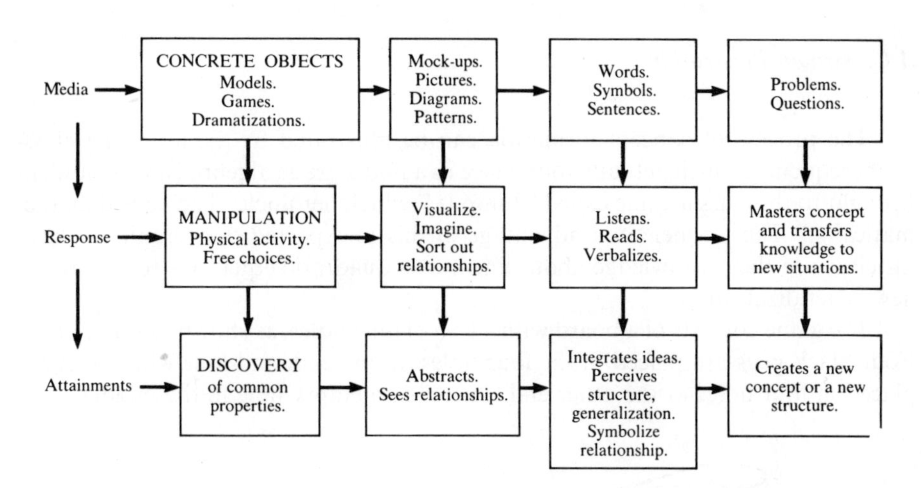

Figure 11.4 Model of Learning

Games and dramatizations provide for interaction and lend informality to learning. Tossing dice, one black for positive numbers and one red for negative numbers, gives informal drill in adding a positive and a negative number.

Mockups, pictures, diagrams, and patterns give the learner an opportunity to visualize and think. This is the way a house plan directs the thinking of the builder. Visualization makes it possible for the student to *see* relationships rather than simply to hear about ideas from the teacher. For this reason the number line is especially useful for learning how to operate with directed numbers.

Words, symbols, and sentences are the tools for thinking and communicating. With these tools, it is possible to relate new concepts to experiences, ideas, or structures that are already known. With these tools, a new concept can be discussed, questions can be raised, and independent reading undertaken. When we can verbalize a generalization, when we can defend a cause-and-effect relationship, when we can describe a structure, and when we can organize a course of action, then we are operating above the plane of rote memorization and mechanical manipulation of symbols. As a child matures, he should also become more reliant on reflection and less dependent on immediate perception in learning new concepts and in building structures with these concepts. Thus, the need for using concrete objects or visual representations is lessened in the upper grades of secondary-school mathematics.

Finally, questions are raised and problems are posed that require the concept to be transferred to a new situation. These probing questions and challenging problems require the learner to combine concepts to form a new concept or a new structure. When a concept is truly mastered by a talented student, he is able to use it to create an original structure. After learning how to add and multiply directed numbers, the student may be able to invent a new numeration system in which the base is a negative number—thus applying his knowledge at a creative level.

A Classroom Illustration

The process of concept formation can be illustrated by the following classroom sequence, which actually took place in a ninth-grade algebra class. A student brought to the class a game called "Jump It," which her uncle "had solved mathematically." The teacher took advantage of this unexpected opportunity to have his class use their knowledge about linear and quadratic equations to discover a new generalization.

The game consists of a board with a row of nine holes, as shown in Figure 11.5. Four black pegs are placed in the four holes at one end, and four white pegs are placed in four holes at the other end. There is an empty hole in the middle.

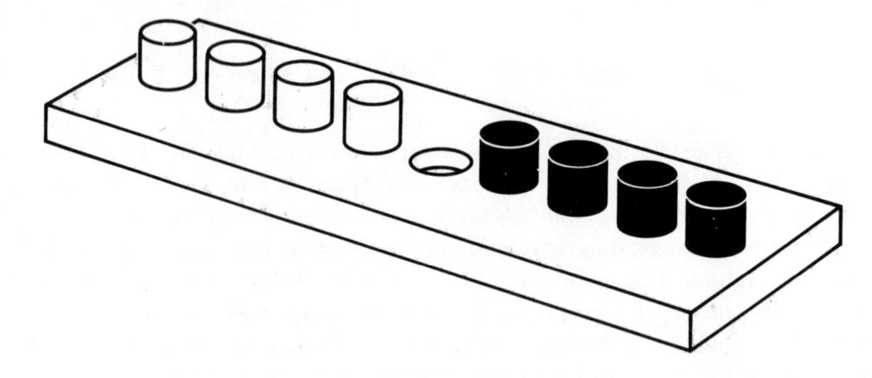

Figure 11.5 Jump-It Board

The object of the game is to exchange the positions of the black and white pegs. A peg may be moved from one hole to another empty hole. A peg may be moved from one hole to the next one or by "jumping" over at most one peg.[1]

1. The learner *confronts* a situation, a question, an idea, or a structure to be explored. The teacher asks Jane, who brought the game, to explain the rules to the class.

2. The learner *accepts* the exploration of the new situation or ideas as being worthwhile. The class is challenged by the uncle's "solution" to the puzzle, and students are eager to try the game.

3. The learner *reacts* to the learning situation by participating in discovery activities. He needs to see, hear, speak, think, write, or do something about the problem presented. Students make replicas of the original game with paper counters and try it themselves. *When they are able to solve the puzzle, they are urged to find a minimal solution using the least number of moves.*

[1] Gerald Rising, "Some Comments on a Simple Puzzle," *The Mathematics Teacher*, 49, 4 (April 1956), 267–269.

4. The learner may use *trial and error* in his search for patterns or relationships from which to abstract a generalization. Most students start with an unorganized attack on the problem, making several false starts and having to begin again. Eventually, they develop a more deliberate strategy. They reduce the puzzle to three holes and two pegs, then five holes and four pegs, then seven holes and six pegs. As they watch the pattern of successful moves, they get a clue to the solution with eight pegs and nine holes.

5. The learner *associates* his previous experiences, memories, skills, or knowledge with the elements of the new situation. Several students relate the game to checkers, while others associate it with related peg games. One student suggests that the game could be played with cards. Another suggests that two sets of different coins could be used.

6. The learner *needs an awareness of his progress* to attain accuracy or reject errors in a generalization. Soon all students have completed the puzzle at least once; they check each other, and several students have what they believe to be the minimal solution.

7. The learner is able to *differentiate* between the properties of the objects, events, data, or ideas involved. Students are urged to focus on the numerical aspects of the puzzle (rather than the moves required to complete the solution), the *number* of pegs and the *number* of moves in the minimal solution.

8. The learner is able to *abstract* the common elements of the situation involved. Bill suggests recording the number of pegs and the number of moves to discover a relationship. The class constructs a table relating these variables.

9. The learner becomes *aware* of the generalization or structure—the concept involved. Sam suggests that the table is exactly like the ones for which the class had to construct equations.

10. The learner is able to *express* the generalization visually, verbally, or symbolically. The class develops the formula relating the variables by the techniques introduced earlier, arriving at the equation $N = n(n + 2)$. (See point 13 below.)

11. The learner is able to use the generalization in a new situation. The teacher poses several questions to ensure understanding: "How many moves would it take to complete the puzzle with 10 counters?" "How many counters would there be if the completion of the puzzle required more than 35 moves?"

12. The learner attains *insight into the implications* of his generalization. The class discusses the generality of the results. The students first consider whether or not the results may be extrapolated; then they test one or two predictions.

13. The learner develops an *understanding* of the explicitness, completeness, precision, and logic of the generalization. The students examine the limits of the generalization (in this case, to positive integers) and begin to examine the way it is developed. They focus on the increase in moves when additional pieces are added. First and second differences are obtained to determine whether the solution is linear or quadratic. The data give the set of ordered pairs shown in Table 11.1. Since second differences are constant, the relation is quadratic, or of the form $y = ax^2 + bx + c$.

Table 11.1

Number of black pegs (x)	Minimum Number of moves (y)	First difference	Second difference
1	3		
		5	
2	8		2
		7	
3	15		2
		9	
4	24		2
		11	
5	33		

$$\text{Since } x = 1 \text{ and } y = 3, \text{ then } 3 = a + b + c.$$
$$\text{Since } x = 2 \text{ and } y = 8, \text{ then } 8 = 4a + 2b + c.$$
$$\text{Since } x = 3 \text{ and } y = 15, \text{ then } 15 = 9a + 3b + c.$$
$$\text{Then } a = 1, b = 2, \text{ and } c = 0.$$

Thus, the general formula for this game is $y = x^2 + 2x$ or $x(x + 2)$.

14. The learner is able to justify the generalization by a *deductive proof*; he then develops a proof by *mathematical deduction.*

15. The learner uses the generalization to *discover or create* a broader concept of which the original generalization is a part. Several students set out to analyze additional games of the same type (e.g., Tower of Hanoi), referring to Hoyle as a principal resource; others attempt to generalize the games to two dimensions.

Basic Principles for Teaching Concepts

The way in which concepts are formed suggests the following:

Concepts cannot be given to the learner. He must construct them out of his own experiences and thoughts. Therefore, effective teaching consists in providing learning experiences for every student.

Concepts are formed as part of a growth process. Wider implications and deeper meanings develop from a variety of experiences; therefore, mathematical maturity is nourished by a spiral approach.

Any concept becomes more meaningful and more useful when it is related to the total structure of which it is a part. Therefore, the mathematical concepts learned each day should be woven into the mathematical structure involved in that day's lesson.

Concepts are best developed by varied experiences rather than by repetitive presentations. Therefore, problem solving, discovery activities, and a varied routine are more effective than monotonous repetition in learning mathematics.

The level at which a concept should be introduced in a given lesson depends on the readiness, motivation, and ability of the learner. Therefore, provision must be made for individual differences and motivation in each day's lesson.

Concepts are more likely to be formed when the learner actively operates on his environment and restructures his own thinking than when he simply carries out instructions in a teacher-directed situation. Therefore, the mathematics lesson usually should not be a lecture by the teacher but a group activity in probing an idea.

Action, manipulation, and imagery precede verbalizing, and verbalizing precedes writing. The learner should try things out to see what happens. He should manipulate objects, drawings, words, and symbols. He should pose his own questions and seek his own answers. He must feel free to ask questions, to make mistakes, to use crutches as he searches for the generalization to be stated.

The Spiral Approach to Learning

A method of teaching that develops a concept from the intuitive level to the analytic level, from exploration to mastery, by spacing instruction is called spiral teaching. It is often emphasized in connection with modern mathematical pedagogy.

In Figure 11.6, the solid line represents the sequential curriculum, the topic-by-topic study of mathematics, while the dotted line represents contact with a given

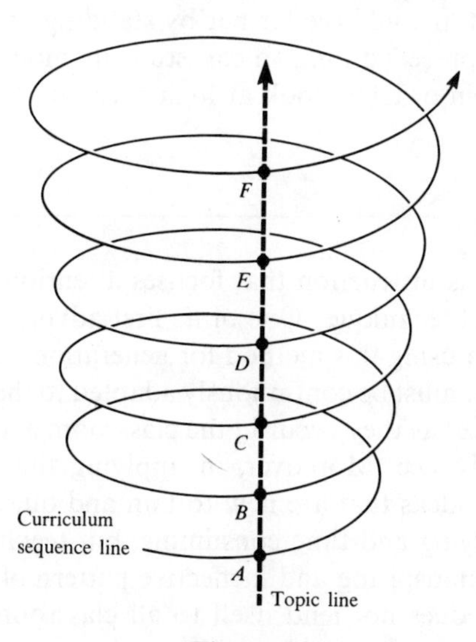

Figure 11.6

concept. The overall vertical ascent indicates an increasingly higher level of maturity and sophistication on the part of the learner as he meets a concept in a new setting at a higher level.

Spiral instruction involves the teaching of a given topic with its related concepts at several widely separated times, with each new exposure to the topic including both new approaches and a higher level of sophistication. Thus, spiral teaching is not at all the same as mere repetition. The spiral approach requires careful planning by both the curriculum developer and the classroom teacher.

The spiral approach may be illustrated by the concept of graphical representation. In the elementary school, pupils read graphs of statistical data and construct bar graphs and distributive graphs (*A* on Figure 11.6). Later, students plot graphs of points on the number line to represent the truth sets of equations and inequalities (*B*). Soon two-dimensional graphs are drawn for linear equations with two variables (*C*). At a more mature level, the intercepts, slopes, and intersections are made meaningful (*D*). Later, these graphs are related to quadratic equations and to periodic and exponential functions (*E*). At a very mature level, graphs are used to discuss continuity, limits, and probability (*F*).

What concepts does the learner have to know in order to learn the next concept? As illustrated in the flow chart of subconcepts of Figure 11.7, each successive step in the hierarchy of the concept depends on other simple subconcepts. The student must have mastered all the concepts at lower levels of the chart in order to understand fully the higher-level concepts. Failure to master an idea at any point in the chart makes the understanding of higher entries questionable at best. To learn a new concept then requires learning and recall of all the supporting subconcepts.

This is, of course, exactly the kind of hierarchical structure described in Chapter 3. Mastery of prerequisite skills provides the opportunity for moving to new levels. Just as Newton could see farther by standing on the shoulders of the intellectual giants who preceded him, so can students mount the skills and concepts already mastered in order to look at higher vistas.

Discovery Teaching

Discovery teaching is instruction that focuses attention on the student. This is not a new pedagogical technique: One of its first advocates was Socrates, and good teachers have been using this method for generations. However, it is not an easy technique because it must be continuously adapted to the students' responses, questions, and experiences as they occur in the classroom and therefore cannot be highly structured in advance. Moreover, in applying this method the teacher himself may "discover" ideas that are new to him and questions that he cannot answer. It is patience trying and time consuming, but teachers who use it never want to go back to an uninspiring and ineffective pattern of recitation-lecture.

Discovery teaching does not lend itself to all classroom activities. It is, for example, difficult to imagine a student discovering a definition. Conceptual

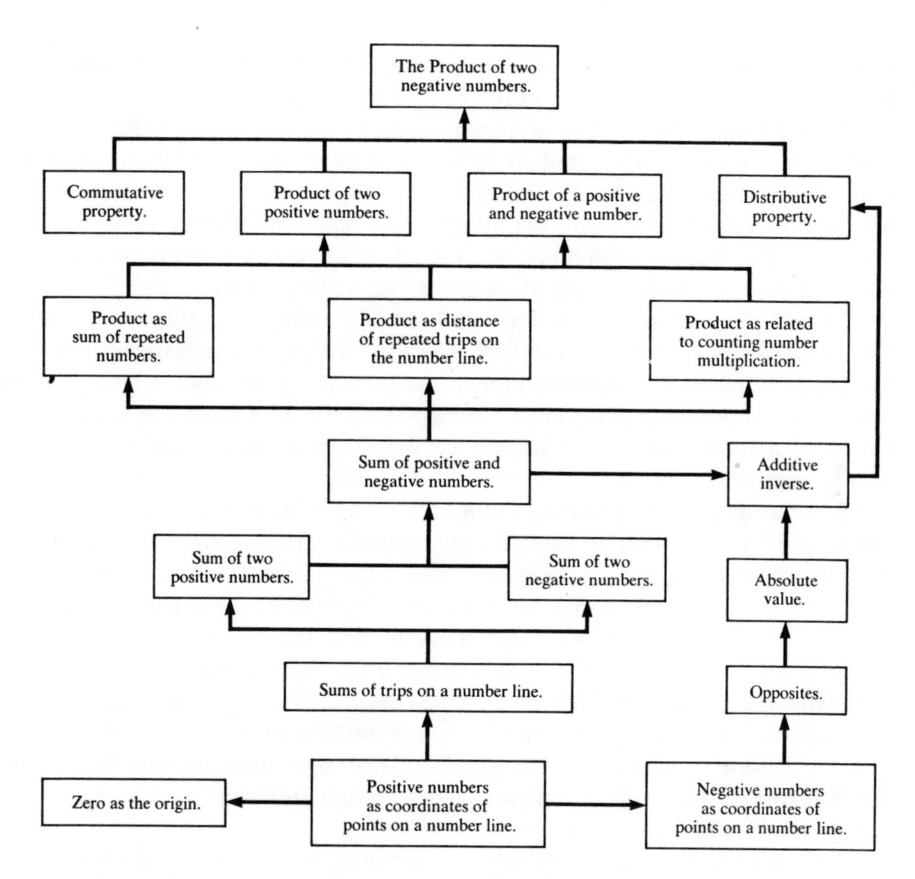

Figure 11.7 ***Flow Chart of Subconcepts that Form the Basis for Multiplication of Negative Numbers***

development, on the other hand, is especially open to development by student discovery.

How Discovery Promotes Learning

The basic objectives in using the discovery method are to present mathematics so that it makes sense, so that it has significance to the learner, and so that it can be a pleasant experience.

Since our individual perceptions depend on our past experiences, interest, emotions, and imagination, the teacher cannot control the perceptions of his students and he cannot "give" meaning to the learner. The teacher's role is to guide the learner to connect new ideas to his storehouse of past experiences and memories. In teaching the elementary ideas about vectors, the teacher may make many references to travel in order to stimulate the sense of the topic's relationship to everyday situations. Then these ideas can be connected to the number line, ordered pairs, and translations on the coordinate plane.

Guesses, conjectures, trial-and-error experiments are used in the discovery method to search for ideas and to relate these new ideas to previous concepts. Thus, the number line quickly relates negative numbers to the corresponding positive numbers, and trips on the number line represent additions.

When the student's reactions are verbal or written, he needs the give-and-take of discussion to clarify his ideas and to guide him toward fruitful investigations. Only when he has attained a high level of mastery of mathematical vocabulary should he be required to state mathematical ideas in correct language.

Discovering an idea independently gives a sense of confidence, which strongly motivates continued explorations. Finding the pattern of results in folding and unfolding a hexaflexagon has stimulated many reluctant learners to try every possible sequence of folds.

Since we remember about one fifth of what we hear, one half of what we see, and three-fourths of what we do, discovery activities are essential to the process of learning. By stressing out most retentive senses they promote a flexible, investigating, creative response to solving problems, and this flexible approach is essential for the transfer and application of knowledge to new situations. Discovery fosters desirable attitudes, for it encourages curiosity for further learning. Discovery is one of the best means of building attitudes of appreciation, enjoyment, and loyalty.

In planning a discovery lesson the teacher outlines a series of questions, problems, or laboratory exercises. The lesson might begin with an introduction, so that the student has a clear idea what he is to explore, what facts he has at his disposal, and what methods seem appropriate.

After the teacher poses a problem, he stimulates the thinking of the students by asking open-ended questions. Student responses should be encouraged by such statements as "That's almost right" or "That's a good idea" or "Keep talking" or even the brief "I see" or "Oh?" Comments such as these reduce the students' fear of being wrong or being embarrassed by the rejection of a poor suggestion. The skillful teacher will use half-formed ideas as stepping stones to correct ideas. It is the teacher's role to keep the investigation going at a challenging rate and in useful directions, for even a correct statement does not always mean that the student understands the idea. The teacher must ask questions that force the student to test his answers, find contradictions, identify special cases, or state a generalization. His role is *not* that of a prosecuting attorney eliciting "yes" or "no" answers. His role is that of guiding the student up a stairway of ideas to the generalization at the top. He guides thinking by helping students to block their own blind alleys and to concentrate on productive avenues.

To illustrate this teaching style here are some questions teachers use in discovery lessons:

"Give me another example."

"Do you believe that, Bill?"

"Will that work with fractions too?"

"How do you know that?"

"Are you sure? Let's test a zero."

"Can anyone find a case for which John's rule doesn't work?"

"Why do you and June disagree?"

"Well, what do you think?"

"How many agree?... Can we decide on the basis of this vote?... Alice, can you convince this majority that they're wrong?"

"That seems to work. Will it always?"

"Have we forgotten any cases?"

"What do you mean by that?"

"Say that another way."

"How can we simplify this?"

"Can you make a rule that a sixth grader could follow?"

The reader should note that these questions have several aspects in common: (1) They say to the student, "Keep thinking!" (2) They are encouraging. (3) They are equally applicable to right and wrong answers. (4) They treat each student as a partner in the learning process. (5) They encourage interaction between students.

A Discovery Lesson

Let us consider a specific discovery lesson to see the interaction of student and teacher:

Teacher. We have learned how to find solution sets for equations. Today we will consider another important relationship—inequalities. First, let's see if we can't devise some informal rules for operating with them. I've written $x > y$ on the board as our basic relationship. Next let's consider statements of the form $x + a$ and $y + a$ which relate addition to our inequalities. Can we say anything about these expressions?

Dorothy. $x + a > y + a$.

T. Oh?

D. Because it's like $x = y$. We add the same thing to both sides.

T. I see.

Bill. That doesn't necessarily work.

Lon. But it does this time. If I take $5 > 4$ (*at this point the teacher offers chalk and Lon writes his mathematical statement on the board*) and add 2 to each side, I get $7 > 6$, a correct statement.

T. Do you agree, Jane?

Jane. Yes.

T. Do you think it will always work?

J. Yes.

T. Why?

J. It just seems that way.

T. Oh?

J. If one thing is bigger than another and you increase them the same amount, the bigger stays bigger.

T. That's good Jane. Can anyone say this another way?

D. Yes. It's like piles of flour on a table. If you add the same amount to each pile, the smaller piles can't get larger than the large piles.

T. Can anyone think of a case where this wouldn't hold true?

D. What if a is negative?

T. Aha!

D. It still works. Subtracting 2 from each side of $5 > 4$ gives $3 > 2$.

T. Can anyone suggest an example when we might run into trouble?

L. How about subtracting 6 from each side?

T. Okay.

L. You get $-1 > -2$. Still okay.

T. (*After a pause*) I guess I'll agree to this until we need to write a more formal proof. Can anyone state this as a theorem?

Konrad. If one thing is greater than a second, the same thing added to each will leave the first greater.

T. Maybe symbols would clarify that.

L. If $x > y$, then $x + a > y + a$. (*Given chalk, he writes this on the board.*)

T. Fine. We'll return to prove that later. Now how about products such as ax and ay?

K. $ax > ay$

T. I see. Everyone agree? (*Writes "If $x > y$, then $ax > ay$" on the board.*) How about some examples?

L. If $5 > 4$, $10 > 8$.

T. How did he get that, Jim?

Jim. Doubled.

T. Okay. How about fractions?

D. If $5 > 4$, $2\frac{1}{2} > 2$. Still okay dividing.

T. Always?

D. Yes. It's like the piles of flour.

T. I see, but I don't know that you've convinced me.

Henry. Hey, it doesn't work.

T. Well, Henry has a counterexample. Before he tells us, can anyone else find one? (*Walks over to Henry's desk to see what he has written.*)

J. $0 = 0$.

T. How does that apply here?

J. Multiply each side by 0.

T. Can you add to my statement to exclude this (*pointing to board*)?

P. Just put "when $a \neq 0$" (*writes with offered chalk*).

T. I don't think that that was your case, Henry.

H. If you multiply by -2 you get -10 and -8. -10 is smaller instead of larger.

T. Now what? Do we discard everything?

D. Change $a \neq 0$ to $a > 0$.

T. (*frowning*) Anyone agree?

L. Yes. If you keep them positive, it's okay.

T. What about starting with $5 > -2$?

L. Still okay. They stay on opposite sides of zero.
T. I'm satisfied then, but we've still got Henry to contend with.
H. How about "If $x > y$ and $a < 0$, $ax < ay$"?
T. Let's test some cases before we decide.

And so the lesson progresses with the active involvement of the students and the teacher.

One method of promoting discovery is to present a lesson in silence. George Polya, a master of discovery teaching, likes to teach the Pythagorean theorem without words, using these diagrams shown in Figure 11.8. Polya works on this lesson by drawing the diagrams on the chalkboard and by having students write statements on the chalkboard. Although neither teacher nor student talks, the relationship $a^2 + b^2 = c^2$ is quickly established.[2]

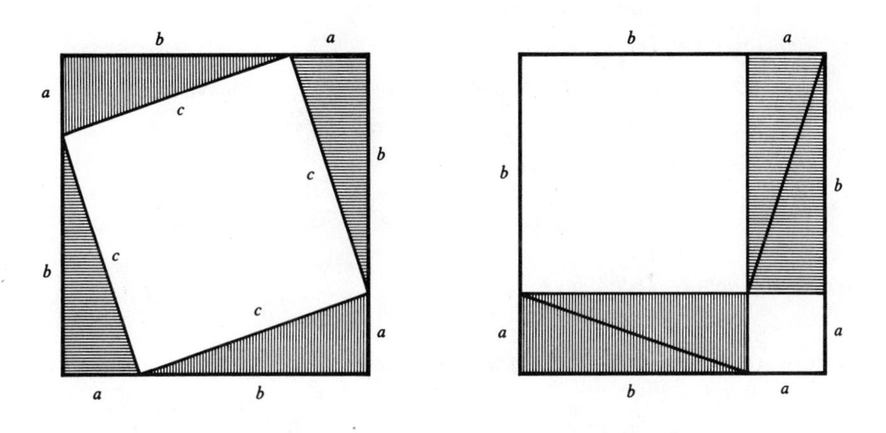

Figure 11.8

Of course, it is quite difficult to have more than one student discover an idea. As soon as the first student sees the pattern, he tells the teacher and, at the same time, the rest of the class. To avoid this, the teacher should not ask for either the solution to the problem or the method of solution, but instead for the answer to a related problem. This question ensures that the student has mastered the technique and, at the same time, allows others to discover the technique for themselves. Another advantage of this method is that other students may produce different and viable methods of solution. Too often, we fix a single pattern when many patterns are possible.

Consider a simple illustration. The game Nim is played by two persons, each in turn removing one, two, or three counters from a pile. The winner is the player

[2] The film *Let Us Teach Guessing* (16 mm., color, 2 reels, Modern Learning Aids) is a delightful illustration of how Polya gets his class to discover a generalization without dialogue.

who takes the last counter. This game has as a pattern of winning (W) and losing (L) (related to the number of remaining counters):

$$\begin{array}{cccccccc} 1 & 2 & 3 & 4 & 5 & 6 & 7 & 8\ldots \\ W & L & L & L & W & L & L & L\ldots \end{array}$$

Students are first asked to continue the fairly obvious pattern. Then they record the numerical pattern of winning counters: 1, 5, 9, . . . and extend that. Given an opportunity to record the rule they followed to get new numbers in this sequence, they may offer all of the following:

1. Every other odd number starting with one.
2. Start with one, keep adding four.
3. Start with one, skip three numbers, say the next, skip three, say the next.
4. Numbers that leave a remainder of one when divided by four.
5. One more than the four-times table.
6. $4n + 1$, for $n = 0, 1, 2, \ldots$

Discovery Teaching Techniques

Here are some specific ways in which a teacher may stimulate learning through discovery:

Pose a question, problem, idea, procedure, example for investigation. Then be sure the student has a clear idea of what he is to explore and how he should go about exploring it.

"How is the value of a rational number affected when you add the same number to numerator and denominator?"

Stimulate the thinking of the students by dialogue. Guide their thinking by asking questions, giving occasional hints, forcing them to relate to known ideas, and requiring reasons. Ask questions that demand student participation.

"Give me some examples."

Plan investigations, exercises, or activities that produce patterns. These patterns may involve numbers, measurements, answers, designs, graphs, number pairs. The patterns are then examined for a generalization that will generate the pattern.

"Each of you answer this question for your example and we'll compare results."

Arrange laboratory periods in which measurements are made, data are collected, equipment is used, models are constructed, charts are drawn, or computations are performed. The results are then examined for a generalization such as a formula.

"Form groups of four to check your results. Be ready to report to the class."

Provide reading, research, or investigations in which ideas are explored independently. Many students discover new concepts directly by reading and thinking about higher relationships rather than by a concrete or intuitive approach.

"Can anyone find a counterexample to our result? It seems to me that we may have stated our rule in too general a form."

Use test items that measure a student's ability to discover new ideas. This will emphasize to the student the need for learning how to discover ideas.

3. Solve for c:

$$\frac{a + c}{b + c} > \frac{a}{b} \quad \text{and} \quad b > a.$$

Examples of Ideas to be Discovered

Almost any topic or lesson in mathematics can be developed through discovery activities. However, to help you get started, the following specific ideas can be used to develop class dialogue.

The difference between the prime numbers 5 and 2 is 3. Why do no other prime numbers have this property?

What is a shortcut for squaring numbers ending in 5? Why does it work?

What are some unusual properties of 1 and 0?

What do we know about sums and products of odd integers?

When is the square of a number less than the number?

In how many ways can a sandwich be cut into two equal parts?

If $ad = bc$, then $a/b = c/d$. When is this not true?

When is $a/b + c/d = (a + c)/(b + d)$?

Why is $1.9999.... = 2$?

Why are all regular hexagons similar?

What is the distance between two skew lines?

What are the properties of the rational numbers of the form $0 < a/b < 1$?

What kinds of numbers can be used for a base of a numeration system?

How do we know that there is an infinite number of millions?

Here are some topics to be investigated by examining patterns:

What is the maximum number of pieces of pie if a round pie is divided by seven cuts?

What is the formula for the relationship between the number of vertices, edges, and regions of a closed network (Euler's formula)?

How are the slope and intercept of a line related to the equation of the line?

What relationships exist among the numbers of Pascal's triangle?

What is the pattern for multiples of 9? 8? Do these patterns occur in another numeration system?

Here are some questions to investigate through laboratory activities:

How is the perimeter of a right triangle related to its area?

If the rays of an angle intercept arcs on a circle, how are the intercepted arcs of the circle related to the measure of the angle formed by the rays?

What is the number of subsets of a set?

What are different ways of illustrating the Pythagorean theorem?

What geometric relations can be discovered by paper folding?

How can the formulas for the areas of geometric figures be related to the area of a rectangle?

What equality properties apply to inequalities?

Show that the sum of the first n odd integers is n^2.

Cautions for the Discovery Lesson

Since there are many pitfalls in using the discovery method, some cautions should be kept in mind.

Be sure that correct generalizations are the end result. Wrong discoveries are difficult to correct. Errors in associations result in chaotic chains of mistaken inferences and deductions; and, of course, failure to discover the generalization is discouraging and confusing.

Maintain a high level of intellectual curiosity even though it may not always be directed toward the idea you wish to attain. Curiosity is often satisfied by incorrect or partial information; so do not expect curiosity alone to generate the discipline, attention, and motivation needed to master difficult ideas. Without judicious guidance and logical restraints, discovery activities may lead to "dense fogs of frustrating perplexity." If a student does not perceive a possible route for his thoughts to follow, discovery activities merely lead to a dead end.

Do not expect everybody to discover every generalization. Even the high-ability student may lack the curiosity or flexibility to discover ideas. Each student will vary in the time needed to investigate an idea, and the entire class cannot be held back until the slowest discoverer has arrived.

Do not plan to discover all the ideas of your course. Discovery of some ideas is too inefficient. Sometimes students do not need an intuitive, empirical, discovery approach in order to understand an idea. Not all learning fits into the discovery pattern.

Expect discoveries to take time. Even though discovery learning is time consuming, the improved meaning and the increased retention that usually result indicate that it is efficient.

Do not expect the generalization to be verbalized as soon as it is discovered. Unverbalized awareness of the generalization should be the first stage in learning. When the generalization is first verbalized, expect the statement to lack precision and completeness. Words often get in the way. Even though they are extremely important for communication, they can actually obstruct the discovery process.

Avoid overstructuring experiences. Permit much individual initiative and originality in exploring ideas. Keep a balance between freedom and direction, reflection and knowledge, uncertainty and memory, creativeness and conformity, exploration and generalization.

Avoid jumping to conclusions on the basis of too few samples. The exploration should be an example of how a rational person explores ideas and problems.

Do not be negative, critical, or unreceptive to unusual or off-beat questions or suggestions. However, incorrect responses must not be accepted as true; and disruptive, nonessential explorations must be eliminated. Students should know that their status is not threatened by incorrect answers.

Keep the student aware of the progress he is making. He should expect difficulties, frustrations, failures. If possible, have crucial ideas "discovered" repeatedly or by different methods.

Finally, each student must recognize why his discoveries are significant and how the ideas are incorporated in the structure involved. This consolidation of the new insight is as important as the discovery itself.

The current stress on pattern development can mislead as well as lead. It is well to have an example or two of patterns that mislead in order to force students to go beyond superficially attractive and intuitively "clear" results. Consider, for example, two problems generated by chords drawn in a circle (see Figure 11.9):

1. How many chords may be drawn joining n points on a circle?

2. What is the maximum number of regions into which a circular disc is subdivided by chords joining n points on the circle?

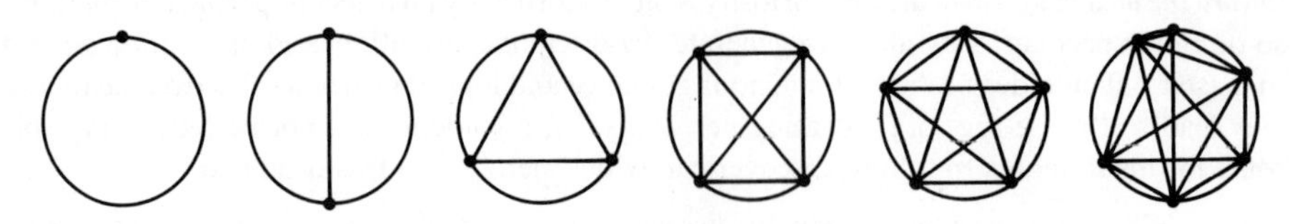

Figure 11.9

The first question is amenable to straightforward pattern construction. But the second is quite different. (See learning exercises 1 and 2.)

The Importance of Concept Learning and Discovery

Concepts are the mortar that holds together all learning, and discovery is the craft of the master mason in applying the cement. There are, of course, many other activities of the mathematics classroom that are important, but none are more fundamental than concept development. Rote learning—learning without conceptualization or structuring—can provide some short-term advantages in performance speed, but it cannot compare with concept learning in retention and in providing the basis for further learning.

Learning Exercises

1. Use the diagrams of Figure 11.9 to generate a pattern for question 1 on this page. Use the pattern to develop a formula, $c = f(n)$. Prove your formula by using mathematical induction. Prove your formula by using a combination formula.

2. Use the first five diagrams of Figure 11.9 to generate a pattern for question 2 on this page. What formula $r = f(n)$ does this pattern suggest? Check your pattern against the sixth diagram. What is it about the sixth diagram that is different?

3. Determine the following: (a) The maximum number of regions into which n lines separate the plane. (b) The maximum number of regions into which n planes separate 3-space.

4. Plan a guided discovery lesson around the exercises 3(a) and 3(b).

5. For one of the following topics, list five or more prerequisite concepts or skills:

(a) Factoring
(b) Indirect proof
(c) Similarity
(d) Logarithms
(e) Systems of linear equations
(f) Coordinate geometry

6. Examine the presentation of a specific concept in three different textbooks. Compare the steps they use in presenting the concept.

7. Examine the development of trigonometry in textbooks for grades 8, 9, and 10. Is this an example of spiral presentation?

8. Plan a discovery lesson with a sequence similar to that shown in Figure 11.10.

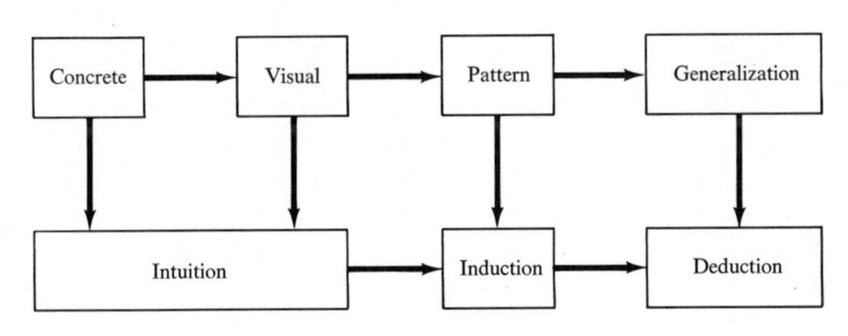

Figure 11.10

Suggestions for Further Reading

Bittinger, Marvin L., "A Review of Discovery," *Mathematics Teacher*, 61, 2 (February 1968), 140–146.

Boeckmann, Hermann, "The Discovery Approach Strategy for Mathematics Teachers," *School Science and Mathematics*, 71, 1 (January 1971), 3–6.

Boyd, Henry, "Developing an Algebra by Discovery," *Mathematics Teacher*, 64, 3 (March 1971), 225–228.

Brown, Stephen G., "Signed Numbers—A 'Product' of Misconceptions," *Mathematics Teacher*, 62, 3 (March 1969), 183–195.

Henderson, Kenneth B., "Concepts," in *The Teaching of Secondary School Mathematics*, Yearbook 33, National Council of Teachers of Mathematics, Washington, D. C., 1970, pp. 166–195.

Hendrix, Gertrude, "Learning by Discovery," *Mathematics Teacher*, 54, 5 (May 1961), 290–299.

Higgins, Jon L., "A New Look at Heuristic Teaching," *Mathematics Teacher*, 64, 6 (October 1971), 487–495.

Jackson, Robert L., "The Development of a Concept: A Demonstration Lesson," *Mathematics Teacher*, 54, 2 (February 1961), 82–84.

Jones, Phillip S., "Discovery Teaching—From Socrates to Modernity," *Mathematics Teacher*, 63, 6 (October 1970), 501–508.

Mann, John E., "Polygon Sequences—An Example of a Mathematical Exploration Starting with an Elementary Theorem," *Mathematics Teacher*, 63, 5 (May 1970), 421–428.

Snyder, Henry D., "An Impromptu Discovery Lesson in Algebra," *Mathematics Teacher*, 57, 6 (October 1964), 415–416.

Thompson, Richard B., "The Special Case May Be the Hardest Part," *Mathematics Teacher*, 63, 3 (March 1970), 249–252.

Wills, Herbert, "Generalizations," in *The Teaching of Secondary School Mathematics*, Yearbook 33, NCTM, pp. 267–290.

Teaching Structure and Logic 12

One of the primary aims of mathematics instruction has always been to teach logic and proof. However, in secondary mathematics courses—except for tenth-grade geometry—very little emphasis was formerly given to proof and deduction. Now all school mathematics emphasizes logic and structure. Even symbolic logic and truth tables are taught as early as the junior high school. In the new school mathematics, students are expected to prove theorems of algebra and number theory, as well as those of geometry.

The emphasis on the structure of mathematics is now begun in the elementary school, where the commutative, associative, and distributive properties of operations are investigated. In the very early grades, the properties of 0 and 1 are used as the basis for rationalizing certain algorithms; and relationships between operations and their inverses are used to find truth sets. In junior high school mathematics, sets, operations, and properties are used to illustrate the mathematical structure of a group: the counting numbers, the integers, the rational numbers, and the real numbers—each is seen as a set of numbers with certain properties for the operations of addition and multiplication; and finite number systems (modular systems) and finite geometries are used to build finite mathematical systems. In senior high school mathematics, fields of mathematics such as geometry are examined as complete mathematical structures; in the process, the nature of proof and the proofs of theorems are studied intensively.

Why do the new mathematics courses place so much emphasis on the structure of mathematics? For several reasons:

When one learns the structure of mathematics, he should have an easier time learning new topics or even new branches of mathematics. Attention to the structure of a subject adds "wholeness" to the learning; and learning in terms of wholes is more effective than attention diverted to isolated elements. Thus, attention to structure is particularly appropriate for continued study or independent learning in the future.

When certain basic properties (e.g., the distributive property) are known, they can be used to explain a variety of new situations. The principle $a(b + c) = ab + ac$ can furnish a rationale for a factoring problem, $3x + 6y = 3(x + 2y)$; for addition of fractions, $2/a + 3/a = 5/a$; even a simple multiplication problem, $3 \times 21 = 63$ is seen as $3(20 + 1)$.

When one discovers the structure of mathematics, he should gain interest in the study of mathematics. Mathematics is a unique, elegant creation of the human mind. Its uniqueness is related to its deductive structure. Consequently, many insightful students, when properly guided, enjoy viewing mathematics as a powerful deductive science.

When one masters the structure of mathematics, he should be able to transfer his knowledge to new problems. Experiences in creating new mathematical structures should build a flexible, insightful approach to problems. Knowing basic procedures should improve skill in analyzing a new situation. Emphasis on the reason for a result such as $(^{-}3)(^{-}4) = {}^{+}12$ should indicate a way of rationalizing other operations.

When one understands the structure of a subject, he should improve his retention of ideas. We are more likely to remember a few big ideas than a multitude of independent facts. Thus, the identity element for multiplication, $a \cdot 1 = 1 \cdot a = a$, can provide one with a basis for reconstructing many individual algorithms. Among the questions resolved by use of this one axiom are the following:

$$\frac{a}{b} = \frac{?}{bc}$$

$$\left(\frac{a}{b} = \frac{a}{b} \cdot 1 = \frac{a}{b} \cdot \frac{c}{c} = \frac{ac}{bc}\right)$$

$$\frac{32.79}{6.3} = ?$$

$$\left(\frac{32.79}{6.3} = \frac{32.79}{6.3} \cdot \frac{10}{10} = \frac{327.9}{63}\right)$$

$$\frac{2}{3} \div \frac{5}{6} = ?$$

$$\left(\frac{2}{3} \div \frac{5}{6} = \frac{2/3}{5/6} = \frac{2/3}{5/6} \cdot 1 = \frac{2/3}{5/6} \cdot \frac{6}{6} = \frac{4}{5}\right)$$

When one learns certain structures, he may be able to invent new structures and thus exercise creative talent. A challenging experience for any person is the opportunity to invent a new operation, write new symbols, select a set of elements, and then use these to build a new mathematical system. Teachers often find it a wholesome experience to have students develop and prove theorems that are original for them. For example, suppose $a * b$ means $ab + b$. For the set of counting members is the operation commutative or associative? Is there an identity element? If not, is there a way to circumvent this problem? Is the set of counting numbers closed with respect to this operation?

The Structure of Mathematics

Mathematics is frequently described as a structure consisting of a collection of mathematical systems (or topics), each of which has the typical structure of a deductive science (see Figure 12.1). What is the structure of a deductive system? A deductive system begins by selecting some undefined terms, called *primitive terms.* These terms are needed to supply basic words for communication. Thus, in geometry we use the term "point" as an undefined term in all statements involving points. The use of the word "undefined" here is true only in the strict sense, since the undefined terms like "point" and "line" are actually defined in context by the definitions of other terms, by the facts that are proved about them and by their relationships. Thus, axioms like "Two points determine one line," "Two distinct lines intersect in at most one point," and "Three noncolinear points determine a plane" characterize or in a very real sense define these "undefined" terms. These primitive terms are then used to state precise definitions of new terms.

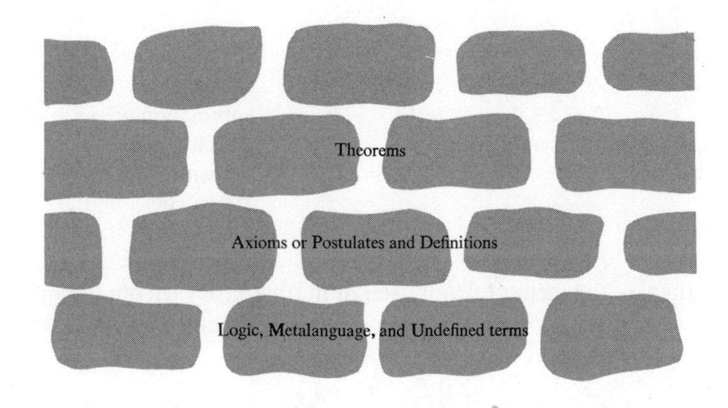

Figure 12.1 Building the Structure of Mathematics

Next, certain basic assumptions, called *axioms*, are stated. These axioms are usually chosen because they seem to agree with our experience. They are the statements that provide basic relationships between the fundamental elements of the system. Finally, certain theorems are stated and are proved by a sequence of statements. Each of these statements is justified by a definition, an axiom, or a previously proved theorem. Let us apply this procedure to one of the basic mathematical systems, the theory of *sets.* In this system, the only undefined terms are "set" and "element." With these terms the operations of union and intersection of sets are defined. Then it is assumed that these operations have properties such as commutativity and associativity. Next, theorems about sets are proved by means of these undefined terms, definitions, and assumptions.

In order to prove a statement by logical reasoning, the following examples illustrate sequences that may be used.

If I get a job, then I will earn money.
If I earn money, then I can buy a bicycle.
If I can buy a bicycle, then I can go to the football game.
Hence, if I get a job, then I can go to the football game.

For beginning experiences, this format may be used:

If □ + 5 = 9, then □ = 9 − 5.
If □ = 9 − 5, then □ = 4.
Therefore, if □ + 5 = 9, then □ = 4.

Notice that these statements are in an "if-then" form, that one statement leads to another. Here is an example of how this reasoning is used in algebra:

Statement	*Reason*
1. If $3x + 7 = 19$, then $3x = 12$.	1. Subtraction property of equality.
2. If $3x = 12$, then $x = 4$.	2. Division property of equality.
3. Hence, if $3x + 7 = 19$, then $x = 4$.	3. Transitive property of implication (statements 1 and 2).

This, then, is the form of proof used in mathematics—from the informal proof of elementary school to the more elaborate and demanding proof of a theorem in advanced mathematics.

Groups, Rings, and Fields

The new school mathematics is concerned with mathematical systems called groups, rings, and fields. These systems are concerned with certain sets, operations, and properties of numbers.

Let us consider a simple mathematical structure called a *group*. To illustrate the nature of a group we will use the set of rational numbers represented by fractions and an operation such as addition. What are the properties of the rational numbers as related to addition?

Are the rational numbers closed with respect to addition? Yes. Every sum of two rational numbers is another unique rational number. For example:

$$2/3 + 1/5 = 13/15$$
$$7/4 + 3/3 = 11/4$$

Does the associative property apply to the addition of rational numbers? Yes.

For example:

$$(2/3 + 4/5) + 7/9 = 2/3 + (4/5 + 7/9)$$

Does the set of rational numbers have an identity element for addition? Is there a number, n, such that $a/b + n = a/b$? Yes. It is 0. For example:

$$3/4 + 0 = 3/4$$
$$a/b + 0 = a/b$$

Does every rational number have an inverse with respect to addition? Is there a number, n, such that $a/b + n = 0$? Yes. Since $a/b + {}^{-}a/b = 0$, the additive inverse of a/b is ${}^{-}a/b$.

Although we have not proved that these properties always hold true (an example does not, of course, constitute a proof), we can easily prove that they do by noting similar properties of the integers. Thus, if we wish to prove that the rational numbers are closed with respect to addition, we note that in the sum $a/b + c/d = (ad + bc)/bd$, both $ad + bc$ and bd are integers because of the closure properties of integers. This means that $(ad + bc)/bd$ represents a rational number and completes the proof. Other properties are proved by similar means.

We now have the six requirements for a mathematical group:

Requirements	*Example*
1. A set of elements.	1. The set of rational numbers.
2. An operation.	2. Addition.
3. The set of elements is closed with respect to the operation.	3. The sum of any two rational numbers is another rational number.
4. The operation is associative.	4. The grouping of the rational numbers in addition does not change the product.
5. There is an identity element for the operation.	5. The identity element for addition is 0: $a/b + 0 = a/b$.
6. There is an inverse, with respect to the operations, for each element of the set.	6. The multiplicative inverse for each rational number is its negative, ${}^{-}a/b$.

In a similar way other sets of numbers (such as the counting numbers, the integers, and the real numbers) are examined with respect to addition and multiplication to see whether they are examples of mathematical groups.

When a set of numbers involves two operations, such as addition and multiplication, we may have mathematical systems called rings or fields. The requirements of rings and fields are summarized and related to groups in Table 12.1.

Table 12.1

Property	Operation	Group	Ring	Field
Closure	$+$	G_1	R	F
Closure	$\times$	G_2	R	F
Associative	$+$	G_1	R	F
Associative	$\times$	G_2	R	F
Identity element	$+$	G_1	R	F
Identity element	$\times$	G_2		F
Inverse	$+$	G_1	R	F
Inverse (non-zero elements)	$\times$	G_2		F
Commutative	$+$		R	F
Commutative	$\times$			F (except skew fields)
Distributive	$(\times)$ over $(+)$		R	F

If we examine the properties of the set of rational numbers or real numbers or complex numbers (in each case excluding zero for multiplicative inverses) with respect to addition and multiplication, we find that we have examples of fields.

Finite Mathematical Structure

A good way to study mathematical structures is to select a finite set of elements and then select operations to apply to these elements. The results of the operations are examined to determine whether or not this finite system is a group, a ring, or a field.

As an example of a finite mathematical structure, consider the set $\{0, 1, 2, 3, 4\}$. Instead of a number line, this set is best illustrated by an integral number circle on which a finite number of elements are used, as in Figure 12.2.

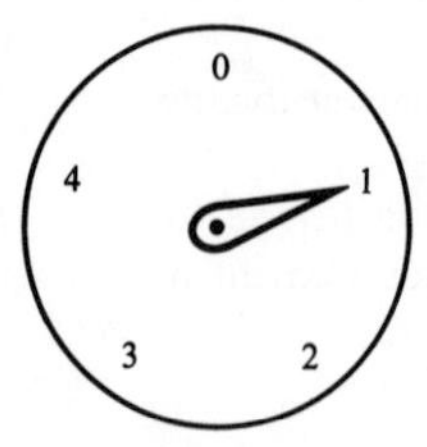

Figure 12.2

Let us select the operation "addition," which we will consider a clockwise counting—beginning with zero. We will use the symbol $\oplus$ instead of $+$ to show this circular "addition." We call this finite system a "modular arithmetic" or a "mod-5" system. Our regular clock is a good illustration of a modular arithmetic. On our clock $8 \oplus 7 = 3$ rather than 15. Similarly, for our mod-5 system we get these results: $2 \oplus 2 = 4$, and $3 \oplus 3 = 1$.

For reference purposes we make an "addition" table (Table 12.2) for all the possible addition combinations.

Table 12.2 **Addition Table (Mod-5)**

$\oplus$	0	1	2	3	4
0	0	1	2	3	4
1	1	2	3	4	0
2	2	3	4	0	1
3	3	4	0	1	2
4	4	0	1	2	3

Is our mod-5 system closed with respect to $\oplus$? Yes. Every result in our table is another element of the set $\{0, 1, 2, 3, 4\}$.

Is our mod-5 system associative with respect to $\oplus$? Does $(2 \oplus 3) \oplus 4 = 2 \oplus (3 \oplus 4)$? Yes. Thus, we assume that the associative property applies to all "additions" in this system. How many checks would have to be made?

Is there an identity element for the operation $\oplus$? If $3 \oplus n = 3$ and $4 \oplus n = 4$, what is n? It is 0. Hence, our identity element for "addition" is 0.

Is there an additive inverse for every element of our set? Is there a number, n, such that $x \oplus n = 0$? If we examine our addition table, we find that $1 \oplus 4 = 0$, $2 \oplus 3 = 0$, $3 \oplus 2 = 0$, $4 \oplus 1 = 0$, and $0 \oplus 0 = 0$. Hence, every element of our set has an additive inverse.

As we check the characteristics of this finite system, we find that it satisfies all the characteristics of a group.

If we establish another operation, $\otimes$, we can then determine whether this set, with the operations $\oplus$ and $\otimes$, is a ring or a field. The operation $\otimes$ can be illustrated by the example $2 \otimes 3 = 2 \oplus 2 \oplus 2$. Now determine whether or not the mod-5 system is a ring or a field.

How to Organize Logic and Structure

There are several ways in which the logic and structure of mathematics can be organized for instruction:

As a separate course or unit, in which principles of reasoning, finite structures, Venn diagrams, symbolic logic, truth tables, and the structure of mathematics are presented as a complete, unified topic.

As a complete structure for one subject or field of mathematics (e.g., high school geometry).

As a part of every mathematics course. Then the emphasis on structure becomes a unifying idea, and the proofs become means of teaching problem solving as well as logical reasoning. Frequently this treatment becomes too rigorous and formal for many students.

No matter what organization is used, the teacher must realize that the logic and structure of an abstract subject such as mathematics must be presented in a well-organized program. This program, like any other, must evaluate what students know and build carefully on this knowledge. It must introduce ideas in a framework understandable and interesting to students and must use all available techniques of reinforcement.

How to Teach Logic and Structure

Reasoning and logical inference are complex mental processes. We need to use every means at our disposal for assuring success and maintaining interest in the process. There are many ways to do this:

Relate the chain of if–then statements to the logic used in daily decisions. Start with simple, concrete situations and lead to logical conclusions by questions and answers. Implications can be used to illustrate answers to any number of questions such as "Why do you study mathematics?" "If I learn mathematics, then I can become an engineer." "Why do you want to buy a new coat?" "If I buy a new coat, then I will look more attractive. If I look more attractive, then Bill may ask me to the party."

Before the formal structure of mathematics can be understood, the student must have available to him a reservoir of experiences and concepts to which operations such as those of logic are applied. Only within the framework of experiences of this type do the structure and logic make sense to the student. Thus he must be aided to relate the formal ideas to concepts and experiences already familiar to him, whether these concepts and experiences come from arithmetic and algebra or the out-of-school informal and (to him) nonmathematical world. This is why the formal geometry of tenth grade is preceded by informal and intuitive experiences with geometric objects and relationships in the elementary and junior high school. Given this background, the proofs of high school are rooted in earlier experiences.

Illustrate the reasoning process visually by Venn diagrams or Euler circles. Why do girls study mathematics? Because:

1. All girls (G) are intelligent (I).
2. All intelligent persons study mathematics (M).
3. Therefore, girls study mathematics.

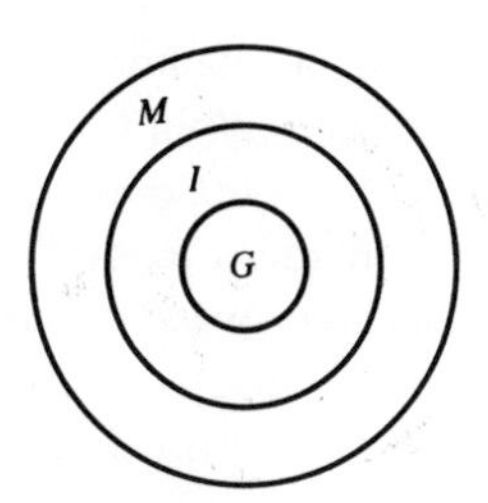

Figure 12.3

Figure 12.3 shows that all girls are a subset of persons who study mathematics.

Complete proofs in algebra and number theory—proofs that are simple and verify experimental results. The product of two even numbers is always another even number. By working out many multiplications, we find that this statement seems to be true. But when we represent the even numbers as $2a$ and $2b$, with a and b integers, the product is $2a \cdot 2b = 2 \cdot 2 \cdot a \cdot b = 2(2ab)$. Since $2ab$ is an integer, $2(2ab)$ is of the form $2n$ for an integer, n; and $2a \cdot 2b$ is therefore an even number.

The Pythagorean theorem is another illustration of a theorem that can be "discovered" by drawing and measurement. We can extend this discovery to a formula by finding the pattern of the triples of measures of sides of right triangles. Later an algebraic proof is given, and finally several geometric proofs are possible.

As much as possible, avoid proving statements that are trivial or meaningless to students. It is difficult to motivate students to learn proofs of statements that are obvious. Only when such statements have meaning and importance to students—as when they are provided in answer to a student question or when they are absolutely necessary to further development—should they be introduced.

Avoid the proof of too many theorems. We cannot take the time to prove every theorem. We prove enough to learn the method of proof and to see the structure of a topic. In particular, theorems with complicated proofs that occur early in geometry are usually better postulated, their proofs delayed until students are better equipped to understand them. For example, students can learn techniques of proof when they use the congruence theorems to write simple proofs of congruency. Once the technique of proof is mastered, the better students may re-examine the proofs of the more difficult congruency theorems themselves.

Never require the memorization of proofs. Proofs, like problems, are useful primarily as exercises in reasoning. Many teachers reason incorrectly that memorization of proofs leads to later understanding. Quite the contrary, there is reason to believe that memorizing the steps in a proof actually blocks understanding. It is certainly better that the student's energy be expended on learning the technique of proof. Better to write two or more different proofs for a theorem than to memorize one.

Give students confidence by providing them with opportunities to write short simple proofs before progressing systematically to more difficult proofs. A student cannot prove that the medians of a triangle are concurrent if he cannot prove triangles congruent. At other times, a teacher can "soften" proofs by providing some of the steps of the proof and leaving blanks for the student to supply reasons or statements. Beginning work with proofs can be made easier if the teacher asks *groups* of students to combine their ideas to find a complete proof.

See to it that students know and understand the cornerstones of proof: the axioms, postulates, and definitions. Failure to master these basic tools makes later work virtually impossible. Systematic review of them should be a continuing part of the classroom program.

Illustrate proofs through the use of symbolic logic. Some people believe that the use of symbols makes proof difficult. But consider the role of symbols in algebra. Until Diophantus introduced the use of a symbol for the variable in an open sentence, little progress was made in developing algebra. Until that time, problems were solved by means of written statements; and an elegant solution was one that could be stated in rhyme. Now we solve these problems by writing equations. When symbols are used for variables in an equation, even young children can find the solution set. Thus, $3\square + 5 = 29$ is now a fourth-grade problem. Only two centuries ago it was a problem for college mathematics. Similarly, symbols such as p, q, and r can be used to represent statements in proofs, just as x, y, and z are used to represent variables in equations. Experimental teaching has found that sixth graders and slow ninth graders can use this type of symbolism correctly and with enthusiasm.

Apply logic to the solution of puzzles and paradoxes. There are a great many interesting situations described in books on puzzles and fallacies. Many of these furnish problems to be solved by the use of logical reasoning. Frequently, students can make up situations to be submitted for solution by the class. These so-called "brain busters" are stimulating, especially to the higher-ability student.

After studying logic and proof by means of truth tables or Venn diagrams, relate the proof to an electric circuit. Simple demonstration boards consisting of circuits with switches in series or parallel illustrate the truth values of compound statements. These circuits are dramatic demonstrations of truth tables and lead to a consideration of the logic circuits of electronic computers.

A simple card device for testing syllogisms is described by Martin Gardner in the March 1952 issue of *Scientific American.* Other simple logic machines and kits for building logic circuits are available at low cost.

Give students opportunities to write proofs in various forms. Actually, students might learn proof more readily by writing five different proofs of a given theorem than by proving five different theorems. Proofs can be done in these different ways:

Double-column "T" form
Paragraph form
Diagrammatic format
Flow chart
Symbolic-logic sequence

Paragraph proofs and diagrammatic proofs are useful as alternate vehicles for demonstration. These alternate forms help students see that there is no single acceptable style for proof, and they relate more directly to forms of proof used in debate and legal reasoning. A paragraph proof in particular allows greater latitude and encourages extra explanation. Consider the following example:

> *Theorem:* Between every two rational numbers there is at least one rational.
>
> *Proof:* We will show that the average of the two rationals lies between them. If A and B are any two rationals, we must show $A < C < B$. Let $C = \frac{1}{2}(A + B)$, a rational number by the properties of closure for addition and multiplication for rationals. To show $A < \frac{1}{2}(A + B)$, we choose $A < B$. By adding A to each side of this inequality, we have $2A < A + B$. Multiplying by $\frac{1}{2}$ gives $A < \frac{1}{2}(A + B)$. Both of these steps are a result of axioms of order (addition and multiplication). By a symmetric argument, it may be shown that $\frac{1}{2}(A + B) < B$. Thus, the rational number $C = \frac{1}{2}(A + B)$ lies between A and B.

On the other hand, a diagrammatic proof forces students to examine in greater detail the elements of their proof. Because reasons for statements would necessarily confuse a diagrammatic proof, they are usually omitted; however, students should be required to produce these reasons on demand. Here is an example of a diagrammatic proof:

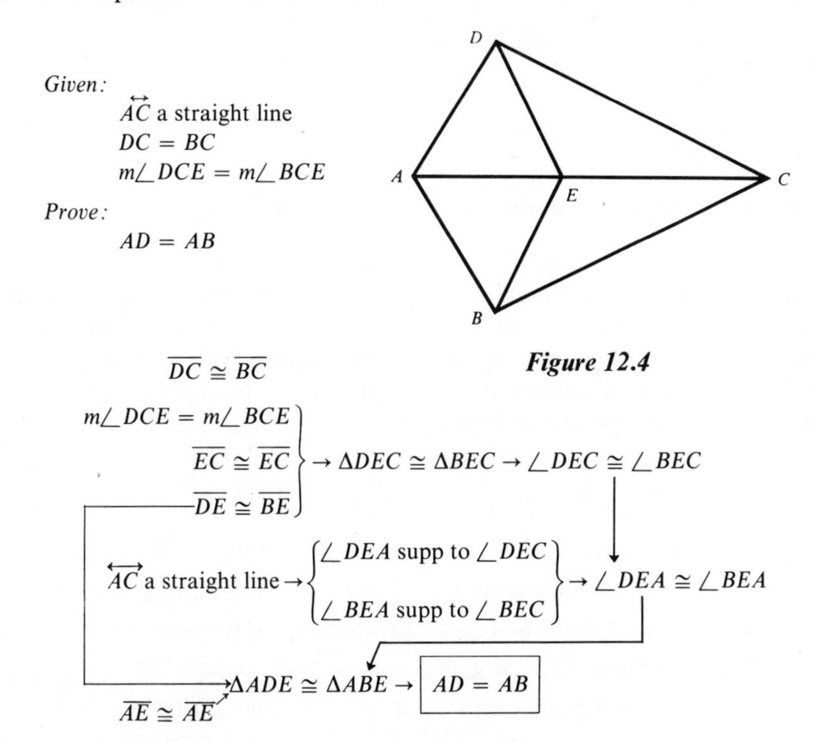

Figure 12.4

A simple flow chart of this proof is shown in Figure 12.5.

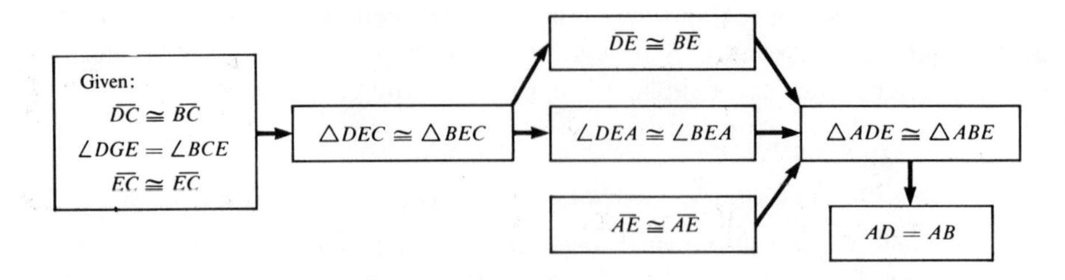

Figure 12.5

In symbolic logic we use the following symbols:

$p =$ "$\overline{DC} \cong \overline{BC}$" $\qquad t =$ "$\overline{DE} \cong \overline{BE}$"

$q =$ "$\angle DCE \cong \angle BCE$" $\qquad u =$ "$\angle DEA \cong \angle BEA$"

$r =$ "$\overline{EC} \cong \overline{EC}$" $\qquad v =$ "$\overline{AE} \cong \overline{AE}$"

$s =$ "$\Delta DEC \cong \Delta BEC$" $\qquad w =$ "$\Delta ADE \cong \Delta ABE$"

$x =$ "$AD = AB$"

$$(p \wedge q) \wedge r \Rightarrow s$$
$$s \Rightarrow (t \wedge u) \wedge v$$
$$t \wedge u \wedge v \Rightarrow w$$
$$w \Rightarrow x$$

By the transitive property of implications, $(p \wedge q) \wedge r \Rightarrow x$.

Help students to develop analytic as well as synthetic proofs and urge them to use combinations of the two approaches in working out a proof. Students who have written synthetic proofs (the standard form) usually think of analytic proofs as "backward" proofs. Consider the proof of the last example in synthetic form. (Reasons are omitted but may be required.)

1. $\overline{AD} \cong \overline{AB}$ because $\Delta ADE \cong \Delta ABE$ (2).
2. $\Delta ADE \cong \Delta ABE$ because $\underline{\overline{AE}} \cong \underline{\overline{AE}}$, $\underline{\overline{DE}} \cong \underline{\overline{BE}}$ (5) and $\underline{\angle DEA} \cong \underline{\angle BEA}$ (3).
3. $\angle DEA \cong \angle BEA$ because they are *supplements* of (4) $\underline{\angle DEC} \cong \underline{\angle BEC}$ (5).
4. They are supplements of these angles because $\overleftrightarrow{AC}$ is a straight line.
5. $DE = BE$ and $\angle DEC \cong \angle BEC$ because $\underline{\Delta DEC} \cong \underline{\Delta BEC}$ (6).
6. $\Delta DEC \cong \Delta BEC$ because $DC = BC$, $EC = EC$, and $\angle DCE \cong \angle BCE$.

(Note that in this proof a technique that is also useful in synthetic proofs is used. In each step an idea that must be established is underlined; when it has been established, the number or numbers of the steps which provide this basis are written in parentheses after the phrase.)

It is useful for a teacher to suggest the different approaches to proof, analytic and synthetic, by means of an unsophisticated example. One such example is that of finding a way of crossing a debris-strewn stream. The synthetic method of planning a crossing would be to think, "I can leap from this bank to that stone, then step onto that log, then . . ." until you chart a route. The analytic method, on the other hand, would be to think, "I can get to the far bank from that island, and I can get to the island from that boulder, and I can get to that boulder from . . ." until you are able to chart a reverse course. With this analogy developed, you can discuss what happens when the student tries to leap from A to B in Figure 12.6, and compare this with gaps in proofs. It is also apparent from the diagram that no matter what form the planning takes, the crossing is in synthetic form.

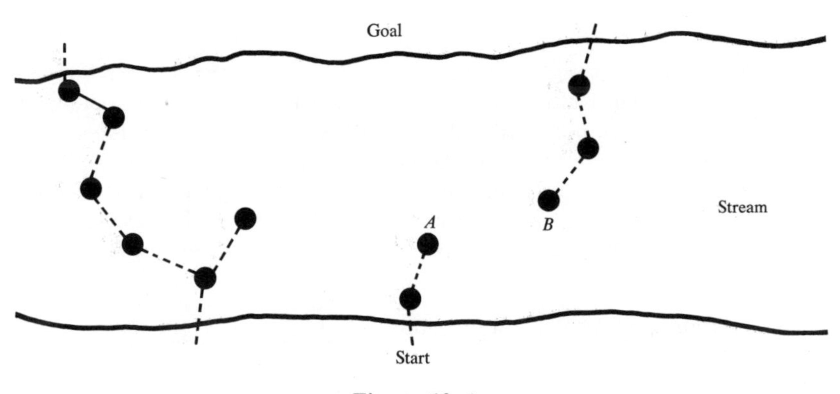

Figure 12.6

A combination of these two methods is usually the preferred procedure. This method, that of "narrowing the gap," is as useful in problem solving as in working out proofs. Working forward from what is given and backward from what is to be proved (by the synthetic and analytic means already pointed out) usually isolates the key difficulty in the proof and makes this difficulty easier to attack. A typical example of this procedure is the solution of identities in trigonometry.

Be sure to provide plenty of experience with indirect proof, a very important form of proof often applied—and misapplied—in modern society. Too often, teachers use indirect proof for only one or two theorems and fail to give students any real understanding of the techniques involved.

The kind of indirect proof used in geometry takes the following form:

1. One of the following (say, A, B, C) is true.
2. All but one (say, A and B) are false.
3. The remaining one (in this case C) is necessarily true.

Students usually see this method applied only to the following theorem:

> If two lines are cut by a transversal to form equal alternate interior angles, the lines are parallel. (1) In this proof there are two alternatives: the lines are parallel, or they are not parallel. (2) That they are not parallel is disproved. (3) The remaining case must then be true: They are parallel.

A number of other theorems, however, can be proved by this technique, in particular theorems involving inequalities. Consider the proof of the following theorem:

> If one angle of a triangle is greater than the second, the side opposite the larger angle is greater than the side opposite the smaller. If the converse of this theorem has already been proved, this theorem is easily established by indirect reasoning: (1) Three cases are apparent: the sides are equal, the side opposite the larger is smaller, or the side opposite the larger side is larger. (2) The first of these cannot apply because equal sides would force the angles to be equal. The second contradicts the converse. (3) The remaining possibility is established, and the proof is complete.

The logic used in discharging the unwanted cases is of the form:

$$\begin{array}{c} p \rightarrow q \\ q \text{ false} \\ \hline p \text{ false} \end{array}$$

This form of proof is often called *reductio ad absurdum.* It should not be confused with:

$$\begin{array}{c} p \rightarrow q \\ p \text{ false} \\ \hline \end{array}$$

In this instance, no conclusion is possible, since q is either false or true (as may be seen later in the truth table for implications). For example, consider the false statement $2 = 3$. By adding,

$$\begin{array}{r} 2 = 3 \\ 2 = 3 \\ \hline 4 = 6 \end{array} \quad \text{and} \quad \begin{array}{r} 2 = 3 \\ 3 = 2 \\ \hline 5 = 5 \end{array}$$

It is rather easy to locate examples of indirect reasoning in newspaper, radio, and television advertising. Students delight in locating sources of misleading statements: in the form of the implication, in failure to list all the possible cases to be examined, or in failure to discharge all alternatives satisfactorily.

Especially at the junior high school level, lay the groundwork for the later, more formal study of logic by encouraging bright students to analyze paradoxes and games and to solve puzzles using logical attacks.

Use good logic yourself and demand its use by your students—not only in formal proofs but in statements and activities. Set a high standard of thoughtful and consistent reasoning in your dealings and relations with students.

A Topic as a Complete Mathematical System

One way to emphasize structure is to organize each topic as a structured unit. Here is how this may be done with a unit on area.

Undefined terms: None are needed because area is treated as a complete topic with a host of already accepted terms.

Definitions:

1. A polygonal region is the union of a polygon and its interior.
2. The area of any polygonal region is a positive real number.
3. The area of a square whose sides are one unit in length is 1.

Assumptions (postulates):

1. If two polygons are congruent, then the areas of the polygonal regions are equal.
2. If the intersection of two polygonal regions does not include any interior points of the regions, then the area of the union of these regions is the sum of their areas.
3. The area of a rectangle is the product of the measures of two adjacent sides.

Theorems:

1. The area of a parallelogram is the product of the measure of any base and the measure of the corresponding altitude.
2. The area of a triangle is the product of half the measure of any side and the measure of the altitude to that side.

Postulate: The area of a circular region of radius r is $K = \pi r^2$.

Other theorems and problems are developed from these.

Venn Diagrams

A good way to illustrate logical reasoning is to use Venn diagrams. These are visual presentations of relationships that are easy and interesting for junior high students. We must remember, however, that such diagrams do not constitute a formal proof.

Venn diagrams arc based on the Greek logic, which stated proofs in the form of syllogisms. These syllogisms consist of three statements: a major premise, a minor premise, and a conclusion. Since these statements use quantifiers such as "all," "none," or "some," we can represent the statements by sets and the sets by geometric diagrams. Then the union and intersection of sets will describe the relationships involved. Here are some examples:

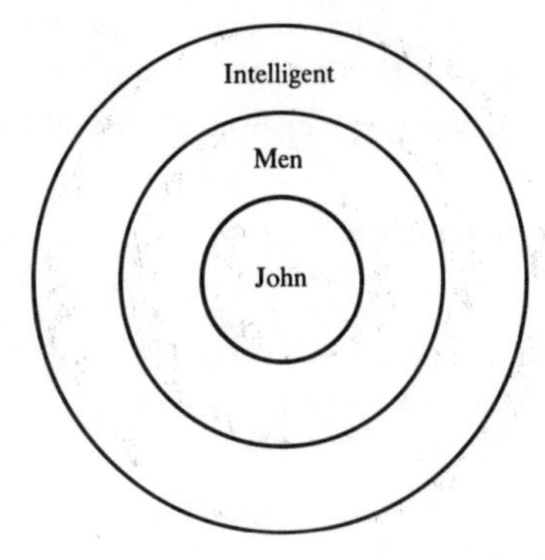

Figure 12.7

1. All men are intelligent. (Men are a subset of the set of intelligent people.)
2. John is a man. (John is a subset of the set of men.)
3. Therefore, John is intelligent. (John is a subset of the set of intelligent people.)

The diagram for the syllogistic reasoning above is often given the name Euler circles. This representation is used only for the special case of set inclusion.

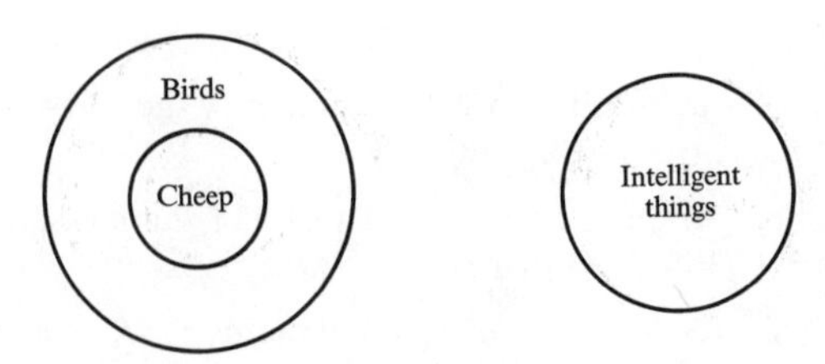

Figure 12.8

1. No birds are intelligent.
2. Cheep is a bird.
3. Cheep is not intelligent.

When we have the quantifier "some," we have difficulty selecting the correct Venn diagram because there are several possibilities.

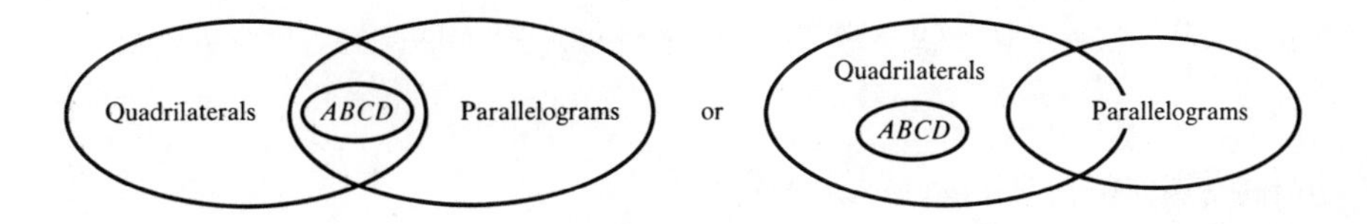

Figure 12.9

1. Some quadrilaterals are parallelograms.
2. *ABCD* is a quadrilateral.
3. *ABCD* is a parallelogram.

These alternate possibilities suggest that Venn diagrams are not adequate to prove all relationships. Fortunately, mathematics has more refined ways of establishing the proofs.

Symbolic Logic, Truth Tables, and Electric Circuits

In mathematics we often use open sentences called equations to solve problems. In the equation $x + 5 = 13$, x is a placeholder for some number of the domain, $+$ is a symbol for an operation, 5 and 13 are numerals for specific numbers, and $=$ is a symbol of comparison. Similarly, in logical reasoning we use sentences such as $(p \wedge q) \rightarrow r$. Here p, q, and r are variables that represent statements, $\wedge$ is the symbol for the conjunction "and," and $\rightarrow$ is a symbol relating $(p \wedge q)$ to r. Just as the equation $x + 5 = 13$ is an easy way to solve a problem, so $(p \wedge q) \rightarrow r$ is a convenient way to represent a logical relationship.

In symbolic logic then we use letters as placeholders for simple declarative sentences such as:

$p =$ "It is raining."

$q =$ "It is warm."

$r =$ "We will go swimming."

Next we combine these sentences to form compound statements such as:

Statement	*Statement in symbols*
It is warm *and* we will go swimming.	$q \wedge r$
It is raining *or* we will go swimming.	$p \vee r$
If it is raining *then* we will go swimming.	$p \rightarrow r$
If it is raining *and* it is warm, *then* we will go swimming.	$(p \wedge q) \rightarrow r$
It is *not* raining.	$\sim p$

In these symbolic sentences we use:

$\wedge$ to represent "and," called "conjunction"
$\vee$ to represent "or," called "disjunction"
$\rightarrow$ to represent "if–then," called "implication"
$\sim$ to represent "not," called "negation"

Many other symbols are used in logic but these are enough to illustrate symbolic logic.

A chain of reasoning uses a series of statements like this:

1. If it is raining, then it is not warm.
2. If it is not warm, then we will not go swimming.
3. Therefore, if it is raining, then we will not go swimming.

We can write this sequence of statements in symbols like this:

1. $p \rightarrow \sim q$
2. $\sim q \rightarrow \sim r$
3. Therefore, $p \rightarrow \sim r$

The basic pattern for a logical argument follows patterns like this: If $a \rightarrow b$ and $b \rightarrow c$, then $a \rightarrow c$. This particular pattern is called the transitive property of implication. This pattern is the one used in the preceding argument.

Let's illustrate how this pattern is used for a proof in algebra.

Theorem: If $3x + 5 = 29$, then $x = 8$.

Proof:

Statement	*Reason*	*Symbolism*
1. $3x + 5 = 29$	1. Given	p
2. If $3x + 5 = 29$, then $3x = 24$	2. Subtraction property for equality	$p \rightarrow q$
3. If $3x = 24$, then $x = 8$	3. Division property for equality	$q \rightarrow r$
4. If $3x + 5 = 29$, then $x = 8$	4. Transitive property of implication	$p \rightarrow r$

We can also prove the converse: If $x = 8$, then $3x + 5 = 29$. This is the way we check our equation solution.

Whenever $p \rightarrow q$ and $q \rightarrow p$, we have a biconditional statement (represented by $p \leftrightarrow q$). This is the format for "if and only if" conditions. Thus the theorem above can be restated to be "$3x + 5 = 29$ if and only if $x = 8$." These statements also have the transitive property—namely: "If $p \leftrightarrow q$ and $q \leftrightarrow r$, then $p \leftrightarrow r$." This relationship is needed especially in algebra, where we often start with an open sentence and end up with an equivalent sentence whose truth set is obvious.

We can also use this example to show the relationship of the contrapositive statement to the implication: "If $(p \rightarrow q)$, then $(\sim q \rightarrow \sim p)$" or "If $x \neq 8$, then $3x + 5 \neq 29$." Whenever an implication is true, the contrapositive statement is true and vice versa. This relationship between the contrapositive and the implication is the basis for indirect proof.

In order for the conclusion to a chain of statements to be true, the statements used must be true. Consequently, the statements used are usually definitions, postulates, or previously proved theorems. Hence, it is of importance to consider the truth of simple statements and compound statements. This is often done by the use of truth tables. Here is a comparison of the truth of p and $\sim p$:

p	**~p**
T	*F*
F	*T*

If it is true that it is raining, then it is false that it is not raining.

Next consider the truth table for the compound statement $p \wedge q$.

p	**q**	**p ∧ p**
T	*T*	*T*
T	*F*	*F*
F	*T*	*F*
F	*F*	*F*

The table illustrates that the conjunction ($p \wedge q$) is true only if both p and q are simultaneously true.

A different result occurs for the "or" ($p \vee q$):

p	q	p ∨ q
T	T	T
T	F	T
F	T	T
F	F	F

This compound statement is false only when both statements are simultaneously false.

The truth table for the implication $p \rightarrow q$ is a little tricky. Recall that a statement can be only true or false.

p	q	p → q
T	T	T
T	F	F
F	T	T
F	F	T

The implication "If it rains, then it is warm" is false only if it rains and it is *not* warm. Thus the statement "If $1 + 1 = 3$, everything is possible" is a true statement.

These truth tables can be extended to many more complex combinations of statements. One of the dramatic applications of truth tables is the translation into electric circuits. If a lighted lamp represents a true statement then the following simple circuits of Figure 12.10 apply:

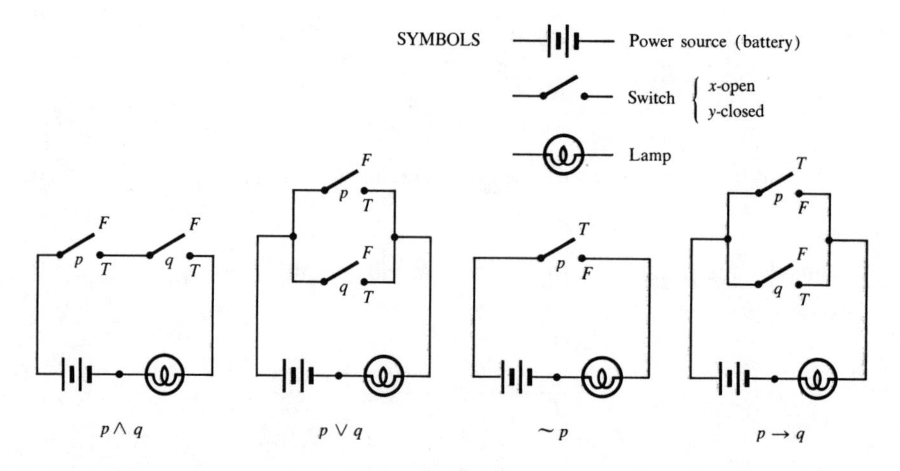

Figure 12.10

Hence the study of logic in terms of truth tables leads to the use of the computer in analyzing a proof.

Puzzles and Paradoxes

Another way to enliven the study of logic is to use it to solve puzzles and paradoxes.

To do this we will accept as valid these arguments:

$$\begin{array}{ll} \text{Given}\begin{cases} p \to q \\ p \quad \text{(is true)} \end{cases} \\ \hline \text{Then} \;\; q \quad \text{(is true)} \end{array} \qquad \begin{array}{c} p \to q \\ \sim q \\ \hline p \end{array} \qquad \begin{array}{c} p \to q \\ q \to r \\ \hline p \to r \end{array}$$

But we will not accept as valid these arguments:

$$\begin{array}{ll} \text{Given}\begin{cases} p \to q \\ q \quad \text{(is true)} \end{cases} \\ \hline \text{Then} \;\; p \quad \text{(is true)} \end{array} \qquad \begin{array}{c} p \to q \\ \sim p \\ \hline \sim q \end{array}$$

Here is a fascinating puzzler from Lewis Carroll:

> *Theorem:* If John wears kid gloves, he is not an opium eater.
>
> The following conditions are given:
>
> 1. If John goes to a party, he does not fail to brush his hair.
> 2. To look fascinating, it is necessary to be tidy.
> 3. If John is an opium eater, then he has no self-command.
> 4. If John brushes his hair, he looks fascinating.
> 5. John wears kid gloves only if he goes to a party.
> 6. Having no self-command is sufficient to make one look untidy.

Use the following placeholders for statements.

p = John goes to a party.
q = John brushes his hair.
r = John looks fascinating.
s = John is tidy.

$t =$ John is an opium eater.
$u =$ John has self-command.
$v =$ John wears kid gloves.

Then the given statements are the following:

1. $p \to q$
2. $r \to s$
3. $t \to \sim u$ or $u \to \sim t$
4. $q \to r$
5. $v \to p$
6. $\sim u \to \sim s$ or $s \to u$

And we are to prove: $v \to \sim t$

By rearranging the given sequence of statements we find this chain of arguments.

$$\begin{aligned} & v \to p \\ & p \to q \\ & q \to r \\ & r \to s \\ & s \to u \\ & u \to \sim t \\ \therefore\ & v \to \sim t \end{aligned}$$

Logical puzzles like this can readily be constructed by students. To do this the students first write a chain of statements for a logical proof. Then the conclusion is stated as a theorem and the steps in the proof are mixed up. Here is an example:

1. If John is well, he hasn't eaten a green apple.
2. If John is not well, he calls a doctor.
3. If John's life is safe, he will not call a doctor.
4. Hence, if John eats a green apple, his life is not safe.

Let p represent John eats a green apple
q represent John is well
r represent John calls a doctor
s represent John is safe

Then the implications are:

$$\left.\begin{array}{l} 1.\ q \rightarrow \sim p \\ 2.\ \sim q \rightarrow r \\ 3.\ s \rightarrow \sim r \end{array}\right\}\text{given}$$

4. $p \rightarrow \sim s$—to be proved

$q \rightarrow \sim p$ implies $p \rightarrow \sim q$ (contrapositive)
$p \rightarrow \sim q$ and $\sim q \rightarrow r$ imply $p \rightarrow r$
$s \rightarrow \sim r$ implies $r \rightarrow \sim s$ (contrapositive)
$p \rightarrow r$ and $r \rightarrow \sim s$ implies $p \rightarrow \sim s$. Q.E.D.

Teachers should provide a stimulating setting for learning logic and structure. To do this, it is necessary to show the need for this structure. Perhaps the best way to illustrate this need is for the teacher to provide many counterexamples and contradicting illustrations, forcing students to fall back on fundamental principles. "Proofs" such as "all triangles are isosceles" and "all numbers are equal" or even "all billiard balls are the same color" force students to examine closely what appear to be examples of correct logical reasoning. This should be a continuing class activity. Examine, for example, proofs of two of those three statements:

I. All triangles are isosceles.

Given:
ΔABC

Prove:
$AB = BC$

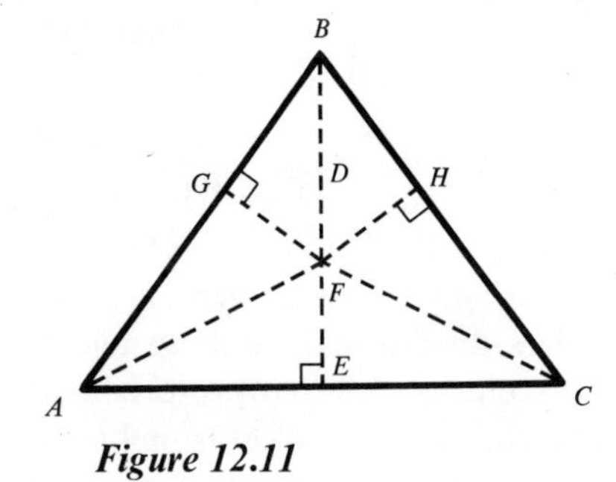

Figure 12.11

1. Construct $\overline{DE}$, the $\perp$ bisector of $\overline{AC}$.
2. Construct the bisector of $\angle ABC$ meeting $\overline{DE}$ in F.
3. Construct $\overline{FG} \perp \overline{AB}$ and $\overline{FH} \perp \overline{BC}$.
4. Draw $\overline{AF}$ and $\overline{CF}$.
5. Since $\Delta BFG \cong \Delta BFH$, $BG = BH$.
6. Since $\Delta FEA \cong \Delta FEC$, $AF = FC$ and $\Delta GAF \cong \Delta HCF$.
7. Since $\Delta GAF \cong \Delta HCF$, $GA = HC$.
8. Adding these, $BG + GA = BH + HC$, or $BA = BC$.

A more complete consideration of this false proof is found in Y. Dubnov, *Mistakes in Geometric Proofs.*[1] In that treatment, five other cases are considered, making the proof seem even more acceptable. The error in this proof lies in the drawing—often a source of erroneous thinking. If in the original triangle $BC > BA$ (as drawn), F will fall outside $\triangle ABC$, G will fall on the extension of $\overline{BA}$ through A, and H will fall on $\overline{BC}$ between B and C. In that case, no congruency argument may be constructed.

II. All billiard balls are the same color.

By induction, certainly true for $n = 1$ (any one ball is the same color as itself). If true for all $n \leq K$, then true for $n = K + 1$, because of the overlap as shown in Figure 12.12.

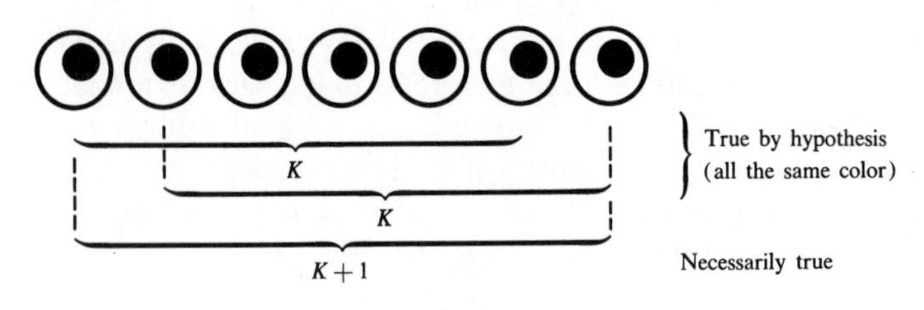

Figure 12.12

Real insight is required here. The induction principle is met *except* for passing from $n = 1$ to $n = 2$, where the overlap illustrated does not exist.

Students are interested in such "proofs." They like to see how closely mathematics borders on illogical thinking and how careful development and use of structural principles is all that protects us.

Every teacher should collect from his own experience less formal illustrations to support this rule. These illustrations are most easily found in newspaper and television advertising, but other sources are available. Here are two examples:

1. The "Peanuts" cartoon in which Charlie Brown is asked by Lucy how he is doing in school.

> "I'm doing quite well in everything but arithmetic."
> "I should think you would do well in arithmetic. It is a very precise subject."
> "That's just it. I'm at my best in subjects that are mostly a matter of opinion."

[1] D. C. Heath & Company, Lexington, Mass., 1963, pp. 9–11, 24–25.

2. Mark Twain in *Life on the Mississippi*:

Please Observe:

In the space of 176 years the Lower Mississippi has shortened itself 242 miles. That is an average of a trifle over one mile and a third per year. Therefore, any calm person, who is not blind or idiotic, can see that in the Old Oölitic Silurian Period, just a million years ago next November, the Lower Mississippi River was upward of 1,300,000 miles long, and stuck out over the Gulf of Mexico like a fishing-rod. And by the same token any person can see that 742 years from now the Lower Mississippi will be only a mile and three-quarters long, and Cairo and New Orleans will have joined their streets together, and be plodding comfortably along under a single mayor and a mutual board of aldermen. There is something fascinating about science. One gets such wholesale returns of conjecture out of such a trifling investment of fact.

Paradoxes have been of interest to mathematicians for centuries. The classic is the statement "If the barber shaves everyone in town who does not shave himself, who shaves the barber?"

There is a new twist to this.

Take a sheet of paper; on one side write:

"The statement on the other side is false."

Turn the paper over and write:

"The statement on the other side is true."

What is your conclusion?

These are both examples of self-contradictory logical systems and are representative of a type of problem that caused real difficulty in the study of the foundations of mathematics, particularly in set theory.

Special Contributions of Structure and Logic

Today's world is one of inconsistency. Students are faced with inconsistency everywhere: their parents alternate from permissiveness to extreme strictness; their teachers, seeking to treat each child as an individual, do not discipline students equally; adults talk of the need for moral, law-abiding conduct but at the same time dodge income-tax payments and exceed the speed limit; politicians often make utterances simply because their constituents want to hear them.

The youngsters of today, brought up in this world of inconsistency, seek stability everywhere. Master teachers know that students react favorably to even the most demanding standards when they know that the standards are the same for all. These same students can, and under the guidance of good mathematics teachers often do, find this same stability in mathematics. John W. Dickey and Edna B. Taylor, in their article "Mental Hygiene and Arithmetic" in *Instruction in Arithmetic*,[2] point out the therapeutic values of mathematics in greater detail.

The cohesiveness of mathematics is made apparent through careful study of its structure; and the integrity of mathematics is made apparent through careful study of the binding logical principles. These two important threads should be throughly developed in the classroom, not only to give students deeper insights into the subject but also to give them a model for their personal behavior and beliefs.

Learning the mathematical method and mathematics as a system of thought should be considered far more important than learning specific mathematical techniques; for all learning has within it similar but less apparent structures of sequence, emphasis, style, logic, and problem-solving procedures. It is in this sense that mathematics is universal in its applications—to all learning, to society, to the individual.

Learning Exercises

1. Collect five examples of misused logic from advertising, political speeches, or editorials.

2. Construct a criminal situation in which indirect proof may be used to determine the guilty person.

3. For the following geometric theorem write the proof in the forms listed below. Supply reasons only for part (a).

Theorem: If a tangent and a secant are drawn to a circle from the same external point, the square of the tangent is equal to the product of the secant and its external segment.

(a) In synthetic form
(b) In analytic form
(c) Diagrammatically
(d) In paragraph form

4. Read the debate about mathematical logic between Robert Exner and Peter Hilton in the May 1971 *Mathematics Teacher* (see references). Indicate points of agreement and disagreement with each author. Summarize your own attitude toward logic in school mathematics.

5. Read the Hesser article in the reading list. Plan a lesson or series of lessons using postulational systems. Are such lessons justified? Why or why not?

6. Some topics contrast intuition and discovery with structure and logic. Identify some of these topics in a mathematics text and suggest ways to transfer the intuitive discovery to the deductive proof of the generalization.

7. Examine a chapter in a contemporary high school algebra text. What is the role of structure and logic in this development?

---

[2] Yearbook 25, National Council of Teachers of Mathematics, Washington, D. C., 1960, pp. 179–201.

8. Look at early chapters of a high school geometry text to determine how proof is developed. How would you adapt this development for teaching?

9. Construct a logic problem along the lines of the Lewis Carroll example on p. 211.

10. Write several proofs of the Pythagorean theorem to illustrate intuitive, deductive, algebraic, and geometric proofs.

11. PROJECT. Investigate Boolean algebra and discuss its use in the logic circuits of electronic computers.

Suggestions for Further Reading

Buck, Charles, "What Should High School Geometry Be?," *Mathematics Teacher*, 61, 5 (May 1968), 466–471.

Cummins, Kenneth, "Mathematics 'In Statu Nascendi,' " *Mathematics Teacher*, 63, 7 (November 1970), 567–570.

Exner, Robert M., and Myron F. Rosskopf, "Proof," in *The Teaching of Secondary School Mathematics*, Yearbook 33, National Council of Teachers of Mathematics, Washington, D. C., 1970, pp. 196–240.

———, and Peter J. Hilton, "Should Mathematical Logic Be Taught Formally in Mathematics Classes?," *Mathematics Teacher*, 64, 5 (May 1971), 389–401.

Farrell, Margaret A., "Geometry—A Psych-o-Logical Learning Strategy," *School Science and Mathematics*, 71, 2 (February 1971), 139–142.

Ginther, John L., "Strategies for Teaching Concepts by Using Definitions," *Mathematics Teacher*, 59, 5 (May 1966), 455–457.

Hesser, Sister Francis Mary, "The Land of the Gonks: An Original Postulational System for High School Students," *School Science and Mathematics*, 66, 6 (June 1966), 527–531.

Klingler, Donn L., "Structuring a Proof," *Mathematics Teacher*, 57, 4 (April 1964), 200–202.

Poincaré, Henri, "Mathematical Definitions and Teaching," *Mathematics Teacher*, 62, 4 (April 1969), 295–304.

Robinson, Edith, "Strategies on Proof," *Mathematics Teacher*, 56, 7 (November 1963), 531–534.

Sitomer, Harry, and Howard F. Fehr, "How Shall We Define Angle?," *Mathematics Teacher*, 60, 1 (January 1967), 18–19.

Smith, Stanley A., "What Does a Proof Really Prove?," *Mathematics Teacher*, 61, 5 (May 1968), 483–484.

Tenney, Arthur E., "Another Format for Proofs in High School Geometry," *Mathematics Teacher*, 56, 8 (December 1963), 606–607.

Thorsen, Carolyn C., "Structure Diagrams for Geometry Proofs," *Mathematics Teacher*, 56, 8 (December 1963), 608–609.

Van Engen, Henry, "Strategies of Proof in Secondary Mathematics," *Mathematics Teacher*, 63, 8 (December 1970), 637–645.

Veblen, Oswald, "The Modern Approach to Elementary Geometry," *Mathematics Teacher*, 60, 2 (February 1967), 98–104.

Wiseman, John D., Jr., "Complex Contrapositives," *Mathematics Teacher*, 58, 4 (April 1965), 325–326.

———, "Scrambled Theorems," *School Science and Mathematics*, 64, 5 (May 1964), 423–427.

Developing Computational Skill

13

Mathematics is something like a game: it has rules, goals, and players, and it requires certain skills. Just as games are fun, so the "game" of mathematics can be played for intellectual satisfaction or for the attainment of skills. But games are enjoyable only for skillful players, and often acquiring skill isn't much fun at all.

To learn mathematics and to use it requires a mastery of computation. To master a skill such as computation requires practice, repetition, and drill. However, this practice does not need to be given by countless time-consuming and boring exercises. When practice becomes a meaningless activity, it causes unfavorable attitudes and habits to develop. Even mathematicians dislike rote computations. Consequently, our first responsibility in teaching computational skills is to make the practice as palatable as possible.

We need to inspire our students to practice computation with purpose and energy. To do so, we must (1) make use of a variety of techniques and materials; (2) make sure that the learner knows the purpose of practice—he should recognize that competence in computation will make other activities such as learning new concepts, playing games, calculating odds, or carrying out operations necessary to business easier and more pleasant; (3) help the learner understand the need for repetition and know how to practice independently; and (4) make sure he is aware that through this practice he will make progress. However, do not make computation skills the only objective or even the primary objective of any course at the secondary-school level.

Importance of Computational Skill

Although computational skill has always been considered essential for the superior student as well as for the slow learner, employers and teachers often complain that young people in elementary school, in high school, in college, and on the job generally lack this skill. Why? Psychologists and educators give these reasons:

Inadequate understanding of numbers and operations with numbers

Lack of interest in attaining computational efficiency

Lack of ability to cope with the abstract ideas and symbols of computation

Ineffective teaching of computational processes

Does it really matter these days whether people have computational skill? Some people might say that it does not, since calculating machines such as the cash register and the electronic computer have taken over computing tasks. However, computational skills are still essential for the following purposes:

To facilitate the learning of new mathematical concepts. If his computation is efficient, the learner can devote his mental energy to reflective thinking when facing a new problem or when exploring a new idea. If a student has mastered basic skills, he is likely to be confident in his ability to learn more mathematics.

To perform many tasks in the home, on the job, and in recreational activities. Elementary arithmetic examples pervade activities such as shopping, cooking, managing a business, or playing a game.

To promote productive thinking in problem solving, research, and other creative activities.

To provide sources for insight into the structure of our number systems. Carrying out a computation may be a means of understanding place value, the properties of certain numbers, or the operation involved in an algorithm.

This last point is especially important. Many of the great mathematicians of the seventeenth, eighteenth, and nineteenth centuries based their insights into numbers on the simple arithmetic calculations they performed. Without their broad experience with calculation processes, a good part of their conceptual work would probably not have been achieved. Foremost in this regard was the German mathematician Karl Gauss, who developed many illuminating theorems in number theory from ideas suggested by his work with paper-and-pencil calculations. When Karl Gauss was a boy of eight, his class was given the task of finding the sum of all numbers from 1 to 100 inclusive. Gauss soon found the answer, 5,050, while his classmates struggled to add all these numbers. Gauss undoubtedly noted that

$$\begin{aligned} S &= 1 + 2 + 3 + \cdots + 98 + 99 + 100 \\ S &= 100 + 99 + 98 + \cdots 3 + 2 + 1 \\ \text{Hence, } 2S &= 101 + 101 + 101 + \cdots 101 + 101 + 101 \\ \text{Then } 2S &= 100 \times 101 \text{ or } S = 5{,}050 \end{aligned}$$

Perhaps this discovery, which reduced routine computation into a creative experience, inspired Gauss to become a mathematician. Gauss is now considered one of the greatest mathematicians of all time. Some of his greatest discoveries were in the field of number theory.

In much the same way, drill exercises, when they are carefully structured, often can be used to develop a mathematical point. For example, consider the following exercises:

$371 + 2{,}934 + 7{,}066$
$899 + 456 + 544$
$3\frac{1}{7} + 5\frac{3}{5} + 2\frac{2}{5}$

Most students will attack these exercises in a straightforward manner. However, if they are urged to find a shortcut or to find the solutions quickly, they soon notice that they can solve the problems most easily by applying the associative property and combining the latter addends first.

The importance of mathematical skills is verified by the statements of recent studies of the current mathematics curriculum:

> . . . children must know how to do simple and rapid mental computations: they must be accustomed to finding very quickly an order of magnitude for a total or a product.[1]
>
> Mastery of the four fundamental operations with whole numbers and fractions, written in decimal notation and in the common notation used for fractions . . . includes skill in the operations at adult level (i.e., adequate for ordinary life situations) and an understanding of the rationale of the computational processes.[2]

At the same time that new mathematics programs are advocating a high level of computational efficiency, they are emphasizing the need for a new approach to skill learning:

> Lest there be any misunderstanding concerning our viewpoint, let it be stated that reasonable proficiency in arithmetic calculation and algebraic manipulation is essential to the study of mathematics. However, the means of imparting such skill need not rest on methodical drill. We believe that entirely adequate technical practice can be woven into the acquisition of new concepts. But our belief goes farther. It is not merely that adequate practice can be given along with more mathematics; we believe that this is the only truly effective way to impart technical skills. Pages of drill sums and repetitious "real-life" problems have less than no merit; they impede the learning process. . . .
>
> We propose to gain three years through a new organization of the subject matter and the virtually total abandonment of drill for drill's sake, replacing the unmotivated drill of classical arithmetic by problems which illustrate new mathematical concepts.[3]

---

[1] *New Thinking in School Mathematics*, Organization for European Economic Co-operation, Washington, D. C., 1961, p. 66.

[2] *Program for College Preparatory Mathematics*, College Entrance Examination Board, Princeton, New Jersey, 1959, p. 19.

[3] *Goals for School Mathematics*, Cambridge Conference on School Mathematics, Houghton Mifflin, Boston, 1963, p. 7.

Actually, this emphasis on learning through meaningful problem situations has been advocated for a long time.

> Continued emphasis throughout the course must be placed on the development of ability to grasp and utilize ideas, processes and principles in the solution of concrete problems rather than on the acquisition of mere facility and skill in manipulation. The excessive emphasis now commonly placed on manipulation is one of the main obstacles to intelligent progress.[4]

However, the present emphasis in psychology on learning suggests a new approach to computational skills. This emphasis—on structure, on discovery, on participation, on meaning, and on reinforcement—has important implications for skill learning. Similarly, new materials such as programmed texts, teaching machines, games, computing devices, remedial learning kits, individualized instruction materials, and audio-visual aids could be used for skill learning.

Purpose of Practice

Although we cannot improve our skills by practice alone (in fact, repetition can fix wrong ideas just as well as right ideas), practice does provide for refinement of technique and fixation of concepts and procedures in the following ways:

> Practice is essential for *retention.* We remember only a small part of what we read, hear, see, or do only once.
>
> Practice is a means of building *accuracy.* The correct fact should become the only response that is remembered.
>
> Practice is the basis for improving *efficiency.* After learning why an operation works, we eliminate crutches and discover shortcuts through exercises that are designed to promote good procedures.
>
> Practice is one way to establish *confidence.* Success in computing correctly and efficiently improves motivation, participation, and attitude.

But such positive results do not arise from random practice. They demand careful planning and execution. Often, a teacher can change the very nature of an assignment from drill to concept learning by the instructions given. Consider, for

[4] *The Reorganization of Mathematics in Secondary Education,* The National Committee on Mathematical Requirements, Houghton Mifflin, Boston, 1923.

example, the difference between the following sets of instructions:

> Do the odd exercises on page 37.
>
> Do the odd exercises on page 37 until you are able to state the pattern involved in each solution. As soon as you can do this, record the rule you would use to solve the exercises you have done. Then do only exercises that differ from that rule.

In the second case the student has a reason to consider each problem carefully. Once he has found a pattern, he need not write other solutions. He must, however, examine all problems in order to test his method and must work out those that do not fit the pattern.

Basic Principles in Practice

If practice and drill are to be effective, they must be an integral part of mathematics instruction. This practice should be *at the right time, in the right amount*, and *with the right exercises*. The following suggestions will add meaning and interest to practice activities.

Practice (unlike our daily practice of walking, reading, speaking, or driving a car) must be done with the *intent to improve*. Before he can *want* to improve, the learner must believe it worth his while to attain skill in computation. He needs to be aware of the advantages of being skillful in computation and the handicaps that will result from failure to attain it. Whenever possible teachers should show students how skills are used in games, sports reporting, driving a car, buying supplies, computing wages, and other activities that relate to their lives.

Practice should be performed *thoughtfully* and with insight, so that it never becomes mere mechanical repetition. The learner should be able to justify the process, know the properties involved, or relate the process to the definition of the operation. Thus, it is probably better for a teacher to assign a few exercises requiring thoughtful solutions than many exercises calling for automatic responses. This role of reflective thinking in learning a skill is the reason for the current emphasis on practice of computation within the framework of problem solving and applications. Whenever possible, drill and concept-developing exercises should be intermixed.

Practice should follow *discovery* and *understanding*. This understanding, as reinforced and extended by thoughtful practice, is the key to learning mathematics, not skill in computation—even though this skill is helpful.

Practice should involve *correct responses* rather than incorrect responses. Errors should be eliminated and correct responses reinforced by immediate knowledge of the right answer. Thus, whenever he assigns exercises for independent practice, the teacher should furnish answers for students to use in checking their work, like a programmed text.

Practice should be *individualized* according to the needs or ability of the learner. Diagnostic tests, observations, or interviews should be used to identify the need for remedial instruction and the reason for the difficulty. The practice of assigning the same exercises to an entire class or assigning more exercises to the student who finishes early is not reasonable. The skillful teacher recognizes that the bright individual needs a few difficult exercises, while the slow learner needs more easy exercises.

Practice should be *brief* and at *spaced* intervals. Spaced practice seems to produce better retention, and brief practice is necessary to avoid fatigue. Practice should be used when and where needed and not wasted on insignificant skills or on well-learned skills. It is not reasonable, for example, for students to drill on division in nondecimal scales. Many teachers begin a class with a short mental computation exercise period.

Practice should be given in *meaningful exercises*, so that transfer and application are promoted. If specifics are emphasized, they should soon be integrated into the whole of which they are a part. Thus, practice situations should closely resemble the situation in which the skill is used. Textbook exercises are never as meaningful as problems that are related to contemporary events and the local environment.

Practice should emphasize *general principles* rather than tricks or shortcuts. For example, the associative and distributive principles provide a general procedure for a variety of situations such as adding polynomials, factoring polynomials, and finding equivalent equations. These general principles eliminate the need for memorizing mechanical processes or tricks.

The learner should be given instruction in *how to practice*. The learner should know how to use answers for independent learning. He should know what he is expected to write and what he should do mentally. Crutches should be permitted but recommendations made for their elimination by more efficient procedures. (Note, however, that it is better for a weak student to function with a crutch than not to function at all.)

Practice should be given in a *variety of activities*, such as games, contests, puzzles, timed exercises, tachistoscope projections, mental computation, group activities, oral or written exercises. Even a short "warm-up" with rapid-fire exercises for mental computation is an effective reminder of facts and processes.

Practice becomes more effective if the learner is *informed of his progress*. He should know what competence is expected; how he compares with class, school, or national norms; and what progress he has made. A graph of errors and time can be a dramatic way of showing improvement. Then improvement should become a game as it is in golf or bowling.

Practice must *never be a punishment*. Learning mathematics should be a privilege and a pleasant experience. Never assign a set of exercises as punishment for any offense.

By whatever means computation is practiced, the goal is improvement in skills. No teacher should forget this. Many textbook writers recognize this fact and provide sections of review exercises. Whether these are available in the text or not, provision should be made for this review and maintenance in the classroom.

Remedial Instruction

Even though the principles outlined above are followed, individual differences are such that remedial instruction is frequently necessary. These differences may be due to the learner's low ability, inadequate educational experiences, emotional problems, or lack of interest. The learner may be a poor reader or a slow worker, or he may have physical defects. In any case, the first step in remediation is to locate the specific area of difficulty. In order to identify the specific need for remedial instruction, the teacher should collect information as follows:

Use diagnostic tests to identify the process or number set causing difficulty. Any achievement test can be diagnostic if an item analysis is made and the errors analysed. Your objective in diagnostic testing is to determine both what is not yet learned and what wrong learning needs to be "unlearned."

Use evaluative tests to measure the level to which the skill functions when needed. Does the student need an inordinate amount of time to perform? Can he do only easy problems and direct applications of a principle? Varying the time or the difficulty of items, as well as recording types of errors, gives added information on the level of skill development.

Observe the way the individual learner solves exercises. Does he make errors in copying problems? Does he use mechanical processes such as "cancellation" or "transposition" without understanding the processes?

Interview the learner to identify his thought patterns. Use questions like: Can you tell me the answer? Is this answer correct? How did you get the answer? What did you think about first when you saw the problem? What did you think before you told me the answer? Do you like this kind of problem? What kind do you prefer?

After the specific needs are identified, remedial instruction is given. To make this instruction successful, the teacher must accept the learners as they are, show some affection for them, and teach them with patience. He will need to use a new approach with different content to get around emotional blocks and hostile attitudes. He will need to use a variety of materials at appropriate levels—possibly even fourth-grade material for a high school student. He will need to impress on his students that the goal is improvement, not competition with others; that they are working in order to make progress—not just to get a passing grade.

Learning Computational Skill through Games

Games may be an effective means of making the practice of computational skills palatable. They are also means of attaining other objectives. Games can be adjusted to the interest and abilities of small groups of students and consequently can be used to advantage with the slow learner, the average student, or the gifted student. The creative student can even devise a new mathematical game. In addition, a game can be an ideal device to involve the parent in out-of-class learning activities. A mathematical game can be sent home with the student to gain an informal setting where student and parent can work together at learning mathematics.

The success of a classroom game, like any instructional material or technique, is highly dependent on how it is used. If a game is to play the roles described above, the following factors should be considered:

The game to be used should be selected according to the needs of the class. The basic criterion is that the game make a unique contribution to learning—a contribution that cannot be attained as well or better by any other material or technique. The material involved should be closely related to that of the regular classwork. Specifically, the game selected should involve important mathematical skills and concepts; and major emphasis should be on the learning of these concepts or skills rather than on the pleasure of playing the game.

During the game situation *all* students involved must be participating. Even though only one person is working on a certain problem, every team member also must be responsible for its solution. Games must also avoid extreme embarrassment for the person who cannot solve a problem. Whenever possible, students should compete with other students of equal ability.

The game should be used at the proper time—that is, usually during the regular class period when the ideas or skills are being taught. There are other times, however, when games promote learning in an otherwise difficult environment. Many teachers prefer to use games on the day before a vacation, or during days of heavy absence due to athletic games, storms, concerts, or excursions. Usually, games should be relatively short so that pupils do not lose interest.

The game must be carefully planned and organized so that the informality and excitement of the setting does not defeat its purpose. Before the game begins, the participants should be briefed on the purpose of the game, the rules, and the way to participate. Often the students can establish ground rules, so that everyone (including the teacher) may enjoy the activities. "Coaching" or "kibitzing" should not be allowed. The loss of points for breaking rules is usually sufficient to maintain appropriate behavior.

The participants in the game must accept the responsibility of learning something from the game. Follow-up activities such as discussions, readings, or tests will emphasize this responsibility. The teacher will need to evaluate the results by asking himself how successful the game was in promoting desired learning.

Games appropriate for learning mathematics are limited only by the ingenuity of students and teacher. They can convert almost any practice lesson into a learning game by choosing teams, participating as individuals or teams, and keeping score. Most of the common parlor games or athletic games can be adapted for use in a mathematics class at any grade level.

For example, Bridget is a game similar to bridge. There are four suits (four colors). Each card has an algebraic expression such as x^3 or $(x^2 - 2x)$. Value cards designate the numeral to substitute for the variable. The player whose card gives the highest value of four cards played takes the "trick." Another example is given by mathematical baseball. Here the "pitcher" proposes problems to the "batter." When the batter solves the problem, he makes a "hit." The number of "bases" of the hit depends on the difficulty of the problem. Even if the "hitter" does not solve the problems, he gets on base if an opposing player cannot solve the problems. Quiz games, card games, word-association games, mathematical bingo, spell downs, and identification contests have been successfully used by many teachers.

A card-game version of bingo is popular with many teachers. The playing cards consist of an array of numbers, algebraic expressions, terms, or geometric figures, depending upon the topic or course involved. The call cards consist of problems, definitions, equations, or algebraic expressions. Player cards are distributed to the members of the class. A student caller selects a call card at random. (Sometimes call cards are large enough to display the problem. At other times the problem of the call card is written on the chalkboard.) Players compute the answer and cover the appropriate space on their playing card.

Identification games may be played like Twenty Questions. A term, number, or principle is selected to be identified in fewer than 20 questions. Students or teacher may select the term or number to be guessed. For example, any natural number less than a million can be identified in 20 or fewer yes/no questions. In a similar way, any set of exercises can be used for a relay contest, quiz game, or card game.

Printed cards containing numbers (for instance, Rook, Flinch cards, or flash cards) are often useful. Dice or spin dials can also be used to supply numbers. As an interesting variation for dice, the students can make and use regular polyhedra such as octahedrons, dodecahedrons, or icosahedrons. Numbers, algebraic expressions, or problems can be written on the faces of these polyhedra depending on the game involved.

These games can be useful as a means of learning new ideas as well as mastering key ideas and skills. They are effective in lending variety and competition to classroom activities. The teacher is no longer the judge or leader to be outsmarted. Instead, the students vie with each other, work together as a team, and accept responsibility for doing their best. This is why the Minnemast Project and the Cambridge Conference recommend games for learning mathematical concepts.

James Coleman[5] of Johns Hopkins University suggests some additional features of games: attention focusing, a balance of chance and skill that gives slower learners an opportunity, a broader range of skills than many drill activities, and a closer approximation to the complexity of the real world.

Many mathematics games can now be purchased ready for use in the classroom, from kindergarten through high school. Commercial games for the mathematics class are listed in Appendix A. Many games that can be prepared by students or teachers are described in *Games for Learning Mathematics*.[6] Making games

[5] "Learning through Games," *NEA Journal*, 56, 1 (January 1967), 69–70.

[6] Donovan Johnson, *Games for Learning Mathematics*, Walsh Publishing Company, Portland, Me., 1960.

based on current classwork may be an excellent learning experience. Students at one grade level also enjoy constructing games for the students of a grade below. The materials you need are simple—paper, pencil, 3 × 5-inch cards or cardboard. With the help of students and the use of some imagination, you are all set to start an adventure. Don't be surprised if your students ask for mathematics games to play at noon hours or at home.

An example of a game that provides a challenge combined with concept reinforcement has been developed by Charles Linn of New York State College at Oswego. For this game fractions are printed on 2 1/2 × 3-inch cards (3 × 5-inch cards cut in half). Five cards are dealt to each of two to five players who lay out the cards as in Figure 13.1. The object of the game is to reorder the fractions left to right, in order of increasing value. The player who first completes the task correctly wins the deal (giving other players a head start in the next game).

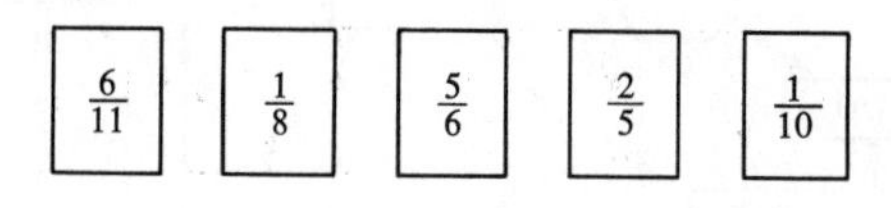

Figure 13.1

This game would be difficult to monitor by the teacher unless some quick checking procedure were available. There is. The reverse side of the cards are numbered with natural numbers as shown in Figure 13.2. When the fractions are correctly ordered, the natural numbers on the reverse are correctly ordered. (The figures illustrate this. You may wish to try the task yourself.) Now no teacher guidance at all is necessary once students know the rules of the game.

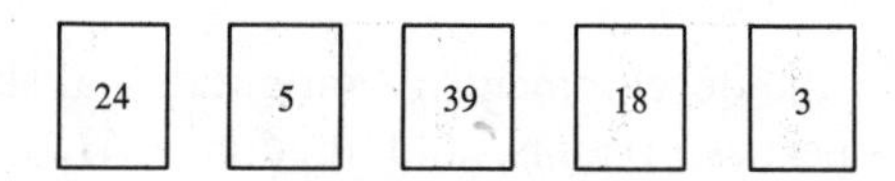

Figure 13.2

The game has a number of additional advantages. The cards illustrated are from a 66-card deck containing all positive fractions with denominators 12 or less whose value is less than 1. The numbering on the back represents the order for the complete deck. Any subset of the deck then retains the order of the full deck. This is true of five-card hands, and also of easier decks formed by retaining only fractions with denominators up to, say, 6. This allows a progression to harder and harder decks—and games. The 66-card deck is challenging to most adults.

A suggestion: there is, of course, a fundamental relationship that provides an easy comparison of two rational numbers. Do not tell this strategy to students. Some will discover it for themselves. Others will use less efficient, but equally effective strategies like changing to decimals.

Computational Skills and Calculating Tools

It would seem that using computers or desk calculators would keep students from practicing skills. In many ways, however, the result is exactly the opposite. Students are intrigued by the machines and gain greater understanding of numerical relationships. When the calculators are used to check computation exercises, they can improve their skills in the process. Interest is probably the key here.

Flow charts provide excellent drill-generating formats. Max Sobel has suggested a flow-chart format like that of Figure 13.3.[7] These ideas are extended in Chapter 25.

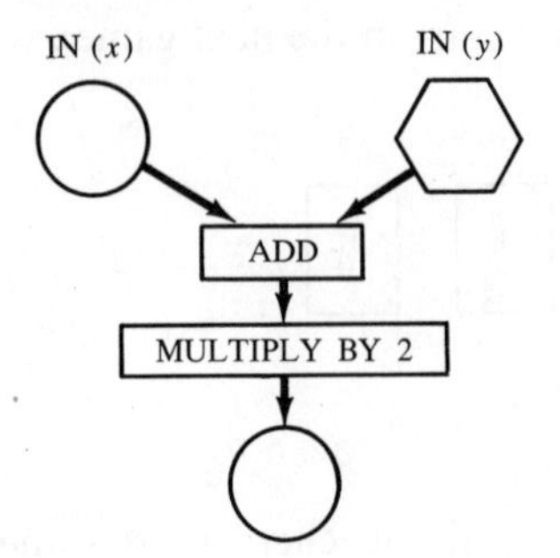

x	y	$x+y$	$2(x+y)$
3	7	10	20
5	2		
−3	4		
−2	−3		

Figure 13.3

Learning Fundamental Mathematical Statements

Closely related to the development of computational skills is the mastery of basic definitions, properties, axioms, and theorems. Here, the student has to memorize the statements (sometimes in a particular order), understand them, and know how to use them to support the statements of a problem or a proof.

To help students learn these fundamental statements, the teacher has to ask himself a number of basic questions: What definitions, rules, constants, and properties are of such importance that they should be overlearned? How can I best help students memorize necessary definitions such as those for π, sine, adjacent angles, linear equations, and rational numbers? How can I help them recall from their "dictionary" of memorized statements the ones appropriate to a particular problem situation? For example, what theorems should I consider when I set out to prove that a given quadrilateral is a rhombus? How can I help students use

[7] *The Teaching of Secondary School Mathematics*, Yearbook 33, National Council of Teachers of Mathematics, Washington, D. C., 1970, p. 301.

informal shortcuts but at the same time be cognizant of the mathematical principle involved? Compare, for example, transposition with the addition property for equations.

Selection of Statements

As a first step in teaching for mastery of fundamental ideas, the teacher must *select the statements, formulas, or procedures* that are of sufficient importance to warrant the time and effort needed for mastery. Wherever possible, the learner should participate in this choice, but he will seldom have had the experience necessary to make choices. The reason for each choice should be made clear to the learner so that he is motivated to master the idea. Consequently, the teacher may want to demonstrate immediate applications or show how future developments depend on the statement. He may, for instance, point out the value of definitions to students who must write proofs: in return for memorizing the definition of a word, the student is allowed to replace a longer phrase with the shorter word or words. It is certainly shorter to write or say "π" than it is to write or say "the ratio of the circumference to the diameter of any circle." Memorizing the formula $A = \pi r^2$ or $\sqrt{2} \doteq 1.414$ will save time and effort in innumerable situations. In the same way, committing to memory a proven theorem substitutes for repeating the steps of that proof again and again. In short, if we use an idea often, we need a definition for economy; if we seldom use the idea, it is more economical to repeat the idea than to memorize the proof. One appropriate technique to encourage students to learn a particular theorem or definition is to examine a development where it is used and see how the same development would have to be written or explained without the theorem or definition.

Mastering Fundamental Ideas

As a first step in mastering a fundamental idea, the learner must understand the idea: that is, he must be able to use it in statements, illustrate it with examples, and use it to solve problems or complete proofs. To build this mastery, teachers need to show in as many settings as possible how the statement is applied. The addition axiom is used in arithmetic, in algebra, and in geometry to operate with the lengths of lines, the measure of angles, the area of polygons, and the volume of polyhedra. These varied settings not only underscore the need for the axioms; they give students exposure to varied practice in the use of the axiom. This varied application will extend understanding and memorization better than mere repetition.

The examples below illustrate the importance of understanding each phrase of a definition:

Why are the angles shown in Figure 13.4 not adjacent angles?

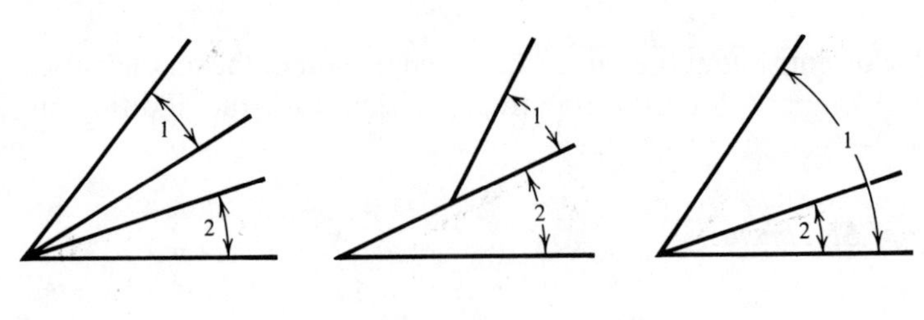

Figure 13.4

What is wrong with the following general definition of a circle: "A circle is the set of points at a fixed distance from a fixed point." (Does a sphere fit this definition?)

Memorizing Fundamental Ideas

When helping students to memorize fundamental ideas, the teacher can use a variety of approaches:

Since memorizing requires repetitive experiences, he can utilize the same principles detailed under the section on practicing a skill (page 222). All these are appropriate here.

He can provide both simple and complex settings for application of the basic ideas to be memorized.

When several statements are related, he can help students see and use these interrelationships to cut down on the amount to be memorized. A geometer of international reputation, Nathan Altschiller Court, has suggested the theorem "When two parallel lines are cut by a transversal, the angles that look equal are equal and the angles that don't look equal are supplementary." This informal statement, used with appropriate discretion, summarizes a number of independent theorems. It does not replace them, but it reduces their individual difficulty and it provides the student with an overall understanding of the ideas lost in long lists of statements.

He can seek out appropriate procedures to provide quick review and applications. The chalkboard, posters, or flash cards can be used for ready reference by the teacher. References for the student can be provided by underlining the text or recording key ideas in a notebook. Games, contests, mental drills, or quizzes are other settings for quick review.

He can encourage students to list available and appropriate ideas, formulas, and definitions when attacking a problem. Such a summary is especially important for an attack on a nontrivial problem for which an answer is not immediately forthcoming.

He can provide mnemonic devices when they help students to learn otherwise burdensome lists of definitions. These learning techniques must be used carefully because they make no pretense at being mathematical in nature; but where definitions are arbitrary and have little relation to the word being defined (or if the word itself is unfamiliar), this approach is helpful.

Shortcuts and Informal Statements

The use of informal statements for mathematical terms and shortcuts is widespread in mathematics. Although informal statements have considerable value (they save time, they save space, they communicate general ideas much as do variables and symbols of operation), there is a basic danger in using them. Students may fail to understand what underlies the informal statement or procedure and may use it without this necessary background of understanding. Informal statements and shortcuts, therefore, should be used with some caution; but they cannot and should not be rejected, since much of mathematics is virtually dependent on such shortcuts in procedure and in organization of information. (For example, what would we do if we could not use shortcuts in carrying out such a basic procedure as multiplication?)

When informal statements or shortcuts are introduced, a single teaching procedure should be constantly utilized. *Students should be asked repeatedly to provide the reasoning underlying shortcuts or informal statements. Unless they can provide such explanations, they should never be allowed to use the shortcut or informal statements.* Since working problems the long way requires more effort, the motivation to learn the basis for informal statements or shortcuts is built in.

Summary

Drill activities have often been overemphasized in the mathematics classroom. Practice and drill are carried to extremes when computational skill is considered sufficient to gain the goals of mathematics. Routine learning of skills results in poor retention, little understanding, and almost no application in daily problems. It is evident that the key to learning skills is through *meaningful* experiences, discovery, and applications.

Teaching the meaning of numbers, the understanding of a process, and the mathematical structure involved precedes practice. Practice, then, is the part of the learning process that builds accuracy, efficiency, and retention. If we provide the proper amount of practice at the appropriate time, it is likely that many of our students will have the mathematical competence that business, science, industry, and colleges are now demanding.

The effective teacher should consider skills as an important part of his program. He should require all students to meet at least minimal standards, but he should also be willing to spend extra time with students who have difficulty. In other words, he should mount an effective program to achieve skills competency just as he does to achieve conceptual understanding. Descending standardized test norms indicate that this is not a generally recognized responsibility. Computational skills should be one of the easiest things to teach, yet we do not do such teaching effectively.

Learning Exercises

1. Attitudes are mixed toward mnemonic (memory) devices such as "Some old hen caught another hen taking oats away," to associate the trigonometric functions with their ratios (sine, opposite: hypotenuse; cosine, . . .) Do you think that they should be used? Support your argument.

2. Make a deck of cards for the game described on page 227. Test yourself with some hands. What is "best" strategy? Try the game with some adults. What are their strageties?

3. One teacher who uses the fraction-ordering game secretly tells only his weakest student the strategy. What do you think of this as a teaching technique? Indicate what might be merits and demerits of such a pedagogical technique.

4. Make up a set of computation exercises covering fundamental arithmetic operations that you feel is attractive, motivating, and creative. Compare your exercises with those of other teachers.

5. Develop another flow chart that would function exactly like the one on page 228. How could the two be used together in the classroom?

6. The teacher writes this on the chalkboard □□ × □□ → 850, asking each student to copy it on a piece of paper. The problem is to use the four digits provided by the teacher to make a product (of two-digit numbers) as close to 850 as possible. The teacher then calls out four random digits, pausing after each because the students must decide then where to enter the digit. They cannot wait until all four are given. The winner is the class member whose product comes closest to 850. What are the values of this class game?

7. How could you use this textbook to generate digits that are nearly random?

8. Devise other patterns like those of exercise 6 that would apply to:
   (a) Addition
   (b) Subtraction
   (c) Division
   (d) Addition of fractions
   (e) Multiplication of fractions

PROJECTS. 1. Diagnose computational deficiencies and provide remedial instruction for an individual student who is having difficulty in this area. Retest to determine progress.

2. Interview several students to determine how they think as they perform a standard computation like long division.

3. Set up a finite mathematical system such as a rotation group. Work a series of exercises with the operations involved. Search for relationships and patterns. Work another comparable set of exercises and compare results.

Suggestions for Further Reading

Forsythe, Alexandra, "Mathematics and Computing in High School: A Betrothal," *Mathematics Teacher*, 57, 1 (January 1964), 2–7.

Hannon, Herbert, "The Role of Meaning in Teaching the Fundamental Processes," *School Science and Mathematics*, 58, 2 (February 1958), 83–89.

Sobel, Max A., "Skills," in *The Teaching of Secondary School Mathematics*, Yearbook 33, National Council of Teachers of Mathematics, Washington, D. C., 1970, pp. 291–308.

Sueltz, Ben A., "Drill—Practice—Recurring Experience," in *The Learning of Mathematics: Its Theory and Practice*, Yearbook 21, National Council of Teachers of Mathematics, Washington, D. C., 1953, pp. 192–204.

Van Engen, Henry, "Rate Pairs, Fractions, and Rational Numbers," *Arithmetic Teacher*, 7, 8 (December 1960), 389–399.

Wendt, Arnold, "Per Cent without Cases," *Arithmetic Teacher*, 6, 6 (October 1959), 209–214.

Teaching Problem Solving

14

Learning to solve problems (that is, finding an appropriate response to a situation which is unique and novel to the problem solver) is perhaps the most significant learning in *every* mathematics class for several reasons:

It is a process whereby we learn new concepts. A problem may be a setting for the discovery of a new idea. These problems may be a question for class discussion, the topic of an entire unit, or an assigned exercise. An excellent way of promoting much independent individual discovery is to assign in exercise form the concept to be discussed in the next lesson. Thus, an appropriate assignment for discovering the method of finding the truth sets for inequalities might be a series of inequalities to be solved.

The skillful teacher makes such assignments carefully. Most students have become accustomed to homework that includes only content previously discussed in the classroom. Parents also expect this type of assignment and occasionally are upset by homework problems on material not already taught. "Are you asking me to teach it?" is their response. The key to using problems to learn new ideas is clear identification of the *type* of problem the new material presents and, when necessary, guides for the student.

Here is an elementary example of the use of problems for learning a new idea or preparing for the next lesson. The day's lesson had been concerned with finding unions and intersections of sets. The assignment consisted of finding unions and intersections by tabulation. The learning problem assigned was: Does the distributive property apply to unions over intersections?

> Does $A \cup (B \cap C) = (A \cup B) \cap (A \cup C)$? (Suggestion: Select sets $A, B,$ and C, which have some common members. Compare results by tallying the sets involved. Now test your results with Venn diagrams.)

Here is a similar example from geometry:

> 1. Draw two large quadrilaterals (sides each greater than 2 inches) on your paper. In each, bisect the four sides and connect the midpoints consecutively. What kind of quadrilaterals seem to be formed? What is the ratio of the area of the first quadrilateral to the second? Test your results with a third quadrilateral. Can you prove your conjecture?

2. Draw another large quadrilateral. Bisect the sides and join the opposite midpoints. Measure the four segments formed. Form a conjecture. How is this conjecture related to the result of exercise 1?

3. Draw another quadrilateral. Trisect the sides and join corresponding points, one on each side. Compare the measures of the sides and the areas of the two quadrilaterals. Form a conjecture. Find a generalization that applies for any number of division points on a quadrilateral.

Exercise 3 will probably be a problem-solving experience for you, the reader. Here is an example from an eleventh-year algebra assignment:

We have studied graphing quadratics in some detail. Now let's look at a few simple functions of higher degree in order to determine whether some patterns are evident. We will first consider only equations of the form $y = x^n$ for positive integral values of n in the range $0 \leq x \leq 1$.

On your graph paper mark a large square with the left and lower sides the y and x axes. Mark scales in tenths on each axis, as in figure 14.1.

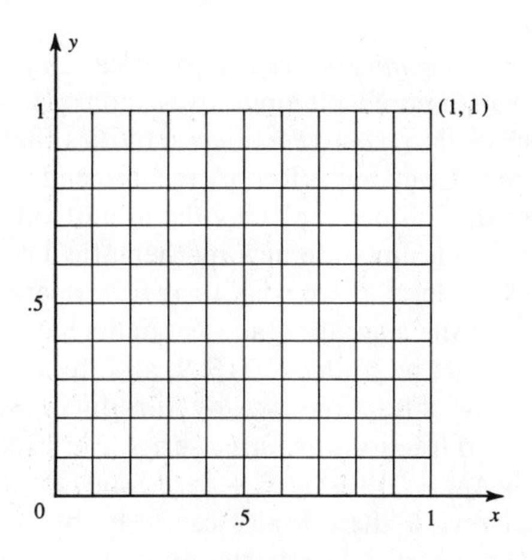

Figure 14.1

1. On this square, carefully plot two graphs with which you are familiar, $y = x$ and $y = x^2$.

2. On a separate sheet of paper make a table of values to be used for additional graphs. Use the form shown in Figure 14.2.

3. Fill in the table for x^3 and x^4. (Do you have to start with x for each computation?)

4. Plot the points for $y = x^3$ and sketch the graph as a smooth curve through these points. (You may wish to use additional x values to ensure accuracy.)

5. Do the same for $y = x^4$.

x	x^2	x^3	x^4
0	0		
.3	.09		
.5	.25		
.8	.64		
1	1		

Figure 14.2

6. What appears to be happening as n increases for $y = x^n$? What would you expect to be the limiting position for $y = x^n$ as n increases, $0 \leq x \leq 1$?

7. FOR THINKERS. Without making a table or plotting points, extend the ideas developed here to the domain $-1 \leq x \leq 0$ for the same graphs. Use your knowledge of symmetry and exponents. Be careful: two different patterns emerge.

8. FOR DEEP THINKERS. What would be the pattern for $y = x^{1/n}$ for positive integral n? (It is not necessary to extract roots to answer this.)

Problems may be a meaningful way to practice computational skills. The rote learning of skills through purely manipulative exercises has been found inadequate. Bright students especially resent repetitive activities that demand little thought. Hence, the Cambridge Conference Report recommends the "virtually total abandonment of drill for drill's sake, replacing the unmotivated drill of classical arithmetic by problems which illustrate new mathematical concepts."[1]

Suppose, for example, that we wish to review operations with fractions and fractional equations in the eleventh grade. Students have already been extensively exposed to fraction units in grades 6, 7, 8, 9, and, to a lesser extent, 10. We may wish to motivate review of fractions now by introducing some problems related to sequences or continued fractions or Farey series. For example, what is the sum of the series $1 + 1/2 + 1/4 + 1/8 + 1/16 + \ldots$? Students who do not have computational skills must review these topics carefully, but in this way they gain an intensive rather than superficial understanding. At the same time, any work students do in these subjects has to involve practice with fractions. Note, however, that the practice contributes to the more valuable problem-solving objectives and does not stand alone, unmotivated and unwanted.

By solving problems we learn to transfer concepts and skills to new situations. We know that the transfer of knowledge from the classroom or textbook situation to new applications is too limited. The amount of this transfer—a fundamental necessity for learning—is highly dependent on the emphasis on transfer. Problems of varied types, which illustrate varied applications, are a means of giving practice

[1] Cambridge Conference on School Mathematics, *Goals for School Mathematics*, Houghton Mifflin, Boston, 1963, p. 7.

in transfer. Furthermore, a general skill such as problem solving is more permanent and more transferable than the knowledge of specific facts or concepts.

The thoughtful teacher continually extends the textbook problems to new situations. The mixture problems of algebra can be extended to the proportion problems of chemistry. The problems about digits have unusual applications to number tricks. The problems involving quadratics have current applications to space travel. The study of relations between the vertices and edges of regular polyhedra can be extended to problems of symmetry and duality.

A student looking for transfer might well see the common basis in the following two problems without having it pointed out to him.

1. In how many ways can you spell the word *mathematics*, starting from the top and working down through the array shown in Figure 14.3?
2. A car drives into a city at A and leaves at B (see Figure 14.4). How many "*direct*" routes are there through the streets?

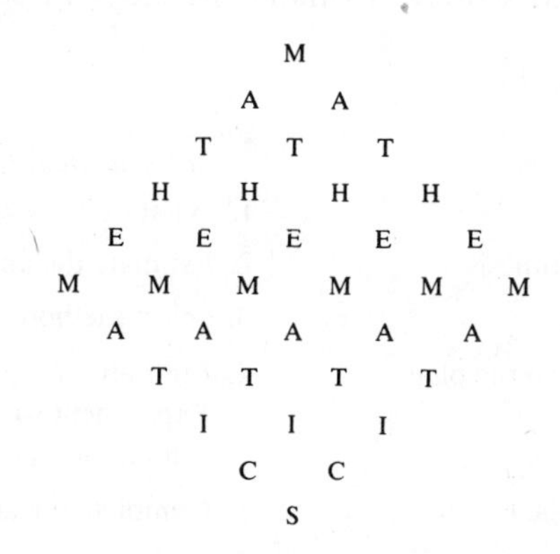

Figure 14.3

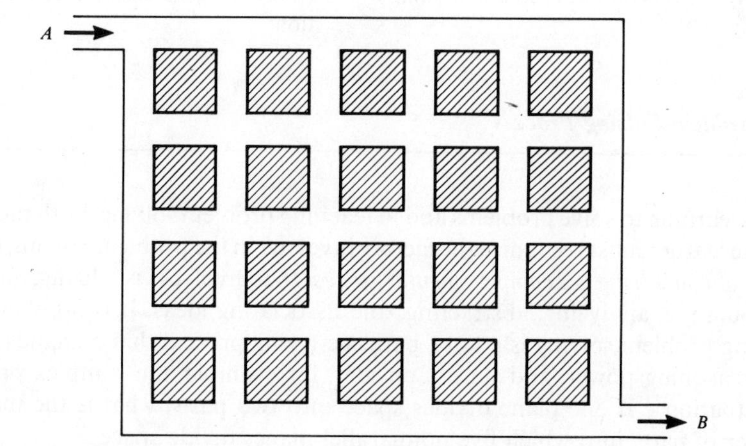

Figure 14.4

Problem solving is a means of stimulating intellectual curiosity. A problem, like a puzzle, brings into play inventive, creative responses. Most people have enough intellectual curiosity to enjoy solving puzzles and problems. The curiosity and interest aroused by the brain busters or challenge problems of many contemporary texts demonstrate the pleasure students get from solving challenging problems. Some teachers use a "problem of the week," posted on the bulletin board, to whet the intellectual appetites of their students. If every problem assigned to the mathematical students could be an intelligent challenge of this type, it is likely that the students would grow greatly in problem-solving skill.

Problem-solving skill probably develops in direct proportion to amount of practice in solving problems. For that reason, the more the better. Selection should, however, be done carefully to avoid too much frustration and wasted effort. A number of sources for problems are listed in the bibliography in Appendix C.

New knowledge is discovered through problem solving. The research of the scientist and the mathematician is largely one of solving problems. Even the steps of the scientific method closely resemble the steps in solving a mathematical problem.

Scientific Research	*Mathematical Problems*
1. Define the problem.	1. What do we want to find out?
2. Suggest a tentative hypothesis.	2. Estimate the answer or solution.
3. Plan a course of action.	3. Select methods, formulas, relationships.
4. Collect facts according to the plan.	4. Find all facts given and compute others. Experiment with the data and conditions. Search for a pattern.
5. Organize and analyze the facts.	5. Complete the analysis or computation.
6. Draw conclusions, state generalization.	6. State answer or generalization.
7. Apply the generalization to a new situation.	7. Apply the generalization to a new situation.

The Problem-Solving Process

In learning to solve problems and in teaching problem solving, both the teacher and the learner must recognize the factors involved in the problem-solving process:

Problem solving is a complex mental process that involves visualizing, imagining, manipulating, analyzing, abstracting, and associating ideas. It is no wonder that learning problem solving is a long, patience-trying process that demands motivation, reasoning power, and time. Consider, for example, the complex process in this situation, "If one plane divides space into two parts, what is the maximum number of parts into which five nonparallel planes divide space?"

Problem solving requires novel, original, unique, and varied responses. The best problem solver is likely to be a nonconforming, creative person who is highly flexible in his responses to the problem situation. The following problem, for example, is possible only if the problem solver switches to thinking in three dimensions: How can six matches be used to form four equilateral triangles if the side of each triangle has a length equal to that of the match?

Problem solving involves a background of knowledge of concepts, facts, and structures so that the learner can recall selected ideas and thus discern relationships and structure. Only when the learner can select the structure of the relationship or operations can he express the situation in symbols and thus find answers to questions. For example, in order to prove that the product of two negative factors is positive, the student needs to know the format for proof, the meaning of additive inverse, and the distributive property.

Problem solving often requires skill in reading as well as computing and ability to state associations. Through reading, the learner usually identifies the problem. To understand all the conditions of the problem situation, he must be able to read in an understanding manner.

Consider the following problem: "Suppose I have two American coins in my pocket whose total value is 60 cents. If one of them is not a dime, what coins do I have?" One must read carefully to discern that one of the coins is a half dollar and the "other one" is a dime. A learner also needs to compute accurately if he is to find correct answers to quantitative questions. He needs a storehouse of words, symbols, and formulas so that he can express associations in symbolic form.

Problem solving involves motivation and curiosity. Only when the problem is accepted as worth time and effort is it a problem for the learner. One way to get a student involved is to have him estimate the answer. Once he has done this, he has committed himself. It has been suggested that problem solving is one third mathematical knowledge, one third desire or curiosity, and one third common sense. Without question, a major factor in learning problem solving is intellectual curiosity. The satisfaction of this curiosity is usually sufficient to supply the incentive needed. This means that the selection of problems of interest to the solver is critical. At the same time, it should be noted that failure to solve assigned problems should not be a threat to the status or acceptance of the student in the mathematics class.

Many famous problemists have recognized that the form of the problem statement itself often increases the level of curiosity. Thus, Sam Loyd poses a problem:

> In describing his experiences at a bargain sale, Smith says that half his money was gone in 30 minutes, so that he was left with as many pennies as he had dollars before, and but half as many dollars as before he had pennies. Now, how much did he spend?[2]

[2] Martin Gardner, ed., *Mathematical Puzzles of Sam Loyd*, Dover, New York, 1959, vol. I, p. 13.

This problem illustrates one other feature of problem posing. For most students it is *not* necessary to take problems from their immediate environment. Often, students are less interested in grocery bills than in cannibals and missionaries, less interested in volumes of oil tanks than in walks through Königsberg.

A related technique used by clever teachers is to state problems in humorous terms. Sometimes interest is attained if class members are named as the persons involved in the problem.

Problem solving requires a procedure, an analysis, and a sequence of steps. In view of the complexity, the individuality, the originality involved, there is no generally accepted procedure or fixed sequence of steps for solving problems. The key points of emphasis are flexibility, originality, and variation. However, since it is the method of solution which is important, the problem solver should analyze his methods by indicating the procedure he used to solve a given problem.

Analysis of Word Problems in Algebra

To many teachers, problem solving means only solving the verbal problems that form an important part of arithmetic and algebra. Many attacks have been made on this particular genre of problem. Some teachers use geometric approaches—in particular, scale diagrams and graphs; others use tables—too often a specific type of table for each type of problem: mixture, distance, coin, age, or income. Even the geometric approaches are somewhat limited by the difficulty of translating many ideas—like price—into reasonable geometric terms.

It is much better to attempt to give the student a general method of attack on such problems; otherwise, he can solve only those problems that fit neatly into the specific patterns provided. It is possible to determine appropriate equations for solving the problem by a direct approach, utilizing steps like the following:

1. Represent what you wish to find, or something closely related to what you wish to find, by x.
2. Represent other unknown values in the problem in terms of x.
3. Seek relationships between the values represented by variables. These statements are often in the form of simple arithmetic relations like $d = rt$, $i = prt$, $S = C + P$.
4. Use these relationships to write an equation that can be solved for x.

Here is an example worked by this procedure, the steps in the solution corresponding to the steps in the given procedure.

Two sums of money totaling $10,000 are invested at 5% and 6% respectively. If the total income for a year is $540, how much is invested at each rate?

1. Amount invested at 5% is x (choice arbitrary!).
2. Amount invested at 6% is $(10{,}000 - x)$.
3. Income for a year is investment multiplied by rate of return, $i = pr$, or $.05x$ and $.06(10{,}000 - x)$. Total income is the sum of the individual incomes.
4. $.05x + .06(10{,}000 - x) = 540$.

Many students have difficulty with this procedure. Another helpful procedure (although more often taught to college calculus students than to high school students) involves the following steps:

1. Represent all the quantities unknown in the problem with letters x, y, z, etc.
2. Taking each phrase of the sentence of the problem statement, translate the problem into equations relating these letters. (Note that if there are n letters, you will need n independent equations. Occasionally, fewer equations will suffice.)

Example:

Jack is ten years older than his sister. In 3 years he will be twice as old. What are their ages now?

1. J = Jack's age now.
   S = Sister's age now.
   j = Jack's age in 3 years.
   s = Sister's age in 3 years.
2. $j = J + 3$
   $s = S + 3$
   $J = S + 10$
   $j = 2s$

Note that the use of extra letters avoids some of the complications of representation. When only the letters J and S are used, the resulting equation,

$$J + 3 = 2(S + 3)$$

confuses the student. When the additional letters are used, the resolution of this additional complication is merely postponed until the equation-solving stage. But most students can perform the simple substitutions and computations readily when expressed as equations.

Another approach that deserves special attention is that utilized in the University of Illinois Committee on School Mathematics (UICSM) program. This program attacks word problems by extending a procedure known to many teachers—e.g., "Write an equation relating feet, f, and inches, i." When using this procedure, teachers usually suggest that the student pick an arbitrary number of inches (or feet), carry out the conversion as he would in arithmetic, and then merely substitute the letters in the appropriate place in the computations.

Since 36 inches is 3 feet,

$$\frac{36}{12} = 3 \qquad \frac{i}{12} = f$$

Extended to problem solving, this method provides an interesting basis for equation writing. The student guesses a solution and checks it, recording his check carefully in equation form. (When he is first learning how to solve word problems, he may be asked to try several trial solutions.) Once he has the equation representing his check, he merely substitutes x for the guessed solution at each point it occurs.

Example:

A train leaves San Fransisco for Philadelphia at 50 miles per hour two hours before another train leaves Philadelphia for San Francisco at 60 miles per hour. If the two cities are 3,400 miles apart and the trains maintain these speeds, how soon will they meet?

Guessed solution: 40 hours

$$40 \cdot 50 + 38 \cdot 60 \stackrel{?}{=} 3{,}400$$

Substituting x for 40 and noting that $38 = 40 - 2$,

$$x \cdot 50 + (x - 2) \cdot 60 = 3{,}400$$

The last example suggests two other features of the solution of word problems —features that teachers should stress. The first of these is best summed up by the famous problemist Georg Polya: "We must teach guessing!" Too many teachers, when they emphasize set procedures, encourage students not to guess. Such teachers fail to recognize that the trial-and-error approach sometimes yields adequate answers. It is true that such methods may fail a student or may at least slow him down, but they should not be totally discarded. The second feature is what may be termed *reduction of parameters*. Reducing the complication of a problem often makes it transparent. In the "train" example, for instance, once both trains start, they approach each other at 110 miles per hour. The second train starts after the first has gone 100 miles. The solution, then, is the result of dividing 3,300 (3,400 − 100) by 110.

In this section on word problems, we have not stressed two features that many teachers are adamant about: form and labeling. For these two aspects of problem solution, we believe, individual preferences and needs should be followed. When students make careless errors because they fail to follow a set form or to label units correctly, they should be asked to modify their procedures. On the other hand, if such formal procedures appear to stifle the creative part of the problem-solving activity, they should be curtailed. To illustrate this, it may be noted that one of the best papers written by a student for a recent advanced placement examination of the College Entrance Examination Board (CEEB) was virtually illegible; another very fine paper was the work of a student who was neat almost to the point of compulsion. Here, as always, the individual makes the difference: Some students are able to display their organization while others are not.

A special word of caution is appropriate for the so-called word problems of algebra. Too often, specified procedures are worked out for each type of problem. These pattern solutions tend to restrict rather than expand a student's competence in problem solving. When a method is prescribed, the word problems are merely exercises in which the student practices a given procedure. It is better to let the student find several ways to solve a given problem and to let him make up original problems based on a given equation than to repeat several stereotype solutions.

Problems as Transformations

One basic skill needs practice, namely, the matter of translating the word sentences to open, symbolic sentences. Translation needs to be done in both directions. First, we should start with the algebraic statement and construct related word problems. Teachers who use this technique report that students produce varied and often quite sophisticated results—not only in constructing interesting word problems but also in solving word problems of the standard form.

Example:

Make up a word problem that would lead to the equations $xy = 50$ and $x + y = 15$.

This two-way approach to problem solving is also fruitful in that it gives students insight into the relationship between the real and the abstract worlds. As Figure 14.5 illustrates, problems from the real world are translated into the

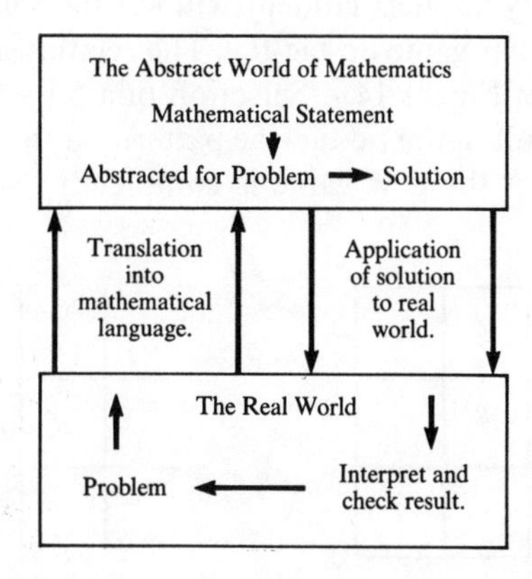

Figure 14.5

abstractions of mathematics. Then mathematical tools are used to produce an answer—the truth set. However, this answer has little meaning unless it in turn is applied to situations in the real world. The abstract world of mathematics provides generalizations to be applied in the real world. Thus, a complete problem solution should include the process, the generalization, and the application.

This concept of transformation is a basic one in mathematics, and its central role in this field should be pointed out to students. Many teachers fail to do this when the opportunity is presented. They miss the forest for the trees. For example, analytic geometry provides the opportunity to solve geometric problems by translating them into algebraic terms (algebraic geometry) and the corresponding opportunity to solve algebraic problems by translating them into geometric terms (geometric algebra). Many students never realize that this is the central characteristic of this subject; they are too bound up in its details.

Here is a way to demonstrate the structural relationship (isomorphism) that occurs here: Present to your students a simple two-person game played with nine cards numbered 1 to 9 as in Figure 14.6. Players alternate turns, choosing one card each time. All cards are displayed face up so that each player knows what cards his opponent has selected and what cards are available to him. The winner is the first player to draw three cards whose sum is 15. Thus, for example, a player who has drawn 2, 3, 6, and 7 wins because $2 + 6 + 7 = 15$.

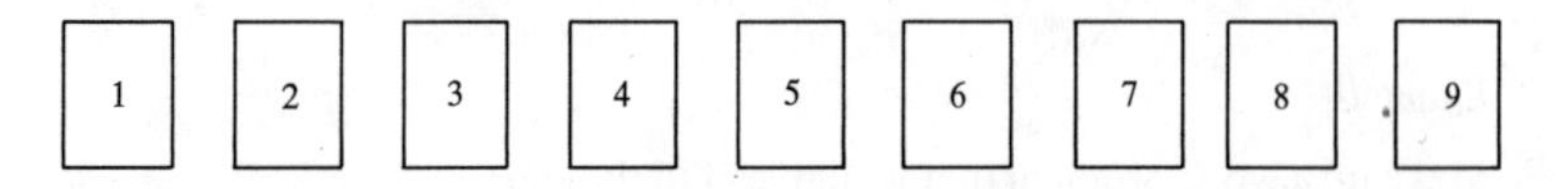

Figure 14.6

As soon as students have played a few games, ask them if they can suggest a strategy to win or at least to guarantee a tie. Most students will give only partial answers, but occasionally an alert student will see the relationship between this game and the more familiar game tic-tac-toe. The relationship is seen in the three-by-three magic square of Figure 14.7. Selection of a 5 by the first player and a 6 by the second would result in the tic-tac-toe pattern on the right. It is immediately apparent to students that the new game is completely "solved" by this transla-

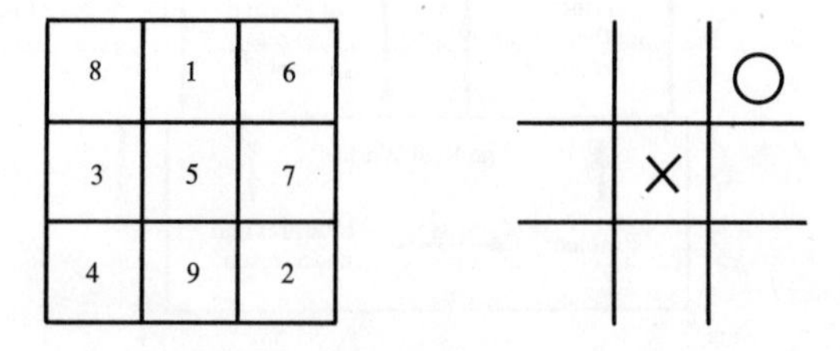

Figure 14.7

tion, for they need only apply the strategy of tic-tac-toe to their play. The power of transformation is evident.

Mind Sets

When teachers follow fixed solution patterns, there is real *danger that students will form mind sets*, which will be obstacles to their further development. Mind set (*einstellung*), or mind fix, is a form of undesirable transfer. Consider Figure 14.8 as an example. Note that there are two *the*'s; one is usually missed on first inspection.

Figure 14.8

Another example: At the time students are studying the Pythagorean theorem, ask them to find the measure represented by x in the diagram shown in Figure 14.9. Thinking of the single method, they fail to notice that the other diagonal of the rectangle is a radius. It is not necessary to use the Pythagorean theorem to find the value of x.

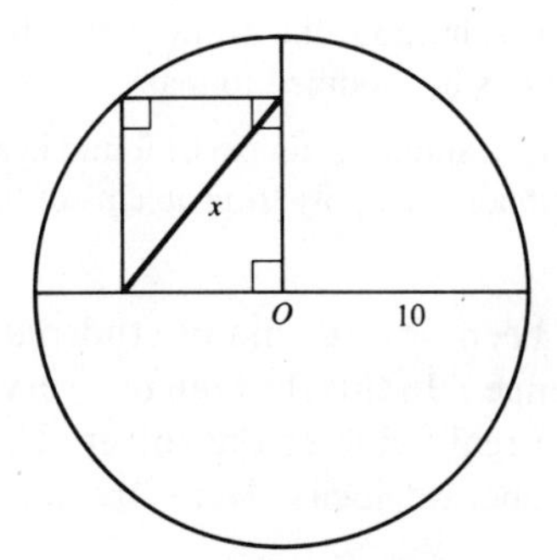

Figure 14.9

To avoid sets of this type, activities designed to force students to keep their minds alert should be continually utilized. For example, students who have solved several equations of the form

$$\frac{x}{a} = b$$

should be asked to solve an equation of the form

$$\frac{a}{x} = b$$

All teachers should be alert to these procedural fixations.

Suggested Approaches to Problems

Often problem solvers get stuck because they have fixed on a single procedure or approach. Many authors have suggested methods to approach problems. From these the following list of suggestions has been gathered:

Identify the question or problem. Do you understand the wording of the problem? (Is there a trick?) Is the problem of a known type? What is given?

Search for related ideas. What type of problem is it? What are the conditions (parameters)? What is an analogous problem?

Simplify the conditions. Can you simplify the conditions by using small numbers or fewer variables? Can you get more information by performing some computations? What details are not needed? Can you use a diagram or an equation to represent the conditions?

Search for a strategy. Can the data be organized into a pattern? Does trial and error point to a solution? What is it about the problem that is different from problems previously solved? Can the conditions be modified to make a simpler problem?

Use reference tools. Can a similar problem be found in a textbook? Do tables provide assistance? What formulas or theorems apply to problems of this type?

Once a problem has been solved, many students and teachers leave it to advance to "another challenge." In this they fail in a very important way. Problems that have been solved have real value to the solver. They are the source of additional problems as they generate ideas. Here are some questions that may be applied at this point:

Can this result be applied to analogous situations?

Under what conditions is the problem solvable? Unsolvable? Meaningless? Trivial?

Is there a generalization of the result?

Is there a more sophisticated solution?

What type of reasoning was involved?

When and why did I have difficulty? How could this difficulty be resolved in the future?

In this regard, it is important to point out to students that generating problems is an art well respected among mathematicians. Encourage them to suggest problems even when the solution is either trivial or too difficult. Developing problems requires careful thought, thinking that is a contribution to growth in problem solving as well. The best time to provide opportunity for such problem generation is when another problem has just been solved.

Steps in Solving Problems

Although there is no generally accepted sequence of steps for solving mathematical problems, psychologists suggest that some common stages or steps operate in most problems.

1. *The problem solver must first know exactly what the problem is.* The situation, the question, or the problem is usually presented in a written statement. To become oriented to the problem, the solver should read and reread the statement. The following questions or activities add meaning to the problem:

What information is given?

What are you trying to find out?

State the problem in your own words.

Write a similar problem about people and places familiar to you.

The following problem illustrates how important it is to relate the ideas about straight lines and directions to geometric relations between lines and angles.

It's as far from *Witt* to *Pitt* as from *Kitt* to *Mitt*.
It's as far from *Sitt* to *Bitt* as from *Ditt* to *Mitt*.
Mitt is on a straight line north from *Sitt* to *Pitt*.
Mitt is on a straight road east from *Witt* to *Bitt*.
Kitt is 8 miles north of *Witt* and 8 miles west of *Pitt*.
Ditt is 6 miles south of *Bitt* and 6 miles east of *Sitt*.
How far is it from *Witt* to *Sitt?*
How far is it from *Pitt* to *Bitt?*

When a drawing is made, it shows that the answers are obtained by means of the Pythagorean theorem.

2. *The problem solver relates the problem to a familiar idea* or a previously solved problem. He must search his memory to recall ideas, facts, assumptions, theorems, formulas, or experiences that are related to the problem. He must ask himself whether or not he has solved an analogous problem. He must reduce the situation to simpler conditions or solve a simpler related problem. He must consider what happens if given conditions vary, and he must estimate the answer.

Here is a problem that requires the solver to search his memory for conditions on the earth's surface:

> An explorer walks 1 mile south, then walks 1 mile east, then turns and walks 1 mile due north. He finds himself back where he started. At what locations (more than one) on the earth's surface is this possible?

Only when he recalls how latitude and longitude are related to directions and locations does he discover the locations at the North Pole and near the South Pole.

3. *The solver must search for a strategy by identifying the structure of the problem.* He must identify the known facts, the conditions, and the variables. By organizing the facts given, he should identify a pattern. Then he selects a search model to represent the structure of the problem—a sketch, an equation or inequality, a graph or flow chart represents the elements of the problem in symbols, and determines what method of proof applies.

What digits may be substituted for the letters in the exercises below? In each separate exercise, each different letter represents a different digit.

(a) addition	**(b) subtraction**	**(c) multiplication**	**(d) division**
SEND	SPEND	SEAM	bfb
MORE	MORE	N	ab)cdeeb
MONEY	MONEY	MEANS	ceb
			gge
			gch
			ceb
			ceb

What strategy should be used for these exercises? These problems require an exploration of many possible combinations to arrive at the equality given. The essential structure here is the specific algorithm.

In the following problem, a sketch would quickly give the clue:

> A chessboard has 64 squares. We have 32 dominoes, each of such size that it covers exactly two chessboard squares. The 32 dominoes may then be used to cover all 64 squares. Suppose that we cut off two checkerboard squares, one at each of two

> diagonally opposite corners of the board. Discard one domino. Is it possible to to place the 31 dominoes on the board so that the remaining 62 chessboard squares are covered? Show how it can be done or prove it impossible.

4. *The solver should use the search model (determined in step 3) to find the answer to the question.* He must perform the computations involved, complete the deductive proof or find the solution set for the equations or inequalities. He must check the answer to see whether the results satisfy the conditions given and state a complete answer to the question of the problem. In this way the search model (sketch, equation, concrete representation) bridges the gap between what is given and what is required.

Make an analysis of this problem:

> Each of two boys had 30 balloons for sale. One boy sold his at the rate of two for a nickel and the second boy at the rate of three for a nickel. At the end of one day their receipts were 75 cents and 50 cents, or $1.25 in all. The next day the boys decided to combine their efforts. They pooled their 60 balloons and sold them at the rate of five for a dime (two for a nickel plus three for a nickel). Upon counting the receipts at the end of the day, they found that they had only $1.20. Why did they lose 5 cents by their merger?

What information about rates is the key to this problem?

5. *The problem solver should interpret the results in the form of a generalization.* He should apply the solution to situations in which the conditions are changed, determine when a solution is impossible, meaningless, or trivial, and write a generalization in terms of a formula, theorem, or principle.

What is the generalization for this problem?

> Philip and his wife, Mary, both work at night. Philip is off duty every ninth evening; his wife is off duty every sixth evening. Philip is off duty on this Sunday evening; Mary is off duty the following Monday evening. When (if ever) will they be off duty the same evening?

What relationship between 3, 7, and 9 is involved in this problem? Under what conditions will they have the same night off?

6. *The problem solver should analyze the method of solution.* He should write the sequence of steps in a logical order, indicate the process whereby information was generated, and identify the type of reasoning involved.

The complete analysis of this problem may be a good example of finding a key idea that simplifies the problem. Try finding several methods of solving it.

A commuter always arrives at his subway station at 6 P.M. each day. His wife meets him at the station to drive him the remaining distance home. One day he arrives at 5 P.M. and begins walking home. His wife meets him on her way to the 6 P.M. train. He gets into the car and they drive home, arriving at home ten minutes earlier than usual. Assuming that the wife drives at a constant rate on the same route, for what length of time did the husband walk before he was picked up?

Programmed Instruction and Problem Solving

Those familiar with programmed instruction may be surprised to see this technique associated with problem solving. There are, however, two very important connections that may be made between the two. The first may be seen best-developed in books by Polya and in the Hungarian Problem Books.[3] All of these use the method of successive hints: a problem is stated, and the student is urged to seek a solution without help. Failing this he is provided with a first hint, which may help him to organize and direct his thinking into productive channels. Subsequently, more leading hints are also provided. Finally, the complete solution is given. At any point, the student may progress to the solution and skip the subsequent hints to check his result. On the other hand, if he was unable to solve the problem at any stage, he was at least forced to examine the problem in greater detail as he progressed through the program. The student who did not solve the problem might be directed to a similar problem in order to help him reinforce his procedure.

Consider, for example, the problem cited earlier: spelling the word "mathematics" in the array of letters (page 237). The following hints might be given:

Hint One: Are you familiar with Pascal's triangle?
Hint Two: In how many ways can you get to each of the A's in the second row?

Note that neither hint takes the student to the answer, but each gives him more direction.

A second application of the idea of programmed instruction comes after a specific problem is solved. Finding the solution to a particular problem, the student should be urged to generalize his result and to develop an algorithm for solving problems of this particular genre. To many who use computers regularly, this programming is the real solution to the problem, anything less having no real application. Such an algorithm may take the form of a carefully structured pro-

[3] Especially Georg Polya, *Mathematical Discovery: On Understanding, Learning and Teaching Problem Solving* I, II, John Wiley and Sons, New York, 1962, 1964; *Mathematics and Plausible Reasoning* I, II, Princeton University Press, Princeton, 1954; *Hungarian Problem Book* I, II, Random House, New York, 1963.

gram, a flow diagram, or a work sheet. In each case, however, the design should be aimed at a layman unfamiliar with the processes involved except for arithmetic computation.

Computer-mediated Problem Solving

As an example of how a problem is attacked by a student with access to a computer, here is a problem assigned to a high school student. He solved the problem by means of a classroom teletype contact with a time-shared computer. Compare this work with the typical format for solving this problem to observe the insights gained.

Problem:

Joe Doakes is looking for a job that pays well. Opportunities Unlimited offers him a starting salary of $1,000 a day with a $100-a-day increase in wages. Double-or-Nothing, Inc., offers him a penny for the first day but doubles his wage each day. Which is the better offer? That is: (1) Would the Double-or-Nothing daily wage exceed the Opportunities Unlimited within a year? (2) If so, after how many days? (3) If so, how long before the total wages were greater?

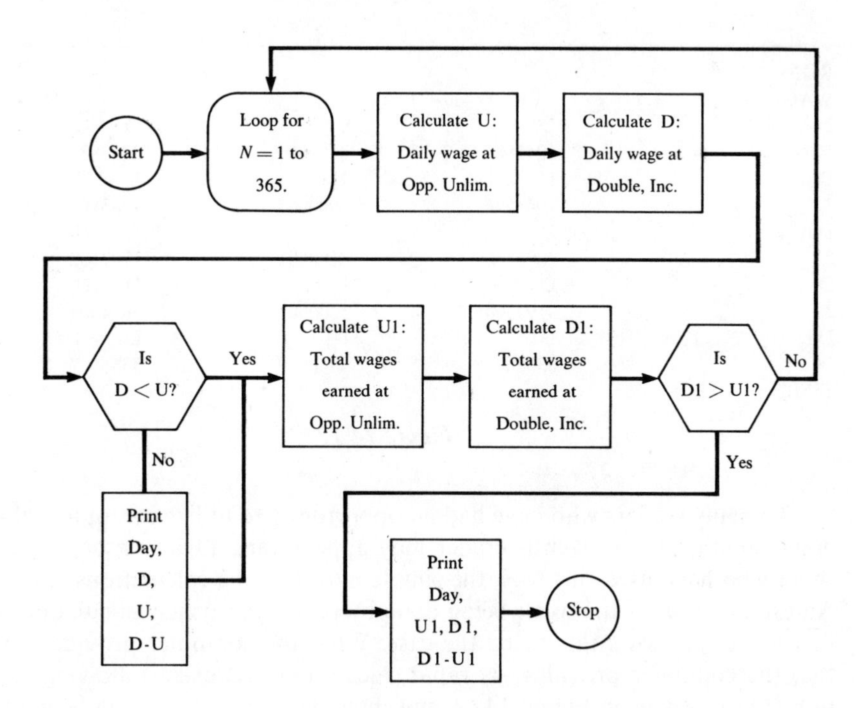

Figure 14.10

First the student prepared a flow chart (Figure 14.10). Next he translated the flow chart into an algorithmic language (BASIC) and ran the program (Figure 14.11) on the machine, which produced the solution shown in Figure 14.12.

```
WAGES              13:56              Fri 06–17–73
10 LET  U1 = 0
20 LET  D1 = 0
30 FØR  N = 1 to 365
40 LET  U = 100000 + 10000 *(N – 1)
50 LET  D = 2 ↑ (N – 1)
60 IF  D < U THEN 100
70 PRINT  “DAY”, “D”, “U”, “D – U”
80 PRINT
90 PRINT  N, D, U, D – U
100 LET  U1 = U1 + U
110 LET  D1 = D1 + D
120 IF  D1 > U1  THEN 140
130 NEXT  N
140 PRINT  “DAY”, “D1”, “U1”, “D1 – U1”
150 PRINT
160 PRINT  N, D1, U1, D1 – U1
170 END
```

Figure 14.11

```
RUN
WAGES        13:57        FRI 06–17–73
Day                       D              U              D – U
20                        524288         290000         234288
Day                       D              U              D – U
21                        1048576        300000         748576
Day                       D              U              D – U
22                        2097152        310000         1787152
Day                       D              U              D – U
23                        4194304        320000         3874304
Day                       D1             U1             D1 – U1
24                        8388607        4840000        3558607
TIME     1 SECS.
```

Figure 14.12

To many readers who have had no opportunity to utilize computers, the foregoing solution by a seventh grader may appear very advanced or unusual. To those who have used this tool, the simple nature of the calculations is apparent. An examination of the work being done by paper-and-pencil calculation for the first few days shows this to be the case. What the computer provides—and all that the computer provides—is rapid calculation and exact following of directions. (That is why, in Figure 14.12, unwanted data for days 21 and 22 and part of 23 are printed.)

On the other hand, the student must be able to analyze the problem carefully; translate the problem into arithmetic calculations; provide a form for sequencing these calculations; give a reasonable interpretation of the data; and spot errors in his analysis, interpretation, or attack on the problem.

Note that this is not the most sophisticated possible program, nor is the print-out as simple as it might have been. Each of these questions offers a basis for further student thought. Progression sum formulas have not been used for either the arithmetic or geometric series; instead the sums are calculated directly (subsequences 100, 110).

In this type of problem solving, the student's thinking is focused on the data from beginning to end. This study often helps him to generate extensions of the problem. To some extent, he has already done so here, printing how much more he would earn as well as the days required. It would also have been possible to print weekly status reports or even the effects of a progressive income tax on the wage comparisons.

This type of activity also forces the student to interpret both question and solution carefully. In this case, for example, he had to use cents as his basic wage unit even though dollars would have sufficed for the first alternative.

Group Attacks on Problems

Every teacher would like to raise the sights of his classes, to encourage student work on more significant problems. But everything seems to work against this. Presented with nonroutine problems, students merely give up. Where no one in a class can make progress, the teacher retreats. He either demonstrates solutions of more difficult problems himself, or he omits them from his program. Textbooks generally reflect this: It is extremely difficult to find textbook problems that require more than a minute's work for solution!

One way to begin to meet this problem is by organizing the students into small work groups. Four on a team seems to be an effective number. Larger groups are unwieldy; smaller groups provide too little interaction. In a study of the results of such teamwork, William Bailey[4] of New York State University College at Buffalo found that students in teams make much more progress on difficult problems than students working alone and that after working in teams students make greater progress on difficult problems than they did before they had this experience. In other words, the experience of working in groups makes the students realize that they can do more than they thought.

The dynamics of team effort are most interesting to observe. Leadership usually develops with brighter students more often taking this role, but when

[4] William H. Bailey, Jr., *A Study of Group Reaction to and Productivity on a Mathematical Task Involving Productive Thinking*. Unpublished Ed.D. thesis, SUNY Buffalo, 1971.

students of nearly equal ability work together one student still tends to be dominant. Generally, all students participate. It is more difficult to justify failing to help your team than to justify failing to help yourself. There is much opportunity to test each other's ideas since the teacher's authority is withdrawn. In fact, the kind of student interaction that most teachers would like to have develop in their classrooms becomes part of the group exchange.

Professor Sherman Stein of the University of California at Santa Barbara and a group of coworkers have developed a complete secondary-school program with team learning in this format the basis for their approach.[5]

Many of the illustrative challenge problems of this chapter could be assigned in this format. One problem that is a useful "starter" for this kind of activity is one suggested by Martin Gardner.[6] The 28 dominoes in a double-6 set are arranged at random to form a seven-by-eight rectangle. The number of spots are recorded as in Figure 14.13. Typical domino tiles are shown in Figure 14.14. The problem: Find the positions of the individual dominoes in the array.

4	1	3	4	3	5	3	3
5	0	4	1	1	5	0	2
0	1	2	0	2	1	6	2
2	5	1	0	6	4	0	0
5	3	5	6	6	6	5	3
6	4	3	0	2	1	5	6
6	2	3	2	4	1	4	4

Figure 14.13

To solve this problem, students must not only apply new patterns of thinking but must also organize their method of attack. Having a set of dominoes available is useful but not necessary. A word or two about the uniqueness of each combination on a one-by-two domino tile may be offered if students are completely unfamiliar with the game.

[5] Houghton Mifflin, Boston, in press.

[6] See "Mathematical Games," *Scientific American*, December 1969, pp. 122–124, for a more complete description of this and related problems.

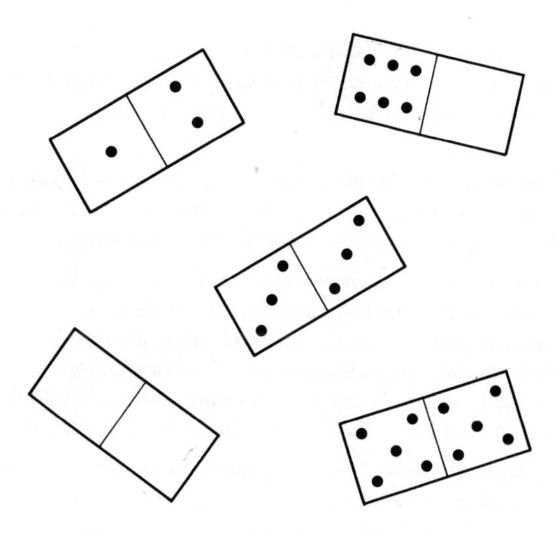

Figure 14.4 Typical Dominoes

Key Ideas in Teaching Problem Solving

Provide a wholesome emotional climate for learning problem solving. Allow ample time for thinking, analyzing, and experimenting. Be receptive to questions. Reduce hostility and fear. Be patient with the student who cannot solve problems.

Establish and maintain students' motivation. Emphasize the importance of learning to solve problems. Provide problems in which some success is assured for each learner. Prepare the learner for necessary difficulties and frustrations. Arouse intellectual curiosity by using puzzle problems and "brain busters." Make reasonable assignments and provide ample time. Concentrate on a few problems and treat them leisurely and thoroughly. Find interesting, unusual, relevant problems.

Provide ways of increasing students' understanding of a problem situation. Show students how to read and reread problems. Help students restate problems so that conditions become clearer and the question more reasonable. Give the learner an equation or geometric relationship and ask him to write a problem that is based upon this idea. Ask questions to ensure that the learner understands the statements, the vocabulary, and the type of problem involved. Have the learner identify key terms or ideas or break the problem down into simpler subproblems. If the solver does not know where to begin, encourage him simply to write the facts of the situation. Have the learner write a related problem, simplify conditions, or reduce numbers. Give a general question and let the learner specify the variables or conditions involved.

Emphasize flexibility and variety in solving problems. Do not prescribe a rigid step-by-step procedure or format. Suggest that students change perspectives when in difficulty. When students make no progress, encourage them to change the search model. Provide some problems with insufficient data and others with extraneous data. Encourage using several methods of solving the same problem. Relate problem solving to solving mystery or detective stories and to criminal investigation.

• *Give instruction in establishing a search model.* Use flow charts to illustrate a solution. Use diagrams, models, or sketches to identify structure of the problem. Establish the variables of the problem. Then use symbols and write equations or inequalities for the relationship involved.

Show the learner how to ask himself questions. Suggest questions like these: What facts are given? What are you trying to find out? What ideas have you studied that are related to this situation? What problems have you solved that are analogous to this one?

Emphasize the method of solution rather than the solution. Instead of assigning numerous problems, ask the student to find several methods of solution for each problem given. Give credit for each part of a correct method rather than only for the right answer. Use test items that demand demonstration of the method of solution rather than merely the right answer. Give opportunities of showing methods of solution that do not require the performance of laborious computations. Provide for an analysis of the analogies used.

Encourage experimentation, trial and error, estimation, intuition, guessing, and hunches to suggest a method of solution.

Concentrate heavily on reading skills. Ask questions designed to identify failure to understand the problem itself. Have students read problems aloud. Have a difficult problem reread.

Provide problems at frequent intervals, so that students get adequate practice in solving problems.

Promote analysis, organization, and communication skill by having students state or write their solutions in logical and orderly form. Outline the logic or structure of the problem. Suggest possible analyses, flow charts, transformations.

Use problem situations to discover new mathematical concepts, principles, or relationships.

Use problem situations as a basis for practice and as a substitute for isolated drill exercises.

Learning how to solve problems can be one of the most satisfying though difficult experiences which we can give our students. It is also probably the most significant learning that we can direct. The key to problem solving seems to be flexibility and reflective thinking. The result of successful learning should be the ability to explore all fields of knowledge independently. To attain this we should emphasize the structure of the problem rather than the computational phase. Then we should emphasize the method of solution rather than the answer. The specific answer to a given problem has limited use. It is the method that can be applied elsewhere.

Learning Exercises

1. What is the difference between solving a problem and learning a concept?
2. State a problem situation that could be a means of practicing computational skill.

3. State a problem situation which could be a means of learning a new concept.

4. State a problem and analyze the background of knowledge and skill that the solver is expected to have.

5. Review an algebra text to determine the approach to word problems. What modifications of their approach would you prefer?

6. Read Part I of Georg Polya's *How to Solve It* (see references). Make up some questions like Polya's that would help students to solve his problem 4 (page 234): "To number the pages of a bulky volume, the printer used 2989 digits. How many (numbered) pages has the volume?" See also his comments on pages 238 and 243.

7. Solve the domino problem on page 254. What were some of the techniques that you used? Do you consider this a mathematics problem? Why or why not?

8. Of two barrels, the first contains water and the second, wine. A dipperful of water is taken from the first and mixed into the second. Then a dipperful of the mixture of wine and water is returned to and mixed into the first. Now consider the resulting mixtures: Is there more wine in the water or water in the wine? Solve this problem. How does it illustrate *einstellung?*

Suggestions for Further Reading

Arnold, William R., "Students Can Pose and Solve Original Problems," *Mathematics Teacher*, 64, 4 (April 1971), 325–327.

Botts, Truman, "Problem Solving in Mathematics, 1," *Mathematics Teacher*, 58, 6 (October 1965), 496–500.

———, "Problem Solving in Mathematics, 2," *Mathematics Teacher*, 58, 7 (November 1965), 596–600.

Brown, G. W., "Improving Instruction in Problem Solving in Ninth Grade General Mathematics," *School Science and Mathematics*, 64, 5 (May 1964), 341–346.

Dahmus, Maurice E., "How to Teach Verbal Problems," *School Science and Mathematics*, 70, 2 (February 1970), 121–138.

Garfunkel, J., "The Recursion Formula," *Mathematics Teacher*, 63, 2 (February 1970), 121–125.

Hadamard, Jacques, *Psychology of Invention in the Mathematical Field*, Princeton University Press, Princeton, N. J., 1949.

Henderson, Kenneth B., and Robert E. Pingry, "Problem-Solving in Mathematics," in *The Learning of Mathematics: Its Theory and Practice*, Yearbook 21, National Council of Teachers of Mathematics, Washington, D. C., 1953, pp. 228–270.

Kinsella, John J., "Problem Solving," in *The Teaching of Secondary School Mathematics*, Yearbook 33, National Council of Teachers of Mathematics, Washington, D. C., 1970, pp. 241–266.

Mercer, Gene B., and John R. Kolb, "Three Dimensional Tic-Tac-Toe," *Mathematics Teacher*, 64, 2 (February 1971), 119–122.

Polya, Georg, *Mathematical Discovery*, vols. I and II, John Wiley & Sons, Inc., New York, 1962.

———, *How To Solve It*, Doubleday & Co., Garden City, N. Y., 1957.

Snyder, Henry D., "Problem Solutions That Ask Questions," *School Science and Mathematics*, 66, 4 (April 1966), 373–376.

Trimble, Harold C., "Problems as Means," *Mathematics Teacher*, 59, 1 (January 1966), 6–8.

Wilson, James W., and Jerry P. Becker, "On the Solution of a Problem," *Mathematics Teacher*, 63, 4 (April 1970), 293–295.

Developing Positive Attitudes and Creativity 15

It is widely recognized that students need to learn something besides facts and skills. They need to develop desirable attitudes and creativity—goals of the affective domain, which are much more difficult to attain than those of the cognitive domain (concepts, facts, and skills).

The Role of Attitudes in Learning Mathematics

Attitudes are fundamental to the dynamics of behavior. They largely determine what students learn. The mathematics student with positive attitudes studies mathematics because he enjoys it, he gets satisfaction from knowing mathematical ideas, and he finds mathematical competency its own reward. Attitudes determine not only his willingness to study mathematics but also his use of mathematics as well. Since our goal is to provide students with knowledge that they can and will use, we should be especially sensitive to attitudes.

The development of positive attitudes toward mathematics is a fundamental concern of the mathematics teacher for a number of reasons:

No student can be forced to learn mathematics that he does not want to learn. He may superficially satisfy you that he is learning by adhering minimally to your classroom demands and standards, but he will carry away virtually nothing unless he is interested in doing so. In this regard all teachers should consider how deeply involved students—even low achievers—become in their hobbies, whether they be stamp collecting, model building, scouting, or sports. Students would make rapid progress in mathematics if they could equal their knowledge of mathematical principles with their knowledge of baseball rules. Such knowledge, which often includes a vast store of memorized detail, is motivated by strong intrinsic attitudes.

Even if students learn the mathematics of a given text, the primary concern of continuing learning is lost when students do not develop positive attitudes. The mathematics of tomorrow cannot be taught today; so an abiding interest in mathematics should be instilled to encourage learning in the future.

Vocational choices are largely dependent on attitudes. Mathematics teachers close doors for students when they allow them to develop poor attitudes toward such an all-pervasive subject as mathematics. Even though mathematics may not be required for college entrance, it is required for many college sequences.

Application of mathematical ideas in large measure depends on a positive attitude. We tend to remember favorable, pleasant experiences and block out the unpleasant. If this leads to blocking out mathematical concepts, they will not be available when needed. Thus attitudes are a major factor in retention.

Many people take pride in professing ignorance of mathematics. Few adults admit that they are poor history students, but the parents of many pupils announce the fact that they "never did understand math." It is in this negative environment that the teacher must work. Hence a major aspect of attitude development is to "unlearn" negative attitudes.

Finally, there is a selfish aspect of attitude improvement: *Positive attitudes reflect favorably on the teacher.* They form an excellent basis for a teacher's rating by students, parents, and colleagues. The mathematics teacher who inspires his students and whose students are enthusiastic about his class is the teacher who is in line for salary increases, institute recommendations, and—best of all—the rewarding experience of success.

The Role of Enrichment in Attitude Building

An effective way of attaining the affective goals of instruction is an enrichment program that is part of the regular day-to-day instruction. This planned enrichment gives breadth and depth to mathematics learning. This program should capitalize on the varying capacity of each student to visualize, to use intuition, to pursue intellectual curiosities, and to be creative. It should encourage each student to develop his mathematical ability, to consider mathematics as being worth studying, and to extend independently the mathematics known to him. Although this program has special implications for gifted students, it should be appropriate for all levels of mathematical ability. In the following discussion, the word "enrichment" is used in its broad sense, not in the narrow sense of "extra materials," to which it has too often been assigned by the teacher.

Consider the role of enrichment in our own daily lives. We frequently hear the expression "he has lived a rich life" or "her life is drab and routine." What are the experiences that contribute to an enriched life? We would likely agree to include the following:

Success in our work and day-to-day activities. Since we spend a major portion of our time on the job, the joy and satisfaction of success is one of the greatest factors for an enriched life. This success gives us a feeling of security that is so essential for participation in a variety of activities.

A *sense of values* that gives meaning to our lives. A life without a purpose or a goal is an empty existence.

A variety of *social activities* at home and elsewhere, such as those we enjoy through travel, conversation, and group organizations.

Participation in *cultural activities*, such as music, art, drama, or literature.

Intellectual activities, such as reading, listening to lectures, and participating in discussions that satisfy our curiosity about life around us.

Recreational activities, such as hobbies and sports.

Creative activities, such as craftwork, painting, photography, writing, gardening, and sewing.

Whether or not one can live an enriched life that involves these activities depends upon the ability and resources of the individual and the resources of the community in which he lives. Enrichment of some type is usually desired by all age groups and at all socioeconomic levels.

In exactly the same way, the learning of mathematics should involve satisfying experiences that enrich the learning process. Activities are needed to satisfy the learner's need for enriched learning. It is the learner's experiences, not the teaching activities, that satisfy this need. Thus, the enrichment of mathematics instruction should provide the learner with the following experiences, which parallel the enrichment factors of daily living:

Success in learning the ideas, skills, and structure of mathematics. Success in the understanding and mastery of mathematical concepts is developed through the learner's participation in discovery, discussion, illustration, problem solving, practice, and application. Thus, instruction that builds confidence, independence, and security in this way enriches the learning of mathematics. This means that adequate time must be devoted to the mastery of ideas. Our task as teachers, then, is to "uncover" material rather than "cover" a specific sequence.

A sense of the value in learning mathematics gives meaning to learning activities. Learning without a goal or purpose is empty activity. Discussions of the role of mathematics in our society, the power of mathematical analysis, the applications of mathematics, and the mathematics needed in different vocations enhance the importance of learning mathematics.

A variety of learning activities can be planned by the use of teaching aids, historical sidelights, dramatic topics, laboratory lessons, or excursions. In these learning activities, there should be much interaction between students as they discover, discuss, and apply mathematical ideas.

The cultural aspects of mathematics are made evident, so that the learner can look upon it as a human invention important in itself. Mathematics has aesthetic aspects—for instance, in the properties of symmetry and reflections—comparable to music or art. And, in turn, music and art have mathematical aspects of their own.

The relation of *intellectual curiosity* to mathematics is unique and exciting. Mathematics is one of the greatest intellectual inventions of the human mind. Its logic, its abstractness, its paradoxes, its study of patterns, its unsolved problems can satisfy intellectual appetites at many levels.

Recreational activities involving mathematics are plentiful. These include tricks, puzzles, games, and stunts. Mathematics is a hobby for many people who are not mathematicians. Unusual properties of geometric figures and number operations provide such entertaining interludes.

Creating new mathematics can be made an intriguing activity. Constructing a model, planning and producing a mathematics assembly program, building a mathematics exhibit—all of these are creative activities. Originality may be nourished by new inventions, an elegant new proof, or a new mathematics structure.

The student's participation in enrichment activities depends upon the ingenuity and background of the teacher, upon the ability and interest of the individual, upon the time available, and upon the resources of the school. When a student's curiosity, concern, and creativeness are aroused, learning becomes automatic.

Student Basis for Positive Attitudes

A positive attitude is indicated when a person moves emotionally toward a subject, activity, or individual. It is behavior that indicates that the person is trying to increase his contact, his experiences, or his knowledge of the target subject. For example, consider the approach responses of a sports car fan:

Talks a great deal about sports cars.

Buys and reads magazines, books, news reports about sports cars, often bringing his materials to the mathematics class to read.

Knows the qualities of sports cars and can recite a great number of facts and figures.

Watches every TV program involving sports cars.

Spends time at a sports car garage and tries to get a job working on sports cars.

Wants to organize a sports car club in school.

Finds friends and even teachers who are interested in sports cars.

In other words, a positive attitude determines how a student will spend his time and money, what he reads and talks about, what facts he considers worth knowing, and even with whom he likes to associate. Thus, the application to learning mathematics is obvious.

It is very important for the classroom teacher to assess his students in order to determine established attitudes toward mathematics and to assure himself of a sound basis on which to build a program of attitude improvement and motivation. While many individual students will bring to the classroom very specific acceptable and unacceptable interests, there are many attitudes that are common to the major segment of the student population. To become involved in emotional dynamics, the classroom teacher must be aware of these and be sensitive to quantitative differences among students. These basic desires include:

Wanting to avoid embarrassment or punishment
Wanting to win approval of teachers and parents

Wanting to gain confidence in his own ability
Wanting to succeed in progress toward goals acceptable to him
Wanting to attain approval of his peers individually and as a group
Wanting to be secure

Many teachers fail to recognize the difficulty they face in developing real sensitivity to these strong drives within their students. It is not easy to be sensitive to the subtleties of student desires and motives. How often has the remark been passed in the faculty room that a student showed quite unexpected concern for something or someone? The fact that such feeling was quite unexpected means that the teacher did not really know his student and that he had therefore underestimated him.

It should be noted that all the desires listed are basically oriented toward the development of positive attitudes. In the beginning they are working for the teacher. Too soon, however, they can all be misdirected. For example, the desire to avoid embarrassment often turns a bright girl into an underachiever when she is criticized by her friends for being the teacher's pet.

This leads to a special point that should be stressed about attitudes. The mathematics teacher cannot choose to avoid the problem of developing positive attitudes toward mathematics. *Whether the teacher likes it or not, each student responds emotionally to him.* Students react strongly, acquiring or rejecting on the basis of this response the attitudes, values, and appreciations of the teachers.

No matter what topic is being taught, much concomitant learning and change in attitude are taking place. For example, when being taught how to solve a mathematical problem, a student may be learning either to dodge responsibility or to be cooperative, to maintain his integrity or to cheat, to trust the teacher or to lose respect for him.

Attitudes may not be taught systematically or directly. Teaching is not a mechanical operation. It is not even a science, although science too has its artistic side. Teaching attitudes is more like painting a picture, playing a musical selection, planting a garden, or writing a letter to a friend. The teacher must put all his heart into his instruction. It is human warmth that adds the emotional vector needed in the development of positive attitudes toward mathematics.

Ways to Build Positive Attitudes toward Mathematics

The basic way to develop positive attitudes toward mathematics is to provide pleasant experiences for the learning of mathematics. We are all willing to give time, effort, and money to those activities that we enjoy. We avoid doing those things that are unpleasant. We usually find that our favorite activities are those that we do successfully and comfortably and that bring us in contact with people we enjoy.

The implication for the mathematics teacher is very clear. However, do not assume that learning mathematics is a task that is unpleasant. The sports enthusiast wanted to work on sports cars, wanted to read and learn facts about sports cars, and wanted to discuss sports cars so that he could learn more and could share his own knowledge with others.

The teacher's appreciation of mathematics as an important, dynamic, remarkable subject must be real and deep, his attitude toward students must be sympathetic and understanding, his interest in learning must be great, his enthusiasm for teaching sincere. If the teacher's attitudes or interests are less favorable or the same as those of the student, no transmission of enthusiasm can take place.

Once the teacher has set his own standards high, he must still establish himself as a person who has the respect and esteem of his students. Maintaining the students' esteem in the face of their constant judgmental response is a sobering problem for the teacher. Students are quick to sense the smallest insincerity just as they note the slightest insecurity. It is extremely important that the classroom atmosphere be friendly, accepting, and supportive, even when it is demanding and challenging. A spirit of security, enjoyment, and loyalty should be the basic goal of classroom organization. The teacher should make his students feel that his attitude will still be friendly regardless of the success or failure of the students' efforts. If we want students to think for themselves, we must allow them to try out their own ideas and answers.

Some of the really important ways in which teachers influence their students' attitudes have to do with pleasant communication habits: the teacher's voice inflection, the way he looks at students, his responses, or even his failure to respond. The classroom itself should be attractive. It should contain books, pamphlets, pictures, and displays reflecting an intellectual atmosphere and providing a proper setting for learning.

Here are some specific examples of things to do to build positive attitudes toward mathematics:

To develop *appreciation* of the elegance, power, and structure of mathematics: (1) Emphasize the nature of mathematics and how it is a model for reasoning—often, we forget to teach what mathematics *is.* (2) Illustrate the harmony, symmetry, and beauty of mathematical patterns. (3) Include current applications of mathematics.

To nourish *curiosity* in mathematical ideas: (1) Give experiences in discovering new ideas. (2) Make each lesson have significance for the learner. (3) Include enrichment topics such as topology, computer programming, and game theory. (4) Assign open-ended questions and problems for thinking and exploring.

To build *confidence* in and *loyalty* to mathematics: (1) Be the kind of person students accept and want to associate with. (2) Work with students with patience and kindness so that *each day each student* has some success. (3) Make learning mathematics a *privilege* rather than a punishment. (4) Be fair in *marking* and in *discipline.*

To make learning mathematical ideas a *pleasure:* (1) Present the material so that it is understood. Be sure that students attain a reasonable level of competence before going on to new topics. (2) Use a variety of materials and methods that provide student participation in discovery, discussion, or laboratory lessons. (3) Make reasonable assignments with which students will experience success.

To nourish *respect* for excellence in achievement: (1) Stress the things a student does well; do not humiliate him because of failure; use errors as a means of learning. (2) Show how mathematical achievement relates to the student's goals. (3) Establish reasonable competition for marks and keep the student informed of his status.

To establish an *optimistic* attitude: (1) Present problems in a way that does not threaten the student's ego. (2) Assign tasks that are within the range of the student's ability. (3) Have a repertoire of illustrations, problems, sidelights, and applications that add variety and sparkle to daily lessons. (4) Be an optimistic, enthusiastic, sincere person.

Using Mathematical Recreations to Build Positive Attitudes

To build positive attitudes, students must have pleasant experiences with learning mathematics. Many teachers have found the use of mathematical recreations in the classroom the key to attitude development. Of course, learning mathematics is not all play, and no student should bypass the hard work of the subject. But much good mathematics can be learned from enjoyable recreations. In fact, enrichment activities of this type have been the source of much high-quality mathematics. Such outstanding mathematicians as Gauss, Leibnitz, and Euler found in such pastimes sources of new ideas and even new fields of mathematics. Two examples of such mathematical topics growing out of recreation are probability theory and game theory.

Recreations may sometimes be brought into the program as optional activities; for instance, many teachers pose a weekly problem for extra credit. On the other hand, it is quite appropriate to incorporate recreational activities that introduce, underscore, or extend the regularly required program.

The possibilities for worthwhile recreational activities in the mathematics class are very great. Most lessons can be dressed up with a puzzle, trick, paradox, or anecdote. The references and books listed in Appendix C will supply you with a wealth of ideas.

Here are some examples of recreations that stimulate students to participate, be observant, learn rules, work independently or cooperatively, and, most of all, find satisfaction in mathematics:

A game like "Battleship," in which teams "shoot" at various-sized ships marked on coordinate grids, provides good practice in locating points on a graph. Another game that provides this same practice is a form of the Japanese game Go-Moku. To play, the class is divided in half, and each side attempts to locate a row, column, or diagonal of four counters, at the same time preventing the opponents from doing so.

Alphametics encourage junior and even senior high students to examine the structure of mathematical operations. In these each different letter represents a distinct digit. Here are some examples:

$$\begin{array}{r} \text{HAVE} \\ +\ \text{SOME} \\ \hline \text{HONEY} \end{array} \qquad \begin{array}{r} \text{SANTA} \\ -\ \text{CLAUS} \\ \hline \text{XMAS} \end{array} \qquad \begin{array}{r} \text{FORTY} \\ \text{TEN} \\ +\ \text{TEN} \\ \hline \text{SIXTY} \end{array}$$

$$2(\text{HOHOHO}) = 9(\text{OHOHOH})$$

$$\frac{\text{EVE}}{\text{DID}} = \text{TALK TALK TALK}\ldots$$

Paradoxes force students to examine more closely the operations they carry out without thinking. For example:

You are as old as I am!

If x represents your age and y mine, then let our average age, $(x + y)/2$, be M. Then $x + y = 2M$. Multiplying by $x - y$ gives $x^2 - y^2 = 2Mx - 2My$, or $x^2 - 2Mx = y^2 - 2My$. Add M^2 to each member and factor:

$$x^2 - 2Mx + M^2 = y^2 - 2My + M^2$$
$$(x - M)^2 = (y - M)^2$$

But this means $x - M = y - M$ or $x = y$. In other words, you are exactly the same age I am.

Number tricks often help students to understand the operations of simple algebra and encourage them to explore the field further:

Think of a number between 0 and 10. Multiply it by 5. Add 6 to your answer. Multiply this answer by 2. Add any other number between 0 and 10. Subtract 5 from this result. The answer can be represented as $10x + y + 7$, which quickly identifies the chosen numbers.

Select a number between 100 and 1,000 that has a first digit and a last digit that differ by at least as much as 2. Reverse the digits of this numeral. You now have another number represented by a three-digit numeral. Subtract the smaller number from the larger of these two numbers. The difference should be another number between 100 and 1,000. Reverse the digits of this difference. Add the difference and the number represented by its reversed digits. The sum is 1,089.

Recreations can show the versatility of logical analysis. The following situation is an example of a problem using indirect reasoning:

A professor wishing to choose an assistant decided to test the mentality of the three top candidates. The professor told the candidates that he would blindfold each one and then mark either a red or blue cross on the forehead of each. He would then remove the blindfold. Each candidate was to raise his hand if he saw a red cross and drop his hand when he figured out the color of his own cross. The professor first blindfolded each candidate and proceeded to mark a red cross on each forehead and then removed the blindfolds. After looking at each other, the prospective assistants all raised their hands. After a short interval of time, one candidate lowered his hand and said, "My cross is red," and gave his reasons. Can you duplicate his reasoning?

Recreations may be appropriate "homework." A puzzle, game, or stunt is an excellent way for student and parents to work together at learning mathematics. Since parental attitudes are a key to student attitudes, this joint enjoyment of "doing" mathematics together can be extremely productive. Such assignments also give an answer to the oft-repeated parent question: "How can we help?"

The Behavior of Students who Have Attained Favorable Attitudes

The student with the proper attitudes will enter wholeheartedly into the learning activities because he is sensitive to mathematics whenever he finds it and derives pleasure from his contacts with it. His conversation, written work, and activities in and out of the classroom will be indicative of his attitudes. For example, in the student's written work his attitudes will be expressed by the pride he takes in completing work, doing extra work, searching for the most elegant exposition or solution, and by the effort he makes to discover new relationships, new forms, new procedures. He may find unique problems or applications to bring to class, or participate in a mathematics club, or read mathematics books. He may keep a mathematics notebook or use ingenuity to create original models—and he will do these things cheerfully and be annoyed by distractions that hinder his progress or that violate good mathematics. These then are the behaviors we observe as a measure of our students' attitudes.

Creativity in Mathematics

When we consider the history of mathematical ideas, we realize that many mathematical concepts have been created by young people. Therefore, continued development of mathematics depends to a great extent on the early identification, stimulation, and education of our mathematically creative youth.

In many ways, mathematics offers unique opportunities for creative and original thinking. Writing and solving original problems, establishing theorems with original proofs, discovering and stating relationships in one's own words are beginning experiences in creative thinking. A further opportunity for originality is found in the communication of mathematical ideas, be it in a demonstration, a proof, an exhibit, a poem, or a research project.

The development of a new numeration system, the building of an original model, or the discovery of new ideas or new applications of mathematics illustrate creative work at an even higher level.

Discovering Creative Students

Research on the problem of creativity indicates that the creative student is the nonconformist, the independent, the offbeat, and sometimes even the unruly student. In locating the specially creative youngster, the teacher must not rely heavily on standard measures of achievement, for the creative student does not always rate unusually high on an intelligence test. On those tests, the ideal performance is conformity to the examiner's criteria and norms. Creativity requires more than high intelligence, special talent, or technical skills—even though creative thinking is related to achieved intellectual skills.

Here are some of the characteristic ways in which creative students perform:

They are unpredictable, flexible, versatile in responses to situations, ideas, problems, adaptive in association, redefinitions, reorganization, and elaborations of ideas.

They are curious about ideas, objects, devices; they are sensitive to problems, relationships, errors, independent and confident in judgment and approach.

They are able to sustain uncertainty and withhold decisions in a complex situation, and able to abstract generalizations from complex situations or apply generalizations to new situations.

Although much recent research, notably by Paul Torrance, has sought objective measures of creativity, the capable teacher attuned to the attributes of creative students can probably locate this talent as well as any testing device. Through individual conferences and out-of-class contacts, through evaluation of student classroom activities, homework, and projects, the teacher will frequently find the student who has special creative attributes.

Often, creative students develop negative attitudes toward the highly structured activities of science and mathematics. They prefer the unusual activities, the unexpected circumstances. Classrooms that offer little opportunity for student participation repress creativity, whereas classrooms that provide for student participation and that encourage original ideas and approaches foster creativity. Within this encouraging environment, the teacher can identify the creative student as the one who takes real advantage of the opportunity provided him.

Creative Activities of Mathematicians

The mathematician, in his search for new knowledge, uses intuition, imagination, recall, and estimation. Often his insight, originality, and flexibility lead to breakthroughs to new ideas. The history of mathematics offers a wealth of anecdotes that illustrate the creative activities of young mathematicians. One of the best examples is Évariste Galois (1811–1832), a great French mathematician. Even though he died at the age of 21, he created the basis of modern group theory—one of the most fundamental concepts of mathematics today. He was inspired by reading the literature of mathematics and encouraged by a teacher who recognized his genius. The creative efforts of Galois and other mathematicians may be illustrated by these historical events:

The determination of the value of π by Archimedes.
The measurement of the earth's circumference by Eratosthenes.
The binomial expansion and Pascal's triangle.

The analysis of the networks of topology by Euler.
The sum of an arithmetic series by Gauss.
Fermat and his unsolved theorem.
The invention of calculus by Newton and Leibnitz.
Goldbach and his conjecture about primes.
Cartesian coordinates as proposed by Descartes.
The musical scale as established by Pythagoras.
The invention of binary numbers by Leibnitz.
The theory of relativity proposed by Einstein.
Cantor and the mathematics of sets.
The non-Euclidean geometry of Riemann and Lobachevski.
The theory of games as built by Von Neumann.

Suggested Activities for Developing Creativity

There are some who say that creative activities must be delayed until the students know all the mathematics already developed. However, numerous times during the course of study a student may depart from the traditional sequence to explore new fields. For example, permutations and combinations are essentially independent of advanced mathematical concepts; and, at an elementary level, properties of numbers can furnish an excursion independent of algebraic concepts. Linear programming, elementary topology, groups, and game theory are other possibilities for original thinking. In geometry, there are extensions to the geometry of a sphere or a cylinder, non-Euclidean geometries, four- or five-dimensional geometry, and projective geometry. At a more advanced level, the creative student might study finite geometry, quaternions, matrices, probability, or transformations. The applications of mathematics in science, economics, genetics, and psychology are also avenues for further original explorations.

For a secondary-school student to discover or develop a new mathematics today would be most unusual, though not absolutely impossible. But for a secondary-school student independently to rediscover or redevelop good mathematics is an extremely valuable experience, as valuable to him as the original act was to the mathematician who first performed it. The mathematician E. H. Moore of the University of Texas recognized the need for redevelopment and rediscovery and made it the basis of his teaching. He provides his students with a set of axioms and definitions and asks them to develop a certain subject independently, without the aid of texts.

Here are a few ideas that may suggest possibilities for creative activities:

Determine why an accepted algorithm or procedure works.

Invent new number symbols and a new numbering system.

Invent new operations or new ways to perform divisions, multiplications, or the finding of roots.

Write an original mathematical poem, essay, or story.

Write and present an assembly program, television program, or classroom dramatization on a mathematical theme.

Discover a new proof for a theorem, such as the Pythagorean theorem.

Construct an original model, such as a device for finding the roots of a cubic equation.

Invent a new scheme of measurement with appropriate units and measuring devices.

Find a new way of graphing data or an equation.

Write and prove original theorems involving sets, non-Euclidean geometry, or game theory.

Extend the theorems of plane and solid geometry to four-dimensional space.

Extend Euler's formula connecting edges, vertices, and faces to four-dimensional tessaracts.

Additional topics are described in greater detail in Appendix D.

Classroom Activities that Promote Creativity

The teacher can encourage investigation and exploration by making available materials, topics, problems, and reading matter. Any discovery by the student in the realm of ideas should be recognized and care taken to avoid discouraging such activities. Often, time and solitude are needed to bring creativeness to the surface; but this freedom for independent work and expression must be combined with sustained effort by the learner, organized presentations of information, and adequate resources for reference work. The psychological climate must be such that the student feels that his qualities are valued by other members of the group as well as his teacher and feels enough confidence in his relations with others so that he can afford to be different and to express his own opinions.

Often teachers reward memory, skills, and information far more than imaginative responses, "irrelevant" questions, and differences of opinion. By requiring specified assignments, courses, and procedures, we tend to explore new ideas in a formal group situation with a textbook containing all the facts and rules. Rather, the teacher should explore new ideas without the aid of the textbook. Finally, test constructors have not built test items that search for creative, unique responses.

If students are given only facts, rules, and drill, then the teacher has no reason to expect creative thinking. If we as teachers think there is only one solution to a problem, then students have little incentive to demonstrate originality. Do not stifle enthusiasm, originality, or creativeness by requiring conformity in analysis, method, or language.

Here is a list of activities that the teacher may check against his current classroom practice to highlight areas which may need improvement. It should be noted, however, that great variation in the number of times a specific activity is used is to be expected in any program. After matching the items in the list against his own classroom practices, the teacher may wish to modify his program to provide greater emphasis in these areas.

The students are actively participating in discovering concepts through reflective thinking, problem solving, experimentation, analysis, or generalization.

The students are encouraged to ask questions, correct errors, propose new solutions of proofs, and introduce concepts that are different from those of the text or class discussion.

The students are required to give reasons for answers, statements, methods, rules, so that they will know the "why" as well as the "how" of what they do.

The teacher is prepared with reading material, applications, illustrations, procedures, and problems that enhance and extend the meaning of a concept.

The students are encouraged to explore topics independently.

The students are given open-ended research projects, reports, creative writing, and supplementary assignments.

The tests used include open-book tests, reading tests, performance tests, or reasoning tests that measure productive thinking.

The teacher shows enthusiasm for and enjoyment of his work and his pupils, and appreciation for new ideas.

Whenever you teach a mathematical topic, ask yourself if it could be explored beyond the textbook. For example, what about a numeration system using a negative base? What unit of measurement could you use for taste? What other figures on the sides of a right triangle give the Pythagorean formula? What happens to a graph if the axes are skew rather than perpendicular, curved, or with irregular scales? Why isn't the sum $a/b + c/d = (a + c)/(b + d)$? Similarly, almost every topic in mathematics can be explored beyond the textbook.

Few of our students will make original, creative contributions to mathematics. Today the horizons of mathematical knowledge are mostly far beyond the limited vision of secondary students. But these students should have many opportunities to uncover things that are new to them. Given proper encouragement and support they will occasionally even discover things that you, their mathematics teacher, do not know.

Learning Exercises

1. Analyze your own attitude toward mathematics. How will this attitude affect your classroom teaching? What are some things you can do to improve your attitude?
2. Identify a student you judge to be creative. This may be a secondary-school or college mathematics student. By talking with him try to determine what are some of the factors in his background and training that supported and encouraged him. Did other factors work against his creative bent?

3. Many teachers feel that the most important time of year for building attitudes is the first month of school. How is this attitude-building period influenced by:
   (a) Extensive review of content from previous grades?
   (b) Concentration on definitions and other foundation learning?
4. Analyze a contemporary mathematics textbook for:
   (a) Interest building activities
   (b) Creative, open-ended projects
5. Detail and justify your attitude toward use of each of the following in mathematics instruction:
   (a) Games
   (b) Puzzles, paradoxes, and tricks
   (c) Historical episodes
   (d) Personal anecdotes
6. The students in a specific class usually have a wide range of interests. How could you manage such a class in order to improve interest levels while not sacrificing already-generated interest?
7. Experiments carried out by Berkeley psychologist David Krech suggest that the brain itself is physically improved by enrichment, and that this improvement is not based on random activities but "species-specific enrichment activities."[1] In the case of rats, Krech found that the best activity was "freedom to roam around in a large object-filled space." How might these findings be interpreted for classroom teaching?
8. PROJECT. Test the following rummy-like game on a small group of junior high school students. Make 42 cards ($2\frac{1}{2} \times 3$ inches) and label them (two each) -10 to 10. Rules: Each player is dealt three cards. His object is to form a sum of zero. As soon as this is done, the player announces "zero" and scores the sum of the absolute values of each opponent's points. The play is as in rummy, each player in turn drawing a card and then discarding. Players may draw one or all cards from the discard pile or one card from the deck. Fifty points accumulated is the goal.
9. PROJECT. Make up a game to reinforce one of the following:
   (a) Locating coordinates
   (b) Determining logical equivalence
   (c) Problem solving skills
   (d) Identifying geometric figures

Test the game with an appropriate group of students.

Suggestions for Further Reading

Bernstein, Allen L., "Motivations in Mathematics," *School Science and Mathematics*, 64, 9 (December 1964), 749–754.

Burns, Richard W., and Barbara M. Ellis, "What is Creativity?," *School Science and Mathematics*, 70, 3 (March 1970), 204–206.

Hartung, Maurice L., "Motivation for Learning Mathematics," in *The Learning of Mathematics*, Yearbook 21, National Council of Teachers of Mathematics, 1953, pp. 42–68.

[1] "Don't Use the Kitchen Sink Approach to Enrichment," *Today's Education*, 57, 7 (October 1970), 31–32, 87.

Hirschi, L. Edwin, "Encouraging Creativity in the Mathematics Classroom," *Mathematics Teacher*, 66, 2 (February 1963), 79–83.

Johnson, Donovan A., "Attitudes in the Mathematics Classroom," *School Science and Mathematics*, 57, 2 (February 1957), 113–120.

———, "Enriching Mathematics Instruction with Creative Activities," *Mathematics Teacher*, 55, 4 (April 1962), 238–242.

———, "A Fair for Mathematics with Marathons and a Midway," *School Science and Mathematics*, 65, 9 (December 1965), 821–824.

Laible, Jon M., "Try Graph Theory for a Change," *Mathematics Teacher*, 63, 7 (November 1970), 557–562.

Malerich, Sister Antone, "A New Look at Enrichment," *Mathematics Teacher*, 57, 5 (May 1964), 349–351.

Marx, Robert, "Mathematics Can Be Fun," *School Science and Mathematics*, 68, 2 (February 1968), 123–129.

Mattson, Robert J., "Mathematics Leagues: Stimulating Interest through Competition," *Mathematics Teacher*, 60, 3 (March 1967), 259–261.

Neale, Daniel C., "The Role of Attitudes in Learning Mathematics," *Arithmetic Teacher*, 16, 8 (December 1969), 631–640.

Nemecek, Paul M., "Stimulating Pupil Interest," *School Science and Mathematics*, 65, 1 (January 1965), 47–48.

Ranucci, Ernest R., "A Tiny Treasury of Tessellations," *Mathematics Teacher*, 61, 2 (February 1968), 114–117.

Read, Cecil B., "The Use of the History of Mathematics as a Teaching Tool," *School Science and Mathematics*, 65, 3 (March 1965), 211–218.

Rosenberg, Herman, "The Art of Generating Interest," in *The Teaching of Secondary School Mathematics*, Yearbook 33, National Council of Teachers of Mathematics, Washington, D. C., 1970, pp. 137–165.

Teaching the Applications of Mathematics 16

Mathematics has often been criticized because it has emphasized logic and structure that seemingly have no physical representation or practical application. This is unfortunate because probably no other subject has greater application than mathematics. It is the prime instrument for understanding and exploring our scientific, economic, and social world. Today, more than ever before, all fields of knowledge are dependent on mathematics for solving problems, stating theories, and predicting outcomes. It is an indispensible tool in creating new knowledge. For mathematics to attain its rightful place in the curriculum, we need continued reference to physical representation and specific applications. It must be relevant!

Sometimes modern mathematics programs overemphasize mathematics as a self-sufficient, self-generating field of knowledge. True, mathematics can be enjoyed as an aesthetic field of knowledge by itself. However, this aspect of mathematics appeals to such a small proportion of our students that it introduces motivation problems. It is the teacher's responsibility to supplement this view.

As mathematics teachers, we are also responsible for the general education of our students. We must build attitudes, patterns of investigation, and problem-solving skills that are broader in usefulness than knowledge for its own sake. We should be concerned about the insights into other fields of knowledge that mathematics may give our students. We should help students become informed about the aid that other subjects can supply in learning the concepts and nature of mathematics. We should encourage them to recognize that mathematics has two contrasting aspects: one the abstract, logical deductive structure that deals with symbols and ideas independent of physical representation, and the other the inductive aspect that deals with observed physical measures and probable conclusions.

Mathematics developed without reference to applications is called "pure" mathematics, while mathematics used to understand our world is called "applied" mathematics. Such a division of mathematics is difficult to make because many important mathematical ideas come from physical representations and often pure mathematics is very practical. Fourier considered mathematics as a tool for describing nature; however, the impact of Fourier series has been felt in the purest branches of mathematics as well as being of crucial importance in science. Cayley, on the other hand, believed that matrices, which he invented, would never be applied to anything useful; today matrices are an essential tool in science, economics, and statistics. Leibnitz invented binary numeration as a philosophical idea but it has become basic in computer science.

Geometry originated in practical problems of land measurement. At one time, geometry was considered a part of physics; but the Greeks developed geometry as an axiomatic system and studied it for its pure logic. Ever since, mathematics has been largely an axiomatic structure. Today intellectual curiosity is the main force behind research in mathematics. Even so some major mathematicians are both "pure" and "applied," such as Hermann Weyl who contributes to the theory of groups as a "pure" mathematical discipline and to the use of this theory in atomic physics as an important aspect of "applied" mathematics.

The history of mathematics makes it clear that mathematics grew out of experience in the real world. Counting, measuring, fractions, geometric relations, equations, probability, and vectors—these and many others had their origin in problems faced by the scientists and mathematicians of the past. It seems reasonable, then, that students can learn mathematical principles in a similar way—in terms of actual problems from the environment in which the student lives.

If we look back at the history of our civilization, we will find that mathematics has always been a major factor in society. As civilization developed it was a means of:

Counting possessions, objects, coins
Measuring property boundaries
Predicting the seasons
Computing taxes and profits
Navigating ships and explorations
Building homes, temples, and bridges
Drawing maps and plans
Developing weapons and planning warfare
Measuring indirectly the motion, size, and location of heavenly bodies
Increasing the power of human reasoning
Analyzing the vibrations of musical sounds
Providing models for the study of patterns in nature
Synthesizing the results of experiments
Measuring forces, time, energy, motion, and locations

During our lifetime, mathematics has provided the tools for:

Discovering new scientific principles
Inventing new machines and new products
Creating computing machines and electronic memories
Developing strategy for games, business, and government
Harnessing atomic energy
Directing traffic and communications
Navigating jets and space vehicles
Forecasting storms, floods, and crop conditions
Determining cause-and-effect relationships
Programming the solution of problems by computers

Testing new vaccines, antibiotics, and drugs
Studying the pollution of air, water, and soil
Discovering new minerals, oil, and new food supplies
Analyzing languages, music, art, and philosophy
Researching new theories in psychology, geology, and biology

Mathematics then grew out of the physical world. For example, geometry is basically a mathematical model of shapes, size, patterns, and motion in two and three dimensions. Similarly, the addition of fractions is based on the physical representation of fractions. If fractions had been considered rate pairs, addition would have been defined as $a/b + c/d = (a + c)/(b + d)$ rather than $a/b + c/d = (ad + cb)/bd$.

Properties of operations are based on results rather than established in advance. If matrix multiplication, to be useful, must be noncommutative, we abandon commutativity. This suggests that even in abstract mathematics we use real problems and real situations to produce mathematical concepts, operations, and theorems.

Actually, new mathematics programs have introduced innovations that should improve the student's ability to apply mathematics. The introduction of sets and set operations should provide a helpful means of classification and description. The emphasis on patterns and discovery should improve the ability to perform experiments in order to determine generalizations. The treatment of rate pairs and proportions provides a better means for using variation to solve problems. The increased content devoted to vectors, probability, and periodic functions improves the understanding of mass, space, time, and motion. The emphasis on problem solving and logical reasoning is a good foundation for the scientific investigation of problems in any field. Related to this is the common use of computer programs for solving problems. Finally, the emphasis on the structure of mathematics should prepare students to see the structure of problems and the structure of other fields of knowledge.

The Role of Applications in Learning Mathematics

The good mathematics teacher develops mathematical ideas and skills so that they can be used for further study, for everyday problems, and for personal satisfaction. If he cannot show what mathematics accomplishes outside his classroom, his students will lose interest and may not transfer their knowledge to a situation in which they need it.

Thus, a major role of applications is to motivate the student. Students frequently ask why they must learn certain facts and skills. When it is shown that these specific facts and skills are used in certain occupations, recreations, or subjects, the students are more likely to exert the effort to learn them. To reply that mathematics is needed for the next mathematics course or for entrance into college is not enough. The argument that scientists and engineers need it or that queer people called mathematicians love it is also not sufficient. Mathematics in total context is a better response.

> There is only one subject-matter for education, and that is Life in all its manifestations. Instead of this single unity we offer children—Algebra from which nothing follows; Geometry from which nothing follows.... Let us now return to quadratic equations.... Why should children be taught their solution?... Quadratic equations are part of algebra and algebra is the intellectual instrument for rendering clear the quantitative aspects of the world.[1]

To ensure the transfer of mathematical ideas and skills, we must do several things. First we must be informed about the possible applications of mathematics. These applications include those in mathematics itself as well as those in other subjects and in the world around us. This enables us to select concepts, structures, and skills that will be needed by our students. For example, we teach theorems about congruent triangles so that this knowledge can be used for indirect measurement as well as to develop ideas about transformation geometry. Knowing these applications gives us suggestions for introducing the topic as well as establishing priorities in deciding what to teach.

To ensure that transfer takes place, we must give our students experiences in applying their knowledge. If we want our students to evaluate the formulas of physics we use formulas such as $s = s_0 + v_0 t + \frac{1}{2}at^2$. We point out that science rarely uses x or y to represent the variables and often uses subscripts as in v_0. If we want our students to use projections for the maps of social studies or the blueprints of the shop, we give examples in the mathematics class. If we want our students to find the relation defined by ordered pairs or a graph, we collect a set of ordered pairs by an experiment such as the stretching of a spring when weights are added. If we want our students to understand vectors as explored in a science class, the mathematical treatment should be related to vectors such as those involved in an airplane flight. If we want our students to interpret correctly statistical graphs in the newspaper, we use data and graphs from the newspaper. If we want our students to use ratios and proportions or equations in solving problems, we include some problems such as these in our mathematics assignment:

1. If 201 grams of mercury and 16 grams of oxygen combine to form 217 grams of mercuric oxide, calculate the percent composition of each element in the mercuric oxide.

[1] Alfred North Whitehead, *The Aims of Education, and Other Essays*, The Macmillan Co., New York, 1929, pp. 18, 19.

2. The volume of a gas is directly proportional to the absolute temperature provided the pressure is constant. If a sample occupies 10 liters at 273° absolute, what will be the volume at 400° absolute?

3. If 12 grams of carbon react with 32 grams of oxygen to form carbon dioxide, how many grams of carbon will react with 640 grams of oxygen?

4. How many millimeters of 0.4 normal potassium hydroxide solution, KOH, will neutralize 100 millimeters of 0.05 normal phosphoric acid, H_3PO_4?

Formula:

$$\frac{\text{millimeters of acid}}{\text{millimeters of base}} = \frac{\text{normality of base}}{\text{normality of acid}}$$

5. A stone is thrown out of a window that is 50 feet above the ground. While traveling 20 feet horizontally, the stone falls 1 foot. If the stone travels in the arch of a parabola, how far away from the building does it land?

6. The equation of the orbit of the earth around the sun may be taken to be $x^2 + y^2 = 9$. There is a comet whose path is almost parabolic. If the comet has the equation $y = x^2/4 - 1$ as it crosses the earth's orbit, what are the points of intersection of the two orbits? (Assume both orbits lie in the same plane.)

7. A certain weight of gas has a volume of 1,200 cubic centimeters at a pressure of 500 grams per square centimeter and a temperature of 21°C. What is its volume at a pressure of 1,500 grams per square centimeter and a temperature of 315°C?

Formula:

$$\frac{P_1V_1}{T_1} = \frac{P_2V_2}{T_2}$$

where T is absolute temperature, 273° more than centigrade.

8. What is the combined resistance of a 9-ohm coil and an 18-ohm coil connected in parallel?

Formula:

$$1/R = 1/R_1 + 1/R_2$$

9. A space vehicle has a circular body with elliptical stabilizers. The equation of the circular body is $x^2 + y^2 = 9$. The equation of the wings is $x^2/16 + y^2/4 = 1$.
   (a) At what points do the wings intersect the body?
   (b) Draw to scale the two curves.

10. In mechanical drawing, a student is faced with the problem of representing a cylindrical can in perspective. This means that he will have to draw congruent ellipses for the top and bottom, whose major and minor axes are 2 inches and 1 inch respectively. Instead of using the usual approximation with circular arcs, the student decides to use a thread to construct the ellipses accurately. Determine the positions of the foci and the length of the thread to do this.

The use of these problems reminds us that we must be familiar with other courses of study, the language, the symbolism, and the topics taught. It is desirable to examine the texts used in other courses and discuss the needs of other courses with the instructors involved so that your students can be prepared with the concepts and skills needed at the time they are needed.

In a study of the mathematics in an engineering physics text for university freshmen, 15,000 instances were listed in which mathematical concepts or processes from secondary-school mathematics were needed.[2] These 15,000 instances indicate that there are ample physical illustrations and applications for teaching mathematics. In doing this, however, care must be exercised to select those formulas, graphs, or measurements that can be intuitively grasped. In the process of using these applications, there will be an opportunity to practice many mathematical skills: computations, evaluation of formulas, solving equations, finding square roots, using measures, and drawing graphs.

Applications are also a means for illustrating the role of mathematics in our society. How has mathematics influenced civilization? It is likely that mathematics has been a greater factor in building civilization than many rulers or battles that are treated in great detail in history courses. A recent book, *Evolution of Mathematical Concepts* by Raymond L. Wilder,[3] explores how mathematics is a major factor in our culture. It describes how mathematics influenced culture and at the same time how culture influenced the development of mathematics. Another book, *Looking at History through Mathematics* by Nicolas Rashevsky,[4] attempts to use mathematical models in studying the processes of history. At the present time the computer is a revolutionary force in our society. Its capabilities and its role should be presented to every high school student. Since business, industry, government, and education are highly dependent on linear programming and the computer in making decisions, every citizen should be informed about the role mathematics is playing in our society.

Finally, the applications of mathematics are a means of teaching problem solving. What better sources for problems are there than those in the world about us? What is the formula for weights on a teeter-totter? How can the distance to the moon be measured? How is an airplane supported in flight? How high will a baseball travel if the batter hits it upward at 45° and 100 feet per second? At what speed will a space vehicle have to be launched to escape the earth's gravitational pull? What is the probability that you will have an automobile accident tomorrow? No matter where you turn in modern life, you will find mathematical problems. Often the best assignment is to have the student himself find such applications in his environment. Then he can decide what data to collect and what relationships to explore. This activity can then be used to relate problem solving to the so-called scientific method of inquiry. This approach to problem solving should emphasize flexibility, originality, selecting relevant data and discarding irrelevant, trying out hunches, probing different possibilities, and discovering generalizations. This type of problem solving is very similar to that of inquiry in laboratory lessons.

[2] Douglas B. Williamson, "The Mathematics Essential to the Learning of Engineering Physics," unpublished Doctor of Education Project Report, Teachers College, Columbia University, 1956.

[3] John Wiley and Sons, New York, 1968.

[4] MIT Press, Cambridge, Mass., 1969.

Some Applications of Mathematics

The most common applications of mathematics are in the subject matter itself. After students have learned the principle of one-to-one correspondence, it is applied over and over in later topics. After they have learned the distributive property, this idea is used to make many operations such as the factoring of binomials meaningful. After they have learned to graph equations, this principle is applied to the study of functions and relations in advanced courses. The operations used to find truth sets for linear equations are applied in solving higher-degree equations and systems of equations. The method of proving a theorem is applicable to a great variety of situations. Frequently, it is these applications of mathematics that are of the greatest significance to the learner. He is not always interested in the applications of the adult world such as taxation, insurance, or banking.

Another major application of mathematics is its use in establishing new facts. Whenever the scientist performs an experiment, he collects data by observing events and making measurements. He then must fit the data into some conceptual scheme called a law, principle, or hypothesis. In his analysis of his experimental data, the scientist searches for a general law that will fit all the data. Usually he searches for a pattern by graphing, by obtaining differences, by testing regression equations—all tools furnished by mathematics. The scientist experiments and observes results enough times to be convinced that a principle has been established. Since he cannot test all possible events, he runs the risk of drawing an incorrect conclusion. He often uses mathematics to give the probability that his conclusion is true.

From special laws and concepts, the scientist attempts to develop a more general theory. To do this he begins with certain assumptions and then uses deduction, the logical reasoning of mathematics. Deductive reasoning—that is, reasoning from the general to the particular—results in conclusions that are accepted as certain to follow from given assumptions and definitions.

There is another significant aspect of mathematics used in the scientist's research. His experimentation usually involves many measurements; the more precise and accurate they are, the better. However, every measurement is an approximation. It is the mathematical aspect of measurement that gives him the necessary means for dealing with these approximations. Fortunately, the computational tools of mathematics are designed to provide for exact data as well as approximations.

If we stop to consider the activities of business, industry, government, science, and education, we recognize very quickly that the following mathematical topics are used daily:

Computation. Despite the availability of calculators or computers, the most common application of mathematics is calculating with counting numbers, rational numbers, and real numbers. Accurately computing additions, subtractions, multiplications, divisions, squares, and square roots is frequently necessary.

Measurement. Making measurements, computing with measures, converting from one unit to another. Computing area or volume of common geometric polygons, circles, polyhedrons, spheres, cylinders, cones, finding the errors of measurement.

Statistical data. Reading tables, charts, and graphs; summarizing data, determining patterns of data; computing means or standard deviations; determining cause and effect and predicting results.

Graphing. Reading or interpreting or drawing graphs; determining trend lines; using nomographs; using graphs for predictions, for solving problems, for showing relationships between variables.

Percent. Reading, converting, computing with ratios expressed as percents.

Equations. Finding truth sets for linear equations, formulas, and quadratic equations. Writing the equation for a pattern of data or a graph.

Geometric relations. Pythagorean theorem; similar triangles; properties of triangles, quadrilaterals, or polygons; angles; relationship of lines or planes; properties of circles.

Ratio and proportion. Writing proportions for direct and inverse variation, solving proportions for any variable, determining constants of variation.

Estimation and approximation. Using significant digits and scientific notation, rounding off numerals, estimating results.

Computing machines. Calculating with slide rule, calculator, flow charts, or by programming a computer.

Probability. Interpreting and computing probability, both theoretical and empirical.

Note that these competencies relate to many of the classical topics of school mathematics as well as the content of the new school mathematics.

Teaching the Applications of Mathematics

In order to teach the applications of mathematics, a teacher should have a background in fields other than mathematics, so that the applications have meaning for him. Because of the number and the significance of applications in science, it has been suggested that every mathematics teacher have a minor in science. Whether or not he has this specific background, a teacher with a broad general education will have had contact with many applications of mathematics in many other fields, and be able to present them to his students.

One of the ways to emphasize applications is to have a variety of them available for use as examples and exercises. Whenever a mathematics lesson involves topics such as equations, graphs, or polygons, this storehouse should be used to supply an appropriate example.

We are fortunate that a variety of books containing applications are readily available. A number of these are listed in Appendix B.

Examine science and social studies texts as well as periodicals and newspapers for applications. You will be pleasantly surprised to find the variety of graphs, tables, formulas, and measurements involved. For example, the Physical Science Study Committee *Physics* looks almost like a mathematics textbook. Give your students open-ended assignments to find patterns, statistics, probabilities, charts, and formulas in their school texts or newspapers.

How applied mathematics and pure mathematics differ can be illustrated by quadratic equations. In the mathematics class, we learn how to solve quadratics by factoring or by formula or possibly by graphing. In no case are we permitted to drop a term or ignore a coefficient. Consider this application:

Determine the depth of a well by dropping a pebble and measuring the time until the sound of the splash returns. Let d represent the depth of the well, ignore air resistance, assume sound travels 1,100 feet per second and that the falling distance is $d = 16t^2$. If the time for the sound to return is 3.1 seconds, then

$$\text{time for stone to fall} + \text{time for sound to return} = 3.1$$

$$\sqrt{\frac{d}{16}} + \frac{d}{1{,}100} = 3.1$$

or

$$\sqrt{\frac{d}{16}} = 3.1 - \frac{d}{1{,}100}$$

and by squaring both members

$$\frac{d}{16} = 9.61 - \frac{2(3.1)d}{1{,}100} + \frac{d^2}{1{,}210{,}000}$$

Then

$$\frac{d}{16} + \frac{6.2d}{1{,}100} - \frac{d^2}{1{,}210{,}000} = 9.61$$

Unless d approaches 1,000 feet the term in d^2 is very small and can be dropped because of the accuracy of the measurements involved. We cannot use more than two significant figures, and the error resulting from dropping this term would be a fractional part of a foot. Solving $d/16 + 6.2d/1{,}100 = 9.61$ gives $d = 141$. Hence the depth of the well should be recorded as 140 feet.

The Apollo Flight to the Moon

One of the greatest technological achievements of man has been demonstrated by his flights to the moon and back. Since this triumph is based on mathematics, we should exploit the interest in it to develop mathematical competence in our students. The basic orbits and trajectories, if we omit perturbing factors, can be computed by high school students.

First, we will simplify the problem to a consideration of a circular orbit (see Figure 16.1). When an object whirls in a circle, the centripetal force F_1 is

$$F_1 = \frac{mv^2}{r}$$

where m is the mass of the object, v is its speed or velocity, and r is the radius of the circle.

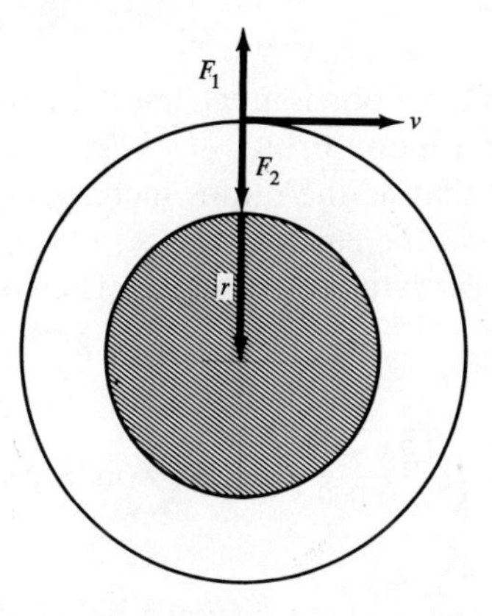

Figure 16.1

When a spacecraft is traveling around the earth, the force toward the center of the earth is supplied by the force of gravity. This gravitational force F_2 is

$$F_2 = \frac{gMm}{r^2}$$

where g is the constant of gravitation (about 32 for the earth), M and m are the masses of the two bodies (M for the earth and m for the spacecraft), and r is the distance between the two centers of mass. If the spacecraft is to stay in its circular orbit, the gravitational force must equal the centripetal force.

If $F_1 = F_2$ then

$$\frac{mv^2}{r} = \frac{gMn}{r^2} \quad \text{and} \quad v_c = \sqrt{\frac{gM}{r}}$$

where v_c = circular velocity in miles per hour.

For the earth

$$gM = 1.24 \times 10^{12} \qquad \text{and} \qquad v_{\text{earth}} = \sqrt{\frac{1.24 \times 10^{12}}{r}} \text{ miles per hour}$$

For a moon orbit

$$gM = 1.49 \times 10^{10} \qquad \text{and} \qquad v_{\text{moon}} = \sqrt{\frac{1.49 \times 10^{10}}{r}} \text{ miles per hour}$$

Note that since the mass of the moon is much less than that of the earth, the velocity is decreased markedly for a moon orbit.

These formulas show that as the radius increases, the velocity decreases. At an altitude 100 miles above the earth's surface, $r = 3{,}960 + 100 = 4{,}060$ miles, assuming the radius of the earth to be 3,960 miles. Then the circular orbital velocity is

$$v_{\text{earth}} = \sqrt{\frac{1.24 \times 10^{12}}{4{,}060}} = 17{,}500 \text{ miles per hour}$$

When $r = 239{,}000$ miles, the average distance between the centers of the earth and our moon, the velocity is about 2,280 miles per hour. This is the approximate velocity of the moon in its orbit around the earth.

In actual practice, circular orbits do not exist. The orbits of spacecraft around the earth or moon are elliptical. In this case we must introduce a new factor into the velocity formula, namely eccentricity. The computation of eccentricity is illustrated by Figure 16.2. If the semimajor axis is a and the semiminor axis is b and the focal point F_1 is c units from the center, then the eccentricity e is $e = c/a$. When $c = 0$ then $c/a = 0$ and the figure is a circle. If $c = a$, then $e = 1$ and the

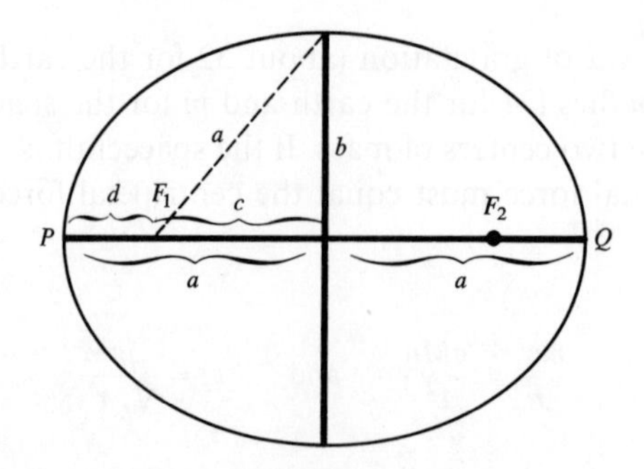

Figure 16.2

figure is a parabola. If F_1 represents the earth, then p is the perigee of this elliptical orbit and q is the apogee. This provides another way to find the eccentricity:

$$e = \frac{a + c - d}{2a}$$

Why must the velocity at the perigee and apogee be different?

$$v_p = \sqrt{\frac{gM}{d}(1 + e)} \quad \text{and} \quad V_q = \sqrt{\frac{gM}{(a + c)}(1 - e)}$$

The minimum escape velocity is related to parabolic orbit and is found by multiplying the circular orbital velocity at the altitude involved by $\sqrt{2}$:

$$v_{\text{escape}} = \sqrt{\frac{2gM}{r}} = \sqrt{2}v_c$$

Actual velocities for moon flight are somewhat less than this in order to assure a "free return trajectory," meaning that the lunar gravity will pull the spacecraft around the moon.

Another useful equation is the formula for finding the sidereal period of a satellite in an elliptical orbit:

$$p = 2\pi\sqrt{\frac{a^3}{gM}} \text{ hours}$$

If the orbit is a circle, then $a = r$. The time computed by this formula gives the time for one complete orbit without regard for the rotation of the earth under the satellite. The time for crossings of a given meridian of longitude is called the synodic period. This period is a difficult concept for high school mathematics, especially if the orbit is inclined to the equator.

Special Teaching Problems

One of the most commonly used mathematical tools is the proportion. It is a means of solving problems involving percentage, chemical composition, mechanical advantage, pressure, volume, light, profits, and a host of others. At the same time, textbooks tend to give limited emphasis to proportionality as an aspect of variation, direct or inverse. This is one topic that should be well taught in the mathematics class by discussion, experimentation, and applied exercises.

Frequently, students will know the technical language and the mechanical manipulation of proportions without really knowing how variation is involved. Ask students to explain direct proportion so that a fourth grader can understand it. What happens numerically to the second variable if the first is doubled? What does it mean to say that distance varies directly as the time? What does it mean to say that light intensity varies inversely as the square of the distance from the light? If the distance is doubled, what happens to the intensity of the light? If the speed of two pulleys connected by a belt varies inversely to their diameters, how does the speed of a 3-inch pulley compare with that of a 12-inch pulley? What is the geometric basis for this relationship? If the volumes of two spheres are directly proportional to the cubes of the diameters, how does the volume of a 2-inch orange compare to that of a 3-inch orange? Which of these cereals is the best buy in terms of the cost per ounce?

Another simple source of difficulty in transfer is the use of different notation and language in courses other than mathematics. For example, the subscript zero is often used in science to denote initial conditions, as in $s = v_0t - 16t^2$. Likewise, exponents are used for dimensions, as 5 in.^{2}. Methods of solving equations and the language used will tend to be that of traditional mathematics. Terms such as "the resolution of forces" or "the constant of variation" or "a percent is a fraction" and "signed numbers" will be foreign to a student in a modern mathematics class. To avoid confusion caused by differences in symbols, methods, and words, the mathematics student should be told of these variations both in the mathematics class and in the science class.

When teaching a mathematical concept, every effort should be made to show how the process is a model for a physical situation. For example, when we deal with relations, we find relationships between slope, abscissa, ordinate, and area. (See Figure 16.3.)

1. Slope $m = \dfrac{y - y_0}{x}$ by definition
2. Ordinate $y = y_0 + mx$ an equivalent equation for (1)
3. Area of trapezoid $A = \left(\dfrac{y + y_0}{2}\right)x$ formula for area of trapezoid
4. Area of trapezoid $A = x \cdot y_0 + \frac{1}{2}(y - y_0) \cdot x$ an equivalent equation for (3)
5. Area of trapezoid $A = x \cdot y_0 + \frac{1}{2}m \cdot x^2$ substituting $y - y_0 = mx$

To show how this relates to science, use the following substitutions: $m = a$, $y = v$, $x = t$, $A = S$, where a, v, t, and S represent values of acceleration, velocity, time, and distance in appropriate units.

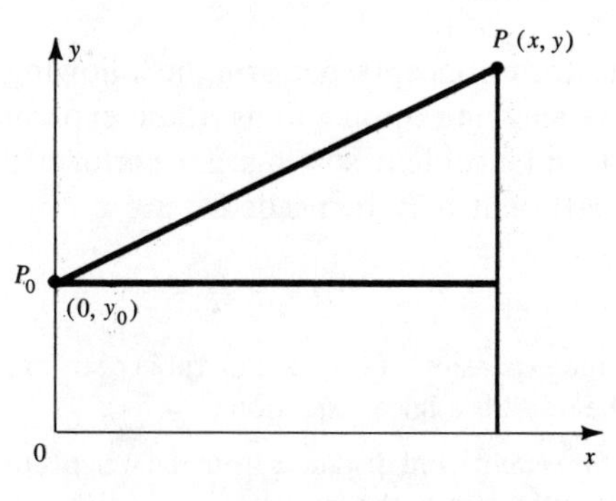

Figure 16.3

Then the formulas above become the formulas for constant accelerated motion:

$$a = \frac{v - v_0}{t} \quad \text{and} \quad S = v_0 t + \tfrac{1}{2}at^2$$

This is the mathematical model of linear motion.

If we make another mathematical transformation, we find another true relation in mathematics:

$$\text{If} \quad m = \frac{y - y_0}{x} \quad \text{and} \quad A = \frac{y + y_0}{2} \cdot x$$

$$\text{then} \quad x = \frac{y - y_0}{m} \quad \text{and} \quad A = \left(\frac{y + y_0}{2}\right) \cdot \left(\frac{y - y_0}{m}\right) = \frac{y^2 - y_0^2}{2m}$$

$$\text{If} \quad A = \frac{y^2 - y_0^2}{2m} \quad \text{then} \quad y^2 = y_0^2 + 2mA$$

Translating to the variables of physics, we find $v^2 = v_0^2 + 2as$, a familiar relation in science. This formula would be difficult to generate by experimentation. This development then shows the student how mathematics as a logical system may serve as a model for a physical system.[5]

Laboratory Exercises for Applications

Another way to treat the applications of mathematics is to perform specific laboratory exercises in the mathematics classroom. In order to collect data for

[5] Fehr, Howard, "The Role of Physics in the Teaching of Mathematics," *Mathematics Teacher*, 56, 6 (October 1963), 394–399.

generalizing or to illustrate concepts, perform the following experiments. Independent of their value in showing applications, these experiments may be valuable lessons in discovery and problem solving. To perform them, some equipment from the science department may be needed.

1. Measure the expansion of a spring or rubber strip related to weights added to the spring. The result is a linear equation ($y = kx$).

2. Measure the weights and distances from the weights to the fulcrum of a balanced lever to discover the law of the lever ($w_1 d_1 = w_2 d_2$).

3. Find the effects of weight, length, and amplitude of a pendulum ($t = 2\pi\sqrt{L/g}$).

4. Measure the distance an object rolls down an incline or falls when dropped in given time intervals ($s = \frac{1}{2}at^2$).

5. Determine the specific gravity of objects by comparing weight in air and weight in water with volume ($s = w_1/[w_1 - w_2]$).

6. Measure the centrifugal force of a rotating object related to the weight of the object and the speed of the object and the radius of the turning ($C = WV^2/gR$).

7. Find the intensity of light related to the distance from the light ($I = k/D^2$).

8. Determine the sequence of the measures of the bounces of a ball.

9. Find the horsepower of a person running upstairs by substituting measures of his weight, distance raised, and the running time in the formula $P = WD/550t$.

10. Discover the law of reflection by drawing and measuring angles and distances between points and images.

11. Compare ratios of gear teeth and gear turns to get examples of inverse proportions ($t_1/g_2 = t_2/g_1$).

12. Use an inclined plane to discover the vectors involved in pulling an object up the incline.

13. Make measurements of a stream to determine its velocity and pollution level.

14. Perform an experiment tossing coins to establish an empirical probability.

15. Deal samples of cards and record results of different-sized samples to discover how to base conclusions on them.

16. Use motion—forward and backward—along a number line to illustrate the meaning of vectors and relate these to addition and subtraction of integers.

Table 16.1 suggests how the real world and mathematics are related. Too often in mathematics we start with the abstraction and then give the application. For more effective learning we should start with the real world that our students have experienced, relate it to the scientific concept that has intuitive meaning, and then give the mathematical model for dealing with it.

Table 16.1

Real world →	Scientific concepts →	Abstract mathematical model
Movement	Displacement	Vector
Push	Force	Sum of vectors
Pull	Work	Scalor
Direction	Velocity	Vector space

Curriculum Based on Applications

It has been proposed that another way to organize the mathematics curriculum might be around applications. Since mathematics is thought of as a sequential field, this notion has received little attention. After attaining basic competence in topics such as measurement, proportion, probability, graphing, statistics, equations, and metric geometry, students could profit from a one-year course devoted to units such as:

Change. Graphs, motion geometry, variation, patterns

Politics. Surveys, sampling, probability, authorities

Pollution. Statistics, formulas, scientific notation

Crafts. Measurement, formulas, projections, geometry

Science. Relations, geometry, measurement, patterns

Economics. Proportions, graphs, probability, statistics

Nature. Measurement, combinations, sampling, symmetry

Reasoning. Sets, Venn diagrams, syllogisms, induction, proofs

Space. Solids, equations, conic sections

Computer science. Inequalities, algorithms

Estimation. Metric systems, linear, area, volume, time, weight

Mapping. Measurement, locations, scale drawing, similarity

These topics suggest one way to direct our curriculum toward an interdisciplinary approach now being advocated.

A mathematics curriculum that grows out of experiences in the physical world has been proposed by Morris Kline, a severe critic of the new school mathematics. His major argument in favor of this curriculum is that mathematics is meaningless and unattractive to the vast majority of our students. He bases his organization on the need to develop mathematics constructively rather than deductively. He also believes that the historical development of mathematics suggests the difficulty level of topics.

Kline's proposal for a ninth-grade algebra course[6] is based on formulas that deal with motion, gravitation, light, temperature, force, elasticity, friction, patterns, and population growth. He begins with the simplest linear function using positive numbers, for example:

> $P = 4L$, where P is the perimeter and L is the measure of a side of a square.
>
> $d = 30t$, where d is the distance and t the time traveled
>
> $h = 2t$, where h is the height of the water in a swimming pool t hours after the water is turned on.
>
> $v = 32t$, where v is the velocity of a dropped object t seconds after it is dropped and 32 is the acceleration of the earth's gravity.
>
> $S = \frac{1}{4}w$, where S is the elongation of a spring when the weight w is attached and $\frac{1}{4}$ is a measure of the spring's stiffness or elasticity.

These formulas are used to illustrate the meaning of formulas, the evaluation of them by substitution and the graphs of the results that show the changes taking place. He also distinguishes between mathematical and physical results. Suppose the swimming pool has a maximum depth of 6 feet. The formula $h = 2t$ gives the mathematical result after 4 hours, namely, 8. However, the physical result is 6 feet.

These formulas then lead to the solution for x of the general equation $y = ax$.

The next linear functions are of the following type:

> $P = 200 + 20t$, where P is the population of a town which has 200 people now and is growing at the rate of 20 persons per year.
>
> $v = 32t + 100$, where v is the velocity of a ball t seconds after it is thrown downward with a speed of 100 feet per second.
>
> $F = \frac{9}{5}C + 32$, the Fahrenheit–Centigrade conversion formula.
>
> $L = \frac{1}{4}W + 3$, where L is the length of a spring which is 3 inches to start with and stretches $\frac{1}{4}$ inch for each pound of weight w attached to it.

First, he proposes physical representations of these formulas, then solutions by substitution, and next graphs of the results. This leads to the method of solving the general equation $y = ax + b$ for x as well as y.

Negative numbers are introduced and operated on in terms of temperature readings where temperatures are rising or falling. Then negative values are used in the previous equations to check the operations with positive and negative numbers.

---

[6] Morris Kline, "A Proposal for the High School Mathematics Curriculum," *Mathematics Teacher*, 59, 4 (April 1966), 322–330.

Next, quadratic functions are treated in the same fashion. This introduces the need for factoring, square roots, radicals, literal fractions, inequalities, and systems of linear equations.

The general pattern then is to develop concepts by these steps:

1. Introduce the function with a physical problem or phenomenon.
2. Graph the function and answer questions about it.
3. Solve equations resulting from values substituted for the dependent variable.
4. Then invert the function and find a general method of solution.
5. Extend and practice the general technique.

In a similar fashion, Kline introduces geometry by showing how it can be used to measure indirectly, for example, the distance across a highway. Next, the proof of the relationship involved is explored. Then the theorems are applied to real problems, such as behavior of light, location of schools, satellite motion, geography, vectors, inclined plane, and the mass of the earth. This infers the study of conic sections and reflections. He also suggests that historical events should be included to add interest.

Supplementary Units

A third way to emphasize the applications of mathematics is to prepare and teach units as supplements to the regular text. Topics that are suitable for application units include the following:

Probability and insurance

Projections and mapping

Vectors and mechanics

The slide rule and surveying

Space travel and conic sections

Perspective and painting

Linear programming and economics

Proportions and music

Installment buying and interest formulas

Chemistry and proportions

Carpentry and the Pythagorean theorem

Astronomy and indirect measurement

Machine work and geometry

Navigation on sea, in the air, and in outer space, and similar triangles

Decision making and probability

Sometimes these topics can be dealt with as student projects to be shared with the class. In this case, the projects should be presented to the class by student demonstrations, reports, or bulletin-board exhibits.

An excellent source of applications topics from which to generate mathematical organization and thinking is *The Man Made World*,[7] a publication developed by the Engineering Concepts Curriculum Committee for use in schools. Some of the topics included are decision making, optimization, modeling, logic, feedback, and stability. The applications are extremely varied: coding, machine design, medicine, traffic flow, transportation, communication, computers, ecology, accident research, bat navigation, space travel, population dynamics, and serving lines are a few of the topics.

A collection of examples of statistical applications for use in schools[8] has been developed by the Joint Committee on Teaching Statistics in Schools of the American Statistical Association and the National Council of Teachers of Mathematics. The chairman of that committee is Harvard statistician Fred Mosteller.

All teachers should obtain reference copies of books like those described in this section even when they are not used as basic texts. They provide activities that can enrich any program whether it be as extra content for classroom development, assigned reading for better students, or a source for projects.

Applications in Special Courses

A fourth way to treat applications is to offer specific applicational or "practical" courses. Consequently, schools may have special courses such as Shop Mathematics, Consumer Mathematics, Basic Mathematics for Science, Business Arithmetic, or Computer Programming.

Textbooks used in applications courses do give emphasis to the applications. However, few college-bound students take these courses, and they need to know applications too. Moreover, frequently little mathematics is taught in these courses; they tend to concentrate on remedial arithmetic. Even worse than this, the courses may emphasize shortcuts and meaningless manipulations, some of which are not mathematically sound.

However, this allocation of applications to courses for low achievers is a tip-off to the low regard for the use of applications. This should not be the case. If we unconsciously avoid or depreciate applications, we contribute to the difficulties of teachers in other fields.

[7] McGraw-Hill Book Company, New York, 1971.

[8] Addison-Wesley Publishing Co., Inc., Reading, Mass., 1973.

In order to live in our society, a person needs to know how to analyze a situation in the real world, relate it to a mathematical model, and translate the results into conclusions about the real world. Then he tests his inferences by comparison and observations again to see if they work; if they do not, he must try another analysis of the situation. In mathematics, we concentrate almost exclusively on the mathematical model. We rarely concern ourselves with where the problems came from or what anyone would do with the answers.

Probably one of the best ways of insuring transfer to the application of a given field is to coordinate the topics of a mathematics course with the mathematical needs of another course. Before the science class uses proportions for problems dealing with simple machines, the mathematics class should study proportions. Before the physics class begins its study of vectors, the appropriate mathematics class should master the mathematical aspects of vectors. If the social studies class is to study map projections, the mathematics class should ensure the necessary background in geometry. Perhaps in the future, team teaching, integrated courses, and flexible scheduling will make it possible for mathematics teachers to work closer with staff members of other fields. Until then, every opportunity should be taken to ensure interdepartment cooperation. Examination of other school courses shows that the mathematical content is very substantial. It would seem appropriate for the mathematics department to teach these mathematical concepts to a level of mastery high enough that the non-mathematics teacher does not need to reteach them. The science teacher will need to review mathematical concepts and skills and assist the student in transferring this knowledge, but he should not need to teach the mathematical topics involved.

A point of special concern to science and mathematics teachers is the matter of dimensional analysis. For years this was a sensitive issue, with scientists promoting incorporation of labels in mathematical statements and purists among the mathematicians throwing up their hands in shock. The issue has been essentially settled, however, by Princeton mathematician Hassler Whitney who has provided for dimensional analysis a full mathematical structure.[9] All the advantages of dimensional analysis for checking labels and procedures are now available to teachers of both science and mathematics.

Here are some examples of dimensional analysis:

$$2 \text{ cows} + 3 \text{ cows} = (2 + 3) \text{ cows} = 5 \text{ cows}$$

$$25 \text{ ft} = 25 \cancel{\text{ft}} \times \frac{12 \text{ in.}}{1 \cancel{\text{ft}}} = 300 \text{ in.} \qquad (\text{since } 12 \text{ in.}/1 \text{ ft} = 1)$$

$$30 \text{ mph} \times 4 \text{ hr} = \frac{30 \text{ mi}}{1 \cancel{\text{hr}}} \times 4 \cancel{\text{hr}} = 120 \text{ mi}$$

[9] "The Mathematics of Physical Quantities. Part I. Models for Measurement," *American Mathematical Monthly*, 75, 2 (February 1968), pp. 115–138; "The Mathematics of Physical Quantities, Part II. Quantity Structures and Dimensional Analysis," *American Mathematical Monthly*, 75, 3 (March 1968), pp. 227–256.

The volume of an 8-ft 2 × 4-in. board:

$$8 \text{ ft} \times 2 \text{ in.} \times 4 \text{ in.} = 48 \text{ ftin.}^2$$

Since ftin.^{2} is not a standard unit, it may be converted to in.^{3} or ft^3 as follows:

$$48 \cancel{\text{f}}\text{tin.}^2 \times \frac{12 \text{ in.}}{1 \cancel{\text{ft}}} = 576 \text{ in.}^3 \qquad (\text{since } 12 \text{ in.}/1 \text{ ft} = 1)$$

$$48 \text{ ft}\cancel{\text{in.}^2} \times \frac{1 \text{ ft}}{12 \cancel{\text{in.}}} \times \frac{1 \text{ ft}}{12 \cancel{\text{in.}}} = \frac{1}{3} \text{ ft}^3 \qquad (\text{since } 1 \text{ ft}/12 \text{ in.} = 1)$$

Note in these examples how the multiplicative identity is used each time it is necessary to convert units from one form to another.

The argument about dimensional analysis is essentially settled in its favor; decisions about its use are not. Mathematics teachers should find out about dimensional analysis from their friends in the science department; but their use of this interesting tool should be judged carefully. Many teachers will choose to leave this topic to the science classroom. Many others will note its value and use it when it has a special contribution to make to problem solving. A few purists may go too far, forcing it into unnecessary settings and overcomplicating simple problems. For example, consider the following exercise and solution:

A picture has length 2 inches more than width. With a 3-inch frame, its area is 224 square inches. Find the dimensions.

Let width be x.
Length: $x + 2$ in.
Area with frame: $(x + 6 \text{ in.})/(x + 2 \text{ in.} + 6 \text{ in.}) = 224 \text{ in.}^2$

$$\begin{aligned} (x + 6 \text{ in.})(x + 8 \text{ in.}) &= 224 \text{ in.}^2 \\ x^2 + 14x \text{ in.} + 48 \text{ in.}^2 &= 224 \text{ in.}^2 \\ x^2 + 14x \text{ in.} - 176 \text{ in.}^2 &= 0 \\ (x - 8 \text{ in.})(x + 22 \text{ in.}) &= 0 \end{aligned}$$

$$x - 8 \text{ in.} = 0 \qquad x + 22 \text{ in.} = 0$$

$$x = 8 \text{ in.} \qquad x = -22 \text{ in. (reject)}$$

There is no question that the technique works (some readers may be surprised at that). Here, however, the question is one of gains and losses. The labels confuse an otherwise easy solution. In this instance, it would probably be better to abstract one step further and leave the labels out of the computation. This is another situation in which the pedagogical decision must be made separate from the mathematical validity of a procedure or concept.

Another variation between science and mathematics teachers involves definitions. The mathematics teacher requires precise statements of definitions, whereas the science teacher uses abbreviated definitions such as the following:

velocity = distance/time

force = mass × acceleration

power = work/time

work = force × distance

acceleration = change in velocity/time

Conversations between the science and mathematics teachers are needed so that each department supports the needs of the other department.

Mathematics can exist as a completely self-sufficient, self-generating theoretical science, but its purposes at the secondary level are best served when it is related to applications or physical representations in the real world. That some mathematical concepts have no physical equivalent should not prevent a study of them since there are criteria other than utility for the selection of content. Some concepts are needed for subsequent concepts, and some topics give aesthetic enjoyment of an elegant idea. Mathematics as an elegant tool is important as is the elegance of a mathematical structure.

Learning Exercises

1. Convert 30 mph (30 mi/1 hr) to feet per second using dimensional analysis. Can you make this conversion without using this technique?
2. Look at the word problems of an algebra text. How many do you find that students would perceive as applications to their world? Copy the best one you find to share with others.
3. Find six applications that tie directly to secondary-school mathematics content, at least one of which is geometric, in each of the following:
   (a) A contemporary high school physics text
   (b) A contemporary high school chemistry text
   (c) A contemporary high school biology text
   (d) A contemporary high school social studies text
4. Use appropriate reference tools to determine:
   (a) How Eratosthenes determined the circumference of the earth
   (b) Which planet was located by mathematics before it was first sighted and how it was found.
5. How would exercise 4 support the ideas of this chapter?
6. A practical problem for pilots: Do you need the same amount of gasoline to fly a round trip at constant air speed in a prevailing wind as in still air? Use your knowledge of vectors to explore this problem.
7. Bergmann's law states that individuals of the same species that live farther from the equator are larger. For example, a mallard duck in Alaska would be larger than a mallard

duck living in the contiguous United States. Give a plausible geometric basis for this law in terms of heat loss through surface area and energy through volume. (It may help to think first of a sphere.)

8. How do exercises 6 and 7 differ from most textbook applications problems?

9. PROJECT. Analyze the mathematical competencies required for use of a specific secondary-school science text.

10. PROJECT. Investigate ways in which a computer may be used to help integrate mathematics and another high school course.

Suggestions for Further Reading

Ahrendt, M. H., *The Mathematics of Space*, Holt, Rinehart and Winston, New York, 1960.

Bell, Max S., "Mathematical Models and Applications as an Integral Part of High School Algebra," *Mathematics Teacher*, 64, 4 (April 1971), 293–300.

Bowen, John J., "Mathematics and the Teaching of Science," *Mathematics Teacher*, 59, 6 (October 1966), 536–542.

Brown, Richard, "Predicting the Outcome of World Series," *Mathematics Teacher*, 63, 6 (October 1970), 494–500.

Bryan, William W., "Some Modern Uses of Mathematics," *School Science and Mathematics*, 63, 2 (February 1963), 133–139.

Chi-Ming, Chow, "The Relation Between Distance and Sight Area," *Mathematics Teacher*, 58, 4 (April 1965), 298–302.

Delman, Morton, "Counterpoint as an Equivalence Relation," *Mathematics Teacher*, 60, 2 (February 1967), 137–138.

Dixon, Lyle J., "Approximations and the Teaching of Mathematics and Science," *School Science and Mathematics*, 70, 9 (December 1970), 826–832.

Fehr, Howard, "The Role of Physics in the Teaching of Mathematics," *Mathematics Teacher*, 56, 6 (October 1963), 394–399.

Fischer, Irene, "How Far Is It from Here to There?," *Mathematics Teacher*, 58, 2 (February 1962), 123–130.

Gosbin, Douglas W., and Edward W. Kleppinger, "An Algebraic Approach to Balancing Redox Equations," *School Science and Mathematics*, 67, 1 (January 1967), 9–10.

Hannon, Herbert, "Label $\times$ Label $=$ Label^2—A Point of View," *School Science and Mathematics*, 68, 6 (June 1968), 506–510.

Kleber, Richard S., "A Classroom Illustration of a Nonintuitive Probability," *Mathematics Teacher*, 62, 5 (May 1969), 361–362.

Kline, Morris, "A Proposal for the High School Mathematics Curriculum," *Mathematics Teacher*, 59, 4 (April 1966), 322–330.

McClain, Ernest G., "Pythagorean Paper Folding: A Study in Tuning and Temperament," *Mathematics Teacher*, 63, 3 (March 1970), 233–237.

McFee, Evan E., "Education in Decimal Currency and the Metric System," *School Science and Mathematics*, 69, 7 (October 1969), 644–646.

Mosteller, Frederick, "Progress Report of the Joint Committee of the American Statistical Association and the National Council of Teachers of Mathematics," *Mathematics Teacher*, 63, 3 (March 1970), 199–208.

Pollack, H. O., "Applications of Mathematics," in *Mathematics Education*, Yearbook 69, Part I, National Society for the Study of Education, University of Chicago Press, Chicago, 1970, pp. 311–334.

Röde, Lennart (ed.), *The Teaching of Probability and Statistics*, John Wiley & Sons, Inc., New York, 1970.

Schaaf, William L., "Scientific Concepts in the Junior High School Mathematics Curriculum," *School Science and Mathematics*, 65, 7 (October 1965), 614–625.

Siemens, David F., Jr., "The Mathematics of the Honeycomb," *Mathematics Teacher*, 58, 4 (April 1965), 334–337.

Souers, Charles V., "An Integrated Math-Science Activity for Process Teaching at the Junior High School Level," *School Science and Mathematics*, 66, 1 (January 1966), 3–5.

Thompson, Robert A., "Using High School Algebra and Geometry in Doppler Satellite Tracking," *Mathematics Teacher*, 58, 4 (April 1965), 290–294.

White, Louise G., and Virginia H. Baker, "Systems in Nonmathematical Disciplines," *Mathematics Teacher*, 62, 3 (March 1969), 171–177.

Wick, John W., "Physical Mathematics," *School Science and Mathematics*, 63, 8 (November 1963), 619–622.

Williams, Horace E., "Some Mathematical Models Used in Plastic Surgery," *Mathematics Teacher*, 62, 5 (May 1971), 423–426.

Teaching the Methods of Learning Mathematics 17

Surely the most important single goal of all instruction is to teach *how to learn independently.* Obsolescence does not affect just machines; it is unfortunately a characteristic of people too. We cannot hope to teach anyone all that he will need to know in his lifetime, so we must try to give him skills that will help him to add to his own knowledge, and particularly to his understanding of new and changing concepts. This is the only buffer we can provide our students against their own obsolescence.

Particularly since most students lack skills in listening, reading, and studying, which are all essential to the independent learning of mathematics, we must teach these skills.

Learning through Listening

One of the most effective ways of learning mathematics is for the student to listen and participate in class discussions. Here are some suggestions that may help your students:

> Listen to the statements made and try to correct or improve them.
>
> Participate in the discussion, and ask questions when ideas are not clear.
>
> Try to anticipate what comes next in a discussion or explanation. Study the pattern, sequence, and facts to discover relationships involved.
>
> Take notes of significant ideas or key examples, but don't concentrate on writing so that you miss the idea being presented.
>
> Reflect on what has been discussed both in and out of class. Relate what is discussed to previously learned ideas. Discuss the concepts with other students.

Learning through Reading

With renewed emphasis on reading material in contemporary mathematics textbooks, supplementary books, and programmed texts, the mathematics teacher must know not only the sources of difficulty his students encounter in reading mathematical material but also ways of overcoming these difficulties.

Mathematical statements are unique in several respects. First, mathematical writing is *compressed and concise*, and unfamiliar words may not be skipped in reading mathematical narrative. Second, the vocabulary of mathematics is *highly specialized and technical.* Even in primary school, terms such as "associative" or "operation" are used to express basic ideas. Consider as a case in point the definition of "adjacent angles" in plane geometry: "Adjacent angles are two angles with a (1) *common vertex* and a (2) *common side* (3) *between them.*" Each of the three italicized phrases is necessary for the definition. Examples can be given of angles that fulfill any two of the three phrases that are not adjacent angles. There is, in other words, no "fat" in a mathematical statement. There is no room for carelessness in mathematics.

Third, *the ideas involved in mathematics are abstract.* The reader may be familiar with an elementary example of this confusion: failure to recognize the difference between the abstract idea of number and the symbol representing it; that is, a numeral. When asked why he said that half of 8 is 3, a student answers, "It's the right half." Presumably the left half is Ɛ.

Students of mathematics must be able to relate the abstract ideas to concrete examples, but they must always recognize the differences and the losses that come with the gains when such examples are used. A dot (a pencil or chalk smudge) is a useful representation of a point, but it immediately introduces contaminating aspects of dimension to the pure idea.

A related difficulty is that the ideas frequently lack immediate application outside the mathematics classroom. Although students may learn how to divide one fraction by another fraction, the possibilities for real situations that require this computation are very limited. Often, the applications of mathematical ideas are beyond the scope of the course or the level of understanding of the learner. Complex numbers have wide applicability in electrical theory, but the physics background demanded for such applications is not yet available to the high school student.

Fourth, familiar words (*root, base, irrational, real, log, point, opposite, similar,* and *function*) frequently have a *special meaning in mathematics*; and some mathematical words have several different meanings. For example, the word *root* may refer to the truth set of a linear equation or to the value of a radical. And some words that are often taken as synonyms (for instance, *root* and *radical*) turn out to have different meanings. A square root of 16 is -4, but -4 is *not* a value of $\sqrt{16}$ (radical 16).

A final source of reading difficulty is *the frequent use of symbols.* Often these symbols represent complex ideas, such as

$$\int_a^b \frac{dx}{x}$$

At other times a given symbol has several meanings. Thus, "$-$" may mean "subtract," or "negative," or "opposite," all in addition to meanings of the dash in Standard English. At other times, symbols are so similar that meanings are confused. For example, within a given problem x^1, x', and x_1 would all have different meanings. At other times symbols are omitted. This X represents $+1X^1$. Sometimes several arrangements of symbols will mean essentially the same thing —for instance, $\frac{a}{b}$ or $a:b$, or a/b, or ab^{-1}, or even (a, b).

Factors in Teaching the Reading of Mathematical Material

In deciding what instructions to give a student regarding a given reading assignment, the teacher must consider several factors. First of all, he should consider the *purpose* of the reading assigned. If the purpose is to learn a new, complex idea, then the reading must be careful, slow, and deliberate; each word and symbol must be noted and, frequently, paper and pencil should be used to respond to questions, examples, or computations. In addition, the material should be reread to be sure the ideas are thoroughly clear to the reader. If the material assigned is read merely for the main ideas, then such an intense reading would not be required.

Second, the *difficulty* of the ideas must be taken into consideration. If the reader is not expected to digest all the ideas in the first reading, then there should be instructions for looking up words, working out problems, making drawings, and applying the ideas to specific situations. After that, a second reading would make further clarification more likely to occur.

Third, the teacher should stress that the reading should be done with *intent to learn*, for it is the teacher's role to stimulate interest in the readings and to ensure the maintenance of this interest.

A fourth factor relates to the student's background—his storehouse of ideas and vocabulary and his reading habits. It is essential that the student be *prepared* for the ideas, words, and symbols involved in a new assignment by reviewing background material and defining new words.

To enhance the student's success in reading mathematical material, the teacher should take the following steps:

Establish the purpose of the reading, so that the student approaches the reading with interest.

Be sure the student is ready for the material involved.

Teach the necessary vocabulary and symbols.

Give instruction in the type of reading which is appropriate.

Let the student read the selection silently and independently.

Have some sections read aloud.

Discuss the purpose of the material read.

Have the student reread the material if necessary.

However, all of these steps will be of no avail if they are not followed by a test of achievement of reading skill. The teacher should design and administer reading tests that measure comprehension, rate, and retention. Many schools today have reading specialists on their staffs who are ready and eager to help classroom teachers identify and respond to both individual and group reading problems. Such a resource person can be most helpful in developing a continuing testing program.

To ensure learning through independent reading, the student should be urged to follow these steps:

Be sure the purpose is clear in your mind.

Read the entire passage to get an overview of the ideas involved.

Look up all unfamiliar words and symbols.

Reread the passage very carefully to see how the details fit together.

Work through each step of examples given. Try to work the example by yourself.

Explore the relationship of this material to previously studied ideas. Anticipate where these ideas will probably be used in the future.

Whenever footnotes, suggestions, or hints occur, be sure to check them.

Whenever drawings, tables, or charts are used, compare them with verbal statements.

Write a summary statement in your notebook.

By providing students with a copy of these instructions and by constant reference to them, their importance can be stressed. (Some teachers also provide parents with copies of these and other study suggestions.)

Learning Vocabulary

The student's vocabulary must be built by direct, planned instruction. New words and symbols should be noted on the chalkboard or screen and then pronounced, defined, and discussed. Whenever possible, words should be related to previous experiences or previous vocabulary. Thus, *binomial* may be related to *bicycle*, *bimonthly*, or *bigamy*. Whenever appropriate, words should be illustrated by visual or graphic illustration. For example, the teacher might make the term *sample space* more meaningful to the student by illustrating the events or "sample points" of the tosses of dice. A new word should be applied to a variety of situations to broaden the concepts it represents. Thus, *irrational* is illustrated by numbers such as $\sqrt{3}$, π, and e, and also incommensurable line segments in geometry.

Word mastery grows in the following way: First, words are heard with comprehension; then the words are read; then they are used in speaking; and, finally, they are used in writing. If students are required to write precise statements before they have mastered the ideas, their learning of the ideas may be hampered.

New vocabulary will have to be reviewed frequently in order to maintain vocabulary mastery, and vocabulary tests will reinforce the retention of technical terms.

Learning through Independent Assignments

If independent assignments are to be effective, the teacher must first recognize the role they play in the learning of mathematics. Assignments are significant if they stimulate independent thought, clarify and extend new ideas, and build skill through meaningful practice. These assignments should increase mastery and retention of ideas, and provide a means for individualizing learning through enrichment or remedial instruction. By so doing they will build habits of organized, clear communication of ideas.

The teacher should always keep in mind that assignments are learning exercises, not measures of achievement. This means that students must be made aware that the assignment has been given to assist them in learning, rather than for grading them or punishing them.

As mathematics teachers, we tend to have too much confidence in the necessity of homework. Although research on the benefit of out-of-class assignments has been limited, evidence suggests that these assignments contribute little to achievement in mathematics—at least at the junior high level. However, before we abandon assignments, we must recognize that they are ineffective largely because they are improperly planned. To be effective, learning assignments should have the following characteristics:

The assignments should always be given in writing so that there is no question about what is to be done.

The assignment should be given for a specific purpose, and the student should know what this purpose is.

Whenever necessary, specific instructions should be given on how to prepare the assignment, how to find the necessary information, what materials are needed, and what difficulties are anticipated.

If possible, the assignments should be differentiated to provide for varying ability to handle abstractions. Capable students need to work a few challenging exercises, while slow students need easier exercises.

The assignment should be reasonable in terms of the time and facilities which the student has available. Whenever possible, make assignments for several days in advance.

Be alert to the "grind" aspect of daily assignments.

Include material such as answers to problems assigned so that students are able to check their progress.

Give the assignment at a time when it will contribute the most to the learning at hand.

In no case should the assignment be given after the period has ended. Assignments given after a passing bell suggest a low priority for this activity.

Whenever possible, use assignments to relate the daily lesson to local situations and up-to-date events.

Use assignments as a way of discovering new ideas rather than for mere practice.

Perhaps most important, show your students that you are making assignments part of the required learning activities of your course. Correct papers and return them. Include in your tests ideas developed in homework and not in class.

Learning How to Study Mathematics

The teacher should use some class time for specific instruction on how to study mathematics. He should help students plan a study schedule and acquire the resources needed for home study through individual conferences and he should check their progress by tests and diaries of study activities.

Here are some easy suggestions for the students:

Know exactly what your assignments demand of you. Record your assignment in writing.

Budget your time so that you do not feel rushed or distracted.

Have all the necessary materials at hand, and, if possible, arrange a study place that is quiet, well lighted, and adequately spacious.

Begin work promptly, concentrate on your lesson, and avoid interruptions.

Understand what you are reading or computing, and reflect as you read. When you get lost, go back to the point where your difficulty started.

Organize the new material by outlining the major ideas. Underline main ideas of your text.

Review frequently. Stop to recall what you have been studying. Look for new applications of past lessons. Summarize key ideas and memorize commonly used facts.

Check the accuracy of your work. Estimate answers to see if your result is reasonable, and use these answers to help locate difficulties (like misplacement of a decimal point) but not to suggest methods.

Complete your written work in an organized, neat manner. Use ample writing space to avoid confusion.

Work independently. We learn best and remember longest those ideas we have discovered by ourselves.

Be optimistic about your progress. Enjoy overcoming obstacles. Expect problems to be frustrating.

Accept the responsibility for learning. No school book, course, or teacher ever gave anyone an education; they only provide the opportunity for one. Your teacher cannot do your learning; only you can do that.

Try to learn the material in such a way that you could teach it to your friends. In fact, explaining a concept to a friend is an extremely useful exercise.

Teacher Attitude and Learning Skills

Concentration on the learning of skills and concepts in the day-to-day classroom setting should not prevent a teacher from continually seeking to make students capable of translating their classroom learning into knowledge and attitudes viable in the modern world.

Teachers must recognize that the specific problems, theorems, computation techniques, and even concepts learned in the classroom will be largely forgotten by students, but that classroom learning techniques—good or bad—will be largely retained. Students who indiscriminately take notes and never make any real use of them will probably continue to do so; students, on the other hand, who develop thoughtful approaches to learning will probably continue to do so. These latter students have two great advantages: (1) they are learning more efficiently now, and (2) they are building a basis for learning new material in the future.

Here are some questions for teachers to consider in regard to their own thinking and classroom instruction:

Am I considering *learning efficiency* in my instruction? Early in the year do I spend the time necessary to establish good study and learning skills or do I race ahead only to have poor learning habits catch up with me later in the school year? Dog owners know the advantages of extra time spent early on house breaking, time rewarded by hours and tempers saved later.

Do I encourage students to extrapolate ideas, to apply the concepts learned to broader concerns? Do I encourage them to apply mathematical principles to the sciences and to the arts? Do I seek to direct students to explore these relationships independently?

Do I work cooperatively with other teachers to coordinate efforts to attack this problem of learning how to learn? Occasionally, teachers can work together on a unit. If, for example, students are reading *Gulliver's Travels* for their English class, their science and mathematics teachers can discuss the problems of size and proportion.[1]

Do I work individually with both strong and weak students to help them develop better learning skills and habits? Do I attempt to encourage depth of interest in my students by all the means at my command?

[1] For a delightful presentation of these ideas see Peter Weyl, *Men, Ants and Elephants: Size in the Animal World*, The Viking Press, Inc., New York, 1959.

Do I seek the cooperation of the parents in encouraging student independence and self-reliance? Unwitting parents often undermine school functions with statements like "I hated school too" or "Math was my poorest subject and look at me now." Brought into contact with schools and made aware of your broad objectives, especially as related to their children, parents can be very supportive and can contribute a great deal to your program.

Is it possible for me to bring in a mathematician to tell about his work or arrange a field trip to a research center to see mathematics being applied?

Do I continue to learn and to treat the acquisition of knowledge as a lifelong opportunity, and transmit this enthusiasm to my students? Or have I based my teaching of mathematics on what I learned and the way I learned it years ago?

It has been said that the successful teacher is one who makes himself increasingly unnecessary to his students. The class that continues to learn when the teacher is called from the room or when a visitor interrupts class activities is often demonstrating the results of superior instruction. To attain this level of independent study, students require instruction in the process of learning and encouragement to stimulate their activities.

An excellent source of information is the NCTM booklet *How to Study Mathematics* by Henry Swain. Use of a classroom set of these pamphlets would give strong support to a program designed to promote independent learning.

Learning Exercises

1. It has been said that the best goal of an instructional program is to bring the student to the point at which he can study independently and successfully content on the same intellectual level as his most recent classroom course. For example, the test of whether a high school student has been taught algebra and geometry well would be his ability to study analytic geometry or trigonometry independently. For what proportion of our high school students do you think that this goal is achieved? Do you think that the goal is a reasonable one? How would accepting this goal change your instructional program?
2. Analyze your own skills as they relate to:
   (a) Learning through listening in the mathematics classroom
   (b) Learning through reading mathematics
   (c) Learning through completing mathematics assignments
   (d) Learning through independent study of mathematics
3. How does your analysis in exercise 2 apply to your teaching of others? What specific things would help you to improve your skills? Would they apply equally to your students?
4. Locate at random three consecutive pages of a high school mathematics text. Determine, by counting, the proportion of the words that are technical. How must the technical vocabulary of this subject affect the teaching program as compared to instruction in English?
5. PROJECT. Read enough of a mathematics library book that you think would be of interest to secondary-school students to write an annotation describing the book in a way that might encourage students to read it.
6. PROJECT. Read *Men, Ants and Elephants: Size in the Animal World* (see footnote 1) and the first two books of *Gulliver's Travels.* Describe in some detail how this material could be used in both science and mathematics courses.

Suggestions for Further Reading

Brannon, M. J., "Individual Mathematics Study Plan," *Mathematics Teacher*, 55, 1 (January 1962), 52–56.

Cohen, Donald, "On Organizing a Mathematics League—A Report," *School Science and Mathematics*, 63, 2 (February 1963), 145–146.

Earp, N. Wesley, "Problems of Reading in Mathematics," *School Science and Mathematics*, 71, 2 (February 1971), 129–133.

Fehr, Howard, "General Ways to Identify Students with Scientific and Mathematical Potential," *Mathematics Teacher*, 46, 4 (April 1953), 230–234.

Grossman, George, "Advanced Placement Mathematics—For Whom," *Mathematics Teacher*, 55, 7 (November 1962), 560–566.

Johnson, Larry K., "Organizing and Sponsoring a Mathematics Club," *School Science and Mathematics*, 63, 5 (May 1963), 424–432.

Lloyd, Daniel B., "Ultra-Curricular Stimulation for the Superior Student," *Mathematics Teacher*, 46, 7 (November 1953), 487–489.

Mann, John E., "Polygon Sequences—An Example of a Mathematical Exploration Starting with an Elementary Theorem," *Mathematics Teacher*, 63, 5 (May 1970), 421–428.

Randall, Karl, "Improving Study Habits in Mathematics," *Mathematics Teacher*, 55, 7 (November, 1962), 553–555.

Rollins, Wilma E., *et al.*, "Concepts of Mathematics—A Unique Program of High School Mathematics for the Gifted Student," *Mathematics Teacher*, 56, 1 (January 1963), 26–30.

Smart, J. R., "Searching for Mathematical Talent in Wisconsin, II," *American Mathematical Monthly*, 73, 4 (April 1966), 401–406.

Swain, Henry, *How to Study Mathematics*, National Council of Teachers of Mathematics, Washington, D. C., 1955.

Trimble, Harold C., "The Heart of Teaching," *Mathematics Teacher*, 61, 5 (May 1968), 485–488.

Wirszup, Izaak, "The School Mathematics Circle and Olympiads at Moscow State University," *Mathematics Teacher*, 56, 4 (April 1963), 194–210.

Developing Instructional Techniques

18

One of the basic requirements for success in teaching mathematics is the development of a repertoire of special techniques to meet particular instructional problems. While many individual teachers have the intellectual resources to develop truly creative approaches to lessons on their own, they too often fail to take advantage of methods developed by others.

One of the best ways for a teacher to find alternate and creative teaching techniques is by direct reference to the extensive literature of mathematics and mathematics education, and to the better texts and teachers' manuals used in classrooms. Naturally, a compendium of "tricks of the trade" would be virtually endless, but the teacher must select from the various techniques those that will work for him, that fit his goals and style of presentation, and that are interesting to him and his pupils.

Many teachers do read extensively and make a real effort to learn from fellow staff members in local discussions and from a broader segment of teachers at state and national conferences. Too often, however, the excellent assistance derived from such interchange is lost if the teacher has made no record of the new ideas. For the teacher quickly forgets his newly found technique and therefore has no way of retaining it. In many cases the same thing happens even to self-developed techniques. A failure to record such activities for future reference means a loss to students and a duplication of effort on the part of the teacher.

To avoid the dilemma of not having various techniques at hand the beginning teacher should develop files of *instructional techniques*, and continue using them for the rest of his teaching career. Such a file would be organized by courses and within courses by topics. The file may at first require only desk-drawer space or a notebook but a continued development would almost certainly create the need for a filing cabinet with a drawer devoted to each subject area and an additional drawer for general techniques.

Types of Material for the Instructional File

Here are some suggestions regarding the kinds of ideas that could be included in the Instructional Technique File:

A unique way of developing the mathematical content of a topic. A typical example is the use of a finite number system as an introduction to a mathematical system.

Historical background related to the development of a topic. For example, the story of Eratosthenes and his method of measuring indirectly the circumference of the earth.

A striking problem to be used to introduce a topic, to motivate the discussion, or to stimulate discovery. For example, locating the path of a basketball can be the basis for an introduction to quadratic equations. The formula $S = S_0 + V_0t - 16t^2$ for given values of height of the ball at release (S_0) and initial velocity (V_0) may determine such things as the two times at which the ball will be at the height ($S = 10$) of the basket.

Alternate strategies for teaching a topic, as described in Chapter 10.

A fresh and stimulating way to review a topic, such as that provided by an identification game.

A game, trick, fallacy, or puzzle related to a topic. The game of Nim can be an excellent setting for examining number patterns and arriving at a generalization.

A specific pitfall to avoid in teaching a topic with some suggestions of how to avoid this pitfall. For example, a common error in dealing with algebraic fractions is to "cancel" as follows:

$$\frac{3x + \cancel{a}}{2y + \cancel{a}} = \frac{3x}{2y}$$

Some good questions that lead into a topic, point out special aspects of a topic, or otherwise direct the thinking of students toward significant results. A typical example might be to ask under what conditions does

$$\frac{a}{b} + \frac{c}{d} = \frac{a + c}{b + d}$$

A useful diagram or a special technique for drawing a diagram. For example, correct drawings of three-dimensional objects are easily done with a stencil.

A logical fallacy that forces students to look deeper into a subject for answers. The proof that arrives at the conclusion that every triangle is an isosceles triangle may be used to challenge the students to find the loophole in the proof.

A laboratory exercise that promotes learning—for example, the exploration of algorithms by the use of calculators.

A project idea or a resource for an enrichment exercise, such as an exploration of recreational aspects of topology.

An application to a local-interest angle, such as the mathematical relationship between students' achievement in the classroom and car ownership.

An interesting extension of a topic to higher dimensions or a more general case—for example, the distance formula and n-dimensional space.

A geometric analysis for an algebraic or arithmetic problem or an algebraic analog for a geometric problem. One example is an algebraic proof of the Pythagorean theorem. Another is the geometric illustration of the arithmetic, harmonic, and geometric means.

The rationale for a computational procedure or an arithmetic shortcut such as the algorithm for the square root. This algorithm has an algebraic explanation in terms of $(x + y)^2$ as well as a geometric illustration.

A new technique to represent a mathematical idea. In this regard, paper folding or curve stitching could illustrate conic sections.

Examples of Special Problems

Here are some examples of material for the Instructional Techniques File, organized under subject and topic.

Arithmetic

Graphing. The use of figures or pictures to represent the quantities in a bar graph is widespread and may be misleading—sometimes, as in advertising, purposely. In the example shown in Figure 18.1, the vertical scale represents the number

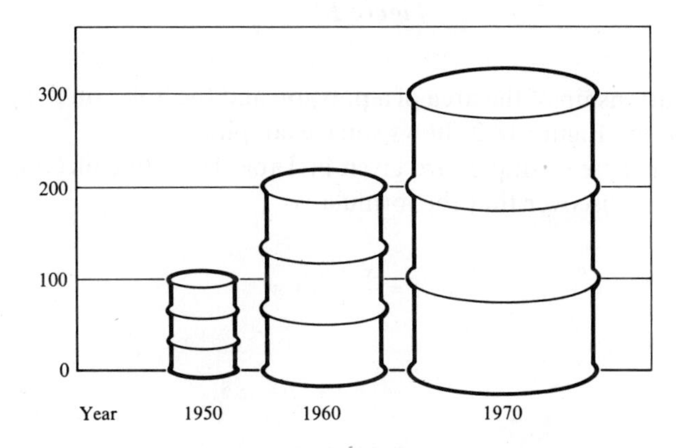

Figure 18.1 ***Number of Barrels***

of barrels of oil, showing that the production in the three recorded years is 100, 200, and 300 respectively. The figures, however, are representations of volume instead of length. If the first represents 100 gallons, the second would represent eight times as much, 800 gallons; and the third, 27 times as much—2,700 gallons. The unsuspecting reader sees a representation of a cubic rather than a linear growth. A variety of similar examples can be found in *How to Lie with Statistics.*[1]

Area. How is the number of holes in a pegboard related to the area of a polygonal region of the pegboard?

Assume that there are holes at the vertices of each unit of area. Also assume that the polygons involved will all have vertices at points with holes. Thus, there will be holes at vertices, holes on the perimeter, and holes inside the polygon.

[1] Darrell Huff, *How to Lie with Statistics*, W. W. Norton & Company, Inc., New York, 1954.

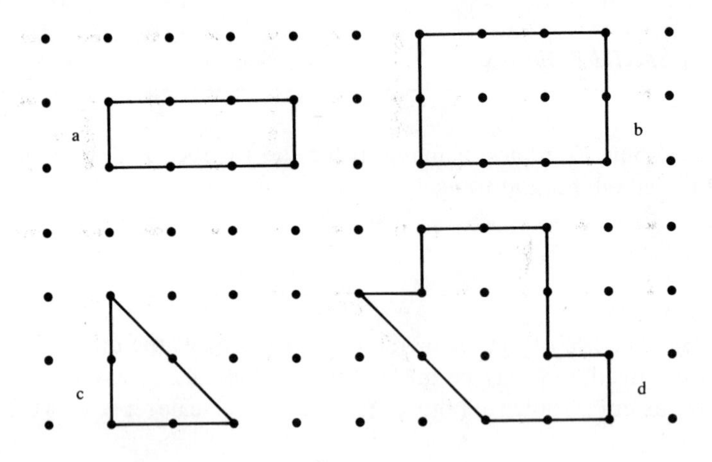

Figure 18.2

What is the relationship of the area of a polygon and the holes on the perimeter or inside the polygon? Figure 18.2 shows some examples.

The data for these examples are given in Table 18.1. By collecting additional data, students can discover that the formula is

$$A = \frac{x - 2}{2} + y$$

Table 18.1

	No. holes on perimeter (x)	No. holes in interior (y)	Area (A)
a	8	0	3
b	10	2	6
c	6	0	2
d	12	2	7

Then conditions should be varied. Consider polygons with common vertices. Consider regions that are not convex polygons. Finally, state the generalization that takes care of all conditions.

Geometry. Geometric relationships give students another application of commutative and associative laws. For example, for Figure 18.3, they might note (1) several ways the area formula for a triangle may be justified from the corresponding formula for a parallelogram and (2) the algebraic relations.

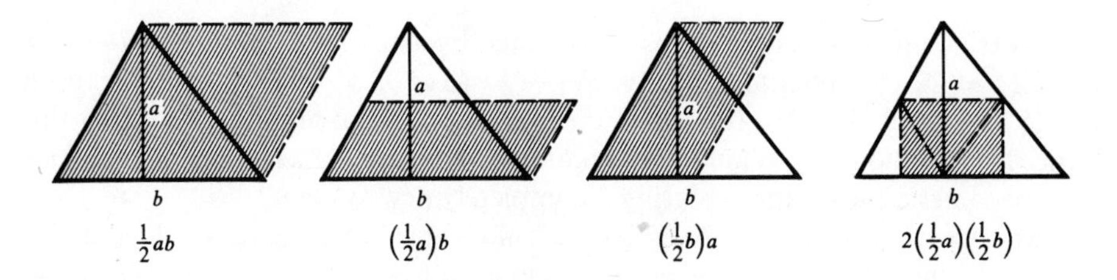

Figure 18.3

Algebra

Patterns. The patterns of certain number relations are discovered rather quickly.

Sum of consecutive counting numbers beginning with 1:

$$S = 1 + 2 + 3 + 4 + 5 + 6 + 7 + 8 + 9 + 10$$

$$S = \frac{10 \times (11)}{2} = 55 \qquad \text{or} \qquad S = \frac{n(n + 1)}{2}$$

Sum of consecutive odd numbers:

$$S = 1 + 3 + 5 + 7 = 16$$
$$S = 4 \times 4 \qquad \text{or} \qquad S = n^2$$

Sum of consecutive even numbers:

$$S = 2 + 4 + 6 + 8 + 10 = 30$$
$$S = 5 \times 6 \qquad \text{or} \qquad S = n(n + 1)$$

Differences of the squares of two consecutive numbers:

$$4^2 - 3^2 = 7 \qquad 5^2 - 4^2 = 9$$
$$9^2 - 8^2 = 17 \qquad 12^2 - 11^2 = 23$$

$$\text{If } x^2 - y^2 = n, \text{ then } n = x + y$$

Sum of rational numbers of the form $1/n(n + 1)$:

$$n = 1 \qquad S = \tfrac{1}{2}$$
$$n = 2 \qquad S = \tfrac{1}{2} + \tfrac{1}{6} = \tfrac{2}{3}$$
$$n = 3 \qquad S = \tfrac{1}{2} + \tfrac{1}{6} + \tfrac{1}{12} = \tfrac{3}{4}$$

$$\text{Thus } S = \frac{n}{n + 1}$$

Maneuvers on a lattice. An imaginative operation has been proposed by David Page. He suggests that a lattice such as the following be used:

51	52	...							
41	42	43	44	45	46	47	48	49	50
31	32	33	34	35	36	37	38	39	40
21	22	23	24	25	26	27	28	29	30
11	12	13	14	15	16	17	18	19	20
1	2	3	4	5	6	7	8	9	10

Next he introduces operations represented by arrows: (→) (↑). Then 7 → = 8 and 12 ↑ = 22. Combinations such as 32 ↑ → = 43 or 32 ↗ = 43. What is the result for 25 ↑ → → ↓? What is 10 → ? Is the operation ↑ closed with respect to this lattice? What about →? What about commutativity and associativity? This lattice then becomes a means for exploring a completely new system.

Many extensions of this basic idea should suggest themselves to the creative teacher. Using a lattice with seven entries in each row or weaving back and forth are possibilities. Applying the operation to a calendar and interpreting the symbols (for example, ↑ is "a week ago") is only one of many others.

Shortcuts for squaring numbers. These shortcuts are based on knowledge of the squares of numbers from 1 to 25.

For numbers from 26 to 75. Example: Square 46.

(1) First two digits	$N - 25$	$46 - 25 = 21$
(2) Last two digits	$(50 - N)^2$	$4^2 = 16$
(3)		*Answer:* 2,116
(4) Why?		*Answer:* $N^2 = 100(N - 25) + (50 - N)^2$

For numbers from 76 to 100 (best 91–100). Example: Square 93.

(1) First two digits	$N - (100 - N)$	$93 - (100 - 93) = 86$
(2) Last two digits	$(100 - N)^2$	$7^2 = 49$
(3)		*Answer:* 8,649
(4) Why?		*Answer:* $N^2 = 100\,[N - (100 - N)] + (100 - N)^2$

For numbers from 101 to 125. Example: Square 106.

(1) First three digits	$2N - 100$	$212 - 100 = 112$
(2) Last two digits	$(N - 100)^2$	$6^2 = 36$
(3)		*Answer:* 11,236
(4) Why?		*Answer:* $N^2 = 100(2N - 100) + (N - 100)^2$

Why were examples chosen within 10 of 50 and 100? *Answer:* To avoid regrouping, but the rule still works for other numbers.

Exponents. Beware the temptation to prove $a^0 = 1$ and $a^{-n} = 1/a^n$ from the definition of a^n. Instead, define these terms separately.

If we define

$$a^n = \underbrace{a \cdot a \cdot a \cdot a \ldots \cdot a}_{n \text{ factors}}$$

then we assume that n is a natural number (a positive integer). We then can establish rules like

$$\frac{a^m}{a^n} = a^{m-n} \quad \text{and} \quad \frac{a^m}{a^n} = \frac{1}{a^{n-m}}$$

only for $m > n$ in the first case, $n > m$ in the second, because we base our argument on counting and applying the definition. For example:

$$\frac{a^m}{a^n} = \frac{\overbrace{a \cdot a \cdot a \cdot a \ldots \cdot a}^{m \text{ factors}}}{\underbrace{a \cdot a \cdot a \ldots \cdot a}_{n \text{ factors}}} = \frac{\overbrace{(a \cdot a \ldots \cdot a)}^{n \text{ factors}} \cdot \overbrace{a \cdot a \cdot a \ldots \cdot a}^{m - n \text{ factors}}}{\underbrace{a \cdot a \cdot a \cdot a}_{n \text{ factors}}} = \overbrace{a \cdot a \cdot a \ldots \cdot a}^{(m-n) \text{ factors}}$$

Now we ask what would we like to have a^0 equal in order to fit (and not contradict) our definition.

Since $a^m/a^m = 1$ for any m, to fit our rule for division, we would like

$$1 = \frac{a^m}{a^m} = a^{m-m} = a^0$$

To make this the case we must *define* $a^0 = 1$. This development also suggests *how* mathematicians arrive at many definitions naturally rather than arbitrarily.

The preceding lesson is a good one for discovery techniques and shows how the brakes must be occasionally applied to avoid trouble.

Division of polynomials. Use $1/(1 - x)$ as an example.

$$\begin{array}{r l} & 1 + x + x^2 + \ldots \\ 1 - x) & 1 \\ & \underline{1 - x} \\ & \quad x \\ & \quad \underline{x - x^2} \\ & \qquad x^2 \\ & \qquad \underline{x^2 - x^3} \end{array}$$

But do not stop there. Raise some questions about the resulting statement:

$$\frac{1}{1 - x} = 1 + x + x^2 + x^3 + \ldots$$

Is it a true statement for all values of x? Students recognize the fact that it isn't true for $x = 1$ (dividing by zero) but usually look no further.

What about $x = -1$? $\frac{1}{2} = 1 - 1 + 1 - 1 + 1 - 1 + \ldots$
What about $x = 2$? $-1 = 1 + 2 + 4 + 8 + \ldots$

Now they wonder if there are any true values.

What about $x = \frac{1}{2}$? $2 = 1 + \frac{1}{2} + \frac{1}{4} + \frac{1}{8} + \ldots$

At this point students should be ready to do some exploring on their own and are more willing to question apparently satisfactory computations. Some students may wish to look ahead to an independent study of sequences and series.

Historical note: The resulting equation for $x = -1$

$$\frac{1}{2} = 1 - 1 + 1 - 1 + 1 - 1 + \ldots$$

led some mathematicians (even Euler temporarily) to accept this result. Such unsatisfactory results led to a more careful logical development of infinite sequences and the calculus.

What about the sequence $1 - 1 + 1 - 1 + \ldots$?

If $S = 1 - 1 + 1 - 1 + \ldots$
then $S = (1 - 1) + (1 - 1) + (1 - 1)\ldots$ or $S = 0$
If $S = 1 - 1 + 1 - 1 + \ldots$
then $S = 1 - (1 - 1) - (1 - 1)\ldots$ or $S = 1$

Which of these is correct?

The Hindu method for deriving the quadratic formula. Once you know the statement of the quadratic formula, you can reconstruct its derivation "from the inside out" merely by reconstructing the statement by "bombardment."

$$ax^2 + bx + c = 0$$
$$ax^2 + bx = -c$$

Complete the term containing c by multiplying by $4a$.

$$4a^2x^2 + 4abx = -4ac$$

Add b^2.

$$4a^2x^2 + 4abx + b^2 = b^2 - 4ac$$

Square roots:

$$2ax + b = \pm\sqrt{b^2 - 4ac}$$

$$2ax = -b \pm \sqrt{b^2 - 4ac}$$

$$x = \frac{-b \pm \sqrt{b^2 - 4ac}}{2a}$$

Geometry

Equality of area, perimeter, and volume. When will the area and perimeter of a triangle by numerically equal? The first example to come to mind is the 6–8–10 right triangle. What other right triangles with sides of integral measures have this property?

If the sides of a right triangle are x, y, and z, then $x^2 + y^2 = z^2$. The perimeter is $x + y + z$, and the area is $\frac{1}{2}xy$. Thus

$$\tfrac{1}{2}xy = x + y + z \qquad \text{or} \qquad \tfrac{1}{2}xy = x + y + \sqrt{x^2 + y^2}$$

Then

$$y = \frac{4(x - 2)}{x - 4} \qquad \text{if} \qquad x \neq 4$$

The graph of this equation is shown in figure 18.4. We also know that $x + y > z$ and $x > 0$ and $y > 0$. It is obvious that $y > 0$ if $x > 4$ and that $y < 0$ for $2 < x < 4$ in figure 18.4.

The isosceles right triangle results when $x = y = 4 + 2\sqrt{2}$ or $x \approx 6.8$ and $y \approx 6.8$. Thus the interval $4 < x < 6.8$ represents all possible measures of the short side of the right triangle having area and perimeter equal. The only integers in this interval are 5 and 6.

Hence, the only right triangles with sides having integral measures and the perimeter equal to the area are (6–8–10) and (5–12–13).

Under what conditions are the area and the perimeter of any triangle numerically equal? The result is that the radius of the incircle is 2. When any triangle is

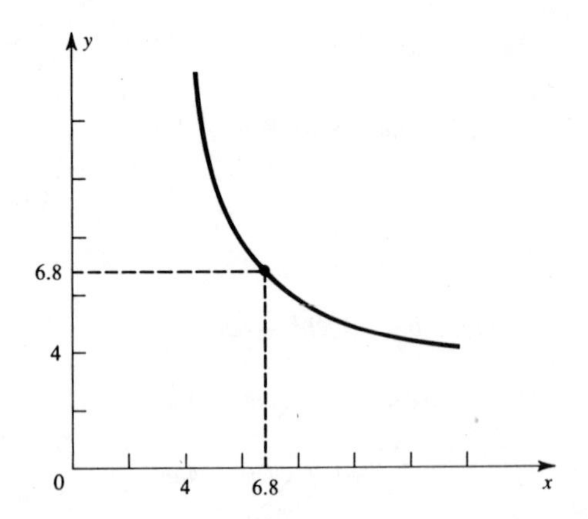

Figure 18.4 ***The graph of*** $y = 4(x - 2)/(x - 4)$

extended to three spaces, the measures of lateral surface and volume are equal when the radius of the insphere is 3. An extension to n-space gives $r = n$.[2]

A comparison of means. Suppose we wish to show geometrically the comparison of the arithmetic mean, geometric mean, and harmonic mean of a and b. (The relationship of these means is illustrated by (Figure 18.5.)

$$\text{Arithmetic mean} = \frac{a + b}{2}$$

$$\text{Geometric mean} = \sqrt{ab}$$

$$\text{Harmonic mean} = \frac{2ab}{a + b}$$

On a horizontal axis locate X and Y so that $AX = a$ and $XB = b$ and Y is the midpoint of $\overline{AB}$. At A and B construct lines $\overline{AR}$ and $\overline{BT}$ perpendicular to $\overline{AB}$ and such that $AR = a$ and $BT = b$. Draw $\overline{AT}$ and $\overline{BR}$ which intersect at S. Then $\overline{XS}$ is parallel to $\overline{AR}$ and $\overline{BT}$. Why? With $\overline{AB}$ as a diameter and Y as the center draw a semicircle. Draw $\overline{YM}$ as a radius perpendicular to $\overline{AB}$. Draw a circle tangent to $\overline{AB}$ with center at S. Extend $\overline{SX}$ which intersects the semicircle (Y) at G. The semicircle (Y) and circle (S) are tangent.

Since $\overline{YM}$ is the radius of the semicircle whose diameter is $a + b$, YM is the arithmetic mean of a and b. The half-chord $\overline{XG}$ in the semicircle (Y) represents the

[2] For details of this extension see Leander W. Smith, "Conditions Governing Numerical Equality of Perimeter, Area, and Volume," *Mathematics Teacher*, 58, 4 (April 1965), 303–306.

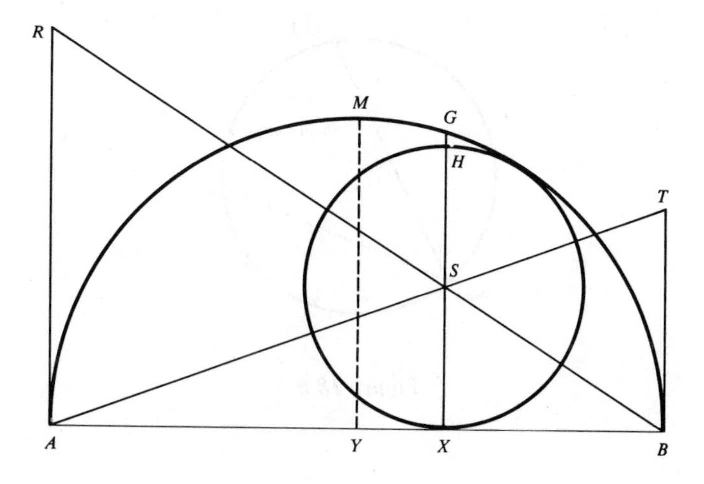

Figure 18.5

geometric mean. In the circle (S), the radius XS is the harmonic expression

$$\frac{a \cdot b}{a + b},$$

and the diameter XH is the harmonic mean. Since circle (Y) and circle (S) are tangent, $XH < XG < YM$. (If $a = b$, the same diagram applies except that $X = Y$ and the three points M, G, and H coincide. In this instance, the three means are equal.)

Suppose AY and XY represent air speed and wind speed, respectively. Then AX is the ground speed with tail wind, BX is the ground speed with head wind, and XH is the average speed for a round trip.

Angle measurement on the circle. With the resurgence of interest in small boats, application of theorems about inscribed angles to navigation will be of interest to many students.

Figure 18.6 shows a chart indicating shoal waters within a circle passing through two navigation lights. Boats in the area avoid the danger by keeping the angle between the two lights less than 42 degrees.

To establish the rationale for this type of navigation, we need only apply an inscribed-angle theorem and the theorem relating an exterior angle of a triangle to nonadjacent interior angles.

In each instance, $m(\angle b) = 42°$. In Figure 18.7a, $m(\angle a) = m(\angle b)$; in Figure 18.7b, $m(\angle a) > m(\angle b)$; and in Figure 18.7c, $m(\angle a) < m(\angle b)$.

This is a good example of a rather trivial application of a geometric idea that still gives students a feeling of the power of mathematics in a situation they can understand and appreciate.

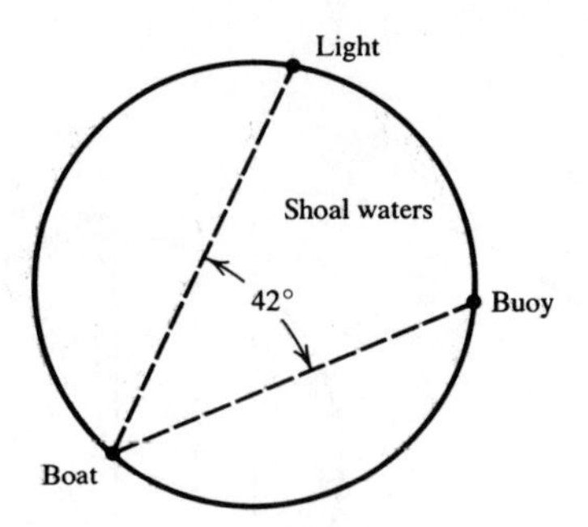

Figure 18.6

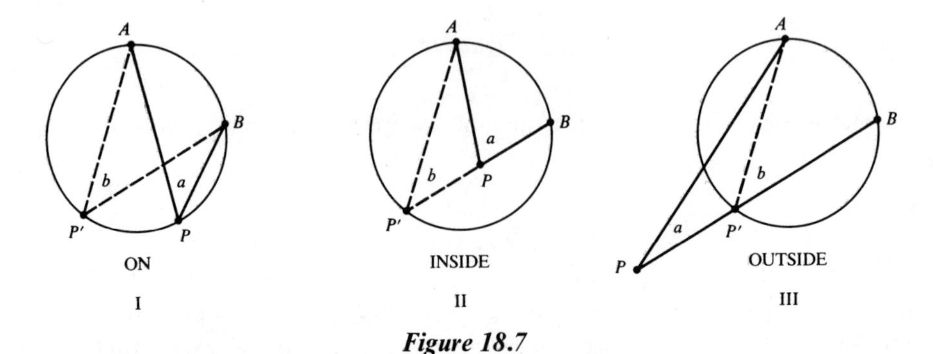

Figure 18.7

Indirect proof. Start with a detective story, a familiar setting for the application of indirect proof.

Four people, alone on an island, have a violent argument. The group separates, each going his own way. When three of them return to the original place later, they discover the fourth seriously wounded. The wounded man is not able to identify his assailant. The hero of the story, knowing that he didn't commit the crime, eliminates himself as a suspect. He eliminates a second suspect who was within his view during the entire time the group was separated. He eliminates the wounded man because the wounds are of a type that could not be self-inflicted. He knows the criminal is the fourth man.

Students can discuss the reasoning contained in this example and may provide better examples themselves. Such a discussion is important because some students have a difficult time accepting this type of proof. Once the discussion has given students a chance to explore informal reasoning, they are encouraged to abstract from their discussion the technique of indirect reasoning: (a) State all possibilities, *one of which must be true.* (b) Eliminate all but one possibility by showing that they contradict given facts. (c) Accept the remaining possibility as proved.

Now return to the story to see how these steps are carried out.

Ruler and compass constructions. Through a point inside an angle but not on the angle bisector, construct one of the two circles tangent to the sides of the angle. (See Figure 18.8).

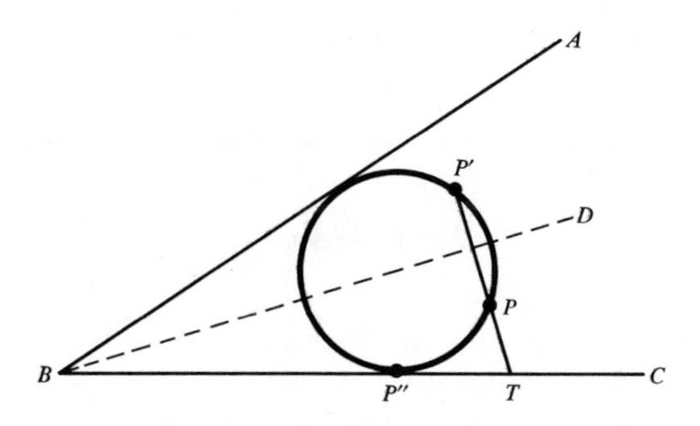

Figure 18.8

Given: $\angle ABC$ and point P.
Solution:

1. Construct $\overrightarrow{BD}$ the bisector of $\angle ABC$.
2. Locate P' symmetric to P on the opposite side of $\overrightarrow{BD}$.
3. Extend $\overline{P'P}$ to T.
4. Construct the mean proportional $\overline{TP''}$ to $\overline{TP}$ and $\overline{TP'}$.
5. Mark off $\overline{TP''}$ on $\overline{BC}$.
6. Construct the circle through P, P', and P''.

The basis for this proof is the theorem "The tangent to a circle is the mean proportional between the secant and its external segment."

Center of gravity. Many geometry books state that the center of gravity of a triangle is the point of intersection of the medians. This is a physical interpretation with no counterpart in abstract geometry because it assumes a uniform weight distribution. Even with such an assumption, it contradicts the usual modern interpretation of a triangle as one-dimensional (like a coat hanger). The following argument illustrates the fact that the intersection of the medians is the wrong answer. On Figure 18.9, let the weight of the sides be distributed evenly, with the unit of weight equivalent for each unit length. The weight of the sides, a, b, and c, then is also a, b, and c. This weight may be considered concentrated at the midpoint of these sides, at R, P, and Q respectively. The center of gravity of lines b and c together, then, is a point, S, on segment $\overline{PQ}$. Unless b and c are equal, S will *not* be the midpoint of this line. The center of gravity, G, then will lie on the line $\overline{SR}$ and not at R, which is generally the only point of intersection of this line with the median, $\overline{CR}$.

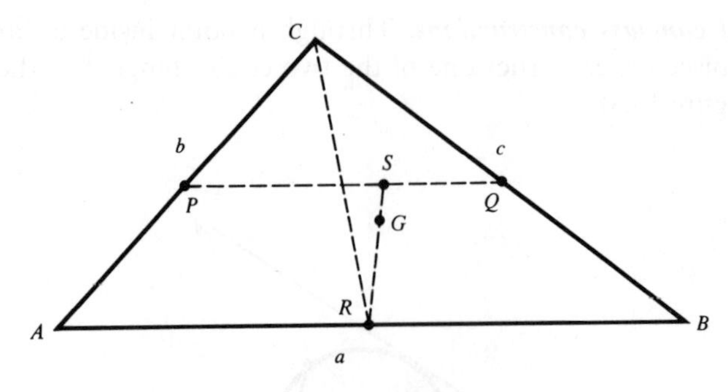

Figure 18.9

In fact, it may be shown that the equilateral triangle is the only case in which the center of gravity of the one-dimensional triangle and the point of intersection of the medians are the same. Students may wish to test this fact by locating the center of gravity of a wire triangle model with threads and attempting to balance the hanger from this point (see Figure 18.10). A further exploration may be made to

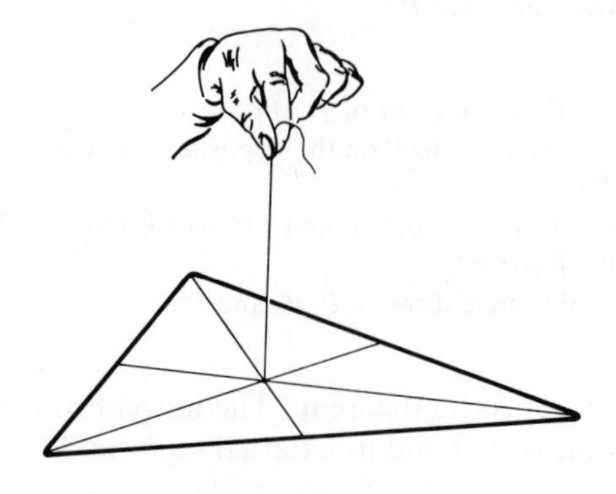

Figure 18.10

discover whether the zero-dimensional (3-point) triangle and the two-dimensional triangle (region) have their centers of gravity at the point of intersection of the medians.

Trigonometry

Wrapping functions. As an introduction to the idea of wrapping functions, graph the length of the string wrapped counterclockwise from the origin as the independent variable, the distance from the x-axis as the dependent variable on the diagram of Figure 18.11. A physical model of this with a graph paper background for graphing and a square block of wood $\frac{3}{8}$ inch to $\frac{5}{8}$ inch thick with twine wrapped

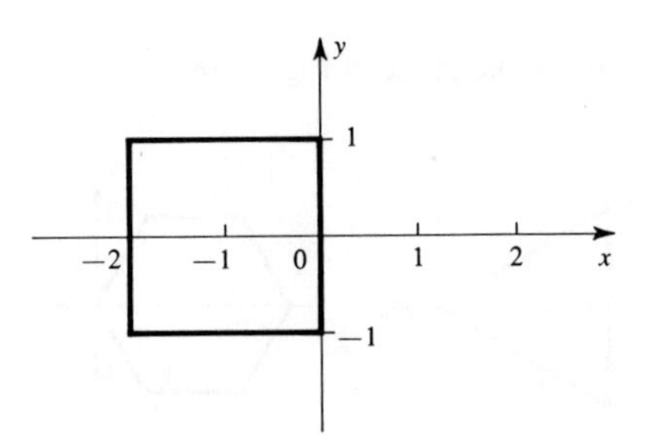

Figure 18.11

around it helps students to see what you are doing. The graph looks like Figure 18.12. The line y_1 shows how a point is located. The arc shows how the string is unwound from P along the x-axis to give x_1; y_1 is the ordinate to P.

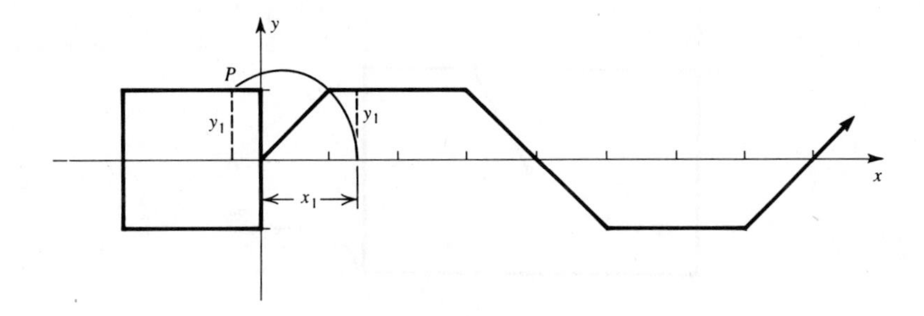

Figure 18.12

Now develop informally some theorems about the function $S(x)$. For example:

$$S(x) = S(x + 8)$$
$$S(x) = -S(x + 4)$$
$$S(6 + x) = S(6 - x) \quad \text{for } x < 6$$

This provides background for the more complicated wrapping functions on a circle. It may be a source of enrichment activities after a study of the circular functions. Questions for research then might be formulated:

$S(x)$ acts somewhat like sine x. What are some similarities? Differences?
Define a function, $C(x)$, that corresponds to cosine x on the square. Graph it and state some theorems about it.
State some theorems relating $C(x)$ and $S(x)$.
Is then a $T(x)$ like tangent x?

Try some other figures like those in Figure 18.13.

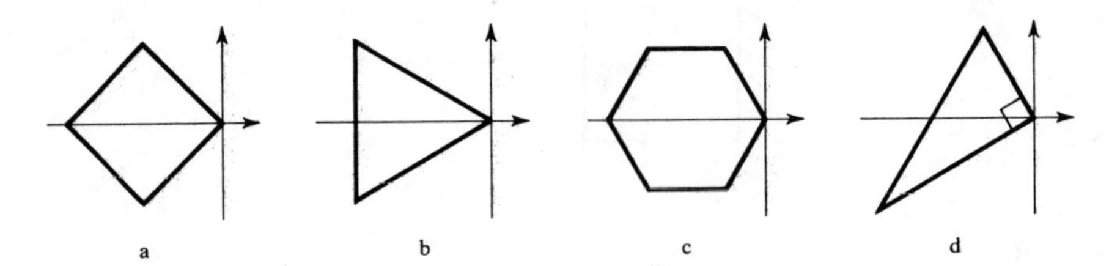

Figure 18.13

Graph the distance of a point on a rectangle from the center of the rectangle as a function of this distance of the point from a given starting point. (See figure 18.14).

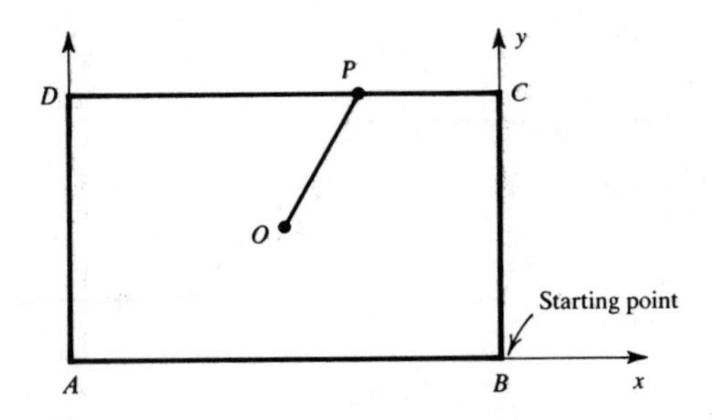

Figure 18.14 *Graph $y = \overline{OP}$ against $x = \overline{BCP}$*

Advanced Mathematics

Analytic Geometry. A problem that draws together many seemingly unrelated parts of mathematics is always useful. Here is one such problem, illustrated in figure 18.15:

> A fixed circle of radius 3 has its center at the point (3, 0) on the x-axis. A second circle has its center at the origin. Its radius is made to approach zero. Consider the line joining two points on this second circle: the intersection with the y-axis and the intersection with the other circle. How far from the origin will this line intersect the x-axis?

Most students will agree that as the variable circle shrinks, points of intersection of the line with the x-axis go farther and farther out. So far so good. But students also believe that the lines move out an unbounded distance. This is not

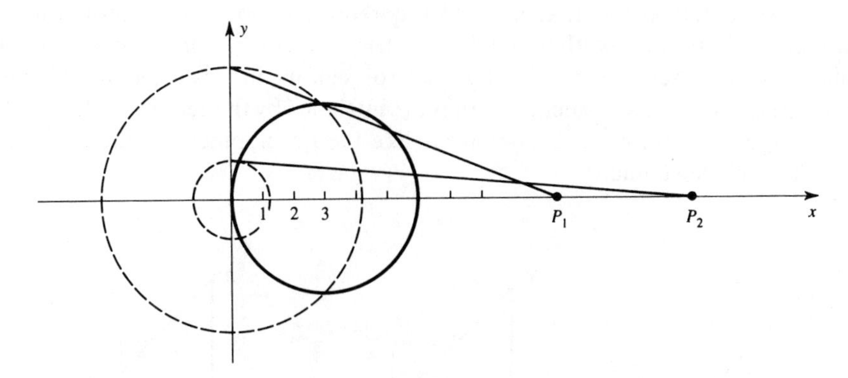

Figure 18.15

the case as they can show by developing the following:

Letting the radius of the variable circle be r, the equations of the two circles:

$$x^2 + y^2 = r^2 \qquad \text{and} \qquad (x-3)^2 + y^2 = 9$$

The coordinates of the two points of intersection described in the problem: $(0, r)$ and

$$\left(\frac{r^2}{6}, \frac{r}{6}\sqrt{36 - r^2}\right)$$

the latter by solving the equations of the two circles simultaneously.
The equation of the line through these two points:

$$y = \frac{\sqrt{36 - r^2} - 6}{r}x + r$$

The x-value of the intersection of this line with the x-axis ($y = 0$):

$$x = \frac{r^2}{6 - \sqrt{36 - r^2}}$$

The x-value rationalizing the denominator of this expression:

$$x = 6 + \sqrt{36 - r^2}$$

Once this expression is reached, this x-value cannot be made greater than 12, no matter how small r becomes—an entirely unexpected result.

This problem is also one that gives students, informally at least, additional insight into the limit process.

Limits. Some students become so enamored of limit processes that they want to use them without restraint. An example that should caution them is the following:

We set out to prove that favorite theorem of junior high school students: Since in a right triangle (with hypotenuse c; legs a and b) $c^2 = a^2 + b^2$, we can take square roots to get $c = a + b$. High school seniors in advanced mathematics know the error in this statement, but they are intrigued by the geometric demonstration of figure 18.16 (which is somewhat like the development of the circle area formula in plane geometry).

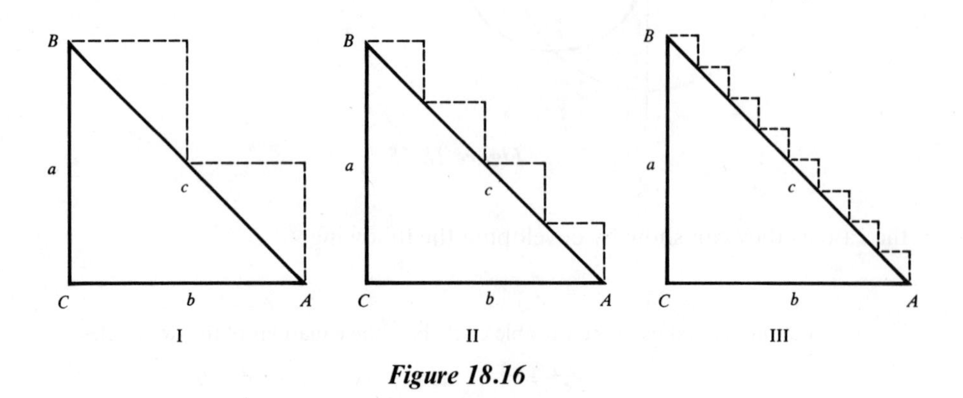

Figure 18.16

In Figure 18.16a, $c(= AB)$ is bisected and the dotted lines parallel to a and b are drawn. The distance along the dotted line is $a + b$. The line c is further subdivided in Figures 18.16b and c, but still the length of the dotted line is $a + b$. Obviously it will soon be impossible to detect any difference between the dotted line joining A to B and the line AB. In other words we may say

$$\lim_{n \to \infty} \dddot{AB} = AB$$

where n represents the number of subdivisions and $\dddot{AB}$ the length of the dotted line joining A and B.

This statement is, of course, untrue. Even though the difference cannot be detected by the eye, it is always there and is in fact,

$$a + b - \sqrt{a^2 + b^2}$$

The existence of irrational numbers. Often the first proofs offered to students are more difficult because they do not have the background for simpler, more evident approaches. It is easier, for example, to demonstrate that $\log_{10} 2$ is irrational than to show that $\sqrt{2}$ is irrational when students have the necessary background of logarithms. Here is the proof that $\log_{10} 2$ is irrational:

Assume $\log_{10} 2$ is rational:

$$\log_{10} 2 = \frac{a}{b}, \text{ for } a, b \text{ positive integers}$$

By definition of logarithms:

$$10^{a/b} = 2 \qquad \text{or} \qquad 10^a = 2^b$$

10^a always has a 0 units digit
2^b has 2, 4, 6, or 8 as units digit, never 0.

Summary

The basic goal of a file of teaching pointers is improvement of one's own instruction. There is, however, a secondary but important goal. Once a technique is developed by the teacher himself or even taken from a suggestion by another teacher and refined through use several times, it should be shared with others through publication in a professional journal. Such articles are always in great demand by editors who know that their readers prefer them to all other types of exposition.

The possibilities for creative teaching techniques are limitless: Tesselations of the plane, graphs with triangular coordinates or on triangular grids, distorting figures by transformations, Fibonacci series, paper folding, optical illusions, special curves like cycloids, devices like pantographs and other linkages, gear trains to show ratios. There are thousands: some showy, others less exciting but equally important to classroom instruction; some providing a few seconds of exposition, some activities or procedures that are repeated almost daily. Some may not even work for you. The important thing is to start collecting them immediately. You will never regret it.

Learning Exercises

1. Start your own instructional techniques file by locating an idea attractive and useful to you from each of the following:
   (a) This chapter
   (b) Another chapter in this book
   (c) A recent journal article
   (d) A mathematics textbook
   (e) Your past experience as a student
2. Outline a lesson that utilizes one of your techniques in exercise 1.
3. Find three or more proofs or demonstrations of the Pythagorean theorem. (They may be located in textbooks, but a good source for literally hundreds of proofs is E. S. Loomis, *The Pythagorean Proposition.*[3]) Which would you use in junior high school? Which in tenth-year geometry?

[3] National Council of Teachers of Mathematics, Washington, D. C., 1968.

4. Add to your instructional techniques file at least one of each of the following:
   (a) A game, trick, or puzzle
   (b) A good examination question
   (c) An enrichment project outline
   (d) A technique that will help to get a class started
   (e) A discovery-generating problem
   (f) A laboratory technique
   (g) A grouping procedure
   (h) An historical reference that would add to teaching a particular topic
   (i) A remediation technique
   (j) A mathematical extension of an algebraic topic
   (k) A mathematical extension of a geometric topic

5. Descartes used the following graphical technique for solving quadratic equations of the form $ax^2 + bx + c = 0$: (1) Locate the point $(-b/a, c/a)$. (2) Join this point to the point $(0, 1)$. (3) Construct a circle with this segment as diameter. (4) The roots of the equation are the points of intersection of this circle with the x-axis. Use Descartes' method to find the roots of the equation $x^2 - 5x + 6 = 0$. Why does the method work? How could this technique be used in a second-year algebra class?

Suggestions for Further Reading

Brendau, Brother T., "How Ptolemy Constructed Trigonometry Tables," *Mathematics Teacher*, 58, 2 (February 1965), 141–149.

Garfunkel, J., "A Project in Mathematics," *Mathematics Teacher*, 61, 3 (March 1968), 253–258.

Harris, Edward M., "Geometric Intuition and $\sqrt{ab} = (a + b)/2$," *Mathematics Teacher*, 57, 2 (February 1964), 84–85.

Kaner, Samuel, "A Compass-Ruler Method for Constructing Ellipses on Graph Paper," *Mathematics Teacher*, 58, 3 (March 1965), 260–261. See also Henry D. Snyder, "Deductive Proof of Compass-Ruler Method for Constructing Ellipses," *Mathematics Teacher*, 58, 3 (March 1965), 261.

Leake, Lowell, Jr., "Distribution of the Mean," *Mathematics Teacher*, 64, 5 (May 1971), 441–447.

Maskowitz, Sheila, "The Crossnumber Puzzle Solves a Teaching Problem," *Mathematics Teacher*, 62, 3 (March 1969), 200–204.

Merchant, Charles J., "An Extension of the Averaging Method of Computing Square Roots to the Computation of Roots of Any Order," *School Science and Mathematics*, 65, 2 (February 1965), 143–144.

Rising, Gerald R., and Richard Wiesen, *Mathematics in the Secondary School Classroom—Selected Readings*, Thomas Y. Crowell Company, New York, 1972.

Schor, Harry, "Altitude, Medians, Angle Bisectors, and Perpendicular Bisectors of the Sides of Triangles," *Mathematics Teacher*, 56, 2 (February 1963), 105–106.

Simon, Julian L., assisted by Allen Holmes, "A New Way to Teach Probability and Statistics," *Mathematics Teacher*, 62, 4 (April 1969), 283–288.

Smith, Leander W., "Conditions Governing Numerical Equality of Perimeter, Area, and Volume," *Mathematics Teacher*, 58, 4 (April 1965), 303–306.

Twaddle, Richard D., "A Look at Base Negative Ten," *Mathematics Teacher*, 56, 2 (February 1963), 88–90.

Teaching Every Student

Individual Differences in the Classroom 19

Human beings are complex, unpredictable, and unequal. No two individuals are exactly alike in appearance, in ability, in personality, or in any other trait. Hence, as soon as we have a class of two or more students, we are faced with the problem of providing for individual differences. And when we have a mathematics class of 30 unlike, active young people, we have a situation that demands variation in content, method, and materials to meet the needs of each individual.

When a retiring teacher was asked how present-day high school students differ from those of a generation ago, he said, "Students today are more extreme than they were when I started to teach. The top students are way out front. The slow student is way behind his counterpart of a generation ago. This seems to be true of behavior as well as of intellectual attainment." Thus, the patterns of individual differences between students in today's classroom are likely to be greater than those of a generation ago.

How Do Individuals Vary?

Before considering ways of providing for individual differences, let us consider the differences that are of significance in learning mathematics. These include:

Mental ability, ability to reason or think reflectively, ability to solve problems. Variability in these traits is usually determined by a so-called intellectual aptitude test and is reported in terms of an IQ. In a typical school we expect IQ's to vary from 75 to 150. Ideally, measures of ability should indicate differences in the learning rate as well as in quality of response. Many teachers and administrators fail to recognize one result of the difference in learning rate—the differences between the students' achievements should be expected to diverge rather than converge. When teachers attempt to have an entire class conform to a set curriculum or when administrators set up classes designed to make weak students "catch up" with their more talented friends, they are working against this factor.

Mathematical ability, ability to use symbols, ability to do logical reasoning, ability to compute. Differences in these traits are often measured by a mathematical aptitude test, a test on quantitative thinking, or a prognostic test. Teacher-constructed achievement tests administered early in the school year give results that compare closely with those of such standardized tests.

Knowledge of mathematical concepts, structures, and processes. This knowledge is related to the previous educational experiences of the learner and largely determines the readiness of the learner for the content of a new course. The mathematical background of a learner is often measured by his past achievement record or by an achievement test. Today this factor is of increasing importance because of curriculum variation from one community to another. Pretests of achievement may be used as general measures, and also as indicators of specific areas of difficulty.

Motivations, interests, attitudes, appreciations. While these are potent factors in the classroom, they cannot be measured well at the present time. Unreliable teacher ratings are frequently used.

Physical, emotional, and social maturity of the learner. The measures of maturity are also at best subjective evaluations of the teacher or counselor. Student variations in cultural, social, and emotional adjustment traits are great and are highly significant for the placement of learners in special classes. A student's real physical disability should be noted and its effect on his emotional reactions should be watched. Minor defects in sensory perceptions, such as in sight or hearing, should also be ascertained.

Special talents or deficiencies such as creativity, lack of reading skill, or retention span. The ability to verbalize is closely related to mathematical achievement. Special talents or special needs are usually identified by observation or interviews, although tests of specific traits are becoming available.

Learning habits, self-discipline, attention and retention span, and organization of written work. Differences in this category bear a significant relationship to the learner's home environment and previous education.

If individuals are unequal in the traits listed above, it is not reasonable to give them all the same mathematics courses, the same assignments, the same instruction, the same allocation of time in class, or the same achievement requirements. Therefore, individual differences should first be provided for by gathering as much information about the student as possible. To do this will mean the use of a variety of tests and a study of the learner's cumulative information folder. It will mean the continued collection of information based on classroom performance, and learner products, and careful observation of and thoughtful conferences with the learner. On the basis of these, the student is counseled into the mathematics program most suitable for him.

Ways of Providing for Individual Differences

Although there are a variety of ways of organizing the mathematics program and a variety of materials for use in meeting individual needs, the teacher is still the key to the success of the program. The teacher's success in dealing with individual differences will be determined by how he accepts the learner, how he uses materials, how he varies his assignments. The teacher must recognize differences and provide for them by offering different experiences for different learners—varying content, language, rate of learning, materials of instruction, and the goals of learning according to individual differences.

Here are some practical ways of providing for individual differences:

A multiple-track curriculum with learners assigned to classes according to certain abilities or interests. Such a program might include:

Accelerated classes for gifted students

Advanced placement classes for gifted students

Correspondence study courses or independent study courses using self-instructional materials for classes with limited registrations in small schools

Remedial instruction courses

Vocationally oriented courses such as consumer mathematics or shop mathematics

Enrichment courses such as probability and statistics or computer programming

Courses designed for students of average ability

Special courses for the low-ability student

Classroom activities modified according to learner needs.

Vary daily learning assignments according to ability or achievement levels. Some system is needed that provides many easy exercises for the slow learner and fewer but more challenging exercises for the talented.

Organize the class into small groups according to ability and then give each group special instruction and assignments. To make this possible it may be appropriate to have capable students act as group leaders. The capable student then receives experience in a leadership role and reinforces his learning by having to teach the material. Such grouping is difficult in a class of more than 25.

Enrich the instruction with student reports, demonstrations, projects, and creative writing. A variety of topics are suggested in Appendix D.

Provide supervised study time so that the work of individual students can be observed and help given when needed.

Involve the students in many of the classroom activities such as writing on the chalkboard, collecting papers, correcting assignments. Each learner needs to feel that he has a place in the class and can participate in some activities without frustration.

Establish a level of participation that is related to the students' ability. Capable students are given difficult discussion questions and are not given undue praise for an average performance. Slow learners are given relatively easy questions and their performance is judged according to their ability. The teacher must seek a balance between coddling students so that they are not challenged and develop unrealistic aspirations and forcing students into a pattern of failure that turns them away from learning activities.

Instructional materials varied according to student ability.

Provide text materials, programmed texts, workbooks, and independent units to supplement the regular text. These supplementary texts should have appropriate readability and interesting content.

Provide manipulative materials, models, devices, or equipment appropriate to the needs and interests of the students. Slow learners need frequent reference to visual and concrete representations. High-ability students need to develop their own challenging devices such as logic machines.

Use visual aids such as overhead projections, films, charts, and bulletin-board displays. (See Appendix A.)

Provide construction material for laboratory work or independent projects. A library of supplementary books and pamphlets for enrichment reading or remedial instruction should be available also. (See Appendix B.)

Evaluation made appropriate to the course or students involved.

Establish reasonable levels of achievement according to the group and individuals involved. Be wary of a continuing high percentage of failing grades: look into the causes of such records.

Establish a marking system in special classes that does not jeopardize the scholastic rank of the capable or frustrate the efforts of the slow learner.

Use standardized tests to determine real level of achievement independent of the group or course involved.

Use common examinations for multiple sections so that the base of comparison is broadened.

Some critics have challenged ability grouping as an undemocratic procedure fostering elitism. An even more serious challenge to grouping policies that has found a positive response in the courts is the charge that such grouping tends to reinforce the barriers between advantaged and disadvantaged students. These sociological and political questions will probably be solved outside of the classroom. Most teachers agree that the bright student gains more and feels less superior in a special class of his intellectual peers than when he easily ranks first in a heterogeneous classroom. They also feel that slow learners are less discouraged when competing in classes in which their work can meet group standards.

The key to all of this is the teacher. Grouping policies do not solve problems: they merely make solutions easier in some cases. The teacher must respond to the group being taught, varying instruction not only to meet group needs but individual requirements as well.

A Multiple-track Program

Many schools, especially those with large enrollments, have developed multiple-track programs in mathematics, designed to serve students of high, average, and low mathematical ability. Whereas such a program is more difficult in smaller schools, occasionally interschool cooperation or even special intraclass divisions make similar programs possible. Table 19.1 shows the tracking for a sample secondary-school program.

Table 19.1 Multiple-track Program in Mathematics

Grade	Track I	Track II	Track III (Honors)
7	Math 7	Math 7	Math 7, 8
8	Math 8	Math 8*	Algebra I*
9	Pre-algebra or lab algebra	Algebra I*	Geometry*
10	Algebra	Geometry*	Intermediate math
11	Pre-geometry or lab geometry; General math	Intermediate math	Senior math*
12	Informal geometry and shop math or consumer math	Senior math*	Advanced placement calculus or modern algebra

* Within starred classes students are grouped according to achievement, not grade level. High-ability sections require extra work.

One of the most important considerations of such a schedule is transfer from track to track as well as from section to section within a subject. Many schools make special provision for late entry into an honors sequence by having students take summer school courses. Although the problem of how to move into an honors track is often considered, how to move out of it is equally important. Students should not be allowed to find "an easy berth" or to repeat courses (with different titles only) for extra credit. And, perhaps most important, courses should be carefully designated on transcripts so that honors students are not punished by grading systems that favor the underachiever.

Selecting Students for Special Classes

It is extremely difficult to select students correctly for special classes or homogeneous groups. No matter how careful the selection process, there are always a few individuals for whom the assignment made is inappropriate. Thus a first principle of selection to adopt is one that makes it possible to change the assignment during the year. This change should allow a student to move into either a lower or a higher group. Of course, the sooner incorrect assignments can be discovered and a new assignment made, the better for everyone involved.

In the assignment of students to special classes, all information available should be used. Intelligence-test scores or aptitude-test scores or achievement records alone are not satisfactory. Teacher ratings of social, emotional, and intellectual maturity help. A rating of the student's interests, attitudes, and motivations would also improve the possibility of an appropriate assignment. Standardized achievement-test scores, especially those that test problem-solving skill or quantitative thinking ability, are also helpful. In the final decision, however, a value judgment must be made by the teacher.

One of the pressures on teachers for student selection for special courses is the parents' desires. To meet this problem, the parent should be informed of the basis for an assignment so that he will recognize it as in the best interest of his son or daughter. Many schools allow parents to change such assignments. When such requests appear unreasonable, special records of the reason for the new assignment should be on hand.

In making decisions some schools establish cut-off boundaries, such as an IQ of 110 or a percentile rank of 50. When this is done without regard to the school population or the community culture involved, these boundaries are likely to be inappropriate. A percentile rank of 50 in an advantaged school may be equivalent in terms of achievement to a percentile rank of 80 in a culturally deprived area.

Individualized Instruction

Although individualized instruction has potential for meeting the problems of individual differences, it also has great dangers. The poor reader, the unmotivated, or the withdrawn student is not likely to be reached by this method. The goals of the affective domain are not likely to be attained since the goals of individualized lessons tend to be narrow and skills-oriented. There is a definite need in every class for interaction between students and between teacher and student. It is in a group situation that a teacher's enthusiasm, humanity, and values stand out. Certainly individual work will in time become monotonous and deadly for most students. However, once you embark on an individualized program, it is difficult to bring the students together again.

Another great danger of individualized programs is their tendency to focus on the things that best fit the format. Thus the programs often stress computation over conceptualization, because conceptual skills are not only harder to teach but also harder to test.

A popular proposal for providing for individual differences is to have students work independently at their own rate and at their own level of achievement. One program, Programmed Learning According to Need (PLAN), organized by the American Institute for Research at Palo Alto, California, has prepared lessons for independent study for several textbook series. The lessons include behavioral objectives so that the student knows what he is to learn. The instruction is given or references are given for the student to study. This is followed by practice and evaluation. When a given lesson has been learned, the student is assigned the next lesson that seems appropriate for his achievement. The progress of students is monitored by a computer.

A related program called Individually Prescribed Instruction (IPI) has been developed by the Learning Research and Development Center at the University of Pittsburgh. This program contains a sequence for individualized instruction designed to improve basic computational skills. The materials consist of pupil booklets, skill sheets, and tests. The teacher evaluates each pupil's records, diagnoses pupil needs, and prescribes programs. Teachers tutor individuals or small groups according to the needs of the students. When a student attains a satisfactory mastery of a given skill as demonstrated on a test, he moves on to the next sequence.

A related program is the Computer Assisted Instruction project of Patrick Suppes at Stanford University. In this program students are given individual drill and practice lessons by a computer terminal. The computer selects the lesson and the questions to be answered according to the response record of the student.

Actually, any teacher can write lessons for individualized instruction. To do this a concept or skill to be taught is selected after which decisions are made about specific objectives related to the lessons. It is also necessary to know what background knowledge is needed by the students to complete the lesson. One format for this self-contained lesson is to begin with a behavioral objective. Then the student is given some instruction such as questions leading to the understanding or possibly even the discovery of a generalization. This might be some sample exercises completed and explained step by step or it might include instructions on where to get information about the process involved. The student is next given some questions, problems, or applications for practice to master the concept or skill objective to the desired competence level. The lesson should also include some test items with answers given so that the student can check whether or not he has answered the items correctly.

The Inner-City Student

Since most mathematics teachers have grown up in a culture different from that of the inner-city student, the teacher of these students needs to develop an understanding and acceptance of their culture. Teachers must also recognize the differing attitudes, aspirations, and pressures on these students. Often these students are assigned to low-achiever classes because the usual mathematics course, text, and instruction turn them off. Many of them have mathematical ability, which they use to solve the problems of their world, but such problems are different from the problems of most texts.

The inner-city student has grown up in a crowded city where housing is substandard, where minority groups are well represented, and where the family is likely to have limited resources. His spare time is spent in crowded conditions at home and on the street. He rarely has known privacy and does not even realize that some people like it, hence he does not hesitate to interrupt school activities. Having rarely done things alone, he often does not do well in individualized activities. Crowding and lack of privacy has also given him information about the inner motives of others; hence he is sensitive to our actions whether approaching or rejecting. To get along in his crowded, rejecting society, he has resourcefulness and creativity that we can capitalize on when activities and assignments are accepted as important.

This student's crowded environment may also make him irritable and touchy. Then classroom tensions may lead to temper outbursts. He has learned to distrust adults, to expect failure in school, and to settle difficulties with his fists. Often he resorts to day-dreaming to shut out the uninteresting activities of the classroom.

The inner-city student may have few possessions of his own; hence he has little concern for school property or the property of others. He may come from a large family in which he does not feel wanted. He does not feel that anyone really cares about him or that he will ever have opportunities to improve his lot. Consequently, he lacks motivation in the classroom and is frequently absent. His communication at home or on the street is not in the language of the school; thus his reading and listening level is low and his vocabulary shocking. The conversation with which he is most familiar is the quarreling, cursing, threatening conversations of the streets and sometimes the home as well. No wonder he tunes out a dull, nagging teacher.

The first step in dealing with the inner-city student is to build his confidence in the teacher. The teacher must show that he accepts the student as he is. The teacher must like him if he is to like himself. He must show faith in the student and in his ability to learn so that the student can have faith in himself. The student must be convinced that he is important and that mathematics will open opportunities for change. If the student senses that the teacher sincerely values him, it will be more difficult for him to reject the teacher and the activities that he proposes.

Classroom Procedures

Individual differences cannot be eliminated or even taken care of by grouping, special courses, or varied material. What makes the difference is how the teacher differentiates his instruction. Here are some suggestions that mathematics teachers report have been helpful:

Class time will be used more productively if every student participates. Teachers who make eye contact with every student and speak to each student every day are often successful. One way for a teacher to achieve maximum participation is for him to ask students the kinds of questions they can answer. To do this, he may have to prepare a lesson plan that lists questions to be directed at specific students. And a teacher should try to ensure that students are not afraid of making mistakes or asking foolish questions. By referring to a student's homework, the teacher can identify areas the student knows well and other areas in which he is having difficulty.

Be sure that each lesson is related to what the student thinks is important. We must exercise caution in imposing our objectives on students of a different culture or of different ability. We must recognize that intellectual goals may not seem worth the effort for a student whose values are different from those of the teacher.

There is no substitute for well-prepared lessons presented with enthusiasm. Neither the gifted nor the culturally deprived will stand for poor teaching.

Encourage students to recognize their ability. Help them set up specific, realistic goals and then hold them responsible for attaining these goals. Building a positive student self-concept is an important affective goal of teaching.

Provide a classroom climate that stimulates excellence and at the same time avoids the frustration of continued failure. Require students to complete work commensurate with their ability. Never permit it to be "smart" or socially acceptable among peers to be stupid or apathetic in class.

Take time to meet individuals or groups out of the classroom, either at noon or after school. It is difficult for a reluctant learner to reject the teacher who shows genuine interest in and concern for his difficulties. Learn to know your students so that you will understand each one better and be able to give them the help and inspiration they need to attain their goals.

Allow class time for motivation, appreciation, and attitude-development activities. Do not confine instruction to the text. Permit individuals to explore topics or skills of special interest. Determine student hobbies, special interests, and activities so that instruction can relate to them.

Give special help in how to study, how to use the text, how to locate information in order to promote individual progress. See Chapter 17.

Demand that classes be of reasonable size. It is virtually impossible to make provisions for individual differences in classes of more than 20 to 25 students of heterogeneous ability.

Continuously evaluate your own mental attitude to see that you enjoy your work and like your students—sometimes in spite of their idiosyncrasies. Students quickly sense your values, your sincerity, your acceptance of them, and your enthusiasm for mathematics.

As we have suggested, individual differences of students pose a constant problem in the mathematics classroom. As teachers improve their instruction and improve their programs, these differences between students will become even greater. This is to be expected because the rapid learner is stimulated to learn to his potential and the slow learner drops proportionately behind. As we recognize how different our students are it seems reasonable that we should vary the goals, the content, the materials, the instruction, and the learning activities. Learning the same content but at a different rate is not satisfactory. Failing a student or dropping him from the mathematics course does not seem reasonable or productive. Mathematics teachers must accept the responsibility for teaching all levels of ability by using proper resources and methods.

Learning Exercises

1. Describe some things you would consider doing to meet the following problems:
   (a) A talented student is scheduled for your weakest class group because of a conflict between his regular assignment and his music lesson.
   (b) A number of students who were only marginally successful in geometry are assigned to your second-year algebra class.
   (c) A very bright group seems to be coasting.
   (d) A girl cannot respond orally in your class although her written work is satisfactory.
   (e) A student does no classwork, no homework, nothing.

2. You have decided to group students in your classroom for a series of lessons. Before assigning group membership you identify students as high (H), average (A), and low (L) in terms of achievement. Indicate some good and bad features of each of the following grouping patterns:
   (a) HHHH, AAAA, AAAA, AAAA, AAAA, AAAA, LLLL, LLLL
   (b) HAAL, HAAL, HAAL, HAAL, HAAL, HAAL, HAAL, HAAL
   (c) HAAA, HAAA, HAAA, HAAA, HAAA, ALLL, ALLL, ALLL
   (d) random assignment to groups
   (e) team captains appointed who then select group members

3. Read the article by Sarah Greenholz and Mildred Keiffer, "Never Underestimate the Inner-City Child" (see the reading list). What is your reaction to this article? What did you learn from it? Do you disagree with any parts?

4. PROJECT. Identify a high achiever and a low achiever in the same mathematics class. Examine their written work, observe their classroom participation, and discuss with them their attitudes and goals. What are some of the similarities and differences between them?

Suggestions for Further Reading

Adler, Irving, "Mental Growth and the Art of Teaching," *Mathematics Teacher*, 59, 8 (December 1966), 706–715.

Brannon, M. J., "Individual Mathematics Study Plan," *Mathematics Teacher*, 55, 1 (January 1962), 52–56.

DeVault, M. Vere, and Thomas E. Kriewall, "Differentiation of Mathematics Instruction," in *Mathematics Education*, Yearbook 69, Part I, National Society for the Study of Education, University of Chicago Press, 1970, pp. 407–432.

Greenholz, Sarah, and Mildred Keiffer, "Never Underestimate the Inner-City Child," *Mathematics Teacher*, 63, 7 (November 1970), 589–595.

Guaru, Peter K., "Individualizing Mathematics Instruction," *School Science and Mathematics*, 67, 1 (January 1967), 11–26.

Harwood, E. Hallie, "Enrichment for All!," *School Science and Mathematics*, 63, 5 (May 1963), 415–422 B.

Matthews, Josephine J., and Harold E. Rahmlow, "Individualized Mathematics the PLAN Way," *Mathematics Teacher*, 63, 8 (December 1970), 685–689.

Schmidt, Roland L., "Using the Library in Junior High School Mathematics Classes," *Mathematics Teacher*, 56, 1 (January 1963), 40–42.

Sott, Joseph J., "Mathematics Enrichment through Projects," *School Science and Mathematics*, 66, 8 (November 1966), 737–738.

A Program for the Unsuccessful Student 20

Many mathematics teachers consider teaching the low achiever an uninspiring and unimportant task. In this we are avoiding just those students who need us most. When we recognize the need for greater mathematical competence for a greater proportion of society, we should see the need for further emphasis on mathematics for every student. Furthermore, as teachers in a democracy that values the worth of each person, we know that we have a professional obligation to teach learners of all ability levels.

> A teacher who accepts the fact that low achievers are teachable; a teacher who has a missionary spirit and a respect for the worth of pupils with limited ability; a teacher who is concerned and interested in individuals; a teacher who can make a pupil feel he not only belongs but also is important; a teacher who can instill a sense of worthiness, responsibility, and desire to achieve; a teacher who cares enough to give his very best to the low achievers will make the program a success.[1]

The low achiever will always be with us. In fact, their number as well as proportion may very well increase. They are found at every socioeconomic level and will have an influence, for good or bad, in our society. If we do something for them in school now, we may be helping them to avoid problems in their adult years. Many of them have latent talents that can and should be developed.

Until recently, very little had been done to improve the mathematics for the low achiever. Classes designed for them have provided a dumping ground for discipline problems. The emphasis and the financial support had been given to educating the potential scientist. For years, little money was spent for equipment and supplies for low-achieving students. Presently, though, attention to the culturally deprived and to vocational training has resulted in financial support for developing new ways of dealing with the low achiever. School board members, representing taxpayers, are pressuring schools to do more for underachievers. They feel that investing in the education of these students can head off later financial support of them when they cannot function in society.

---

[1] U. S. Office of Education and National Council of Teachers of Mathematics, *Preliminary Report of the Conference on the Low Achiever in Mathematics*, Washington, D. C., 1964, p. 20.

Identifying the Low Achiever

Who is the so-called low achiever? Students generally classified as low achievers do not necessarily share the causes for this disability. Special terms are sometimes used in classification to identify the specific cause of low achievement:

The *low ability student*, who has below-average academic ability.

The *slow learner*, who cannot keep up with the normal class pace.

The *underachiever*, who has potential for mathematics achievement but has not realized the potential.

The *reluctant learner*, who lacks interest in the usual school mathematics program.

The *disadvantaged learner*, who has been given little background for adjusting to the usual mathematics program.

The *culturally deprived learner*, who has been reared in a culture with meager educational experiences.

The *disaffected learner*, who has developed a negative attitude toward mathematics.

The *rejected learner*, who has been rejected by teachers, peers, or parents.

The *non-college bound*, who has inadequate ability, interest, or resources for college attendance.

This multiplicity of names smacks of educational gobbledegook or jargon. (Someone has even suggested "deliberate abstractor" as pedagese for the poor student.) But the many names do have a place here. They indicate the wide assortment of problems that come under the general heading of "low achiever." They should point out the need for careful identification of specific problems and individualization of response to those problems.

Another reason for concern for class groups of low achievers is the wide variation in ability and achievement that they represent. A look at the normal curve distribution shown in Figure 20.1 suggests the basis for this problem, which affects

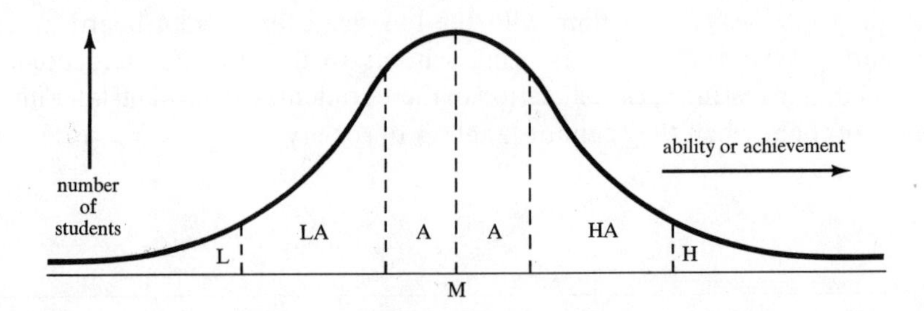

Figure 20.1

both high- and low-ability sections. The area under this curve represents numbers of students. It is drawn for a hypothetical homogeneous grouping into six classes: two average (A), a high average (HA), low average (LA), high (H) and low (L). Notice how the average classes represent narrow ranges of ability whereas classes differing from the mean (M) represent wider and wider ranges. Students in low-ability or low-achievement groups represent a much wider range. (Scheduling misassignments and just plain errors exacerbate this problem).

We will consider the low achiever to be the student who normally ranks below the 30th percentile in achievement due to any of the following factors:

Low mental maturity. Although the student's ability may not be measured accurately by our culturally oriented mental tests, he has little ability to perceive relations, is unable to generalize, and has difficulty transferring knowledge. He usually has an IQ below 90. His reading and mathematical achievement are often two or three years below grade level. He needs to learn from this level of achievement.

Emotional immaturity. The student usually needs and lacks acceptance, affection, security, and success. He comes to school without hope, hostile to the teacher and the school, depressed, and frustrated. His dislike for authority may result in rebellious and delinquent behavior. He is often confused, lacking in self-confidence, yet unaware of the sources of his difficulties. He has probably been deprived of an adequate amount of affection, love, and emotional support. The best way for a teacher to build his self-concept is to see that this type of student has some success.

Social immaturity. The student's social and cultural experiences at home, in the community, and in school are often meager. He may be prejudiced and intolerant. He lacks leadership experience and does not like to cooperate in group activities. He may never have found a person whom he wishes to imitate as he grows up. He is often overly aggressive and physically restless. He has not experienced respect so he does not know how to show respect. He is often absent and is likely to become a dropout.

Physical deficiencies. The student's poor achievement may be due to poor health, insufficient food, or insufficient rest. He may have poor eyesight and poor hearing and lack motor coordination. He may even have a serious disease or be a drug addict.

Psychological deficiencies. The student's attention span and memory are short. His reading level is low, his visualization and imagination meager. Insight, creativity, and problem-solving skills are often almost nonexistent. His motivation for learning is weak.

Limited cultural experiences. The student has had few experiences such as traveling, reading, craft work, or hobby work. Even his conversations have been limited to jargon and cant. The world of literature, art, music, and drama does not seem relevant to his daily life. He may have a record of juvenile offenses. However, mathematics is probably less culturally conditioned than any other subject, and mathematics has fewer derogatory connotations for the culturally deprived than other fields.

Meager educational experience. The student's previous mathematical experiences, his study habits, and his achievement are likely to be inadequate. He is more oriented to vocational than to academic education. Learning for its own sake has little if any meaning for him. He may be anti-intellectual and pragmatic. Talk, reading, and intellectualizing in general have little appeal to him. He has not learned how to ask questions, how to study, or how to take tests. His ability to verbalize is likely to be very low. He is an expert at tuning out action around him. Although tests can be used to determine the extent of some of these deficiencies, these tests themselves require a certain level of reading skill and cultural background. Many of the low achievers are lacking in these qualities, and they may be rated far below their potential. Therefore, conferences, observations, and past performance records need to be considered as a basis for classification.

Problems to be Resolved in Providing for the Low Achiever

There are many problems that complicate provisions for the slow learner:

The teacher problem. Teachers do not want to teach classes of slow learners. They consider it futile and difficult. (Mathematics teachers have generally been spoiled by the natural selection of the academic mathematics program.) At the same time, our society requires that ordinary citizens have enough mathematical competence to be employable. And we would never condone a doctor who refused medical attention to an indigent patient. There is no question but that the slow-learner course needs top-quality teaching. Usually, however, a beginning teacher, a barely competent teacher, or a teacher without seniority rights or without a major in mathematics is given the assignment. The result is chaos and even less respect for the class by teacher, student, and parents. Furthermore, it may mean the loss of a beginning teacher who has the potential to become a highly successful teacher if given a more appropriate first assignment. On the other hand, schools that assign excellent teachers to these sections note real progress on the part of the students. Young people are sensitive to teacher attitudes and often respond remarkably to the accepting teacher. They identify with the teacher who works with them and shows that he enjoys this work.

The curriculum problem. The major modern curriculum-development groups have not selected objectives or designed courses for the low achiever. Hence, there does not seem to be agreement on what the content of courses for the low achiever should be and on which students should take the course. Furthermore, there is little evidence regarding the effectiveness of the course. One recent study showed that a year of reading instruction was just as effective during the ninth grade as a year of instruction in "easy" algebra in terms of later achievement.

The student problem. It is possible for the low achiever to attain significant goals of mathematics instruction—but only if he *wants* to learn. To build this motivation the teacher will need appropriate content, materials, and methods of instruction, and an appropriate method of evaluating the achievement of slow learners in a special class.

The parent problem. Many parents will not accept the classification of their child as a slow learner. They respond negatively to placement in the special class for low achievers, especially if it is for more than one year. They are frequently adamant about their child's taking the mathematics courses that are prerequisite for college entrance.

The materials problem. The slow learner usually needs a variety of experiences with concrete materials. He needs short exposure to topics, with novelty features that will hold his attention for short periods. He needs brief units that are complete in themselves, so that difficulties do not compound. He needs activity-oriented materials that are now available. He needs reading material that is at least one grade level in readability rating below the grade being taught.

Guidelines for an Effective Program for the Low Achiever

Knowing the nature of the slow learner, his needs and potential, and the problems of the current school program, what can be done to build an effective program for the low achiever? The following principles must be considered:

Goals of instructing should be appropriate, realistic, and attainable. Unless he has specific goals in mind, a teacher's instruction will lack direction or significance. Emphasis should be placed on attitudes, appreciations, habits, and values. Attention should be given to reading and other communications skills. The objectives of good citizenship (honesty, responsibility, cooperation, and respect for others) should also be accepted as a major aspect of the course.

Students for courses for low achievers should be selected carefully. Selection should be based on the student's intellectual ability, mathematical ability, mathematical experiences, vocational interests, and emotional maturity. Provision should be made for transfer into another course whenever the achievement warrants a change. At no time should students be assigned to these courses merely because they are discipline problems—students who can achieve but do not try should be excluded. The practice of transferring students from courses in algebra and geometry into a general mathematics class or even allowing a student who has passed algebra and geometry to elect a general mathematics course for a "soft" credit is entirely unwarranted. Since the highest correlation coefficients are found between achievement in algebra and previous achievement as measured by marks in mathematics or a standardized achievement test, previous achievement in mathematics is the best single factor to consider in placement at the ninth-grade level. Algebra-aptitude tests, reading tests, and intelligence tests also correlate well with achievement in algebra: that is, they provide evidence about the student's ability to achieve in this course.

Class size should not exceed 20 students. A classroom of this size will permit laboratory work and individualized instruction and will curtail the number of disciplinary problems. The size of the class is a major factor in the success of a slow-learner course. One justification for having a small class is the wider range of ability to be found in low-ability classes (and in high-ability classes) than in average-ability sections.

Instruction should emphasize student participation in a variety of learning activities such as laboratory work, class discussion, discovery activities, informal practice, games, and the use of teaching aids. Every effort should be made to choose introductory activities that are new and interesting and that are not based on skills that students do not have.

The teacher assigned to the course for the low achiever must be especially well qualified. Besides being skilled as a teacher, he will need to understand and accept these youngsters and their culture, to value the purpose of the course, and to have a warm personality and a reservoir of information about mathematics and techniques for teaching.

Facilities for low-achiever classes must include a variey of instructional aids, such as micrometers, slide rules, models, kits, programmed texts, supplementary books, and games, as well as audio-visual equipment such as projectors, tape recorders, and tachistoscopes (machines that provide timed exposures of drill material in reading, or mathematics).

Evaluation of progress should be realistic—taking into account the purpose of the course and the ability of the pupils. This evaluation should include measures of attitudes, skills, and habits. Wherever possible, marks should record the amount of growth or progress made and the specific levels attained. A continuing record of failure destroys incentive.

Applications should include those appropriate to the modern world, such as computers, space travel, science, technology, sports, and other daily events.

Course content should include new ideas, new treatment of old topics, and new presentation of computation skills. The content should also include much traditional material such as measurement, equations, rational numbers, geometry, graphs, and statistics. In addition, topics such as probability, symmetry, topology, patterns, codes, transformations, and tesselations should be offered.

The individual instructor should be allowed and should take advantage of considerable latitude to modify his program to take advantage of special student interests, to reteach topics not well understood, and to extend topics that he finds provide significant learning and success. While content patterns and units are usually provided for the teacher, he should feel free to explore enrichment topics, applications, or informal work that appeals to his students. In other words, more attention should be given to the needs of the student than to covering the content. The way in which content is taught is more important than the content.

Courses for the Low Achiever

A variety of courses have been proposed for the low achiever.

Remedial mathematics. This course reteaches *computational skills.* Since the slow learner has been unable to learn computation in his previous six to eight years, it is not likely that another year of similar content and method of presentation will be successful. Frustration will continue, hostility will increase, and improvement will be negligible. There is a need for improving the computational skills of the slow learner, but this needs to be done in novel ways such as games, contests, puzzles, laboratory activities, and out-of-door mathematics. In no case should an entire course be devoted to remedial mathematics without any relief from drudgery.

Slow algebra and geometry. This course teaches the mathematics of the college preparatory courses but teaches it at a *slower pace,* with more concrete examples, less stress on precise language, and simpler problems. For example, it is suggested that most of elementary algebra and geometry be taught over a span of three or even four years rather than the usual two. This seems to ignore the needs and interests of the slow learner. Many teachers accept this proposal because they feel more secure teaching this content, and text material of this type is available. Others argue that continuing in the academic sequence keeps more student options open. On the other hand, some teachers argue that the content of a course should be based on the needs and interests of the students. They feel that many topics such as probability, measurement, and statistics are of more relevance to low achievers than pure algebra and synthetic geometry.

Applications. Build the course around the mathematics that the learner will *need* as a citizen, a worker, a housewife. This course is the typical general mathematics of the ninth grade, which covers topics such as installment buying, insurance, taxes, and investments. This has not generally been successful since the topics are too remote for the ninth-grade student. Furthermore, these topics have often merely provided exercises in computation without extending the mathematical understanding of students. Most of the topics of this course are too far removed from the current interests and needs of students. Applications from sports, motorcycles, games, and local events would have more relevance.

Vocational mathematics. This course is organized around a *vocational* area such as shop mathematics, business arithmetic, nursing mathematics, or mathematics for home economics. These courses have the advantage of being related to the student's vocational goals and take care of the problem of transfer to applied problems. Again, however, these courses tend to be narrow, with emphasis on computational drill and it is difficult to find vocations of interest to all students. Furthermore these vocations may not exist when the students are ready for employment. This course, if offered, should involve the use of calculators and computer terminals.

Consumer mathematics. This course deals with consumer problems such as best buys, measures of quality, installment buying, investments, insurance, discounts, and taxation. This may be of greater interest to seniors than to ninth or tenth graders. Again it can be a mere setting for tedious computation. One way to enliven this course is to add computer programming.

Cultural mathematics. This course introduces the student to the nature of mathematics, its varied topics and techniques, its patterns and logic. It may deal with historical mathematics, applications, or new topics such as topology. Its goal is to build an appreciation of and interest in mathematics. In the process it builds mathematical concepts.

Laboratory mathematics. This mathematics course emphasizes experiments, activities, surveys, games, and computing devices. The major concern is that students discover mathematical ideas. These discovery activities not only teach the student how to learn but also add meaning to mathematical ideas. In the process the student performs computations as he needs them.

Topics appropriate for laboratory mathematics include the following:

- Probability
- Calculating devices
- Equations, formulas, and relations
- Rational numbers
- Measurement
- Ratio, proportion, and percent
- Graphs
- Geometry, metric, and motion
- Reasoning, logic, and decision making
- Fact or fancy: statistics
- Design and construction
- Mathematical recreations
- Flow charts, problems, and computers
- Patterns, codes, and topology

All these courses have advantages and disadvantages. Perhaps the best course for the low achiever is one that uses ideas from all of them. To provide such a program, it will be necessary to have more than one year of mathematics for the low achiever in high school. That is as it should be.

Specific Practices for Teaching the Low Achiever

To motivate the low achiever, to gain his acceptance, to help him to identify with the teacher, and to guide him to learn topics that do not always seem relevant to him—all of this requires a teacher with ingenuity and energy. The following activities have been found effective by successful teachers of the low achiever:

Present the material in short, independent topics. Do not insist on complete mastery before moving on to the next topic. Since these students learn slowly, an insistence on their mastery of topics might mean that too limited progress is made and major goals are not attained. Be sensitive to the students' need for a change in topic and teaching technique. At the same time, be aware of the security provided these students when you repeat a successful activity as well as one that needs reteaching. Thus, a topic such as probability could be organized into several independent units. One unit could be permutations, another on probability experiments, and a third on applications of probability. However, if there are several approaches to a new concept being taught, only one should be presented in a given lesson to avoid confusion.

Use a variety of methods and activities from day to day and even during a given period. To meet the short attention span and low motivation, variety of activity is essential. Each lesson should consist of several different activities such as a game, a laboratory discovery period, a short study period, and a short mental-computation drill. Be sure each activity is short, preferably not longer than 15 minutes.

Use a variety of materials to capture the interest and arouse the curiosity of the learner:

Pamphlets, such as *The Amazing Story of Measurement*
Models, such as the dynamic geometry devices
Computing devices, such as the slide rule or calculator
Kits, such as the *probability kit*
Films, such as *Donald in Mathmagic Land*
Construction material, such as cardboard and balsa strips
Games, such as *Prime Drag*
Projectuals, such as the set on Venn diagrams
Programmed topics, such as *measurement*
Charts, such as *optical illusions*
Resource books, such as *Field Work in Mathematics*
Measuring instruments, such as micrometers
Puzzles, such as polyominoes
Remedial instruction kits
Film loops

Use the *discovery approach* by collecting information through laboratory work. Provide background material as needed rather than as an independent unit. Relate ideas to the students' experiences. For example, to discover area formulas, cut out rectangles and transform them by cuts into triangles.

Have the students read about things they see, feel, and do. Use models, paper folding, and measurements that students can manipulate. Stress tangible, not abstract, qualities.

Apply mathematical ideas to local situations. Collect data from class members, parents, or community projects. Capitalize on the curiosity of youth to explore, manipulate, and cope with their environment. Use their names, local situations, local newspaper advertisements, catalogs, radio or television programs for problems.

Give remedial instruction to individuals or groups based on a diagnosis of needs, but do not expect slow learners to attain the normal grade level. Gear learning to the students' readiness to learn. Correct errors and give feedback on progress made. One way to help review is to provide semiweekly quizzes on fundamentals, essentially the same test given twice. Students know that if they can correct their errors they can always score well on the second test. Or provide a sequence of very short problem sheets that students work and correct independently during a 5- to 10-minute period set aside frequently for such a procedure. Only when they score perfectly on one sheet can they go on to the next.

Provide short periods for independent study on activities in which the students can experience some success. Assign short, easy, concrete exercises. Strive for overlearning of basic ideas to ensure retention and transfer. Students grow in security when they have complete mastery of at least a few basic ideas.

Provide informal practice sessions, using games or realistic problems rather than formal drill. Use adequately spaced repetition and review. Give feedback on progress to confirm, clarify, and correct previous learning. A series of easy equivalent tests may be used for students to collect data on their performance; their grades should be noted on a graph. This graph should show improvement in the student's accuracy and speed.

Give specific help in communication: reading, speaking, writing, spelling, neatness, organization, and vocabulary. Be sure to use text materials that students can read. Spend class time in teaching students to state ideas correctly, spell correctly, and write legibly.

Give special help on how to study, how to use the text, how to locate information, and how to check work. Do not expect much to be accomplished by homework. Permit the use of "crutches" but avoid the use of shortcuts. Frequently provide answers so that students can reinforce correct responses and locate their errors more quickly.

Provide the pupil with opportunities to assist the teacher in board work, art work, demonstrations, displays, projects. These activities give the student a feeling of acceptance and security. Never punish by giving extra work.

Enrich the instruction with historical incidents, dramatic topics, unusual problems, puzzles, stunts and tricks, personal experiences, and local applications.

Help pupils select specific, attainable goals and encourage them to recognize their greatest potential. Help them attain these goals. Have good practical reasons for everything done or assigned. Emphasize that the school program is set up to help each student. Do not try to cover up the student's limited ability; he already knows it. By your acceptance of it, help him accept his own limitations.

Create a classroom climate that stimulates interest, curiosity, and participation. A pupil should feel secure and accepted even if his dress, cleanliness, appearance, or response is not what we would desire. Try to have some activity every day which can be done successfully by each student. Be slow to take offense and use a light touch to deflate a potential blow-up. Control expressions of irritation that rebuff or antagonize students. Let them feel secure even when they make mistakes.

Make specific provisions for developing attitudes, interests, appreciations, habits. If a topic is completely lacking in appeal, drop it for a time and move on to a new topic.

Permit pupils to explore topics or skills of special interest. Do not feel obligated to cover all topics of a given text. Allow for class expression and interplay. Tailor your questions to the ability of the student.

Provide adequate time for the pupil to complete assignments. Being hurried produces anxiety, which inhibits learning. Write out the assignments, make them brief, and keep your directions simple. Provide answers for some problems so that the assignment functions like programmed instruction.

Be patient, kind, understanding, and fair. The learner wants the teacher to be genuinely interested in him. He wants to be respected, not patronized. He wants a teacher who will stand by him, someone upon whom he can depend. Don't moralize, preach, or be too sentimental.

Try to make eye contact with or to speak to each student every day. This will tell him that he has status in your class.

Use diagnostic tests to determine needs. Provide programmed texts, workbooks, pamphlets, or supplementary texts to meet individual needs for remedial instruction. Have texts for several grade levels for study. Review frequently to counteract short memory spans.

Contact the parents and recommend cooperative activities for meeting the needs of their sons or daughters. Do not let lack of cleanliness, odd dress, appearance, or socioeconomic status of the parents influence your acceptance of a learner (or his parent).

Collect all the information possible about each student. Use the cumulative folder, the comments of other teachers, and the parents' comments to get a complete picture. Make up your own file with a cumulative folder for each student.

Find some activity in which the pupil can do superior work such as drawing, constructing, or computing. Give genuine praise for every inch of progress. Assume that students can learn, and push them to their limit.

Establish a classroom behavior standard that permits learning to take place. The low achiever usually thrives on specific rules of behavior; and if he respects you and likes you, and if you respect him and like him, he is not likely to be a rebel. But never abdicate your role as the adult in charge.

Try to give some opportunities to *avoid* computation. Provide students with a desk calculator, a slide rule, tables, or a nomograph so that they can attack the organization of a solution. In this way, you overcome some of the hostility generated by errors due to their weak computation skills. Often they find that they can solve problems with these crutches and are then motivated to improve their skills.

There is not just one right technique for teaching the low achiever, although there are many wrong ways. In teaching the low achiever, the teacher needs to be straightforward and specific, stating clearly what is to be done as many times as necessary. At the same time, the teacher should be informal, warm, and down-to-earth. He should encourage the learner and indicate that he expects him to learn. Since low achievers tend to be insecure and defensive, they do not respond well to "challenges."

Once you have established rapport with low achievers, you can teach them many ideas. Your own excitement, enthusiasm, and interest will be contagious if the learner identifies with you.

Reteaching the Low Achiever

One of the greatest needs of the low achiever is reteaching. He does not know very many mathematical facts; his computational skill is poor, as is his understanding of mathematical principles; and he is at a loss in solving word problems. Often the mathematics he does know is *wrong* mathematics. Thus, reteaching includes "unlearning" wrong mathematics or wrong procedures as well as learning correct mathematics. To be effective, reteaching must apply to all of these aspects of performance.

The first step of course, is to identify what the student knows and what he does not know, what he does correctly and what he does incorrectly. Part of this can be done by administering a diagnostic test. Also, whenever a student makes an error on a test, assignment, or class question, you should try to find the reason for the error so that it can be a basis for reteaching.

When a student gives an incorrect response, he may do so for a variety of reasons. He may not know the correct response or misread the problem or have used an incorrect algorithm. He may not have been interested in responding correctly. He may have known the correct response but was careless in responding. He may even know the correct response but be too insecure to give it.

Each of these sources of wrong responses requires different instruction. When the student does not know the correct response, he needs basic instruction beginning with sensory experience so that it will have meaning and better retention. If his error is due to misreading, a careless mistake, or the selection of the wrong algorithm, this should be called to his attention so that he will be aware of his *source of error* and thus proceed more carefully in the future. He may need more practice so that he gains accuracy in responding with what he already knows is the correct response. If he is not interested in responding correctly, we must find his "hang-up" so that we can develop the necessary motivation. Perhaps he is not even ready for instruction on this concept or skill.

If, however, he has learned an incorrect response, then reteaching is more difficult. Since the wrong response may have been used for some time, it may be difficult to replace it with the right response. Jack Forbes of Purdue has suggested that we use retention curves to suggest a strategy for re-teaching.

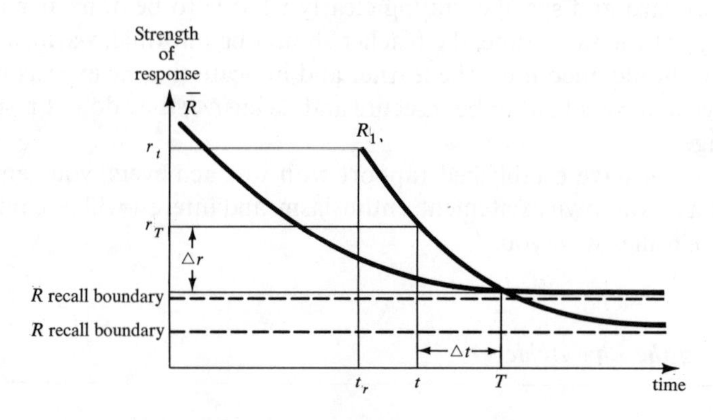

Figure 20.2 Retention Curves

Figure 20.2 displays retention curves for $\overline{R}$, a learned and consistent wrong response, and R, a later-learned correct response. These curves are typical of learning: The strength of anything learned tends to diminish with time to a plateau called a recall boundary at which it remains quite stable. When these conflicting curves are superimposed, the more recently learned correct response, R, is temporarily stronger than the older incorrect response, $\bar{R}$. A key point on the graph occurs at time T, when the old response reasserts itself.

For any time $t(t_R \leq t)$, the retained response to the learning R is designated r. Then two measures are important: $r_t - r_T = \Delta r$ and $T - t = \Delta t$. Given this model, it is evident that keeping Δr positive is the goal of remedial instruction. At the same time Δt (as originally pictured) is the time still left before the old incorrect response extinguishes and replaces the correct response.

The most obvious remedial strategy is to teach R meaningfully and effectively so that its recall boundary is higher than that of $\bar{R}$. Then Δr will remain positive.

Forbes suggests a second strategy, which he calls an extinction strategy. $\bar{R}$ is carefully identified to the student as an incorrect response. Then R is taught. Hopefully, this both reduces the $\bar{R}$ recall boundary and improves the R recall boundary.

A third strategy is called response modification. Rather than eliminating $\bar{R}$, this strategy attenuates $\bar{R}$ until a correct (R) response is attained. This is done by loading the existing $\bar{R}$ with an additional response that produces R. For example, if a student says that $6 \times 9 = 56$, this instructor might respond, "No, how much is 6 times 10?"

"60."

"Good. Now 6 times 9."

"54."

"Good."

Which of these strategies are most effective for "unlearning" a wrong response is unknown. Until research finds an answer, teachers are urged to try different methods to discover what works best for them. The most significant part of this procedure, however, is the identification of the wrong learning of the student so that steps can be taken to "unlearn" it.

A Sample Lesson for the Low Achiever

To illustrate how some of the principles mentioned above operate in the classroom, here is a 45-minute lesson plan:

Introduction (10 minutes). An activity designed for fun is an appropriate class opening, particularly because it encourages promptness. One such activity is the game "What Number?" One student whispers a rule to the teacher like "add 1" or "take half" or "zero for even, 1 for odd." (If the student cannot think of a rule, the teacher provides one.) Without telling what his rule is, the student asks the other class members for a number between 1 and 20. The student then replies by giving the number he arrived at after applying the secret rule. When students think they have deduced the rule from the examples, the leader tests them by giving them numbers to answer. For example, if the rule is "subtract 2," the exchange might be:

Leader. Bill?
Bill. 5.
Leader. 3. Mary?
Mary. 19.
Leader. 17. John?
John. 10.
Leader. 8. Frank?
Frank. I know the rule.
Leader. Okay. Try 15.
Frank. 12.
Leader. Sorry Frank, that's wrong. Bill?
Bill. I know. 13.
Leader. How about 7?
Bill. 5.
Leader. Okay. Anyone else?

Review (5 minutes). Review the short three-problem homework work sheet. Two students write each solution on the board. Other students judge the better solution to each problem on the basis of correctness, presentation, and neatness.

New content (10 minutes). Teacher presentation. With an overhead projector the teacher develops some simple area and perimeter relationship based on counting squares on a lattice, as shown in Figure 20.3.

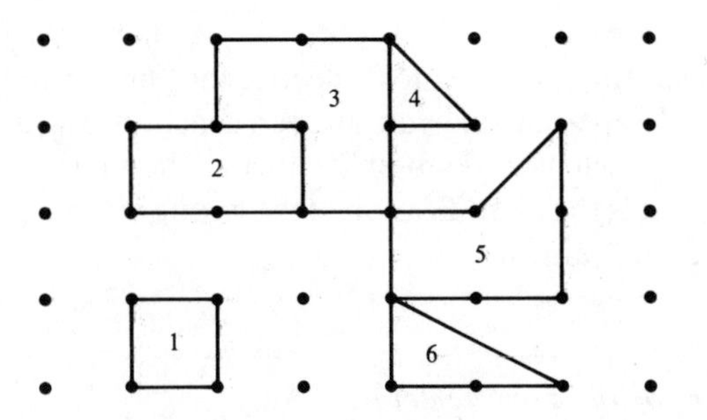

Figure 20.3

Class practice (10 minutes). Students draw dots on their papers and make their own figures. The challenge is to make an odd figure to determine the area and, when okayed, to copy the figure on a transparency to challenge the class.

Individual practice (5 minutes). Start on a worksheet assignment that has one problem based on classwork and two on review material.

Motivation (5 minutes). As a bonus for a good class period, the teacher plays a challenge game of One-Pile-Nim with a student chosen for good participation. The teacher chooses the number of counters in the game, the student tells how many counters may be taken in one turn (no more than 3 or 4 or 5 as his choice) and takes the first turn.

Many readers would challenge the outlined lesson as only nominally mathematical. However, it stresses goals that are even more important: developing interest and participation, good work habits, and successful experiences and providing leadership. The mathematics itself is basic but good. Many algebra students have trouble with the problems of "What Number?" in its abstract form: "Write the equation representing the linear relation between the ordered pairs of which (5, 3), (19, 17), and (10, 8) are examples." In the same way, students of coordinate geometry, who would reject some of these area problems as being too simple, would also have difficulty with them in abstract form.

Summary

The activities suggested in this chapter are based on the following assumptions:

1. It is *important* that we teach mathematics to the low achiever. Our nation needs citizens who have enough mathematical competence to be employable and who can participate in a democratic society.

2. It is *possible* for the slow learner to attain significant goals of mathematics instruction.

3. It is *necessary* for the slow learner to have a desire to learn before learning will take place.

4. It would be *helpful* if the content, grade placement, and ability level needed for progress were established.

5. It is *required* that the teacher of the slow learner be a competent teacher.

6. It is *appropriate* that the content be selected according to the needs, ability, and vocational goals of the learners.

7. It is *essential* that the negative attitudes of the slow learner be changed. This is the key to the problem. To do this the teacher must be convinced that the learner has real potential and is worth time and effort. The teacher must:
   - (a) Respect the student for who he is, so that the student can respect and have faith in himself.
   - (b) Encourage the student to try, and assure him that failure is no crime.
   - (c) Show confidence in the student's ability to learn.
   - (d) Show sincere pleasure at a reasonably good attempt even though there has been no demonstrable progress.

8. It is *important* that class size be kept small to allow for the greatest amount of individualized instruction.

Learning Exercises

1. Are the terms defined on page 340 mutually exclusive? If you answer this question "no," give some patterns of characteristics that overlap.

2. Indicate some differences you might expect to find between a low achiever in a school serving disadvantaged students and a low achiever in a school population of largely advantaged students. How might these differences affect the program of the two schools?

3. Name the five most important characteristics of a class for low achievers. Which of the characteristics on your list are not applicable to classes for students of mixed ability?

4. Often teachers are assigned classes of underachievers at the junior high school level and told that their goal is to "catch them up" with other groups so that they can enter algebra in grade 9. Is this a realistic goal? Why? Indicate some other goals for a group like this.

5. Accountability firms are contracting with schools to achieve exactly the goal of exercise 4: They are paid for each student reaching "grade level" as tested on a standardized examination. Assume that you have been assigned an algebra class of students who have been "brought up to average" by such a crash program. Indicate some concerns you might have for such a class and some techniques you might use with the students.

6. Write a lesson plan for low ability students in one of the following classes:

(a) Geometry
(b) Seventh grade
(c) Algebra
(d) Twelfth-grade general mathematics.

7. Examine a general mathematics text. Indicate some of its strong and weak features.

8. It is the practice in many schools to assign general mathematics or low-achiever classes to beginning teachers who then "work their way up" to college preparatory subjects. What are some things wrong with such a pattern? Are there any arguments in its favor?

9. PROJECT. Make a case study of a specific low-achieving student. Identify some of the sources of his mathematics difficulties. Recommend ways to help him.

Suggestions for Further Reading

Beckmann, Milton W., "Teaching the Low Achiever in Mathematics," *Mathematics Teacher*, 62, 6 (October 1969), 443–446.

Braunfeld, Peter, Clyde Dilley, and Walter Rucker, "A New UICSM Approach to Fractions for the Junior High School," *Mathematics Teacher*, 60, 3 (March 1967), 215–221.

Dodes, Irving Allen, "Some Comments on General Mathematics," *Mathematics Teacher*, 60, 3 (March 1967), 246–251.

Easterday, Kenneth E., "A Technique for Low Achievers," *Mathematics Teacher*, 58, 6 (October 1965), 519–521.

Elder, Florence, "Mathematics for the Below Average Achiever in High School," *Mathematics Teacher*, 60, 3 (March 1967), 235–240.

Fremont, Herbert, and Neal Ehrenberg, "The Hidden Potential of Low Achievers," *Mathematics Teacher*, 59, 6 (October 1966), 551–557.

Greenholz, Sarah, "Successful Practices in Teaching Mathematics to Low Achievers in Senior High School," *Mathematics Teacher*, 60, 4 (April 1967), 329–335.

———, "What's New in Teaching Slow Learners in Junior High School?," *Mathematics Teacher*, 57, 8 (December 1964), 522–528.

Holt, John, *How Children Fail*, Pitman Publishing Corp., New York, 1964.

Keiffer, Mildred C., "The Development of Teaching Materials for Low Achieving Pupils in Seventh and Eighth Grade Mathematics," *Arithmetic Teacher*, 15, 7 (November 1968), 559–604.

Kueitz, Margaret H., and John L. Creswell, "An Action Program in Mathematics for High School Dropouts," *Mathematics Teacher*, 62, 3 (March 1969), 213–217.

Sobel, Max, "Providing for the Slow Learner in the Junior High School," *Mathematics Teacher*, 52, 5 (May 1959), 347–353.

———, *Teaching General Mathematics*, Prentice-Hall, Inc., Englewood Cliffs, New Jersey, 1967.

Weiss, Sol, "What Mathematics Shall We Teach the Low Achiever?," *Mathematics Teacher*, 62, 7 (November 1969), 571–575.

Wells, David W., and Albert P. Shulte, "An Example of Planning for Low Achievers," in *The Teaching of Secondary School Mathematics*, Yearbook 33, National Council for Teachers of Mathematics, Washington, D. C., 1970, pp. 397–422.

Wirtz, Robert W., "Nonverbal Instruction," *Arithmetic Teacher*, 10, 2 (February 1963), 72–77.

A Program for the Talented 21

Students with special mathematical talent are a precious human resource. We must develop procedures to improve the mathematical education of these gifted youngsters who constitute an important national resource.

Identifying the Talented Mathematics Student

For the purpose of our discussion we will assume that the talented mathematics student is simply the person with a high potential for learning mathematical concepts and structures. This student is frequently in the top 10 percent in academic achievement of his age group.

The talented student in mathematics usually shows the following traits:

He is able to abstract generalizations from complex situations.

He is able to transfer and apply his knowledge to new situations.

He demonstrates a high level of intuition, association, insight, creativeness, and original thinking in learning new ideas, solving problems, or finding discrepancies.

He has an extensive vocabulary, which facilitates his thinking, reading, and communication.

He learns new ideas rapidly, easily, and usually with pleasure.

He has a storehouse of knowledge, which often extends to advanced mathematics.

He has an extraordinary memory—sometimes even total recall of what he has read, heard, or seen.

He is curious about new ideas and the world around him.

He maintains initiative for independent work toward goals that may even be remote or difficult to achieve.

He enjoys a wide range of interests in hobbies or special fields such as mathematics or science.

He tends to be physically fit, socially mature, honest, and charitable to others. The usual picture of the gifted student as an awkward introvert is not borne out in studies of the gifted.

He may be a nonconformist, independent, confident in his judgments, and impatient with routine activities, drill, and computation.

Although many talented students have these characteristics, their performance level may vary considerably. Some demonstrate outstanding achievement and interest in mathematics; others have had their initiative blunted by a routine and conforming school program. Some have a record of average or even below average achievement because they are unmotivated, have negative attitudes or conflicting interests, or come from culturally deprived homes or communities.

Generally, taking these characteristics into consideration, a teacher can identify the talented student by specific testing procedures and by careful observation of the student. The following specific tests are helpful:

A mental-ability test. Students with IQ's above 125 are usually talented in mathematics. However, special mathematical talent is not isolated by intelligence tests.

A mathematical-achievement test with published norms. Students ranking in the top 5 percent would likely have special aptitude for mathematics.

A test of reasoning ability, problem solving, or critical thinking. This type of intellectual activity is highly correlated with mathematical talent.

A reading test. The talented student is usually a rapid and avid reader.

A communication test. The talented usually have high verbalization competence.

These test scores combined with achievement records and the judgments of teachers and counselors become a basis for selecting talented students. One large metropolitan school established these standards for their definition of a talented mathematics student:

1. An IQ score of 120 or above.
2. A percentile rank of 90 or better on a standardized mathematics achievement test appropriate at his grade level.
3. An achievement record of A in previous mathematics courses.
4. A strong interest in learning mathematics.
5. A record of good work and study habits.

Although these guidelines make the selection of the talented straightforward and objective, errors in selection still will be made. There is always the risk that some plodding overachievers who are not really talented will be selected, while creative, nonconformist, unmotivated underachievers will be omitted.

Activities for the Talented

Every school should establish special activities and courses for its talented students. The type of program will depend on the size, location, and current resources of the school.

Homogeneous Grouping

The generally accepted method of dealing with the talented student is by homogeneous grouping in special classes. When students are grouped according to ability, the content, materials, method, time, and learning activities can be tailored to the students involved. The talented need open-ended, challenging exercises instead of repetitive reinforcement and review exercises. Talented students need opportunities for enrichment and independent investigations. They need to work at an accelerated pace; and, in a class of his peers, the talented student is stimulated to work more nearly up to his potential. In competing with other students of equal ability, the talented student does not get an inflated notion of his ability. In doing the things he enjoys with others who share his interests, he should be motivated to continue his study of mathematics. However, there are also dangers inherent in ability grouping. The talented group may forget that their special assignment is a privilege and consider themselves an elite group. As a result, they may abrogate their role in leadership and service and may withdraw from student activities. Assigning them to different groups in other subjects and in home rooms is one way of keeping them in touch with the entire student body. Separating the gifted from other students also imposes problems for teachers, who would then need to develop leadership in the average classes to replace that siphoned off by the top sections.

Special Courses

One of the most frequent proposals for dealing with superior students has been to provide them with special courses. These courses permit the extensive study of mathematical topics adapted to the ability of the student and presented at a challenging rate. They provide an organization in which acceleration, enrichment, and independent activities can be coordinated into a balanced program. In small schools, this approach has been modified to provide clubs or seminars, which often meet after school. Currently, we devote a tremendous amount of out-of-school time, provide special coaches, and buy all kinds of equipment for the athletically gifted without concerning ourselves with the probable development of conceit on the part of stars or frustration on the part of the third stringer. We should work for equivalent provision for the mentally gifted.

Scholastic provisions for the mathematically talented students usually take one of the following forms:

1. Acceleration of gifted students into a class normally elected only by students in a higher grade.
2. Individualized instruction using correspondence study, programmed texts, or independent study.

3. A small special-ability group within a regular mathematics class.

4. A special-curriculum track for talented students.

5. Attendance at an optional out-of-school seminar, summer-school class, or camp with an academic program.

6. Registration in a course at another school or college.

7. A special enrichment course such as computer programming, probability and statistics, or game theory.

8. An advanced placement course.

Enrichment

Our description of the gifted has emphasized their possession of characteristics such as intellectual curiosity, ingenuity, independence, high reading interest, imagination, creative talent, leadership, and ability to assimilate and generalize. In order to capitalize on these exceptional traits, any program for the gifted should provide a variety of enrichment activities. A supply of mathematics books, pamphlets, and periodicals should be available to provide literature to satisfy intellectual curiosity and reading interests. Creative talent and independent work habits may be developed through research projects, written reports, or the construction of mathematical devices. The possibilities of using imagination in dealing with infinity, the fourth dimension, non-Euclidean geometry, a new number base, or Boolean algebra are far-reaching. The analysis of original data collected at school or in the community can be a vehicle for social responsibility and leadership, as well as training in applying methods of analysis. To capitalize on the ability of the gifted to assimilate and generalize difficult material, the teacher can direct the student to a much more intensive coverage of a given course. For example, the study of graphing can be extended to topics in analytical geometry such as polar coordinates or three-dimensional coordinate systems, the logic of geometry to statistical inference, the study of variation to limits and elements of calculus. The main point to remember is that the talents of the gifted student must be nurtured; that, rather than narrowing his horizons by a rigid curriculum, the teacher must broaden these horizons by a more flexible program.

Extracurricular activities are another form of enrichment that is attractive to the highly motivated talented student. Mathematics clubs, contests, interscholastic competitions and fairs provide excellent opportunities for these students to extend classroom activities in a new and attractive setting. See Chapter 15 and references to that chapter for more on these activities.

Acceleration

Another frequent proposal is to accelerate the superior student. This method has the advantage of providing challenging material at a pace that does not permit poor work habits and avoids the frustration of lock-step class work. It involves certain risks of social and emotional maladjustment when sooner or later the superior student finds himself studying with older students, who sometimes have greater social maturity. However, several studies have shown that moderate acceleration is desirable—especially when the gifted individual is socially and emotionally mature. When an acceleration is started at a given level such as seventh grade, it should be maintained in the senior high school. Although acceleration is most frequent in the junior high school (e.g., algebra in grade 8), it is more appropriate at the senior high school. At any level, it should not be done so that the student loses out on important mathematics. For example, to have a student bypass an algebra course to study calculus in the high school is unwarranted.

Another danger of acceleration is giving students too much abstraction too fast. Maintaining the balance between providing an exciting and challenging program and demanding too much with too little preparation or motivation is a central concern of the teacher of the talented. Too many of these students drop out of mathematics programs in high school and college, depriving society of still another rich resource. Related to this is the fact that talent is usually not specific. Many of these students are in top sections in all subjects. They are also active in a wide variety of school activities. They need much assistance in defining their own limits and particularly in making choices. The teacher of the talented must be particularly sensitive to characteristics of the overextended student: declining quality and quantity of work, degenerating work habits, occasionally cheating, and often signs of physical exhaustion.

Leadership

The desire for social leadership has often resulted in the gifted student becoming so involved in extracurricular activities that he has lacked time to build his foundation in science and mathematics. Opportunities for social responsibility and community leadership must be assured the gifted, but they should be part of a balanced program planned with the help of well-informed counselors and teachers. This planning should involve the selection of goals as well as the building of interests. At higher levels, it should also explore means of financial support if such help is needed to attain the goal selected. Counselors can also be helpful in establishing the testing program that will identify the gifted, locate their strengths and weaknesses, and measure their progress by contrasting it with the progress made by students both in their school and in other schools.

Counseling is one of the most important aspects of dealing with the talented, for the talented student needs to know what his potential is and how to capitalize on his ability. He needs to know what professions require mathematical talent and what the requirements are for different occupations. He needs also to know what sources of financial support are available to attain the education he needs.

The pamphlets listed in Appendix C (under mathematical careers) are helpful in giving guidance to the talented.

Criteria for Programs for the Talented

All provisions for the talented will fail unless the student is motivated to use his talent. His intellectual curiosity must be stimulated by good teaching, appropriate materials, and recognition of his high accomplishment. In this regard, awards, A grades, scholarships, and contests are probably secondary to intellectual satisfaction. The stimulation of an enthusiastic, well-informed teacher who knows the needs of his students is indispensable.

An effective program for the talented, then, will have these aspects:

The *curriculum* for the talented should be a complete curriculum, extending from grade 7 through 12. This curriculum should provide enrichment as well as acceleration. It may involve correspondence courses, programmed texts, college classes, or independent study. Avenues for late entry into the program should be provided, as well as opportunities for leaving the program.

The *goals* of instruction should include those associated with attitudes, habits, values, creativity, leadership, and communication skills as well as the content goals of the mathematics courses.

The *students* for this special curriculum should be selected with care. Selection should be based on tests of intellectual and mathematical aptitude, vocational interests, and social maturity as well as teacher judgments.

The *teachers* for special classes should be assigned on the basis of qualifications rather than seniority or prestige. The successful teacher of the talented is one who can give the necessary stimulus and guidance for effective and enriched learning. The teacher of the talented must have a strong background in mathematics and, if possible, special training in teaching this type of group. Extra preparation time should be provided for this teacher, especially during his first assignment to this group.

The *resources* for the talented must include special facilities and materials. An independent study room, laboratory, or seminar room is helpful. Supplementary texts, enrichment pamphlets and books, computers, laboratory equipment, and construction materials are needed.

The *evaluation* of the talented must be realistic. A student's class standing or grade-point average must not be jeopardized by his being assigned to an accelerated program. One way to handle this situation is to give an "S" grade in special courses or to base class standing on test scores. Some schools offer two grades for the talented—within the course, a grade that expresses his standing among his peers; and a school grade, which compares him with the entire class. Another procedure that has been used is to assign standard (within course) grades but add a point to these grades when computing grade-point average and class standing. In the majority of cases, average marks are not acceptable in a special class for the talented.

Examples of Enrichment

Since the talented student has such broad interests and the ability to do independent study, enrichment is one excellent way of meeting his needs. A number of enrichment resources are given in the Appendixes of this text. The wealth of materials now available makes it possible to enrich the learning of the talented by a variety of activities such as the following:

Solve challenging mathematical problems like those found in periodicals such as *The Mathematics Student Journal* and *School Science and Mathematics* or in books such as *Mathematical Puzzles for the Connoisseur* or *Chips from Mathematical Logs*. (See Appendix B.)

Read about exciting mathematical topics in enrichment pamphlets such as *Exploring Mathematics on Your Own* or *Thinking with Mathematics*. (See Appendix D.)

Learn new mathematical ideas from books such as *Mathematics and the Imagination*, *The Number of Things*, or *The Education of T. C. Mits*. (See Appendix C.)

Study advanced topics such as limits, infinity, non-Euclidean geometries, finite mathematical systems, computer programming, symbolic logic, or game theory.

Write creatively by preparing scripts, research reports, poems, plays, or essays.

Enjoy recreational reading of science fiction based on mathematical ideas, as found in Fadiman's two books *Fantasia Mathematica* or *Mathematical Magpie*.

Learn to perform tricks, solve puzzles, or analyze paradoxes as described in books listed in Appendix D.

Participate in out-of-class activities such as mathematics clubs, seminars, contests, and fairs.

Build models such as logic machines or tesseracts to represent mathematical ideas.

Find applications of mathematics in science, economics, industry, government, music, or art, as described in Kline's *Mathematics in Western Culture*.

Collect information about opportunities in mathematics and the work of contemporary mathematics.

Study more intensively and deeply the standard topics included in the accelerated course.

Investigate the history of mathematics and biographies of mathematicians.

Prepare lessons or reports on unusual topics such as topology, transformations, symmetry, or space travel. See Appendix D.

Perform laboratory activities such as forming pendulum patterns, determining probabilities, or measuring indirectly the diameter of the sun.

Create an exhibit on a mathematical topic: mathematics in nature or the Möbius strip.

Tutor students who need remedial instruction or who are culturally deprived. (This activity should increase the talented student's sense of responsibility and give him experience in communication. It forces him to reexamine content in a different light, aids his judgment, maintains a good relationship with weaker students, and gives him more appreciation for and perspective on his own gifts.)

Act as an assistant in the mathematics laboratory or in the mathematics class. (The activities assigned should be those that develop the initiative and leadership of the talented. Care must be taken so that this type of assignment does not cause the student to lose status with his peers.)

From the ideas listed above, it is apparent that enrichment requires an extensive library of materials other than mathematics textbooks. If students are to participate in these enrichment activities, they must also be provided with the necessary school time and not be expected to do all their reading outside of school. Of course, the extra work done by the student should be given credit toward his mark, and the work should be recognized as a job well done.

Acceleration of the Talented

Most new ideas in mathematics and science have been created by young people before they reach the age of 26. If we are to help get our talented mathematics students to a level where they too may be able to create new mathematics while they are young, we must accelerate their education.

An easy means of acceleration is the assignment of the talented student to a class one level or more above his own grade. This may be especially appropriate for a small school but even then care must be taken to see that the accelerated student is placed in a class where he is accepted by the rest of the students.

A better method of acceleration consists in the selection of enough able students to form a class. When possible, this should be done at the seventh-grade level so that a complete secondary sequence is available. A possible sequence for acceleration is outlined below:

In grades 7 and 8, the content is that of the modern three-year junior high school mathematics course.

In grade 9, the course content includes plane, solid, and coordinate geometry as well as work with transformations and vectors.[1]

In grade 10, the course is a second year of algebra with emphasis on deduction, proof, and structure.

In grade 11, the course emphasis is on functions, including trigonometric and polynomial functions as well as matrices, vectors, sequences, limits, permutations, and probability.

[1] The algebra-geometry-algebra sequence is suggested in the light of current textual materials. Most second-year algebra texts demand fairly extensive knowledge of geometry for the study of coordinate systems.

In grade 12 there are a number of alternatives: (1) a course in analytical geometry and calculus (usually the College Entrance Examination Board Advanced Placement course); (2) a course in modern mathematics (e.g., linear programming, game theory, probability and statistics, geometric transformations, and computer problem solving); (3) an integrated science-mathematics course; or (4) a selection of semester offerings (e.g., analytical geometry, probability and statistics, and modern algebra), from which the student may elect one, two, or even more. Some schools include among the electives listed in alternative 4 a course geared toward the specific interests of one or two teachers, the concentration being not on the content as such but on student exploration and the development of interest. Such a course might concentrate on inequalities, continued fractions, or computer programming. Several publishers are offering monographs that would be useful in such courses. See Appendix C.

Many high schools today do not have mathematics teachers with adequate mathematical backgrounds to teach these twelfth-grade courses. Hence, teachers and administrators may wish to explore the possibilities of correspondence courses, a televised course, a course arranged cooperatively with one or more neighboring schools, a local college course, or a programmed text for independent study.

Although acceleration is a good way to provide stimulating courses for the talented, it is not appropriate to accelerate students at the expense of depth of treatment and enrichment. Note that in the program suggested, the content in grades 9 through 11 comprises enriched versions of standard courses. This seems the best course of action at all grade levels.

Specific Techniques for Teaching the Talented in Mathematics

Since the teaching of the talented is of such high significance, it is essential that it be appropriate. Too often it is assumed that the talented will learn even though the instruction is poor. In some schools the poorest instructor may be assigned to this group. This is a practice that cannot be condoned. These are some necessary teaching activities:

Know each talented student. Collect information about his ability, interests, hobbies, home environment, school activities, and summer activities. Expect differences among students even when they are grouped according to ability. It should be kept in mind that ability differences in high and low sections will be greater than in average-ability classes.

Know your subject. Present challenging material at an appropriate rate. Select thought-provoking questions and problems that are stimulating. Present mathematics as a subject that is fascinating and of the greatest significance. Be ready to explore unexpected avenues when students initiate unexpected digressions.

Provide time, resources, and stimulation for independent work on topics of special interest to individual pupils.

Make available and encourage the use of a variety of enrichment material—books, pamphlets, periodicals, supplementary texts.

Arrange opportunities for out-of-class activities such as mathematics clubs, contests, fairs, seminars, camps, summer enrichment courses, mathematics bulletins, or mathematics teams.

Encourage leadership and communication skills by having student demonstrations, reports, lessons, exhibits, projects, and research papers. Provide opportunities for the talented to teach topics or give demonstrations to his class.

Enrich your instruction with new topics, historical incidents, biographical sketches, unique problems, current applications, or elegant proofs found in library books.

Share original ideas, elegant solutions, superior achievement of individuals with the class. Provide an environment in which intellectual achievement is valued. Keep the school, community, and parents informed of the program for the talented and of special attainments. The bulletin board is one resource to be used for the display of superior work.

Have frequent conferences with individual pupils. Explore vocational plans, interests, difficulties, needs, and emotional problems. Help the pupil accept responsibility for the maximum use of his talents without conceit. Explore future plans, courses, and scholarships. Investigate the total work load of the pupil and his progress in other areas.

Encourage creative activities such as creative writing, independent and group projects, and class reports.

Provide an emotional climate in the classroom that stimulates interest, curiosity, participation, and security. Evaluate achievement by marks that indicate the standing in an appropriate comparison group.

Communicate with the parents. Inform them of their child's potential and actual achievement and recommend appropriate ways of developing the potential of their talented child. Some parents put too much pressure on their children while others do not understand the talents of their children.

Capitalize on community resources. Local engineers, computing centers, industries, colleges, or research laboratories may have opportunities for work experience, projects, or seminars.

Be receptive to the off-beat questions and suggestions of the creative pupil. Expect the talented to know things that you may not know and to solve problems that you cannot solve.

Require clear, precise statements and neat, correct, well-organized written work. Since talented pupils see results quickly, they sometimes become lazy and slipshod in their work.

Enjoy your association with these students. Show them that you support them in efforts to use their gifts in positive directions. Give them support in dealing with teachers and other students.

Summary

These proposals for a program for the talented mathematics student are based on the following assumptions:

Intellectual talent is our *most precious resource.* Intellectual leadership, insight, and creativeness are indispensable in our current society.

It is possible for the talented to *learn* mathematics *rapidly* and to *be creative at an early age.*

For maximum development of talent, the talented pupil needs *the opportunity, the challenge, and the materials* for learning.

Allow *the talented student to work independently* in his fields of special interest.

It is essential that *the teacher of the talented be a specialist* who is talented in mathematics and in teaching.

It is appropriate that the talented have *courses, materials, instruction, and opportunities that are different* from those for the average or slow learner.

It is important that *adequate recognition be given* to the superior achievement and contributions of the talented.

Learning Activities

1. How does the discussion of distribution of talent on page 340 apply to talented students?
2. Read a journal article on one of the following activities. Discuss the advantages and disadvantages of the activity as a program for the talented:
   (a) A mathematics fair
   (b) A mathematics contest
   (c) A mathematics team
   (d) A mathematics club.
3. Write a lesson plan for a topic for a talented group of students studying one of the following:
   (a) Seventh-grade mathematics
   (b) Geometry
   (c) Algebra
   (d) Elementary functions.
4. Design a presentation for a mathematics club. (*Enrichment Mathematics for the High School*, Yearbook 28, NCTM, is a good resource.)
5. List the five most important techniques for teaching talented students. Which of these differ from instructional techniques for average students?
6. "Boy, do I have an easy assignment this year: all top sections." Comment on this too-common teacher statement. What attitude toward students does it reflect? Which type of section demands the most teacher preparation and effort? Justify your answer with care!
7. PROJECT. Prepare a case study of a talented student. Identify ways his program might be improved to respond to his special characteristics.

Suggestions for Further Reading

Enrichment Mathematics for the Grades, Yearbook 27, National Council for Teachers of Mathematics, Washington, D. C., 1962.

Enrichment Mathematics for High School, Yearbook 28, National Council for Teachers of Mathematics, Washington, D. C., 1963.

Hollingshed, Irving, "Number Theory—A Short Course for High School Seniors," *Mathematics Teacher*, 60, 3 (March 1967), 222–227.

Holmes, Joseph E., "Enrichment or Acceleration," *Mathematics Teacher*, 63, 6 (October 1970), 471–473.

Johnson, Donovan A., "A Fair for Mathematics with Marathons and a Midway," *School Science and Mathematics*, 65, 9 (December 1965), pp. 821–824.

Pieters, Richard S., and E. P. Vance, "The Advanced Placement Program," *Mathematics Teacher*, 54, 4 (April 1961), 201–211.

Rising, Gerald R., and Richard Wiesen, *Mathematics in the Secondary School Classroom—Selected Readings*, Thomas Y. Crowell Company, New York, 1972, "Enrichment" sections.

Rogler, Paul V., "The Mathematics League—for Motivation and Inspiration in Mathematics," *Mathematics Teacher*, 56, 4 (April 1963), 223, 267, 274.

Salkind, C. T., "Annual High School Mathematics Contest," *Mathematics Teacher*, 57, 2 (February 1964), 75–78.

Srinivasan, P. K., "Detection and Care of the Gifted in Mathematics," *Mathematics Teacher*, 61, 4 (April 1968), 396–398.

FIVE

Use of Teaching Aids

The Textbook

22

The mathematics textbook is a major factor in determining what mathematics topics are taught and how they are taught. A textbook has often dictated the scope, the sequence, and even the pace of the mathematics program. Thus, the textbook is a powerful means of determining whether new topics are brought into the schools or whether the old mathematics is maintained. This is all in addition to its basic function as a learning tool in the classroom. Its importance increases when instruction is inadequate. However, the mathematics curriculum should *not* be determined by the text; rather, the text should be selected on the basis of prior curriculum decisions.

Since the mathematics textbook is such a powerful influence in the mathematics classroom it may be an invaluable servant or an intolerable master—depending upon the intelligence with which it is handled. Too often, it is overused. The greater variety of new textbooks available and the continuing development of new school mathematics curricula mean an increasingly frequent change of textbooks is necessary. Some forward-looking schools respond to this need by multiple text adoptions for every course, with a new text added every two or three years.

The mathematics textbook has a unique role in the classroom. The reasons are:

Direct experience, visual aids, and classroom instruction cannot provide all the instruction necessary. Some of this instruction must be covered by reference to a textbook.

Teachers have too many pupils, preparations, and extracurricular assignments to make it possible for them to plan and write complete units and daily lessons without the aid of a text.

Mathematics requires a sequential study treatment, and the textbook provides a useful aid to this approach.

For mathematics teachers with an inadequate background in mathematics and in the methods of teaching mathematics, the textbook is a substitute (*albeit a poor one*) for this background they lack.

Many schools are limited in resources such as library books, concrete and visual learning aids, community resources, duplicating equipment; and so the text provides the basic and sometimes the only resource.

Learning mathematics depends on the mastery of concepts and skills. Students may grow in this mastery by performing the exercises of the text.

Mathematics requires a storehouse of facts, theorems, formulas, and definitions to which reference can be frequently made. In this way, the mathematics text is as necessary as a dictionary or encyclopedia in English or social studies.

The Contribution of a Good Textbook

Thousands of hours of effort by usually highly qualified authors and thousands of dollars in production costs go into the development of textual materials. This very substantial investment often results in books that are attractive, well organized, and well illustrated, thoughtful, and accurate. They can provide a most valuable resource for teaching.

The superior mathematics textbook offers the following aids to teaching and learning:

It provides most of the *content* for a course. As such, it should contain appropriate, mathematically correct topics presented in a readable and orderly fashion.

It presents topics in a manner that builds *understanding* of concepts, structure, problem solving, and computations. In other words, it is a tool to be used in attaining the objectives of the course.

It provides the exercises, the experiences, the directions for attaining *mastery* through practice, review, application, and thought-provoking questions.

It provides a means for *independent study* and, hence, is useful for assignments, make-up work, remedial instruction, and independent study.

It provides a means of making provision for *individual differences.* By giving assignments tailored to different ability, by providing suggested enrichment materials, by permitting independent acceleration, the textbook can be a source of satisfying, challenging experiences.

It provides a compact *reference book* that is useful in building the structure of mathematics. Tables, definitions, formulas, graphs, sample problems, theorems, and proofs are available to make problem solving efficient.

It provides a basis for *achievement testing.* Chapter tests, review tests, practice tests, and accompanying semester tests provide ready-made devices for evaluation of content mastery.

It brings directly to the student the *exposition* of the writer or writers, often major figures in mathematics and mathematics education or master teachers.

It forms the basis for classroom *instruction,* which may and should often follow a different but essentially parallel development. In this way the student is offered various approaches to a single topic.

The Qualities of a Good Mathematics Textbook

If a mathematics textbook is to serve its proper function, it must be a good text. It needs qualities such as the following:

Topics.

The topics are those that will attain the objectives of the course.

They allow selection to fit the sequence: building on the previous course and foreshadowing the course to follow.

They are appropriate in terms of interest, difficulty, and usefulness to the students electing the course.

They are in harmony with current curriculum emphasis.

Mathematics.

The mathematics is correct.

The structure of each topic is clear and concise.

The level of the rigor and precision is appropriate for the course.

The use of symbols is correct but reasonable, accurate, and not overly cumbersome.

Language.

The narrative is readable and comprehensible.

The abstractions and symbols are made meaningful.

The language is interesting and thought provoking.

The definitions and explanations use only those terms that the student can be expected to understand.

Pedagogy.

Material is included to create interest and motivate learning.

Terminology and content is justified in terms the students understand so that they can see how it relates to them.

Material is included to make it possible to meet the needs of different levels of ability.

The strategies used are based on sound learning principles.

Concepts are introduced by providing opportunity for the student to discover ideas through reflective thinking, problem solving, experimentation, analysis, and generalization.

Tests for the evaluation of achievement by student and teacher are included.

Mastery.

Exercises emphasize reflective thinking and problem solving as well as straightforward manipulation.

Adequate exercises of different difficulty levels are included.

Review and remedial materials are included.

Some exercises require the student to generalize, others to consolidate concepts, and still others to improve skills or to apply what is learned to new situations.

Enrichment.

Enrichment topics are included in the text.

Suggestions are given for independent study.

Research topics, projects, and independent experiments are suggested.

References for enrichment reading are included.

Aids to Learning. These include:

A teacher's manual with suggestions for teaching

An answer key with worked solutions

Achievement tests

Overhead projectuals

Accompanying workbooks, laboratory manuals, audiotapes, film loops, and computer supplements.

Physical Characteristics.
The format of the pages is attractive and inviting.
The arrangement, headings, and type make the location of material convenient.
The use of color and illustrations is functional in terms of text content.
The size of the book is convenient.

The Proper Use of Mathematics Textbooks

Too often the text is misused, overused, or, at the other extreme, ignored. In this latter connection, many school administrators sometimes believe that they are introducing a new program by adopting a new textbook—when actually the teaching remains unchanged and the text is misused or ignored.

The following suggestions are given for the proper use of a textbook for the typical mathematics class:

A selection of topics to be taught should be made from the text. *Only in rare circumstances should the entire content of a text be presented in a single course.* Textbook authors purposely include more content than necessary so as to give a teacher the possibility of selection.

A decision should be made as to what topics that are not in the text should be included in the course. *Every textbook needs to be supplemented by up-to-date material* from sources such as library books, pamphlets, and other texts.

The text should be used as a resource and reference book. Rather than repeating the examples of the text or reading the text to the class, *the competent teacher uses different examples and different explanations.*

The text is used by students as well as teachers as a source for questions, exercises, reading material, and reference material. *Students are expected to read the text, to answer the questions thoughtfully, to work the exercises, and to find information.* At the same time the text should suggest problems, topics, or exercises related to independent study and suggest further study in other source books.

The students should be given instruction in how to use the textbook—for example, where to find material, how to read the narrative, how to use chapter summaries and tests, how to review, and how to solve problems.

The narrative or exercises of the text should be assigned with the students' differing abilities and needs in mind. The low-ability student works a greater proportion of easy problems and reviews frequently. The high-ability student works fewer and more difficult exercises. The suggested enrichment material, outside reading, references, or projects are used, according to the interests of the students.

The assignment of textbook exercises is done carefully. *The purpose of each assignment should be to improve the understanding, accuracy, efficiency, and retention of the students.* These purposes must be made clear to the student who should never think of homework as mere drudgery.

Answers to at least some exercises should be provided to students working exercises so that they can know what success they are having. Correct answers then reinforce correct methods while errors suggest further study of concepts and rechecking computation.

The textbook exposition is supplemented by instruction that provides discovery exercises, audio-visual illustrations, references, and local applications.

The textbook exercises also are supplemented with learning activities, reports, projects, games, and surveys.

The textbook tests are supplemented with other tests such as unit tests, reading tests, diagnostic tests, essay tests, performance tests, and open-book tests.

The resources of the text are used to enhance instruction. The projects, references, historical sidelights, and enrichment topics should be recognized to be of as great a significance as the explanations and exercises.

The references support teacher requests for school purchases of additional library books and other supportive material. *Additional textbooks provide the teacher with a valuable resource library* to help him with planning and executing his program. They are a source of different strategies, discovery exercises, enrichment, and test items. Often texts are supplied by textbook companies as a service and in hopes of adoption. However, it is not ethical for the teacher to request "free" books purely for supplementary material. Whatever supplementary materials are needed should be purchased by the school.

The Dangers of "Textbook Teaching"

Many teachers are strictly textbook teachers, more concerned to "cover" the text than to "uncover" ideas. Their focus is on the text rather than on the learner. Some of the following results have been noted:

> The mathematics textbook becomes the mathematics curriculum. To bring the curriculum up to date merely means the adoption of a text with a recent copyright date.
>
> The content of the text becomes the total content of the course, with rate and sequence rigidly prescribed. Some teachers even divide the number of pages of their text by the number of school days in the year to get the daily average rate. Confining the class content to that indicated by the text gives students a limited experience and increases the danger of poor attitudes and little appreciation of the elegance of mathematics.
>
> The tendency of textbooks to emphasize given rules and procedures defeats the possibilities for discovery, independent thought, and intellectual curiosity. Even where discovery questions are included, the answers usually appear on the next page and students find it easier to look ahead than to discover the concept.
>
> Student memorization of the language of the text and stated definitions and rules does not nourish skill in communication or the development of understanding. Students need experiences in stating generalizations in their own words even though these may lack precision.
>
> The constant use of the textbook kills interest by its monotonous, formal treatment. Learning needs a variety of meaningful, interesting experiences.

The blind regimentation of textbook teaching loses the slow learner and bores the rapid learner. It is a discouraging experience to see teachers using the same text examples and exercises for all students in a given class.

The narrow emphasis on the text ignores the importance of objectives such as attitudes, problem solving, creativity, appreciations, or values. These are objectives that seem of greatest importance today. They are seldom attained and rarely tested by the textbook teacher.

From this discussion, it is apparent that "textbook teaching" is highly unsatisfactory. Dependence on the text and only the text is one mark of an unsuccessful teacher.

Workbooks and Laboratory Manuals

Workbooks were very popular a generation ago. They provided the teacher with many drill exercises and saved him time in preparing material. They also saved the student time by eliminating the need to copy exercises. However, the workbooks' emphasis on rote manipulation at the expense of understanding caused them to be discarded.

Workbooks do, however, have many possible uses if they are of good quality and are used properly. Workbooks can contribute to instruction in the following ways:

Workbooks of different levels in a single class provide one way of meeting the problem of individual differences.

They permit individual work at varying rates. Each student can work on lessons selected specifically for him.

Workbooks provide a varied testing program with progress ratings and keyed remedial exercises. This variety is especially important for slow learners who are unmotivated by traditional textbook format.

They supplement the textbook in building meaning as well as skills.

Workbooks often provide discovery activities for each student to complete. He can write his responses in specially provided places. This helps the student to avoid errors in copying problems as well as saving him time.

Workbooks have the advantage of being the student's own property. The student's own work is organized for him, providing him with a valuable tool for review.

Workbooks provide for independent work and orderliness in the overcrowded classroom. They simplify the mechanics of independent assignments.

They provide the teacher with duplicated material written by experts. Even without considering the cost of teacher time, they are usually less expensive than worksheets prepared and duplicated in the school.

In the future, workbook-type activities may be printed by computers. The computer would be programmed so that exercises could be tailored to previous student achievement.

Another type of workbook now on the market is gaining rapidly in popularity. This is the laboratory manual. These manuals provide discovery activities of the kind discussed in Chapter 27. These laboratory activities workbooks provide instructions and data tables for experiments and pose related questions that guide students to discover generalizations. For example, it would seem appropriate to approach the topic of probability by a series of experiments. These experiments would provide the intuitive, empirical experiences to which formulas, definitions, and abstractions would later make reference.

Programmed Texts

A programmed textbook is a book designed in such a way that it guides the student by a series of short steps to understand the material being presented. These short steps are in the form of brief expository statements and questions, to which the reader responds by writing the answers in spaces provided. The reader checks his answers against those supplied by the book (often hidden by a slide or presented on the following page). The questions provide the learner with immediate knowledge of his progress. In some programs, called branching programs, incorrect answers may direct the learner to remedial work, but the most important function of the question-and-answer process is the deeper involvement of the reader in the learning activity. He is virtually forced by the questions to share responsibility for learning.

These texts are being advocated because they require each student to react to each question. In this way he is "led" to discover a generalization. In addition, answers are given to each question or problem so that correct answers are reinforced and errors are corrected.

Programmed texts also are advocated because they permit independent study at whatever rate the individual wishes to establish. This seems an ideal way of providing for individual differences. The slow learner progresses at a rate appropriate to him, and the talented can be accelerated. Also, the text might be used by a student to make up work missed during an absence or for remedial instruction by the student having difficulty.

High hopes were held for these programmed texts when they were first developed. However, research on their effectiveness has shown that these texts have some negative effects. It is much too monotonous for students, especially slow learners, to work independently day after day writing answers to questions, a problem shared with many independent structured programs. To the superior student, the questions also seem trivial and time consuming. To the slow learner, the reading is difficult and his motivation wanes.

These unsatisfactory experiences with programmed texts do not mean that they are not useful. Rather, it means that different ways of using these texts must be devised. Classroom discussion, experimentation, and group activities need to be combined with programmed texts. Perhaps their major role is to supplement instruction by presenting remedial work or enrichment topics. Another possible use is as a text for correspondence lessons or television courses. As is true for all instructional aids, programmed texts do not replace the teacher. Instead, they supply a new teaching tool to be used to improve instruction.

A particularly attractive use of programmed texts and the related audiotapes or videotapes is to provide assistance to students who have been absent. This use alone should encourage teachers to familiarize themselves with available programs so that they can assign specific sections to students who miss classes.

Selecting a Mathematics Text

An assignment that often becomes the responsibility of a classroom teacher is the selection of a textbook. Some teachers view this responsibility with alarm: It is a major, time-consuming job requiring careful examination of and evaluation of many textbooks. Others see the responsibility as an opportunity: They can improve their course by finding the best book available that is suitable to their needs.

Of course both views are correct and both should be taken into account. In particular, an individual teacher making a textbook selection has a very heavy burden. He must be careful not to limit his selection to a narrow or conservative view of his course and yet he must be concerned with the needs of his students. These two concerns pose a difficult balance for one person and suggest the need for a committee to select a text involving not only several teachers of the same subject but also input from teachers of preceding and following grades, a professional mathematician, and the department head or supervisor responsible for the overall program. (Occasionally in a small school all of these responsibilities except those of the professional mathematician still reside in the same person.)

Basically, text selection involves three steps: (1) identification of the goals of the course, especially as they relate to the total mathematics program; (2) location of available texts (too often this step is limited to those books brought to the attention of the selecting individual or group by more energetic salesmen); and (3) evaluation of the texts in the light of step 1.

Many selection committees use checklists related to the individual points noted in the section of this chapter on textbook quality (pages 372 to 373). Such a checklist can be very useful in locating particular values and specific inadequacies in texts, but caution is in order. The final selection must still be a subjective one and not the sum of an objective point system. In particular, the various criteria must be weighted differently. For example, the most attractive text in the world is inappropriate if its mathematical content is wrong or its pedagogical approach is unsatisfactory.

The basic subjective questions are these:

Is the content correct and related to mathematics in general use? Correct content is a must. Definitions and notation not in common use do not necessarily mean rejection, but their use must be justified by strong advantages. For example, an isolated definition of a parallelogram as a quadrilateral with a point of symmetry is quite unjustified by itself in a traditional geometry course. On the other hand, it would be quite appropriate in a special course in transformation geometry. And if the argument for substituting the latter course for traditional geometry is strong enough, as a growing group feels it is, then that may be an appropriate choice. That is an extremely important and difficult judgement for an individual or committee to make.

Is the text appropriate to the audience? The response to this question often leads to selection of different texts for differing ability levels or even more than one text for the same group of students.

Is the text suited to the teaching staff? Edwin Moise of Harvard has commented that many texts are "like the shot heard around the immediate vicinity"; that is, that they may be used with splendid results by the author and a small group of his followers but are quite limited in value to the average classroom teacher. Too many programs rich in potential are changed in the hands of teachers who fail to understand their intent into something far worse than what they previously used. On the other hand, a challenging program with adequate support and preparation may so excite a staff as to improve its quality tremendously.

Summary

The textbook is a tool for teaching, a basic resource but one that must be used thoughtfully and with discretion. At worst, it can serve either as a dust collector and exercise book or on the other hand as *the* course: unmodified and often quite inappropriate. At best, it is an integral part of a total program, drawn upon when its use is most appropriate, but not called upon to perform tasks for which it was not written. The text provides foundations on which student and teacher build.

Learning Exercises

1. What is the best mathematics textbook you have ever used? Why? How are your personal characteristics related to your choice?

2. Many students never read a mathematics textbook. It serves merely as an exercise collection for them. Why does this happen? What are some things that a teacher should do to change this?

3. The interplay between textbook and curriculum is an interesting one. To some extent, it is a "which came first, the chicken or the egg?" matter. Why is a textbook not the basis for a mathematics curriculum? What aspects of the curriculum are provided by the textbook? What aspects are not provided?

4. Make a critical evaluation of one chapter in a contemporary textbook in terms of the qualities listed in this chapter.

5. Choose a topic covered in a contemporary mathematics textbook. What are some of the things that you as a teacher need to add to the text presentation of that topic?

6. Reexamine your answer to exercise 5. Which of these things are really done in the textbook and could be learned directly and independently from the book if you trained students to read the book with understanding?

7. Choose a topic covered in a contemporary mathematics textbook. If you were a textbook author, how would you change the approach to this topic?

8. PROJECT. Compare a topic in a contemporary text with one from 1950 or before. Consider not only content but also style of presentation, stress on exercises, review, and enrichment, and level of difficulty.

9. PROJECT. Write a lesson in programmed format for independent study. Try it with three or more students. What are some of their difficulties?

10. For a specific textbook, indicate content and materials that would be needed to supplement the book. Where would you obtain these things?

11. Make a collection of definitions from various secondary-school texts to compare levels of mathematical sophistication.

Suggestions for Further Reading

Forbes, Jack E., "Textbooks and Supplementary Materials," in *The Teaching of Secondary School Mathematics*, Yearbook 33, National Council of Teachers of Mathematics, Washington, D. C., 1970, pp. 89–109.

———, "Programmed Instructional Materials: Past, Present and Future," *Mathematics Teacher*, 56, 4 (April 1963), 224–227.

Kane, Robert B., "The Readability of Mathematics Textbooks Revisited," *Mathematics Teacher*, 63, 7 (November 1970), 579–581.

May, Kenneth O., "Programming and Automation," in Douglas B. Aichele and Robert E. Reys (eds.), *Readings in Secondary School Mathematics*, Prindle, Weber, and Schmidt, Boston, 1971, pp. 357–370.

Rappaport, David, "Definitions—Consensus or Confusion," *Mathematics Teacher*, 63, 3 (March 1970), 223–228.

Smith, Frank, "The Readability of Junior High School Mathematics Textbooks," *Mathematics Teacher*, 62, 4 (April 1969), 289–291.

Zoll, Edward J., "Research in Programmed Instruction in Mathematics," *Mathematics Teacher*, 62, 2 (February 1969), 103–110.

Models 23

Modern man makes extensive use of models to clarify his thought. The geologist uses maps and charts to locate a new oil field. The chemist uses a Tinker-Toy model to represent a molecular structure. The architect makes a scale drawing of a building design. The physicist or chemist uses a formula to predict the results of an experiment. The economist uses the graphs of inequalities to estimate the most efficient production process. The psychologist uses a flow chart to diagram processes in learning concepts. All these models are constructed to clarify problems.

Most of the great mathematical ideas arose when someone sought a way of dealing with a physical situation. Consider major concepts such as counting numbers, rational numbers, irrational numbers, location, direction, congruence, similarity, symmetry, vectors, probability. All these originated from problems involving objects or situations in man's environment. Hence, it seems logical that if physical situations made it possible for man to invent mathematical ideas, then related physical representation will enhance the learning of these ideas. For example, tossing dice and systematically recording possible results makes the idea of a sample space and a sample point meaningful. Physical representations also are needed to test theories or to provide examples. Thus, for instance, a slated sphere is used for the representation of spherical triangles.

The effective mathematics teacher uses models to help his students think. These models, which may be sketches on paper and chalkboard, concrete devices, or mathematical formulas, furnish the basis for solving a problem, discovering a new idea, or creating a new system. These models are links between the thought processes of man and the reality of nature. They help make transitions from one level of abstraction to another and are a means for expressing ideas and providing stepping stones to new relationships. They add reality to abstract ideas and facilitate creative thinking. They are a means of relating past experience to a new situation.

At the same time, there are inherent dangers in the use of physical models to represent abstract ideas. The concrete representation adds qualities that are not mathematical and by so doing may lead to misconceptions. Compare, for example, "wheel" and "circle." A circle is a mathematical conception. The abstraction cannot be accurately represented by a concrete model, for it is a set of dimensionless points. The wheel can be manipulated as it can be seen as well as felt, rolled, and measured, but while the wheel may clarify certain ideas about circles, it may also mislead the student into thinking of a circle as a disc or a circular region. When teachers utilize models, they must be constantly alert to the dangers of misleading students by false constructions. The model lays the foundation for learning an abstract concept, but the concrete representation does not give a complete conception or definition of the abstract idea.

We usually think of models as three-dimensional, but the mathematics teacher commonly utilizes models that are essentially two-dimensional (linkages or the diagrams accompanying plane geometry proofs) and even one-dimensional (a slide rule or number line). In fact, the teacher may utilize three-dimensional representation of four-dimensional figures, such as the tesseract or pseudosphere, to provide insights into non-Euclidean geometries and geometries of higher dimensions.

The use of such concrete representations of mathematical ideas to teach is far from new. In the first English translation of Euclid's *Elements* published in 1570, pop-up models similar to those in children's storybooks of today were included to illustrate three-dimensional concepts. These and other early models are described by L. G. Simons in "Historical Material on Models and Other Teaching Aids in Mathematics." [1] The longstanding and continuing concern for the use of models in the past is also illustrated by yearbooks of the National Council of Teachers of Mathematics, *Multi-Sensory Aids in the Teaching of Mathematics* (1945), *and Instructional Aids in Mathematics* (1972).

Models are, of course, of special value in the study of geometry, where spatial relationships are the basic concern. However, teachers should not restrict their thinking about models to this area, since models of algebraic and arithmetic relationships can provide students with bases for understanding concepts in these subjects. A good example of the use of a model in algebra is in exploring the terms of the square of a binomial made out of Tinker-Toy parts. In Figure 23.1, it is easy for the student to see that $(x + y)^2$ gives the area of a square. This square is

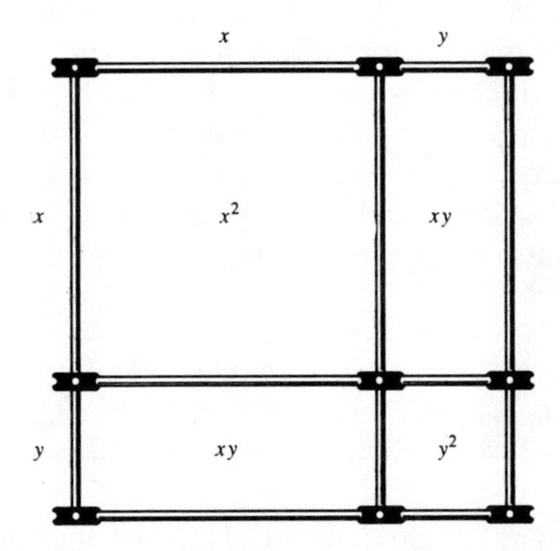

Figure 23.1 Tinker-Toy Models of $(x + y)^2$

[1] *Multi-Sensory Aids in the Teaching of Mathematics*, National Council of Teachers of Mathematics, Washington, D. C., 1945, pp. 253–265.

made up of regions whose areas are x^2, xy, xy, and y^2. Consequently $(x + y)^2 = x^2 + 2xy + y^2$. This is a simple example of how a model adds reality and meaning to an abstract, symbolic representation. Extension of this concept to the cubic model is displayed in Figure 23.2. In this model are one x^3 block, three x^2y blocks, three xy^2 blocks, and one y^3 block.

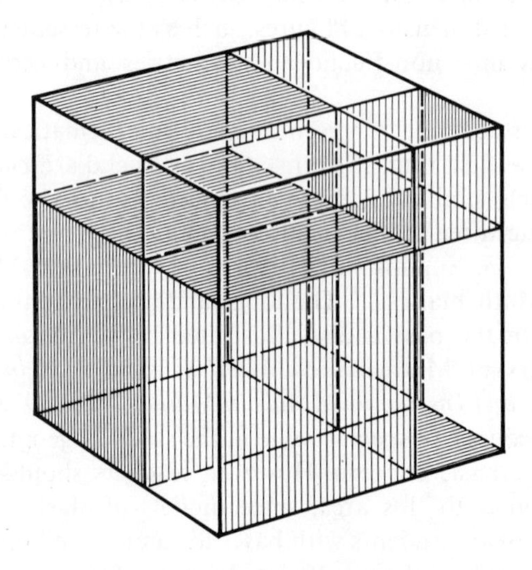

Figure 23.2 *Model of $(x + y)^3$*

One other aspect of the use of models in the classroom should be emphasized: models facilitate one of the central goals of modern mathematics teaching—student discovery. This facet of mathematics instruction, balancing the modern concern with structure and abstraction, is strongly supported by teaching with concrete aids. Successful discoveries prompted by observing, manipulating, or even contemplating physical models give students a sense of individual achievement, a feeling of active participation, and the exhilaration of independent discovery. Through a student's independent exploration, his positive attitudes and intellectual curiosity are stimulated.

Models for Creating Mathematics

In this chapter we usually restrict the word *model* to include only those concrete devices utilized by teachers and students to demonstrate mathematical concepts. Thus, although an equation is a mathematical model of a genus of scientific problems, such models are not the subject of this chapter. We are concerned rather with dynamic devices and demonstration equipment for classroom use. The more general use of the term is, however, central to all mathematics.

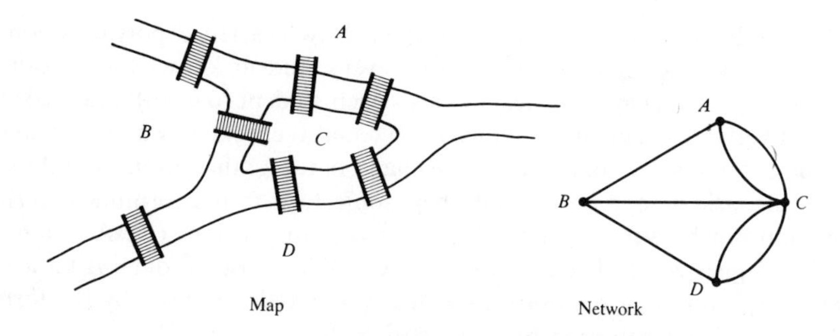

Figure 23.3 The Königsberg Bridge Problem

The invention of topology is a typical example of the development of a mathematical topic beginning with a physical situation. The problem that initiated the study of topology involved the seven bridges of Königsberg (see Figure 23.3). The problem posed was "How can one walk through Königsberg from a given starting point and cross all of its seven bridges once and only once?" No one could find a route that would satisfy the conditions of the problem. To save time someone tried the problem by drawing a sketch of the town and tracing possible walks; but still the problem remained unsolved.

The mathematician Euler was requested to solve the problem. He changed the physical problem to a sketch called a network. He called each bridge an arc and each region of land a vertex. From an analysis of this network, he found generalizations that apply to all networks.

This analysis was extended to a consideration of regions, arcs, and vertices that had no physical counterpart. Today, topology has been extended far beyond this simple beginning into a highly structured discipline.

Similarly, it is presumed that when man invented the concept of number, he first represented this idea with physical objects, such as pebbles. Later, he used marks in the sand or notches on a stick. With the invention of writing, he used tally marks; still later, digits with place value. Soon he learned to use a number line, which visualized the order and magnitude of number. Today, these primitive representations of number have been refined and extended to number systems with properties such as closure and identity elements. And new numbers (such as quaternions) have been invented, numbers that seem to have no application to a concrete situation. This sequence from the physical to the operational to the abstract illustrates the typical sequence of mathematical development and the ideal sequence for learning mathematics.

A Lesson Based on Models

A description of a lesson on Euler's formula for polyhedra illustrates the role of models and visualization in the teaching and learning of a mathematical concept. The teacher begins by having each student draw a variety of polygons, concave as well as convex. After making sure that every student knows the meaning of *vertices*, *edges*, and *regions*, the teacher asks each student to count and record the number of vertices, edges, and regions. Each count is checked by a second student and some of the results tabulated on the chalkboard. As the results are tabulated, students are asked to record what they conjecture is the formula describing the relationship between vertices (V), edges (E), and regions (R). As soon as a formula is suggested, students are asked to check it with their data. After a short period of exploration, a relationship is discovered and expressed by the formula $V + R = E + 2$ (for simple closed curves $R = 2$).

The teacher then asks whether the formula applies to compound figures or networks. The formula is found to satisfy all conditions the student can think of. The teacher then raises a further question: How can this formula be extended to a more complex situation? A student may suggest the formula for polyhedra.

At this stage the teacher distributes models of polyhedra. These can be commercial plastic models, homemade cardboard models, or models constructed by the class out of soda straws and elastic thread or even careful drawings like Figure 23.4. Again the students count and record the number of edges, vertices,

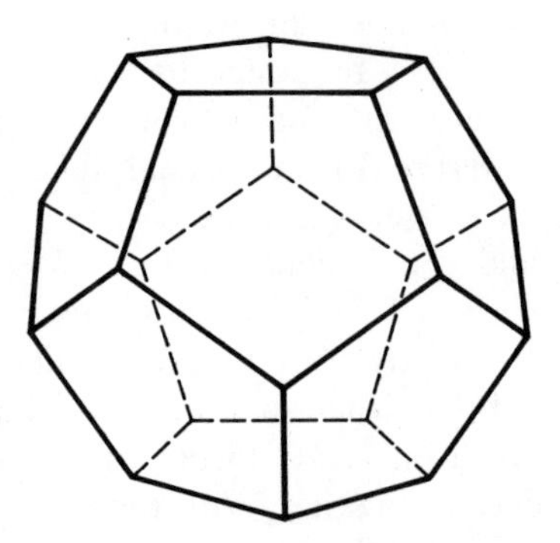

Figure 23.4 ***Dodecahedron***

and faces of each polyhedron. The formula is verified by each of these models. Even so, the formula is only a conjecture arrived at by scientific (as opposed to mathematical) induction. The students explore the formula further by applying it to a variety of objects in the classroom. The next step is for the teacher to guide the students to an abstract proof of their conjecture. The proof is based on systematic dissection of the polyhedron.[2]

The students are now sure that their discovery is correct. A new model is now presented, still a polyhedron model but different from the others in that it has a

[2] See Richard Courant and Herbert Robbins, *What Is Mathematics?*, Oxford University Press, New York, 1941, pp. 236–240, 258–259.

hole through it. A count of faces, vertices, and edges leads to a contradiction of the "proved" relationship. This creates a lively discussion. Is it possible that the same statement can be proved and disproved? The students eventually come to recognize that their formula is true under certain conditions but is only a special case of a more general formula. Their formula is true only for polyhedra without holes—that is, topologically equivalent to (or distortable into) a sphere.

In this discovery lesson, students have shifted back and forth between abstract ideas and concrete representations. They have carried out activities that are similar to the scientific method in their structure. At every stage, the students' abstractions have been tested against the data obtained from the concrete representations available to them. When their tests failed, they were forced to reexamine previously established conclusions and then to derive new ideas to fit varied conditions.

In fact, the students saw in operation a scientific research activity. Faced with a contradiction, they had three avenues open to them: (1) reject their entire effort; (2) limit their theorem in order to exclude the contradiction; (3) extend their theorem in some way in order to include the contradictory case. The first is not very satisfying, especially in the light of what had seemed to be a correct proof. The second is possible. Students should see that the proof they have developed fails when there is a hole in the original polyhedron. The dissection cannot be carried out. They then may correct the theorem by stating it for polyhedra without holes.

Finally, a much more satisfying exploration will show that the formula $R + V = E + 2 - 2H$, where H is the number of holes, takes the additional cases into account. When $H = 0$, the formula reduces to the original form.

The foregoing lesson is an illustration of the role of models in teaching. The model, whether a drawing or a three-dimensional model, helps to make the problem realistic. The models then become the source for data. By exploring the pattern of the data, students receive various clues, which lead eventually to the forming of a generalization. The generalization is checked against a variety of different situations. In addition, the relationship can be applied to everyday situations to increase familiarity. Finally, the generalization is established by a mathematical proof which is independent of the model. The objective is the establishment of the mathematical concept at the abstract level. The models used are tools for achieving that objective. It would be difficult to envision a lesson on Euler's formula without the use of models, drawings, and experimentation.

The Model in the Classroom

Each time a model is used in the classroom, it should play a positive role in providing deeper student understanding of mathematics. The model may illustrate a specific concept (a wooden conic section); it may provide the basis for a development (unit cubes for building solids); it may be used as a vehicle for a student discovery (polyhedra for Euler's formula); in each case the teacher should know the purpose and appropriate use of the model.

The primary purpose of a model is to provide a concrete visualization for thinking about and discussing an idea. For example, a model that allows a student to transform a rectangle physically into a triangle provides a basis for developing the formula for the area of a triangle. The model provides experiences which can then be used for thinking about an idea by providing a frame of reference for sense perceptions and experiences.

Such transformations, from the concrete to the abstract, are often difficult to achieve. They must be supported at all stages by thoughtful teaching. The teacher must recognize and anticipate the level of insights that occur to individual students as a result of experiences with physical models. As rapidly as possible, these students should be weaned away from the concrete model and be made to understand the conceptual model. At the same time, the teacher should recognize a student's need for additional work with a concrete model and should seek to provide him with that assistance. A slower student may be encouraged to work with the model at his desk, for the sense of touch may provide him with the needed bridge to the conceptual level. And, finally, the model should be kept available to reinforce the concept if and when students regress in understanding.

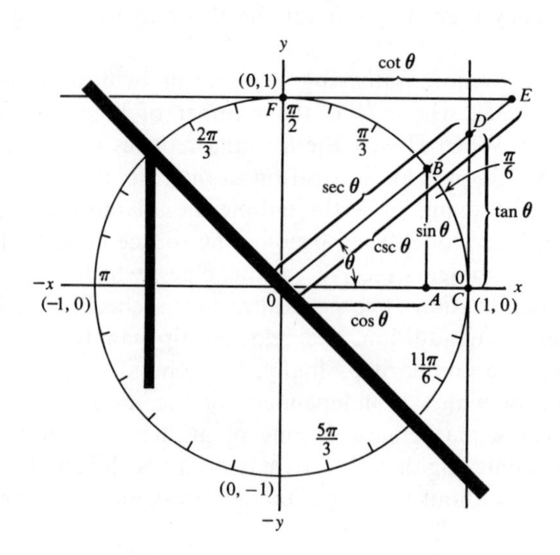

Figure 23.5 Trigtracker

A linkage, sometimes called a Trigtracker (see Figure 23.5) may be used to explore the trigonometric ratios for a unit circle. This device can be made out of cardboard, or a commercial demonstrator may be purchased. By forming different right triangles with the linkage, one can get approximations for the sine, cosine, and tangent of various angles. These approximations show such things as how the values of these ratios increase and decrease with the size of the associated angle and how the maximum and minimum values for the sine and cosine are $^{+}1$ and $^{-}1$. This device is useful for building a table of values for trigonometric ratios. Keeping them available helps to remind students of the relationships examined in the original demonstration.

Many teachers fail to take into account the later uses of models in supporting the initial demonstration. Since the teacher understands the abstractions, he often moves too quickly to the abstract level, carrying with him only the brightest students. In this regard consider this example: a seventh-grade teacher of remedial arithmetic has used felt pie cutouts for several days to illustrate fraction concepts. Students have had experience constructing these models and using them to demonstrate addition problems involving fractions. Presently, the students, most of whom have progressed to the abstract level, are computing sums involving fractions. The teacher tours the room, occasionally asking a student to return to the flannel board to check his computation. The teacher is spotting weaknesses and allowing regression to the use of the concrete devices by students who still need this assistance, thus giving students additional security in dealing with a complicated algorithm.

A description of a lesson on circles may illustrate other aspects of the use of models. This lesson is concerned with the relationship between the measures of the angles formed by intersecting lines and the measures of the intercepted arcs of circles. The class was first requested to make drawings to illustrate all the possible ways in which two intersecting lines could be related to a given circle. From these results, a selection was made of all situations in which each of the two intersecting lines intersects the circle in at least two points (see Figure 23.6).

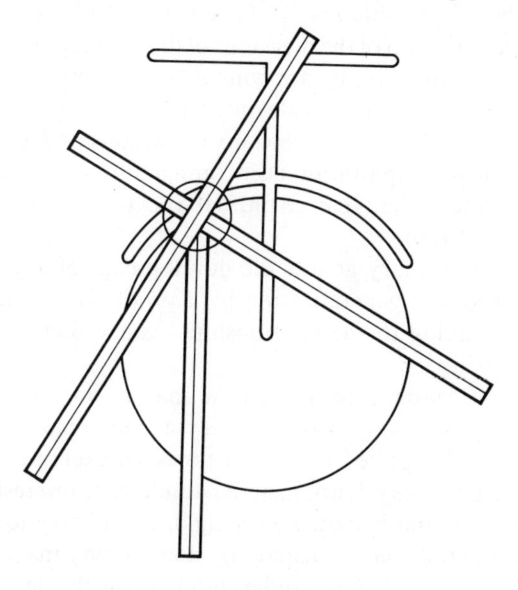

Figure 23.6

The teacher then distributed a device for experimenting and collecting data of the measures of angles and arcs. This device has a circle on its base with marks on the circumference indicating degrees of arc. Two plastic strips are marked off into unit lengths for reading measures of length. The strips intersect at a point which may be placed at the center of the circle, within the circle, on the circumference of the circle, or outside the circle. A protractor measures the angle between the lines represented by the strips. In the different positions the strips may represent radii, chords, secants, or tangents.

The students were then asked to make three different settings for each different type of intersection. They were asked to record the measures of the angles formed by the intersecting lines and the measures of the intercepted arcs for each setting. Each setting was checked by a second student.

For this laboratory lesson some teachers would provide the students with a laboratory guide sheet. This guide sheet might tell the student how to manipulate the device and what measurements to make; in addition, it might suggest sketches of each type of intersection, provide a table for the collected data, and give some hints about the generalization to be discovered. Other teachers would prefer to have the students experiment in a completely unstructured manner, or have students work in groups and permit each group to complete the generalization. Another method would be to make the experiment a class activity in which the data collected would be recorded on the chalkboard.

No matter what method is used, the goal of the lesson is to lead students to discover the formulas that describe the relationship between the measures of angles formed by intersecting lines and measures of intercepted arcs of a given circle. After each of these is formulated, the teacher asks whether any one generalization will describe all possible results. (The measure of the angle in degrees is equal to half the algebraic sum of the measures of the intercepted arcs.) For practice, the class tests the generalization by applying it to a variety of exercises.

The discoveries at this stage provide only a *conjecture*. Once this conjecture is arrived at, students are asked, "Are there any conditions under which the formula does not apply?" Further exploration is encouraged to try to find contradictions. When none is found, the students can go on, in the next lesson, to finding a deductive proof for the generalization.

In this lesson, the primary goal is the development of a generalization. The end activity establishes the generalization by a proof. The models are used as a means for adding meaning to the relationship—as a vehicle for the students to discover the relationship.

Of course, such a lesson takes more time than it does to look at a drawing and read the theorems that apply. Most often these theorems are proved as outlined in the textbook and then applied in a set of provided exercises. But such a short lesson is not likely to produce retention, understanding, or interest. Such essentially sterile teaching presents mathematics as a rigid, completely formulated subject. In this type of lesson students participate by memorizing material developed by others. On the other hand, when he studies models and the development of ideas, the learner is more likely to feel—as he should feel—that he is a creative mathematician.

Note that in each of the previous examples, the model plays an important but *secondary* role to the mathematical concepts at the abstract level. In both cases, efficient advance to the abstract level is the goal: the model is a tool to achieve that goal. The additional expenditure of time in using models is amply compensated by the student's ability to retain the concept and to recognize its utility.

Models are not necessary for all students—some talented students are capable of and even enjoy discovering generalizations through reflective thinking with abstract representations. For other students, even the models are meaningless and fail to help them build an understanding of a concept. It is the responsibility of the teacher to decide *when* models are needed, *what* models are effective, and *how* the model should be used for a given class. Suggestions for models to be purchased are given in Appendix A.

Types of Activities Using Models

The following examples do not exhaust the uses of models in the classroom. Rather, they point up the variety of activities that the use of models can provide:

Models may be a *demonstration aid* to add meaning to a mathematical concept: an area demonstration device adds meaning to area formulas; a large demonstration rule helps one to read the slide-rule scale; insight into concepts of the fourth dimension may be achieved by reference to a tesseract; plastic models illustrate conic sections; Tinker-Toy models demonstrate the terms of polynomial products.

Models are useful *laboratory equipment* to provide a means for discovering new ideas and relationships. For example, a quadrilateral device like the Welch quadrilateral is useful in discovering the relationships between measures of segments and angles. A graphing board may be a means of investigating the periodicity of trigonometric functions; a probability kit can furnish the equipment for sampling experiments.

Models may be used as *practice devices* with which the student builds accuracy, understanding, and efficiency. For example, folding paper provides exercises useful in illustrating relationships between lines; nomographs provide a simple device for checking computations; a percent computer may be a device for showing the relationship between certain ratios as well as for solving percentage problems.

Models may be used as *measuring instruments* for studying mathematical applications within or outside the classroom. A transit can be used to measure indirectly the height of a building; an angle mirror (Figure 23.7) will locate points for measuring an inaccessible line segment; a stadimeter measures distances by means of trigonometric ratios and a reference object; area linkages may be used to measure area relationships; the volumes of polyhedron models can be measured

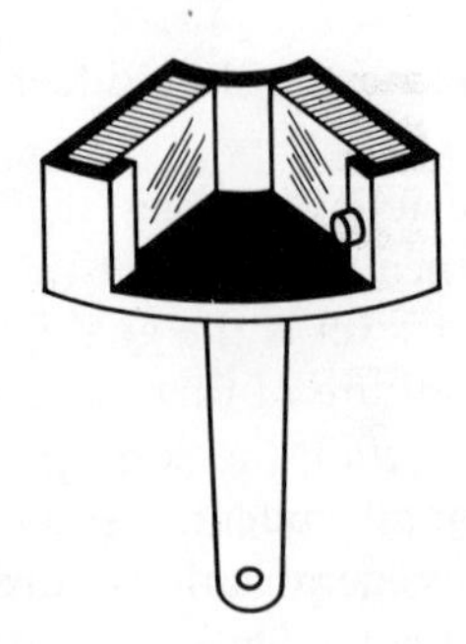

Figure 23.7 Angle Mirror

by liquid displacement; the Pythagorean theorem can be discovered by various models that compare the areas of squares.

Models may be *projects* for independent student work. A student may plan, construct, and demonstrate a model such as a logic machine to generate truth values for various logical connectives, or the dual regular polyhedra—those regular polyhedra (such as the icosahedrons and dodecahedrons) that have the correspondence Faces ↔ Vertices.

Models may be used as *enrichment devices* when they present ideas that are not in the textbook. Möbius strips or hexaflexagons suggest some of the curiosities of topology; the Galton probability board relates the normal curve to binomial expansions; polyhedron models made out of balsa strips may illustrate a finite geometry.

Models as Learning Aids

Teachers who use models are enthusiastic about the contributions they make. They suggest, in summary, that models are effective because they perform the following functions:

They give concrete representation to abstract ideas.

They relate new ideas to previous experience or previously learned ideas.

They enhance active participation of the learner in the learning activity, and thus provide additional motivation for the learning of mathematics.

They concentrate attention on the concepts involved and stimulate interest in these concepts.

They teach how to solve problems and how to explore new ideas.

They speed up communication.

They consolidate details that are related to the generalizations being sought, and thus enhance retention.

They lend variety to classroom activities and provide a useful change of pace.

They provide a program of enrichment and acceleration for individual students.

They provide successful, meaningful activities for the slow learner.

They encourage the participation of many practical-minded students, who need this continuing contact with the concrete world.

If, through the use of models, only some of these goals are attained, it is hard to understand why mathematics teachers do not use them more extensively.

Misuses of Models

Of course, models are sometimes misused or used at inappropriate times. At the same time, it should be noted that the most frequent misuse of models is failure to use them at all. Some of the principal abuses and misuses of models are:

Excessive and indiscriminate use merely for the purpose of using a model. A model is used only where it does something which cannot be done as well without it.

Failure to transfer from the concrete representation to a generalization or abstract representation. The goal in mathematics is to use symbols and abstractions. Mathematical ideas themselves are independent of physical representation or application.

Failure to use the right model at the right time or for a sufficient length of time to establish the concept involved.

Use of an inadequate model—a model that is too small, too crude, or—at the opposite extreme—too complex.

Failure to adapt the model to the needs of the students or to the objectives involved.

Models as Projects for Students

One way to encourage discovery activities in mathematics is through the construction and demonstration of a model or the preparation of a science-fair exhibit. It is usually an individual, out-of-class activity. However, it is appropriate for group work and as a part of the regular classroom activity. In any case, it should contribute to the attainment of the objectives of the mathematics class.

The construction of a model provides an opportunity for independent exploration of ideas usually beyond the scope of the typical textbook. It places major emphasis on creativeness, originality, and craftsmanship. It supplements the usual classroom activities with activities that develop the desire and capacity to think independently.

To be successful, project activity needs careful planning and adequate resources. To plan models, the teacher needs ideas for possible projects, and he needs resources such as reference books and construction materials. One convenient way to have model ideas available is to make a list of possible projects and include them in the Instructional Techniques File. These cards should contain such information as a description of the model, drawings, references, difficulties, and the mathematical principles involved.

The student's first step is the selection of an appropriate model. This choice should be a joint decision of the student and the teacher, with student interest, experience, and aptitude as major considerations. After a choice of model has been made, the teacher furnishes the student with references, materials, and a few specific suggestions. However, the teacher should not give too many hints or suggestions; if he does, the student's opportunity to discover an idea or invent a model may be lost. The student should be allowed—and encouraged—to try out his own ideas. It is advisable to set up a schedule for the completion of the model to avoid wasting time. Additional conferences may be needed to check upon the student and to give added suggestions and encouragement. Do not be too rigid in your requirements.

Some teachers require a model or project from all pupils. Others require a project such as a research report, classroom demonstration, or science-fair exhibit in order to qualify for an A or B grade. In all cases, the completed project should be evaluated carefully and recorded for grading purposes. Projects should be evaluated for originality, completeness, craftsmanship, accuracy, and organization. Of prime importance is the mathematical learning that the student demonstrates as he presents his model or project to the class. Recognition of excellence in project work might be given in a letter to the parents, by a description in the school newspaper, or by an exhibition in the school. Exhibits should always be labeled so that the person who completed the project is given the credit due him. These exhibits should also be properly titled so that the mathematics involved is communicated to the viewer. The informal examination and manipulation of these projects invariably builds curiosity and interest on the part of other students and teachers. Displays in store windows or at PTA meetings gives excellent publicity to the work of the students.

A variety of models for laboratory activities may be made in an ordinary classroom. For more extensive laboratory work, materials and work space (such as those discussed in Chapter 27) are needed.

The following are suggestions for student projects:

Display the best models of previous years to suggest possibilities and stimulate good work.

Use student-made models in the classroom presentation of mathematics.

Provide a guide sheet with suggested models, procedure, time schedule, and references.

Suggest the type, source, and cost of appropriate materials. Wherever possible the school should supply the materials.

Place major emphasis on the mathematical principles involved rather than the entertainment value.

Encourage students to make models attractive by good craftsmanship.

Have the builder demonstrate his model to the class. The demonstration or discussion of the model will give desirable leadership experience to the student demonstrator and will give the class a proper introduction to the topic.

Encourage the use of simple material and simple devices so that construction time is kept to a minimum.

Examples of Discovery Lessons Using Models

Here are some of the concepts, relationships, or structures which can be discovered by using models. The suggestions are appropriate for students in grades 7 through 10:

Seventh- and Eighth-grade Models

The area of triangles and quadrilaterals as related to the area of a rectangle by the use of area boards or geoboards

The meaning of *regrouping*, *carrying*, and *borrowing* by a base 5 or binary abacus

The sums and differences of numbers written in bases other than 10 derived by means of nomographs, Napier's bones, or slide rules

The properties and operations in a finite modular arithmetic, determined by use of a circular number line or dial

The sums of the measures of angles of polygons, found by means of polygon models

The relationship between the squares on sides of right triangles, seen by use of cardboard or graphical models

The formula for the circumference and area of a circle, discovered through use of a circle demonstration board

The probabilities of events, determined by the spinning of dials or the tossing of thumb tacks, coins, or dice

The sample space for the events resulting from tossing two dicc

The symmetries of geometric figures with two- and three-dimensional models

The computation of any one of the three variables in a percentage problem by a percentage computation model

The illustration of the actual conditions in verbal problems by terrain models made of papier-mâché, clay, sponges, and toys

The program for a computer solution to a problem with a computer terminal

The practice of computational skills with games

An indirect measurement of an inaccessible distance with a hypsometer

The properties of optical illusions in two- and three-dimensional objects

The properties of polyhedra in crystals with cardboard models

The comparison of perimeter and area of various polygons by forming different shapes on a pegboard or lattice

The comparison of volumes of prisms, cylinders, cones, and spheres by measuring capacity of models with sand or salt

The properties of similar figures drawn with a pantograph or by graphing patterns on a grid

The discovery of the Pythagorean theorem by use of a model

The transformation of geometric figures from one shape to another shape with equivalent area

The development of formulas by the analysis of puzzles such as Tower of Hanoi or the Oxbow puzzle

The computational algorithms used by a calculator

The drawing of original designs and patterns with ruler and compass

The operation of an electric quiz board for mastering concepts

The representation of units of measure (such as cubic foot) by cardboard models

The application of units of measure to measuring packing boxes such as those for cereal

The comparison of rational numbers by sectors of circles or parts of rectangles

The comparison of percents by a 100 board, pegboard, graph paper, or objects

Algebra

The difference between first-, second-, and third-degree algebraic phrases by models made out of balsa strips, cardboard, or Tinker-Toy dowels

The properties of graphs of linear equations by models using elastic thread between golf tees on acoustical tiles or pegboard

The binomial distribution and Pascal's triangle with a probability board

Practice in finding sums of positive and negative numbers by tossing dice of two colors

The sums and differences of positive and negative numbers with a slide rule

The meaning of formulas such as $V = lwh$ or $S = rt$ by mock-ups

The setting for verbal problems by mock-ups using toys or cardboard models

The application of proportions and variation by using gears, levers, springs, and pendulums

The solutions of quadratic equations by a quadratic slide rule or nomograph

The setting for quadratic equations by mock-ups (e.g., a basketball player "shooting" a basket)

The drawing of conic sections by linkages or paper folding

The terms of a binomial expression such as $(a + b)^3$ by a sectioned cube and $(a + b)^4$ by a tesseract

The terms of a geometric series by a model that measures the bounces of a "super" rubber ball

The sums of arithmetic progressions by cardboard cartoons

The use of the slide rule by student-made rules using logarithmic graph paper

The meaning of the complex roots of a quadratic equation with a three-dimensional graph

The graphs of equations with three variables by using heavy screening for the grids and elastic thread for lines and plastic sheets for planes

The properties of curves such as parabolas, ellipses, and hyperbolas demonstrated by models, drawings, or paper folding

The properties of surfaces such as the hyperbolic-paraboloid by curve stitching

The applications of conic sections by bridge arches, buildings, and space orbits

The permutations and combinations of objects such as cards or blocks

The periodicity of functions by a model of the wrapping function

The values of trigonometric functions demonstrated by a Trigtracker.

The relationship of different number systems by models of Venn diagrams

The meaning of inverse variation by the use of a gear chain

Geometry

The properties of triangles by linkages that form various types of triangles

The properties of quadrilaterals by linkages that form various types of quadrilaterals

The properties of polygons by using elastic thread, golf tees, and protractors mounted on a pegboard

The properties of a finite geometry by constructing models out of sticks or soda straws and elastic thread

The meaning of the fourth dimension by a series of zero-, one-, two-, three-, and four-dimensional models

The use of congruence in indirect measurement in surveying

The use of similar triangles for indirect measurement of inaccessible distances

The Pythagorean theorem and different ways of proving it

The approximate value of irrational numbers by drawing successive right triangles with sides $(1, 1, \sqrt{2})$, $(\sqrt{2}, 1, \sqrt{3})$, $(\sqrt{3}, 1, \sqrt{4})$, $(\sqrt{4}, 1, \sqrt{5})$, and so on

The properties of parallelograms by a linkage model

The meaning of loci and the intersection of loci by combining plastic models

The relationship between the measure of angles formed by radii, chords, secants, and tangents, and the measure of the intercepted arcs of the circle by a dynamic geometry device

The Euler formula and other properties of regular polyhedra by cardboard or plastic models of polyhedra

The truth values as verified by simple electric circuits

The projections of three-dimensional objects on a plane by a projection model or mock-up

The transformations and symmetries of reflections by the use of mirrors and kaleidoscope

The properties of non-Euclidean geometry by a slated sphere and a model of a pseudosphere

The properties of curves (such as the cycloid, spiral, catenary, and cardiod) formed by moving points of light or by curve stitching or paper folding or drawing

The possibility of trisecting an angle with a trisection device

The properties of minimum surface and maximum volume of soap bubbles, honeycombs, and other objects in nature

The properties of three-dimensional symmetries by wire and string models

The combination of vectors by demonstrations with a vector board

The properties of spherical triangles by drawings on a slated sphere

The determination of the diameter of a sphere with a spherometer

The relationship of lines and planes shown by models out of cardboard and balsa strips or a commercial solid-geometry device

The applications of geometry in the home, highway construction, nature, art, advertising, sports, architecture, carpentry, navigation, maps, photography, and machinery

Commercial or Homemade Models?

The variety of commercial models currently available for the mathematics teacher has multiplied tremendously. Some of this increase is due to the emphasis on the discovery method, some of it is due to the support of National Defense Education Act funds, and some of it is due to the normal invention and development of new devices. Therefore, the individual mathematics teacher and the mathematics department head or supervisor will need to render professional judgment in the selection and use of these teaching aids. It is the responsibility of the teacher to know what materials are available, how to select appropriate models, and once selected how the models can contribute best to the learning of mathematics. In order to be informed about available materials, the teacher should send for the catalogs of companies that specialize in mathematical devices (see listing in Appendix A).

In addition, current issues of professional journals should be searched for reviews, announcements, and advertisements of new models. Of special interest is the section of the *Mathematics Teacher* devoted to reviews of new models, games, devices, audio-visual aids, and computers.

When a model is needed, it is necessary to decide whether the school should purchase a commercial product or whether a teacher or student should construct it. In constructing a model, the student or teacher may discover some aspects of the ideas involved that he had not thought of before; and constructing a model may also give some experience in creativeness and craftsmanship. However, it is usually better to use a commercial device whenever a good one is available, for the commercial model is likely to be better constructed than a homemade model. Commercial products are usually designed by experts, produced by specialized machines and finished in an attractive manner. In addition, manufacturers usually modify commercial models on the basis of classroom trial.

The usual argument against commercial models is that they are expensive. This is often true, but the mathematics teacher's time is extremely valuable, and when the time it takes to construct a model is considered, the homemade model usually proves to be more expensive. When a desired model is not available commercially, it should be made locally. Frequently, the industrial arts department of the school will cooperate in its production.

Types of Homemade Models

A variety of models for laboratory work or demonstrations can be made by students or teachers. These models are usually made of simple materials, such as cardboard, balsa wood, soda straws, plywood, plastic, or pegboard. The selecting, planning, making, and using of these models can be a learning experience that justifies the time it takes. A number of the books listed in Appendix D are good sources for ideas about the actual construction of such models. Below are the instructions for sample models which can be constructed in the classroom.

A Percent Computer

The device shown in Figure 23.8 can be used to compute the answer for all three types of percentage problems. The simple materials consist of a piece of string, a sheet of graph paper, and a piece of stiff cardboard. The best graph paper to use is one with ten squares to the inch.

Mount the graph paper on the cardboard. Draw a horizontal line, $\overline{AB}$, at the bottom of the graph paper 100 units long. This line represents a percent scale and should be labeled in convenient units from 0 at the left to 100 at the right.

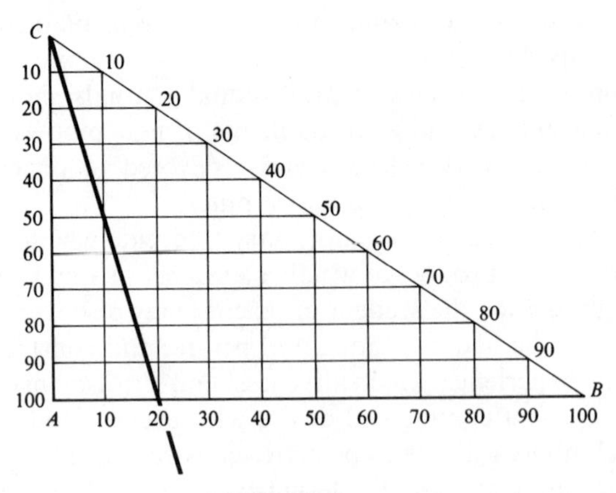

Figure 23.8

At A (the 0 point) draw the perpendicular to $\overline{AB}$. Locate C on this perpendicular so that $\overline{AC}$ is 100 units long. Punch a hole at C and attach a string slightly longer than the diagonal of the sheet. Draw $\overline{BC}$. Draw lines parallel to $\overline{AB}$ at 10-unit intervals such that the segments between $\overline{AC}$ and $\overline{BC}$ will be 10 units long or multiples of 10. Label these lines at each end according to their length. The computer is now ready for use.

To find what percent 8 is of 40, find the horizontal line between $\overline{AC}$ and $\overline{BC}$ that is 40 units long. Locate the point that is 8 units to the right of $\overline{AC}$ on this line and place a pin at this point. Hold the string taut to form the straight line that passes through C and the pin point. The percent is then found where the string crosses $\overline{AB}$, the percent scale.

To find 20 percent of 40 hold the string taut across the 20 point on the percent scale. The point where it crosses the horizontal line that is 40 units long between $\overline{AC}$ and $\overline{BC}$ will be the answer, 8.

To find the number if 8 is 20 percent of it, hold the taut string across the 20 point on the percent scale. Locate the line which the string crosses exactly 8 units to the right of $\overline{AC}$. The length of this line $\overline{AC}$ and $\overline{BC}$ is the answer, 40.

For numbers above 100, it is necessary to increase the horizontal number scales while keeping the percent scale the same. For example, if the scale is doubled, the 10 line becomes a 20 line and each square will represent 2 units. The device will then represent numbers from 0 to 200. Other multiples may likewise be used to extend the scale. Another method of extending the scale is to cut strips of graph paper representing different scales and insert them on the scale at the appropriate place depending on the problem.

The Spherometer

A spherometer is a device for determining the diameter of a sphere. It may consist of a cylinder open at one end such as a tin can with a dowel as a measuring rod at its center. When the open end of the cylinder is placed on a sphere, the measuring rod indicates the height of the sphere inside the cylinder. In Figure 23.9, OR, the height of the sphere in the cylinder, is measured on the measuring rod by the distance DE. The radius of the tin can, OA, is known. The diameter of the sphere, RS, can be computed with the proportion $OS/OA = OA/OR$. What is the basis for this proportion?

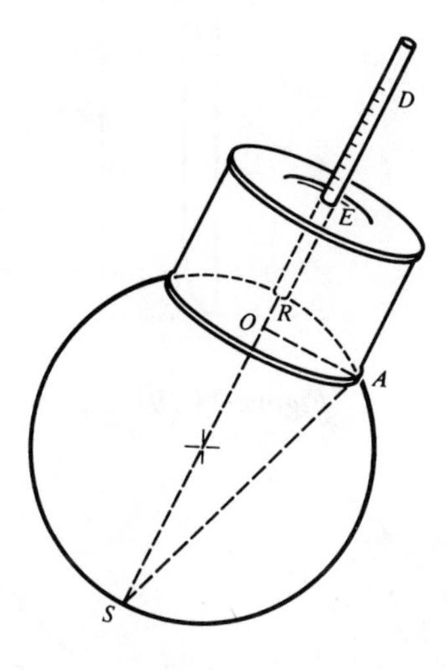

Figure 23.9

The measuring rod is usually calibrated by many measurements like the one above. Then points such as E are labeled according to the diameter of the sphere being measured, so that the diameter can be read without computation.

The Tomahawk Angle Trisector

A simple angle trisector can be cut out of cardboard, plywood, or plastic. The device has the shape shown in Figure 23.10. It is somewhat similar to a carpenter's square. A circle is drawn with the center at T and radius r. Then $\overline{DR}$ is tangent to the circle at D and thus is perpendicular to the diameter $\overline{FTD}$. A convenient length for $\overline{DR}$ is 7 times the length of the radius. The diameter is extended the length of the radius to E. The device is cut along arbitrary curves GH and RE to lend support to it.

The use of this device is shown in Figure 23.11. Place point E on one side of the angle to be trisected. Slide the device until the circle is tangent to the other side of the angle and $\overline{RD}$ is coincident with the vertex of the angle. Then the points C and D locate points of the trisection lines. The proof of the trisection is based on congruent right triangles, $\triangle OAC \cong \triangle ODC \cong \triangle ODE$.

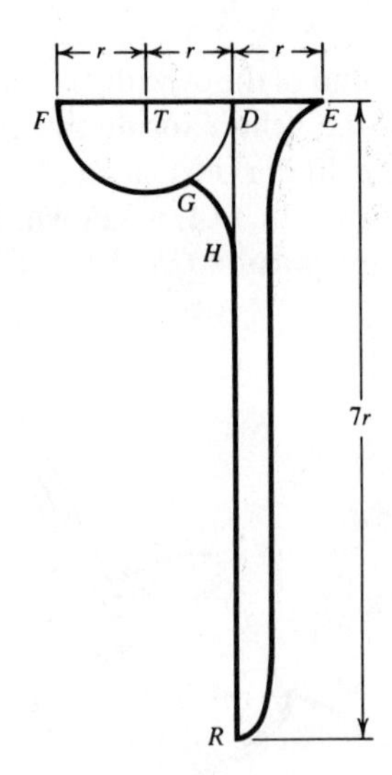

Figure 23.10

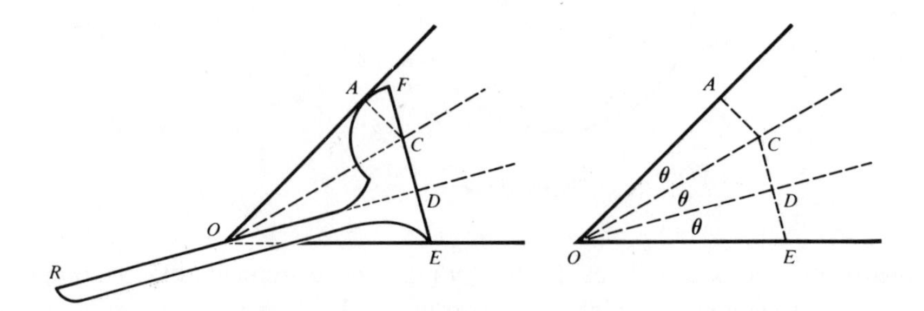

Figure 23.11

Selection of a Model

In selecting a model for purchase or construction, the teacher should keep in mind the objective he wishes to attain and the concepts he wishes to represent.

In evaluating a specific model, one should ask the following questions:

Does the model promote the discovery of an idea?

Does the model present the idea correctly?

Does the model permit the idea to be transferred from the concrete to the abstract?

Does the model add interest to an idea?

Can the model be easily stored, maintained, and repaired?

Is the cost of the model in time or money commensurate with its contribution?

Is the model designed for handling by the students?

Is the model large enough so that its important parts are clearly visible?

Is the idea which the student derives from the model significant?

Does the model do something that cannot be done as well or better by something else?

Mathematics teachers should develop an almost instinctive feel for the use of models—in planning instruction or in answering student questions. When the conditions of a problem are not clear, the teacher should illustrate it concretely with the objects involved. When the relationship between factors, products, and terms of an algebraic expression are not clear, use a model. When a thorem about lines and planes seems difficult, have the students make a model of it. If an application of a quadratic equation does not seem real, toss a chalkboard eraser to a student and ask for the mathematical description of the eraser's path. When inequalities seem dull, set up a linear program for the production of the model toys displayed. In fact, use yourself as a model when you attack a new problem before the class. In these and in hundreds of other ways, models can be used to communicate mathematical ideas in an exciting, understandable way.

Learning Exercises

1. Review your own experience as a student to determine whether any of your teachers made extensive use of physical models in mathematics instruction. Describe some of the models used and their success or failure. How could you improve upon their use?

2. Locate a journal article describing a model to be used to clarify a mathematical concept. Construct the model. Describe how you would use the model for instruction.

3. Review the *Mathematics Teacher* department "New Products" for the past four issues to locate commercial models. Name the three that sound most useful to you. Justify your decision.

4. When weak students are supplied crutches to help them with computation, parental response often takes the form, "Do you want them to carry an abacus around with them?" Give a careful answer to this criticism with a justification of your position.

5. Make a series of two-dimensional models to illustrate the progression in area formulas:

   (a) From rectangle to parallelogram
   (b) From parallelogram to triangle
   (c) From rectangle to triangle
   (d) From triangle to trapezoid (2 models)
   (e) From parallelogram to trapezoid (2 models)
   (f) From rectangle to trapezoid
   (g) From triangle to rhombus
   (h) From triangle to kite

6. Construct a model or models and develop an associated discovery lesson for one of the lessons described in this chapter.

Suggestions for Further Reading

Bell, Max S., "Mathematical Models and Applications as an Integral Part of a High School Algebra Class," *Mathematics Teacher*, 64, 4 (April 1971), 293–300.

Cundy, H. Martyn, and A. P. Rollet, *Mathematical Models*, Oxford University Press, Inc., New York, 1961.

Instructional Aids in Mathematics, Yearbook 34, National Council of Teachers of Mathematics, Washington, D. C., 1972.

Johnson, Donovan A., *Paper Folding for the Mathematics Class*, National Council of Teachers of Mathematics, Washington, D. C., 1957.

Sowell, Katye Oliver, and Jon Phillip McGuffey, "Nondecimal Slide Rules—and Their Use in Modular Arithmetic," *Mathematics Teacher*, 64, 5 (May 1971), 467–472.

Syer, H. W., "Sensory Learning Applied to Mathematics," in *The Learning of Mathematics, Its Theory and Practice*, Yearbook 21, National Council of Teachers of Mathematics, Washington, D. C., 1953, pp. 99–155.

Wahl, M. Stoessel, " 'We Made It and It Works!' The Classroom Construction of Sundials," *Arithmetic Teacher*, 17, 4 (April 1970), 301–304.

Wenninger, Magnus J., "Stellated Rhombic Dodecahedron Puzzle," *Mathematics Teacher*, 56, 3 (March 1963), 148–150.

———, *Polyhedron Models for the Classroom*, National Council of Teachers of Mathematics, Washington, D. C., 1966.

Audio-visual Aids 24

Audio-visual aids are as essential for the mathematics teacher as spices are for the chef. They add the variety, the depth, and the breadth which make the learning process pleasant and meaningful.

The use of visual aids in learning mathematics is not new. Since primitive man first drew pictures on cave walls to communicate the number of animals he had seen, some form of visual referent has been used to represent mathematical ideas. Today, new concepts of learning and new materials have put renewed emphasis on the role of audio-visual aids in mathematics teaching.

The successful, dynamic teacher is always searching for ways to make his instruction more meaningful. He knows that through the proper use of audio-visual materials he may help build in students the number sense, the space perceptions, and the imagination needed to master mathematical ideas.

Although audio-visual materials are a necessary part in the curriculum of mathematics, they are not magic: teachers still have to teach, and students still have to study. But the teacher using audio-visual aids will be teaching in an atmosphere where meanings become clearer and where what the student learns will be more helpful in solving his future problems.

Mathematics by its very nature tends to be abstract, even though some areas, such as geometry, are based largely on visual representation. Despite the fact that the social functions of mathematics have increased in our society, the applications of mathematics are often far removed from the classroom topics. This creates problems of student motivation and learning that may be partially solved by the use of the right audio-visual material, at the right time, in the right way. If audio-visual materials and equipment are used properly, the result should be more correct and richer learning, an economy of time, and improved student retention.

Visual and audio materials, however, can be a waste of time and money if they are not used effectively. The question we must ask when selecting material is, "What *unique* contribution will this particular aid make toward better learning?" Aids should not be used just because they are fashionable, or because they are available, or because they fill in a time gap in a class period. Every teacher needs to recognize the unique contributions that audio-visual equipment and devices can make and on this basis select and use those most suitable in accomplishing aims.

The Function of Audio-visual Aids

Mathematics teachers use films, television, filmstrips, tapes, charts, pictures, projections, bulletin board displays, and sketches to attain the following goals:

Visualize abstract ideas so that they have meaning. The meaning of a number such as π, a word such as *probability*, a process such as division, a theorem such as the right-triangle relationship may be enriched by visual representations. Typical films that serve these purposes are *The Meaning of Pi, Ratio, a Way of Comparing, A Function Is a Mapping*, and *Possibly So, Pythagoras*, and the filmstrip *Thinking in Symbols.*

Illustrate applications of mathematics in our world. Although field trips are best for seeing mathematics in action, they require time and community resources that are not always available. Typical materials which add interest to mathematical ideas by emphasizing applications include the following: the film *Mathematics of the Honeycomb*, the filmstrip *Indirect Measurement*, the chart, *Life Insurance*, and the graph *Population Growth.*

Bring to the school and the classroom important firsthand accounts of new activities in mathematics and mathematics education. By tape recording (or obtaining the tapes of) speeches at state and national conferences, the teacher can bring the thoughts of others, expressed in their own words, to the classroom. Films such as those of the Mathematical Association of America Series bring students into contact with great living mathematicians, such as Richard Courant and George Polya.

Build favorable attitudes toward and interest in mathematics. The uniqueness of mathematics—its power, its elegance, its artistic side—is seldom dealt with in textbooks. Visual materials may be unusually dramatic or attractive as presented in films such as *Donald in Mathmagic Land*, filmstrips such as *Optical Illusions*, or charts such as the *Geometric Progression.*

Present the history of mathematics and other enrichment topics. Many topics and historical incidents can be used to enrich the learning of mathematics. Especially appropriate for enrichment are the films *Mathematical Peepshow, Koenigsberg Bridges, Stretching Imagination*, and *How Man Learned to Count*, as well as the chart *History of Mathematics.*

Illustrate the discovery of relationships or principles. We learn best those things that we discover through our own experiences. Films often show how relationships can be discovered. See, for instance, *Patterns in Mathematics*; *How to Multiply Fractions*; and *Volumes of Cubes, Prisms, Cylinders.* Audiotapes or overhead projectuals with worksheets are also useful for presenting these experiences because they provide activities for the student.

Present dynamic ideas that depend on motion. Motion pictures that use animation or slow motion can give dramatic illustrations of mathematical ideas. Examples of such films are *Locus, A Mathematician and a River*, and *Slide Rule.*

Correlate mathematics with other subjects by presenting supplementary materials. Topics presented in the mathematics textbook are usually limited to definite mathematical principles. A film such as *Curves*, a filmstrip such as *Vectors*, a chart such as *Math at General Electric* have mathematical aspects interwoven with the topic involved.

Provide complex drawings of three-dimensional effects. Visualization and spatial relationships are often difficult to achieve with textbook or blackboard drawings. Three-dimensional stereographs such as *Solid Geometry*; films such as *Rectilinear Coordinates, Dimensions*, and *Triple Integration*; charts such as *Polyhedrons* are highly realistic in portraying three dimensions.

Teach how to solve problems. Since it is extremely difficult to teach problem solving, any supplementary aid should be welcomed. Examples of films that deal with this skill are *Let Us Teach Guessing, Variations: A Lesson in Reading*, and *The Language of Mathematics.*

Introduce a new subject or unit. The beginning of a new course often sets the pattern for the year. Films such as *Mysterious X* and *Language of Algebra* and the filmstrip *Introduction to Plane Geometry* give a broad picture of the nature and use of the subject.

Summarize or review units within a course. At the end of a unit a new view of the entire unit may be possible with a film such as *Measurement* or a filmstrip such as *Deductive Thinking* or a chart such as *Number Systems.*

Show how to teach a topic. Many visual materials picture objects that would have greater meaning if the student could handle the object pictured himself. These materials, though, are useful to teachers in providing teaching hints. For example, the film *What Are Fractions?* shows how one can clarify the meaning of fractions by the manipulation and cutting of objects. Audiotapes or, when available, the much more expensive videotapes of classroom presentations provide the teacher with a wider range of effective teaching techniques and ideas for better instruction.

Provide remedial instruction. Tachistoscopic projections or tape recordings pace drill in computation. Filmstrips and film loops may be viewed individually and reviewed as many times as necessary to learn an idea.

Some Available Audio-visual Aids

For some time in the sixties, federal funds were provided to schools for support of purchase of audio-visual equipment. During that period, a wealth of materials became generally available. Now, despite cutbacks in financing, a wide variety of equipment is available for purchase. The teacher must select carefully, train himself well, and then use appropriately the equipment that best suits his program. Here are some of the major types of audio-visual aids:

The Chalkboard

Next to the textbook, the most commonly used aid of the mathematics teacher is the chalkboard. Mathematics teachers have traditionally used this primary tool in the same manner as Socrates and Archimedes used sketches in sand to solve mathematics problems. The chalkboard provides an immediate and effective mode of presentation of material. Its uses are wide and varied, and teachers should consider these uses in developing and executing their plans. The chalkboard is uniquely suited to make the following contributions:

Provide a medium for the participation of the student in class activities.

Emphasize major points by outlining and summarizing. It permits a point-by-point development and a reference as the lesson progresses.

Present assignments, problems, or discussion questions.

Combine a visual and oral presentation of ideas.

Since many teachers use the chalkboard in a haphazard fashion here are some suggestions for improved uses:

Write clearly, neatly, and correctly. Combine writing and speaking. Do not erase too quickly. Use very large figures so that all class members can see them.

Make all drawings simple and accurate. Use stencils and drawing instruments to give pleasing, accurate figures. Add realism by giving attention to perspective and color. Add interest with captions and cartoons.

Arrange frequent student participation in chalkboard work. Board work has the advantage of requiring physical and emotional involvement as well as mental activity. Select a student to act as class secretary.

Use colored chalk to identify key ideas, to add attractiveness, to emphasize common elements, contrasts, and relationships. For example, use color to distinguish between symbols of operations and signs of positive or negative numerals. In writing numerals in different bases, identify the base by the color of the numeral. In drawings, use color to identify corresponding parts or measures in formulas. Color can be used to distinguish between coefficient and exponent, conditions in loci constructions, characteristic and mantissa of logarithms, factors, and terms.

Have adequate materials available. A variety of available chalkboard tools can be used to improve board work:

Straight edges, meter sticks, or yardsticks
Geometric figures, including triangles, rectangles, and polygons
Stencils for perspective drawings of three-dimensional objects
Stencils for grids for graphing
Drawing instruments, including compasses and protractor
T-square and linkages
Chalkholders

Chalkboards themselves now come in a variety of materials and colors. The new materials include porcelain steel, frosted glass, formica, plastic paint on wood, linoleum, cement asbestos, and silicon carbide as well as the traditional slate. Slate continues to be preferred because of its durability, its texture, and its reflective qualities but it has the disadvantage of being available in only one drab color. Steel backing has the advantage that magnets adhere to its surface.

It is difficult to determine how much chalkboard space is adequate in a mathematics classroom. With modern duplicating facilities and the overhead projector replacing some of the functions of the chalkboard, there is less demand for board space than formerly. And space must also be provided for the wealth of material available for exhibit on bulletin boards, pegboards, book shelves, or exhibit shelves. However, there should be much chalkboard space available at the front of the room for developing daily lesson material. One way to increase the amount of board space is to use sliding panels.

It is likely that every mathematics classroom needs a grid marked on the chalkboard for graphing. If this is painted or scratched on one section of the chalkboard, that section cannot be used for other purposes. It would seem more suitable to obtain a stencil on a plastic cloth that rolls up like a window shade. When this stencil is placed on the board and dusted with chalky erasers, the chalk dust forms a usable grid. This permits the placing of grids on several sections in a very short time. These grids can then be used for class discussion, for tests, or for student board work. After being used, the grid can be erased. Some teachers prefer a painted pegboard for graphing; others use a grid drawn with semi-permanent chalk. Teachers have found a variety of ways of enhancing board work, such as the following:

Use an opaque or overhead projector to give outlines for copying complex drawings from a text or picture.

Draw complex illustrations before the class starts to avoid wasting class time.

Erase or clean the chalkboard carefully so that old material does not distract from the material being presented.

Enlarge drawings to desired size by the use of a pantograph or proportional squares or by projections.

Draw semipermanent lines, using soft chalk soaked in a saturated sugar solution. The lines will not be erased with a chalk eraser but will wipe off with a damp cloth. Use this chalk to draw coordinate axes, number lines, triangles, circles, parallelograms, or other figures used repeatedly.

Use old lipstick holders as holders for colored chalk to reduce finger stains.

Mount drawer knobs or empty spools to meter sticks, T-squares, or templates to facilitate holding them for board drawings.

Use small suction cups to hold fixed points for drawing circles, ellipses, or spirals.

Add realism to drawings representing three-dimensional objects by having the actual objects on display.

Draw horizontal lines from left to right; vertical lines are drawn downward; freehand circles are drawn with a stiff arm so that the shoulder acts as a center.

Use vanishing points to establish proper perspective. Accent the nearest edge of an object with heavy solid lines. Use shading, dotted lines, overlap to accent the third dimension. Use stencils that aid correct representation of three-dimensional perspectives.

Speak to the class, not to the blackboard. Combine writing and talking so that the board work has maximum interest and meaning.

Use stencils, templates, drawing instruments so that your drawings are accurate and attractive.

Dissolve a teaspoon of show-card colors and a tablespoon of Bon-Ami in one cup of water to get chalkboard paint that is colorful and at the same time can be removed easily. Avoid all oil or wax base colors for board work.

Have students with drawing skill place colorful drawings on the board related to the season, holidays, or special events.

The Overhead Projector

The overhead projector (see Figure 24.1) is an audio-visual aid which makes a special contribution to mathematics. Many teachers prefer it to the chalkboard. The teacher can write on blank acetate, projecting his material on the screen, or he can use previously prepared transparencies or commercial projectuals. Overlays allow sequential development of a diagram or dissection of a confusing array to show basic parts. Some teachers even use the overhead projector and chalkboard simultaneously, exposing a complete diagram but tracing on the blackboard only a key feature. Thus, when the machine is shut off, only the desired feature remains. This is an especially useful technique in proving congruency when triangles overlap.

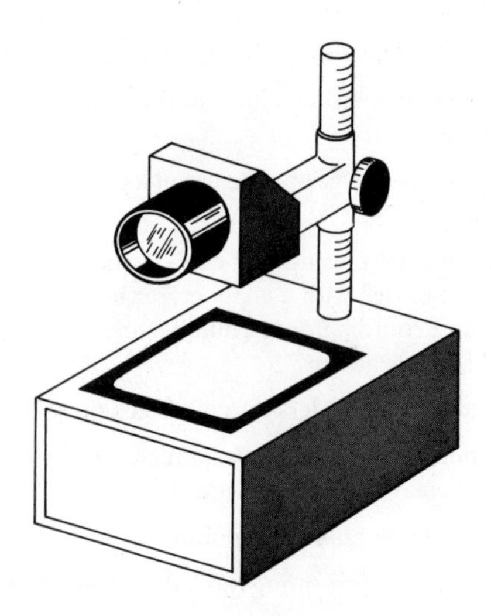

Figure 24.1 An Overhead Projector

While commercial transparencies—including plastic slide rules—are available, teachers can modify them or construct their own visuals to suit their special needs.

There are four primary advantages of the overhead projector over other projected visual aids:

1. The projector is used in front of the class, so that the teacher faces and talks directly to the class. The teacher sees the material projected and, at the same time, maintains eye contact with his class.

2. The projected image is bright enough so that it can be seen under normal classroom lighting conditions.

3. The teacher operates the projector and thus maintains control of class attention. The operation of the projector requires only a flick of a switch.

4. A variety of transparencies is available since they can easily be made by the teacher or purchased from commercial producers. Complete sets of projectuals are available for almost any school mathematics topic. Frequently, text series are accompanied by a set of visuals.

The overhead projector is used in several ways. One involves the use of clear acetate and a nylon pen. The teacher writes or draws on the acetate with the pen much as he would on the chalkboard. The image of his writing is then projected in magnified size on the screen behind him. Thus, lesson development, step-by-step problem solving, or the presentation of an assignment is possible. The material can be erased with a damp cloth and the acetate sheets used for another lesson.

Another use of the overhead projector involves the presentation of material by silhouettes. The shadow of an opaque object or piece of paper will give projections of shapes such as triangles, graphs, or patterns. A variation of this shadow projection is the step-by-step presentation of prepared material by a sliding mask. A sheet of paper will mask material so that the teacher can present material in a proper sequence or direct attention to a given point.

Another major use of the overhead projector is to present transparencies prepared before the class session. These transparencies, called projectuals, may be prepared in several ways. Written material may be typed directly on an acetate sheet or written with a grease pencil or nylon pen. Pictures or text material may be copied by a commercial copying machine. Some of these copying devices will also duplicate material in color.

In mathematics discussions, much use can be made of overlays, which superimpose multiple images upon each other. These overlays consist of a series of transparent sheets attached by hinges so that they can be projected in a prearranged sequence. For example, the intersection of the graphs of a series of inequalities can be built up by first projecting the grid and axes alone, then showing one graph in a specific color, then a second graph in a different color. The final projection now shows clearly the intersection of the truth sets of the two inequalities.

The overhead projector can also demonstrate change when conditions are varied. Suppose that one transparency projects a circle and a second transparency projects an angle. By moving the second transparency over the first, one can show how an angle intercepts arcs of the circle in different positions. This technique can also be used to illustrate the relationships between loci and similar polygons, and the graphs of systems of equations.[1]

Topics for which transparencies have been found particularly useful include the following:

Graphs of equations and inequalities

The number line and nomographs

[1] For further information see the pamphlet by Stephen Krulik and Irwin Kaufman, *How to Use the Overhead Projector in Mathematics Education,* National Council of Teachers of Mathematics, Washington, D. C., 1966.

Venn diagrams illustrating unions and intersections of sets
Geometric figures for two- and three-dimensional space
Reading slide rule scales
Geometric constructions and loci
Flow charts and computer programs
Conic sections—graphs, drawing with a linkage formed by paper folding
Systems of equations and inequalities—intersections
Number patterns such as the sieve of Eratosthenes
Curves of space travel
Measurement—accuracy, precision, angle measurement, areas, volumes
Perspective drawings of three-dimensional objects
The wrapping function
Geometric representation of algebraic expressions
Tables of trigonometric functions, square roots, logarithms
Tests and answers to test questions or assignments
Student papers for illustration of excellent work or common errors

The Bulletin Board

A bulletin board display is one of the least expensive instructional aids available and at the same time one of the most versatile. It has one major advantage—it may present a display for several days. In view of the fact that displays are seen by all students using the room, it is important that the display be appropriate.

There are several ways in which the bulletin board plays a unique role:

It involves the active, cooperative participation of students and teacher in planning and producing attractive displays.

It emphasizes enrichment by presenting further information about supplementary topics, historical events, famous mathematicians, unusual ideas, current applications, and challenging problems.

It is a means of presenting materials to attain objectives that cannot be taught directly—objectives such as values, interests, appreciations, curiosities, and study habits.

The preparation of a display provides a means of stimulating creativity, craftsmanship, and communication skill. By planning, organizing, and completing a display, students learn to work together, to share ideas, and to accept responsibility.

Whether a bulletin board display attains these purposes depends largely on the teacher. Displays should be an integral part of his instruction and students must be aware of their responsibility for learning from the displays. Here are some basic principles to follow in preparing displays that will attract viewers and arouse interest:

> The display should have *simplicity, clarity, and unity.* Proper titles, few key ideas, novel pictures, unusual framing, unique titles, and tricky questions. Background material such as colored construction paper, colored corrugated cardboard, burlap, felt, or cork add color and softness to the display.
>
> The display should have *simplicity, clarity, and unity.* Proper titles, few key ideas, flow lines, and simple arrangements—even blank spaces—tell a story.
>
> The arrangements should be *interesting.* Informal balance, striking contrasts, variations in shapes and colors add interest.
>
> The display should be *timely* and should be changed frequently. An old display fades into the background like wallpaper.

The success of a display is highly dependent on appropriate and clever titles and captions. Titles should be brief, eye-catching, and meaningful. These titles can be formed by commercial letters, cutout letters, stencils, or writing. Titles may be written with a brush pen, string, pipe cleaners, spray paint, or adhesive tape. Three-dimensional effects can be attained by mounting letters and pictures on pins, spools, boxes, or paper cups.

Topics for bulletin boards are many and varied. The headings and titles listed below are merely suggestive.

> Central themes of mathematics, such as "The Number Family," "Basic Mathematical Laws," "Size, Shape, and Similarity"
>
> Mathematical applications, such as "Mathematics in Nature," "Musical Mathematics," "Mathematics in the Machine Shop"
>
> Habits, attitudes, or appreciations, such as "There Is Math in Your Future," "How to Make Mathematics Easy," "What's My Line?"
>
> Enrichment topics, such as "Topology, the Mathematics of the Magician," "The Fourth Dimension," "How to Take a Chance"
>
> Recreational exhibits, such as puzzles, problems, fallacies, or illusions

Bulletin board materials are all around you. In addition to the usual display materials, colored paper, tacks, and pins, such materials as boxes, graph paper, balsa wood, cellophane, and many others should be considered. Usually, the art department will be ready and willing to provide ideas and support for the classroom teacher.[2]

---

[2] For further ideas see *Bulletin Board Displays for Mathematics* by Donovan A. Johnson and Charles Lund, Dickenson Publishing Company, Inc., Belmont, Cal., 1967.

Other Audio-visual Aids

Some of the additional audio-visual aids to instruction are the following:

Motion picture films. The number of mathematics films available has increased dramatically in recent years. Current films are reviewed in *The Mathematics Teacher* and other professional journals. These films are able to show reenacted events, use animation, and present situations that cannot be presented through text material alone. Hence, films should be a unique means of bringing historical events, dynamic variation of conditions, demonstration lessons, or applications of mathematics into the classroom. Publications that list film titles are reported in Appendix A.

Filmstrips. Filmstrips are inexpensive and can be used with great flexibility. The frames may be shown at any desired speed, allowing for appropriate class discussion. Filmstrips are effective in presenting ideas that do not require motion. They seem most effective when they present material in a modified programmed format so that the viewers respond to questions posed by individual frames. Some filmstrip series have been prepared to accompany a specific mathematics text. Individual slides (2 inches × 2 inches in size) allow greater flexibility than do filmstrips, but they must be prepared by the teacher (see Appendix A).

Pictures. Collections of pictures on almost any topic can be gathered quickly from current magazines. Having them mounted and filed makes them readily available for use on bulletin boards or for opaque projections. Students should participate actively in adding to the collection. Writing appropriate captions for each picture adds to their effectiveness.

Charts, maps, graphs. These materials are usually available free or at very low cost from commercial companies or governmental agencies. The weekly graphs of current statistics issued by the Conference Board of New York are available to teachers upon request. Large corporations such as the Ford Motor Company, General Motors, General Electric, Chrysler Motors, and IBM have frequently published free mathematics charts.

Stereographs. To add a three-dimensional effect to a picture, several techniques are possible. Duplicate pictures are taken by two cameras whose lenses are as far apart as our eyes. Then these pictures are printed in different colors or by polarizing the light. When the pictures are viewed through the proper equipment, called a stereoscope, they give the appearance of depth that we perceive when viewing the actual situation. This technique seems especially appropriate in learning to "see" three dimensions in a geometric figure or in learning to represent a three-dimensional object with a two-dimensional drawing. Some new geometry texts furnish views of this type to assist the learner in visualizing space geometry.

Television. Individual commercial or educational television programs occasionally provide an opportunity for students to see good mathematics developed by expert teachers. Closed-circuit television provides an opportunity for the school system to make available unique lessons or courses to many students. Videotape recording of classes, demonstrations, or events extends this use of television.

Tape recorders. Tapes of lessons may provide review material for weak students or original lessons for students who have been absent. The best use of tape is the planned lesson. For this a student is provided with work sheets geared to the specific tape. Earphones allow individualization of lessons even in a crowded classroom, and tape tables—tables seating 4 to 20 students with earphones provided at each station—allow the teacher to carry on two lessons in the classroom at once: taped lessons for some students; for others, direct work with the teacher.

Another format for instruction is the individualization allowed by tape players and individual cassettes with accompanying worksheets. Preparation of taped lessons requires considerable skill and practice on the part of the teacher. It is usually best for the teacher to outline in detail what he wishes to cover in such a lesson, to develop a worksheet from the outline, and to record his instructions as he works the worksheet problems for himself. In this way he can pace his material. The teacher should not be afraid of silent periods (when students are working problems), and he should often provide answers so that students can see their mistakes as soon as possible. Finally, the teacher should personalize his lessons by injecting humor into them just as he would in the classroom.

You need use a taped lesson only once to find how effective it can be. Student interest is concentrated, distracting sounds are shut out by the earphones, and neighboring students are not bothered. And, best of all, each student is working in a one-to-one relationship with the teacher. Taped lessons can extend the individual grouping used so successfully at the elementary school level but essentially lost in secondary schools.

Television tapes and individual film-loop motion pictures. The previous comments about sound-taped lessons extend to the use of such individualizing materials as video television tapes and 8 mm film loops. Each of these allows a student to view a lesson as many times as is necessary to understand the content. Some of the modern schools with this type of equipment have special sections of the library or rooms between classrooms (supervised through windows) where such materials may be used by students. As has been the case before, mathematics is lagging far behind science in the development of films and tapes.

Opaque projector. The opaque projector, or stereopticon, allows projection of the colored image of a nontransparent object. Thus, this instrument allows projection of a student's paper on the screen whereas the overhead projector must utilize transparent projectuals and could not reproduce the paper without an intervening transparency reproduction process. It is especially useful in focusing class attention on student work or on material in current periodicals. Additional uses include enlarging graphs and tables, slide rules and other scales, architectural plans and pictures. Formerly a widely used classroom device, the opaque projector has been largely replaced by the currently favored overhead projector. It is easy to copy materials for use on the overhead projector.

The Proper Choice and Use of Audio-visual Materials

It is obvious from the long list of available materials and equipment that one of the basic problems of the classroom teacher is that of choice. Which mode of presentation will be most effective and will best promote learning? This choice is an extremely important one. Usually, there is no one best way. In fact, use of a combination of devices is occasionally appropriate. The overhead projector may be used to project a complicated geometric diagram on the chalkboard. Special parts of the diagram may then be traced in chalk directly on the blackboard. When the projector is turned off the students' attention may focus on the chalkboard display or on a mock-up of the problem at hand.

It is essential that the mathematics teacher carefully *select* the materials to be used. The excessive use of audio-visual aids may cause the students to become over-dependent on physical representation and unable to work with the symbolic representation of an abstraction, a procedure so necessary in mathematics. The improper use of teaching aids may create confusion and misconception of mathematical principles. The indiscriminate use of these aids may occasionally result merely in entertainment. They should be used only when they are the most effective means available for attaining an objective. Here are some guidelines:

Select the proper material or equipment to be used at a certain time to attain a specified objective. Think carefully about supportive aids in planning units. Make yourself aware of the wide selection. Find titles in catalogs, film guides, reviews in the *Mathematics Teacher*. Select wisely.

Prepare yourself for the use of the audio-visual material. Become familiar with the content of visual material. Preview a film or filmstrip, read the teacher's guide if there is one, try transparencies on the overhead projector.

Prepare the classroom. Every modern classroom needs the facilities and equipment that make possible the optimum conditions for the frequent use of a variety of materials. Before the class meets, check to see that all needed materials are at hand and that equipment is in working order.

Prepare the class. Tell your students the purpose of the learning aid, what it will do, why it is used at this time, what they will learn from seeing it, and how they should be able to apply the information gained. Often a written guide prepared for the students increases the effectiveness of a learning aid.

Present the material. Be sure that each student hears and sees each important item. Discuss the items so that the visual or audio learning is related to the lesson at hand and to the symbolic representation that will later be used.

Provide follow-up activities. Discussions, readings, reports, projects, tests, and reshowings are needed if maximum learning is to result. Wherever possible, provide opportunity to apply the information learned.

Evaluate the effectiveness of the material. A card-file system is helpful. Over a period of time you will accumulate what are for you the best materials for each lesson. This evaluation should include remarks on the quality and content of the material as well as the response of the students to its use.

Deficiencies of Visual Aids for Mathematics

The quality of visual aids now being produced has improved greatly in the past few years. Unfortunately, many of the visual aids now available, especially the older productions, have deficiencies such as the following:

Treatment of topics exactly parallels textbook treatment. Often the language and drawings are identical to that of a text, thus providing no new insights.

Too much material is covered in a single film or filmstrip. Many one-reel, ten-minute films cover a complete topic or unit.

Applications and illustrations are lacking in interest to the student. Installment buying applied to buying a car is more pertinent to high school students than furnishing a doctor's office.

Drawings are too frequently used in place of photographs. Photographs are hard to get for all ideas covered but are much more interesting and realistic.

Commentary and situations are unrealistic. If youths are acting in the picture, the action and language needs to be quite informal and the dress contemporary and acceptable.

Showing cannot be a complete substitute for concrete experience. Observation is never as forceful as actual participation.

Presentations are sometimes used as substitutes for good teaching. Visual aids are sharp tools and require as much planning and organizing as a verbal presentation. They are not produced to replace the teacher.

Available audio-visual aids are not comprehensive for any subject or topic.

Some audio-visual aids contain mathematical inaccuracies. While such inaccuracies may not result in his rejecting an aid, the teacher should be aware of them so that they may be corrected. Many teachers like to use errors in films or texts as the basis for class discussion.

The creative teacher instinctively reaches out for an audio-visual aid to supplement verbal instruction. He knows what aids will add to his lesson so that he has them on tap when needed. The teacher who has a mature feeling for audio-visual aids and who gives careful attention to what works for him will soon learn that these aids should be part of the total approach to a lesson. While the overuse of aids is a danger, it is a far less common problem than failure to use available materials when they can add to a presentation. It is a rare mathematics lesson that cannot be improved with the use of an audio-visual aid.

Learning Exercises

1. Visit the audio-visual center at your school or college. Determine what audio-visual aids and services are available. Is assistance provided in learning to use equipment? Are published lists of films available?

2. Read the *Mathematics Teacher* department "New Products" for the past four months. List and justify what you think is the best of each of the following:
   (a) Films
   (b) Games
   (c) Computers

3. Develop a 5- to 10-minute script for a tape cassette and an accompanying worksheet to present a single topic. Write BELL in your script when the student is to turn off the cassette player and work on a worksheet exercise.

4. Develop an overhead projector transparency and a plan for its use in the classroom.

5. Make a tape recording of your own voice. Can you identify needed improvements in pitch, volume, or clarity?

6. Make a bulletin board display on a mathematical topic (see appendixes).

7. Practice drawing the diagram of Dandelin's Cone (see page 157) on the chalkboard. Hint: Draw the circle and arc first. Try to improve your freehand drawing as you do this.

Suggestions for Further Reading

Anderson, Frank A., "Translation of Axes Discovered through the Overhead Projector," *Mathematics Teacher*, 63, 8 (December 1970), 669–670.

Byrkit, Donald R., "Using Televised and Aural Materials for Mathematics Teachers," *Mathematics Teacher*, 64, 6 (October 1971), 519–524.

Fitzgerald, William M., and Irvin E. Vance, "Other Media and Systems," in *The Teaching of Secondary School Mathematics*, Yearbook 33, National Council of Teachers of Mathematics, Washington, D. C., 1970, pp. 110–133.

Hansen, Viggo P., "New Uses for the Overhead Projector," *Mathematics Teacher*, 53, 6 (October 1960), 467–469.

Hoisington, Robert, "Semi-Permanent Chalk for Teaching," *Mathematics Teacher*, 47, 6 (October 1954), 407.

McCurdy, Sylvia E., "Colored Chalk Techniques for Basic Mathematics," *Mathematics Teacher*, 48, 8 (December 1955), 369–371.

Montgomery, James, "The Use of Closed Circuit Television in Teaching Junior High Mathematics," *School Science and Mathematics*, 68, 8 (November 1968), 747–749.

Osborne, Alan R., "Using the Overhead Projector in an Algebra Class," *Mathematics Teacher*, 55, 2 (February 1962), 135–139.

Vollman, William D., and Philip Peak, *How to Use Films and Filmstrips in Mathematics Classes*, National Council of Teachers of Mathematics, Washington, D. C., 1960.

25

Calculators and Computers

One of the basic issues facing secondary-school mathematics departments today relates to the role of the computer in the school program. Should mathematics courses teach computer programming? Should there be a special elective course on computers or should computer science be included in regular mathematics courses? Should the computer be used as a tool for solving problems? Should data processing be included in courses? Should the mathematics department have desk calculators, a computer terminal, or a low-cost computer? Who should assume responsibility for teaching the role of the computer in society? Should computers be used to monitor instructions?

At the present time, many mathematics departments are installing computers and teaching computer science. Courses and units on computer programming are being introduced without thoughtful consideration or careful evaluation of the outcomes of this instruction. Some schools have found that computer programming can be a means of teaching mathematical concepts and problem-solving skills, while others propose that computers serve mainly in providing the students with a greater motivation to study mathematics. Others are using computer facilities to keep student records, to administer tests, and to provide individualized instruction.

The Computer and Society

The computer is causing a change in our society comparable to the change occasioned by the Industrial Revolution. Automation controlled by the computer is creating an economic upheaval in industry: data processing by the computer is making accounting and record keeping an automatic process; scientific research now has practically instantaneous analysis of data; linear programming by the computer is a means of decision making in government, education, military science, and business. Computer memories are being used to store entire courses for presentation to individual students. Medicine, law, and engineering are using information retrieval by the computer for diagnosis and decision making. And, of course, space travel and commercial plane scheduling are dependent on the computer. Linguists, psychologists, and historians are using the computer for research in their fields. Even music and poetry have been composed by the computer. In mathematics it has even been suggested that numerical analysis and the computer will make calculus, as it is presently taught, obsolete.

At the present time, major computer facilities are being made available for schools through the use of computer terminals such as a teletypewriter. These terminals will make it possible for schools to use a computer for the solution of a variety of mathematical problems. At the same time, school records, schedules, research data, payrolls, and report cards can be processed by the computer. In addition, inexpensive terminals are available for instruction in computer programming. It is increasingly evident that schools must explore and utilize the capabilities of the computer to facilitate and extend educational processes.

Who should be teaching the role of the computer in our society? Logically, this should be the function of the social studies department. However, at this time, only the mathematics teachers are likely to have the background needed for this instruction. Hence, secondary mathematics teachers will need to teach a unit either in the social studies class or in their own mathematics class if this important aspect of current society is to be included in the secondary curriculum.

Computer Programming

Computer instruction today is usually divided into three categories: computer programming, data processing, and computer design. Teaching computer programming should be the responsibility of the mathematics department, while data processing could be taught in the business department. Computer design and maintenance should be the responsibility of the industrial arts department.

Computer programming should be part of the mathematics curriculum for several reasons:

The computer circuits are based on a binary numeration system.

The logic circuits are based on truth tables and Boolean algebra.

Computer programs are frequently based on flow charts like those used in solving problems, so that the study of computers and the study of problem solving are mutually supportive.

The operations of the computer include all the operations of mathematics.

The code language of the computer is similar to the symbolic language of mathematics.

Mathematics teachers usually have the background and the interest necessary for successful instruction.

Knowledge of the techniques for this important tool helps to broaden the instructional base for the individual teacher.

Programming a problem for a computer is a relatively simple process, especially if a simple language such as BASIC, JOSS, or INTERCOM is used. More complex languages, such as FORTRAN, ALGOL, COBOL, or PL/1 are useful for more extensive programs and more complex computers.

The first step in programming a problem consists of writing a flow chart or problem analysis similar to those illustrated in Chapter 12. A second step consists of learning the code language to be used. With these tools at hand, a problem can be programmed for computer analysis and solution.

The computer can perform many mathematical operations. It can add, subtract, multiply, divide, extract square roots, raise a number to a power, find the sine of a number, and so on. The computer performs these computations by evaluating formulas that are supplied in a program. Table 25.1 gives some examples in BASIC language. The typewriter symbol shown in the first column communicates the operation to the machine.

Table 25.1

Symbol	Example	Meaning
+	$A + B$	Add B to A
−	$A - B$	Subtract B from A
*	$A * B$	Multiply B by A
/	A/B	Divide A by B
↑	$X \uparrow 2$	Find X^2
< >	If $A < > B$, then	If $A \neq B$, then
INT( )	INT(A/B)	Integral value of A/B

Let us apply this language to the solution of a system of two simultaneous linear equations:

$$\begin{aligned} X + 2Y &= -7 \\ 4X + 2Y &= 5 \end{aligned}$$

We know that if $AX + BY = C$ and $DX + EY = F$ and $AE - BD \neq 0$, then

$$X = \frac{CE - BF}{AE - BD}, \quad \text{and} \quad Y = \frac{AF - CD}{AE - BD}$$

are solutions of the system. If, however, $AE - BD = 0$, there is no unique solution.

The BASIC program for solving this problem is given in Figure 25.1. Note that the program uses only capital letters, since teletype has only capitals. Also zero and the letter O are distinguished in that 0 is used for zero and Ø for the letter O. Each line of the program begins with a *line number*, which identifies each statement of the program in addition to specifying the order in which the statements are to be

```
10 READ A, B, D, E
15 LET G = A * E - B * D
20 IF G = 0 THEN 65
30 READ C, F
37 LET X = (C * E - B * F) / G
42 LET Y = (A * F - C * D) / G
55 PRINT X, Y
60 GO TO 90
65 PRINT "NO UNIQUE SOLUTION"
70 DATA 1, 2, 4, 2
80 DATA ¯7, 5
85 DATA 1, 3, 4, ¯7
90 END
```

Figure 25.1

performed by the computer. After each line number, an English word denotes the type of statement.

The first statement (10) is a READ statement. It must be accompanied by one or more DATA statements. The DATA statements are 70, 80, and 85. In the program above, the computer will use statement 70 to assign 1 to A, 2 to B, 4 to D and 2 to E (as values for A, B, D, E of the given equation). With these values for A, B, D, and E, the computer computes G (statement 20). If $G \neq 0$, then statement 30 reads the next entries in DATA (80): namely, -7 for C and 5 for F. Statements 37 and 42 provide the computer with instructions for carrying out the rest of the computations. The computer is now ready to solve the system:

$$\begin{aligned} X + 2Y &= {}^{-}7 \\ 4X + 2Y &= 5 \end{aligned}$$

Instruction 55 tells the computer to print the truth set. The addition of DATA in 85 gives the solution for these two additional systems.

$$\begin{aligned} X + 2Y &= 1 \\ 4X + 2Y &= 3 \end{aligned} \qquad \text{and} \qquad \begin{aligned} X + 2Y &= 4 \\ 4X + 2Y &= {}^{-}7 \end{aligned}$$

When these solution sets are printed, the computer has completed the program. Once the given program is printed by the typewriter, the programmer merely waits for a second or so until the computer operates the typewriter so that it prints the solutions.

The problems below illustrate other BASIC programs. The reader should be able to reconstruct the method of attack from the machine instructions.

Problem:
What are the terms of the Fibonacci series and approximations to the golden ratio? (The golden ratio is approximated by the ratio of a term of the Fibonacci series to the following term.) The Fibonacci series is 1, 1, 2, 3, 5, 8, 13 . . . where $F_{n+2} = F_n + F_{n+1}$ for $n \geq 1$.

Program: See Figure 25.2.

```
 5 LET P = 0
 6 LET B = 0
10 LET N = 1
11 LET X = 1
12 PRINT "THIS PRØGRAM PRINTS THE FIRST FIFTY TERMS ØF THE"
13 PRINT "FIBØNACCI SEQUENCE AND GIVES APPRØXIMATIØNS TØ"
14 PRINT "THE GØLDEN RATIØ"
15 PRINT
16 PRINT
17 PRINT "                    RATIØ ØF TERM          RATIØ ØF TERM"
18 PRINT "                    TØ PRECEDING           TØ FØLLØWING"
19 PRINT "TERM                    TERM                   TERM"
20 PRINT "*****               *****************      *****************"
21 PRINT
22 PRINT "1                       —                      1"
23 LET B = N
24 LET X = X + B
25 LET N = X − B
26 LET G = X / N
27 LET H = N / X
28 LET P = P + 1
29 IF P = 50 THEN 35
30 PRINT N,                   G,                     H
31 GØ TØ 23
35 END
```

Figure 25.2

Solution Printout: See Figure 25.3.

```
    THIS PRØGRAM PRINTS THE FIRST FIFTY TERMS ØF THE
    FIBØNACCI SEQUENCE AND GIVES APPRØXIMATIØNS TØ
    THE GØLDEN RATIØ
*****************************************************************************
                      RATIO OF TERM        RATIO OF TERM
                      TO PRECEDING         TO FOLLOWING
    TERM                  TERM                 TERM
    ******          *****************    *****************
    1                      —                   1
    1                   2                     .5
    2                   1.5                   .666667
    3                   1.66667               .6
    5                   1.6                   .625
    8                   1.625                 .615385
    13                  1.61538               .619048
    21                  1.61905               .617647
```

34	1.61765	.618182
55	1.61818	.617978
89	1.61798	.618056
144	1.61806	.618026
233	1.61803	.618037
377	1.61804	.618033
610	1.61803	.618034
987	1.61803	.618034
1597	1.61803	.618034

RAN 3 SEC.
STOP.
STEADY.

Figure 25.3

Problem: Print out all the positive primes less than 100.

Flow chart: See Figure 25.4.

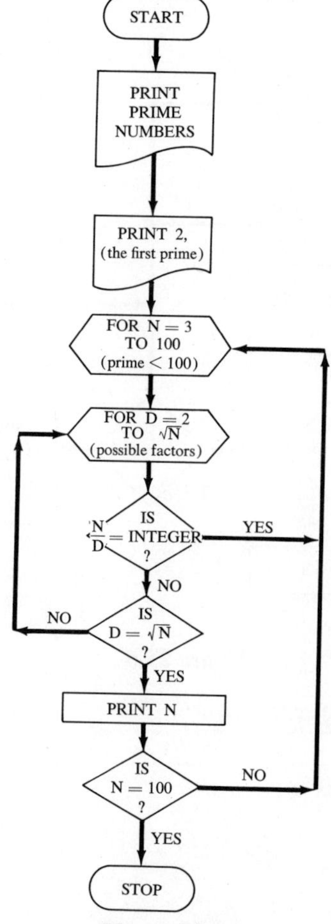

Figure 25.4

```
 5 PRINT "PRIME NUMBERS"
10 PRINT "2";
15 FØR N = 3 TØ 100
20 FØR D = 2 TØ SQT(N)
30 IF N/D = INT(N/D) THEN 60
40 NEXT D
50 PRINT N
60 NEXT N
70 END
```

Figure 25.5

Program: See Figure 25.5.

Problem:
Print out the pairs of inverse elements for "clock 5" addition. (Use zero as the additive identity.)

Flow chart: see Figure 25.6.

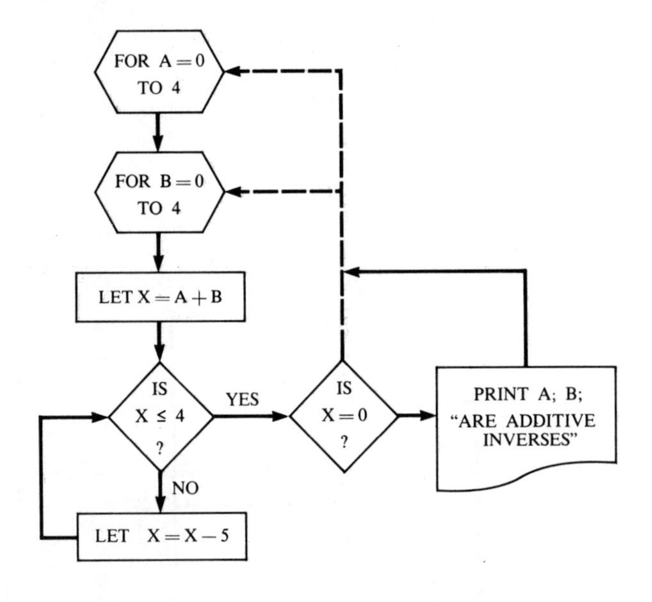

Figure 25.6

Program: See Figure 25.7.

```
10 FØR A = 0 TØ 4
20 FØR B = 0 TØ 4
30 LET X = A + B
40 IF X < = 4 THEN 70
50 LET X = X - 5
60 GØ TØ 40
70 IF X = 0 THEN 110
```

```
 80 NEXT B
 90 NEXT A
100 GØ TØ 130
110 PRINT A; B; "ARE ADDITIVE INVERSES"
120 GØ TØ 80
130 END
```

Figure 25.7

Note: This program would be given after the development of mod 5 addition (which actually is defined by statements 10–60 of the above). Additional levels of sophistication are available: (1) the program for mod k addition; (2) prefacing with the search for the identity element, mod 5; and (3) prefacing with the search for the identity element, mod k.

Solution Printout: See Figure 25.8.

```
        0        0        ARE ADDITIVE INVERSES
        1        4        ARE ADDITIVE INVERSES
        2        3        ARE ADDITIVE INVERSES
        3        2        ARE ADDITIVE INVERSES
        4        1        ARE ADDITIVE INVERSES
TIME:  1 SEC.
```

Figure 25.8

One difficulty beginning computer mathematics students have is understanding a statement like $N = N + 1$. Such a statement means, "Change N to N + 1," and, thus, is not the usual interpretation of "=".

Goals for Computer Programming

Computer programming should be a part of the mathematics curriculum if it is an effective means of attaining goals such as the following:

Learning how to solve problems. In programming, the problem analysis is very complete when it is set up in a flow chart. Furthermore, the problem is usually programmed to give a general solution rather than one single solution, which is rarely the case when students solve a specific problem. Thus, programming focuses on the method of solution and the generalization of a solution to a greater extent than traditional specific problem solving. This should develop greater skill in problem solving and greater competence in transferring the solution to other applications. It should emphasize the condition of the problem and how the solution changes when given conditions change.

Learning how to communicate mathematical ideas. The translation of the problem conditions and operations into a program requires minute attention to the sequence, the operations, and the symbols. Writing programs and interpreting computer solutions is another means of building communication skill. It requires careful attention to order, sequence, and detail. And it involves the use of a symbolic language with new symbols.

Developing desirable attitudes, interests, and appreciations. In view of the tremendous current interest in computers, it is difficult to imagine a more effective device for stimulating student interest. The computer is an exciting, frequently discussed device. It holds the promise of reducing tiresome computations by its instantaneous, almost magical response. Teachers who teach programming testify that students are willing to use their leisure time for computer instruction and operation. They use lunch hours, time after school, Saturdays, evenings, and even summer vacation to run the terminal or computer. If students are willing to participate in a mathematics-related activity to this extent, it seems reasonable to provide them with the needed instruction and facilities.

Teaching mathematical concepts. In programming, a student must design algorithms and then program the algorithms for the computer. In effect, the computer does *only* the "busy" work. When a student writes programs, such as the solution of a system of equations, the solution of quadratic equations, or the properties of a finite system, he is likely to have greater mastery of these ideas than can be provided by the usual paper-and-pencil exercises. Since *sequence* is important in programming, the structure and logic of a given algorithm are emphasized. (It is worth noting that programmers must guard against degenerate cases and zero divisors.)

Developing skill in computation. Usually it is assumed that since the computer does all the calculation, computation is eliminated. While this is true, in programming the computer, the student must have considerable insight into the meaning of each step of an algorithm. Programming demands understanding—the very kind of understanding that develops computational skill. Examples of this include programming the computation of a square root, the computation of the arithmetic average, the evaluation of a determinant, the solution of a right triangle, or calculation of an approximation of the golden ratio. At the same time, programming emphasizes the significance of units, approximations, and limits.

Adapting instruction to individual abilities and interests. Computer programming is of great interest to students of all abilities. For the slow learner, it is a new topic free of many of the frustrations and hostilities associated with topics that depend on concepts they have not yet mastered. For these students it offers relief from the grind of unsuccessful computations. For the culturally deprived, it gives a promise of new and upgraded employment opportunities. Also, problems of different degrees of sophistication are available for the enrichment of almost every topic. For the gifted, it provides a tool for probing challenging problems free of monotonous computation. To meet individual needs, computer programming has been offered as an elective course, as an extracurricular activity, and as a summer school course. If programming attains goals such as those listed above, it seems reasonable to offer it to all students as part of a regular mathematics course.

Whether computer programming actually attains the goals listed above has still to be accurately determined, and a variety of research projects are presently investigating this problem.

Two new areas of computer applications underscore the potential of the computer in the classroom. One is the use of the computer in the analysis of functions. By feeding equations or ordered pairs into the computer, the student can observe the results in graphical form on the teletype printout or on a television screen. These graphs show how lines curve, change in slope, or change in location as they represent graphs of functions.

The second area is simulation. Through simulation the computer can present the student with situations or problems as they might occur in the real world. The student is then asked to make decisions. His decision is fed back into the computer, which then presents a new set of facts. The student then makes new decisions based on the results of his previous choices. This rapid response to student decisions should be highly effective in teaching concepts and problem solving.

Mathematics for Computer Programming

The major aspect of computer programming is the use of this process to deal with a variety of topics. Therefore, a very small part of the instruction in computer programming should be devoted to learning how to draw flow charts and to write the computer program. This should take a short time, particularly if a trainer is available and a simple language is used. Some teachers report that one week at the seventh-grade level is sufficient.

With the introduction of mark-sensed cards to be recorded with a pencil, writing programs becomes independent of a typewriter. These cards also have the advantage of being corrected merely by erasing pencil marks and making new marks. Thus programs may be corrected without using expensive computer or card-punch time. The card in Figure 25.9 is marked "10 LET N = 3."

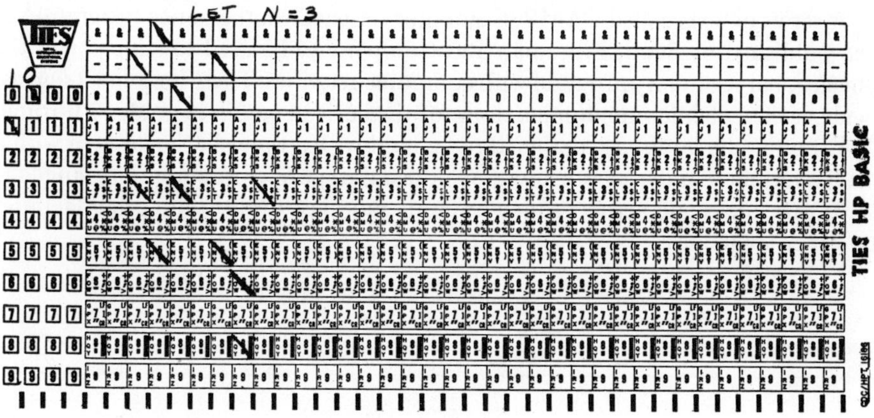

Figure 25.9

Programming is usually introduced by presentation of relatively trivial problems, such as generating the counting numbers. Next, simple equations are evaluated for several sets of values for the variables. These evaluations demonstrate the kind of analysis and formulation needed for programming, give practice in programming, and indicate the potential of the computer. To stop there (as is occasionally done) is not productive, however; additional significant problems should be explored.

Given the opportunity, teachers and students will find a great variety of problems that lend themselves to computer programs. Here are some examples:

> *Arithmetic.* Numeration systems: conversion from one base to another; divisibility of numbers; prime numbers; arithmetic average; properties of numbers; square root; evaluating exponential numerals.
>
> *General mathematics.* Compound interest; modular arithmetic; mortgage tables; salary problems; evaluation of equations and inequalities; graphing; probability.
>
> *Algebra.* Evaluation of phrases; systems of equations; quadratic equations; verbal problems; evaluation of determinants; coordinates, slopes, intercepts of equations; operations with ordered pairs; matrix addition and multiplication; sums and limits of sequences.
>
> *Geometry.* Right-triangle solutions; calculating areas and volumes; calculating approximations to π; finding sums of angles of polygons; measures of triangles related to coordinates of vertices; measures of circles; figures formed by coordinates.
>
> *Trigonometry.* Solving triangles; calculating approximations to e; calculating a table of sine values from series expansion; computing the path of a projectile; testing validity of identities.

A series of texts called CAMP (Computer Assisted Mathematics Program) has been published by Scott, Foresman and Company to be used to supplement the conventional secondary program with computer activities. These textbooks suggest a wide variety of computer applications for the basic program in grades 7 through 12. Although the books are designed to be used with a time-sharing system with a teletype terminal in the school and they utilize the specific programming language BASIC, the ideas and activities may be easily modified for use with other facilities and other languages.

Computer Equipment

To teach computer programming and to use it for developing mathematical ideas requires time and resources. Time should be provided in regular mathematics courses. Resources needed include a teacher with competence in computer science, computer installations or a computer terminal, programming cards, and text material. All these are now available.

Computer equipment for school programs includes the following: (1) adding machines and calculators, (2) desk-top computers or calculators with computer facilities, (3) computer terminals, and (4) electronic computers.

Adding machines and calculators are inexpensive devices for beginning instruction. These machines perform most of the operations of the computer but at a slow rate. Since each operation is transmitted by keys, these machines are manually programmed. They have limited memories and cannot execute a sequence of operations automatically.

Calculators, however, can be used to show the relation of preparing a flow chart to machine solutions. For example, the flow chart for computing an arithmetic average can lead to the machine calculation of the average. Calculators can also be used to discuss various parts of a computer, such as input, control, memory, and output. The keyboard is the input. The inside gears provide the controls that perform the operation desired. One register is the memory and another register gives the output (the result of the calculation).

Each school should have one mathematics classroom or laboratory with enough calculators for an entire class. (Calculators cost from $120 to well over $1,000.) Calculators provide hands-on operation in a manner that is fascinating to students. They illustrate visually place value in base 10 and often add meaning to a given operation. For example, calculators perform multiplication by repeated additions, division by repeated subtraction, recycling one place when the register passes zero. They are probably most appropriate for use in general mathematics classes.

A variety of calculators with limited programming units are also now available. Accepting input from punched or mark-sensed cards, these machines handle many of the shorter programs that would be developed by secondary-school students. More expensive than standard desk calculators, they are still less expensive than even small computers. Many departments purchase a few of these machines to supplement their inventory of calculators.

The computer terminal is usually an electric typewriter connected to a computer by a telephone line or cable. The typewriter can be operated to send a program to the computer or it can cut a tape for transmission at a later time. Cutting a tape permits the examination of a program to identify errors before using valuable computer time for the program. When the computer has received the program and is finished with other work received earlier or of higher priority, it solves the problem and types the solution on the terminal typewriter. Thus, a computer terminal provides the facilities of a real computer. It is relatively inexpensive because the terminal rental is nominal and very little computer time is needed.

A computer terminal eliminates the need for a computer in the school. With a computer terminal there are no problems of computer maintenance, nor is there an investment in an expensive computer that is likely to become out-of-date in a short time. To substitute for hands-on-the-computer, the class should take a field trip to a computer center. After that, the terminal is a sufficiently realistic hands-on experience. Some very large schools or school systems may, however, justify a computer installation on the basis of multiple uses such as facilitating school business, some financial savings if guarantees of updating equipment are obtained, and insurance against the delays occasionally faced in real-time terminals (that is, terminals that share the same computer).

Solving Problems with Flow Charts

Even where no computer or calculator facilities are available, computer-related activities are attractive to students. One of these is flow-charting, a technique developed by computer programmers to facilitate and explain their processes. Often omitted by all but beginning programmers, this technique has valuable applications in the mathematics classroom, especially for slow learners who need help with organizing their approaches to mathematical problems.

These flow charts provide a condensed visualization of a sequence of operations and facilitate analysis much as an equation facilitates the finding of truth sets. These charts are like classification diagrams, organization charts, historical charts, or industrial process charts. Figure 25.10 shows how a flow chart may follow the directions for drawing a common pattern.

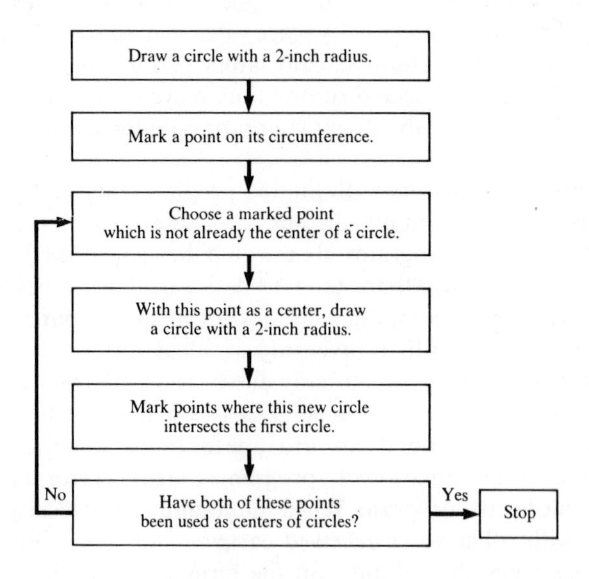

Figure 25.10

Flow charts are the basis for programming a problem for computer solution. A simple flow chart for finding an average is given in Figure 25.11.

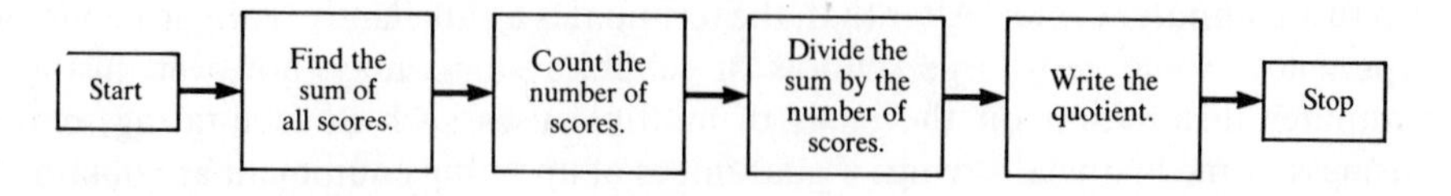

Figure 25.11

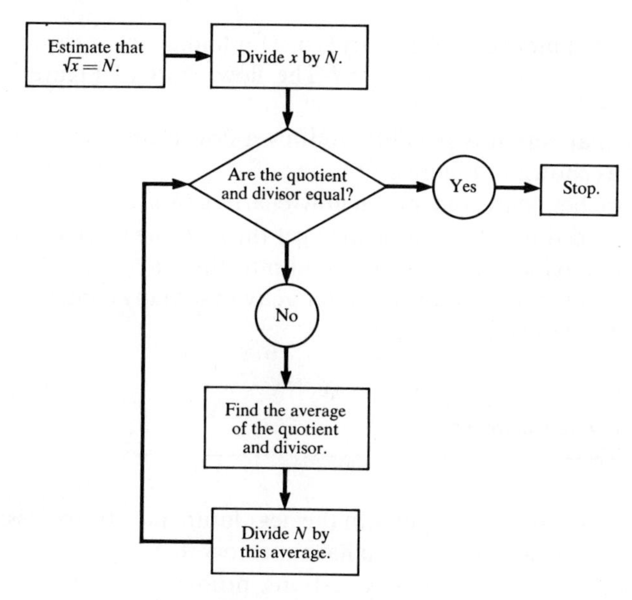

Figure 25.12

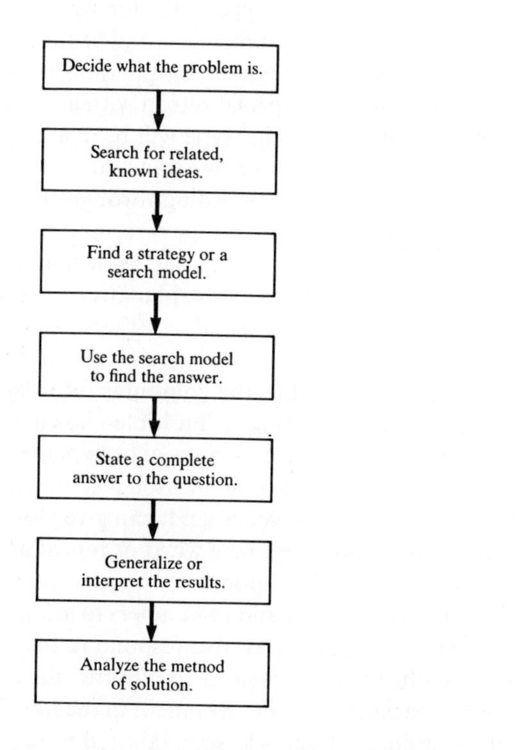

Figure 25.13 Flow Chart for Problem Solving

A somewhat more complex chart is one for finding the square root of a number by Newton's method. What is $\sqrt{x}$? The flow chart of Figure 25.12 gives one algorithm for finding this root.

In a similar way it is possible to draw a flow chart for any problem-solving procedure, as shown in Figure 25.13.

Some teachers start students on flow-charting by asking them to program how they get from bed to school, how they get through the cafeteria line, or a similar sequence of activities. It is easy to build into these activities decisions and even loops that are fundamental to computer activities. Many contemporary textbooks utilize this device as a central feature.

Computer-assisted Instruction

A major role of the computer in the near future may be to "assist" or monitor instruction. It can do this because it has a memory that can store much information. This information may consist of questions, problems, proofs, answers, tests, and facts. Therefore, an entire course similar to a programmed text can be placed in the computer memory. This information can be communicated to the student by terminals, which may include a typewriter for written responses, a screen like a television screen for visual communication, and headphones for sound. The student can communicate with the computer by pressing multiple-choice keys or typewriter keys or by writing on a special screen with a "light" pencil. It is anticipated that in the future a terminal of this type will be available for each student.

The computer can act as a tutor by asking questions or presenting information. It can do this by playing a tape recording through the headphones, by displaying a paragraph or even a photograph on the screen, or by writing with the typewriter. The screen could also present videotapes to accompany the verbal statements. This communication may present an explanation, a series of discovery exercises, a step-by-step development of a proof, a series of practice exercises, or thought-provoking problems.

The student may respond to the computer tutor by answering the questions, working the problems, selecting a multiple-choice response, drawing on the screen, requesting the next step in the proof or explanation, or requesting a repeat of a previous paragraph. The computer lesson is somewhat like that of programmed texts but is usually much more versatile. It can give verbal or pictorial information as requested. It can pronounce a new word or repeat an explanation. At the same time, it records the student's response. If the student responds incorrectly or too slowly, it gives the right answer and remembers to ask this question again as part of another lesson. The computer may also respond to the rapid, correct responder by giving a more sophisticated lesson in the future; the slow responder with many errors is given an easier, slower development in the next lesson.

Thus, the computer presents lessons tailored to the individual student, who is given new material as fast as he can demonstrate proficiency. The student is not bored by repetitious drill on material already mastered. He is not discouraged by new material before he has mastered the previous topic.

Consider, for example, what a computer with a display screen could provide the student in an introductory lesson on slope. The student could be presented with the general equation $y = mx$ and invited to select different values of m (within reasonable limits, say $-5 \leq m \leq 5$). As he chooses different values, the equations and their graphs would be instantly displayed. He would have a chance to observe the regularities that are apparent and would then have an opportunity to generalize in response to subsequent questioning. In this way, the computer can facilitate experimentation and encourage discovery.

The computer can also free the teacher from routine activities such as correcting assignments, reviewing, giving tests, keeping records, giving remedial instruction, selecting individualized assignments. The teacher with a computer-tutor could have time to do creative, professional work. The teacher's role could involve inspiring students, developing attitudes and values, teaching originality and application, and adding the human element in learning. The computer can teach basic facts and skills, but it cannot help the student discuss original ideas, new applications, or new ways of communicating ideas. It can teach basic concepts and provide practice, but it is more limited in developing individual interests, values, and creativity.

The teacher in a computer-tutor classroom could devote his time to these professional activities:

Planning. Selecting or developing new topics, new programs, new sequences, new materials for his classes.

Counseling. With complete information supplied by the computer for every student, the teacher could work with individual students in terms of their needs and abilities.

Evaluating. By analyzing student responses, reacting to student feedback, questioning individual students, the teacher could have a wider basis than before for deciding what to teach and how to increase student achievement.

Teaching. The teacher would still conduct discussions, present new topics, make assignments, encourage independent study, direct enrichment activities. With freedom from routine, he could conduct these activities with finesse, selecting proper materials and tailoring activities to individual needs.

The computer revolution is at hand. The computer is a tool that may have multiple uses in the mathematics classroom. It may be a device for making the learning of concepts, skills, and problem solving more effective than that of traditional practice. The technology is available. The question now is, do we have the teachers and the materials for establishing computer programming in the mathematics classroom?

Computer-assisted instruction can create an almost ideal teaching situation—continuous, direct, individual contact between teacher and student. The implications of such improved instruction speak hopefully for the future.

Learning Exercises

1. Reactions to computers range from extreme enthusiasm to complete rejection with all gradations between represented. Some mathematicians see the computer as a tremendous resource; others think that it has misdirected mathematical priorities and encouraged superficiality. How must you take your own attitude into account in your teaching?

2. It is extremely important for all mathematics teachers to know as much as possible about computers. Propose a program for extending your knowledge.

3. Write flow charts for solutions to these exercises:

(a) $y = ax^2 + bx + c$, for different values of parameters

(b) $y = 1 + \frac{1}{2} + \frac{1}{4} + \frac{1}{8} + \cdots + \frac{1}{n}$ for $n = 1, 2, 3, \ldots, 30$

(c) $y = \begin{vmatrix} a & b \\ c & d \end{vmatrix}$, for different values of parameters

4. Write the computer programs for the flow charts of exercise 3.

5. PROJECT. Investigate availability and cost of calculators, computers, and terminals.

6. PROJECT. Visit a school with a computer teaching program. Describe the program in terms of goals, achievements, and proportion of mathematics students involved. Suggest improvements.

Suggestions for Further Reading

Biddle, John C., "Resource Unit in Computer Programming for Junior High School," *School Science and Mathematics*, 66, 6 (June 1966), 539–550.

Dorn, William S., "Computer-Extended Instruction: An Example," *Mathematics Teacher*, 63, 2 (February 1970), 147–158.

Forsythe, Alexandra, "Mathematics and Computing in High School: A Betrothal," *Mathematics Teacher*, 57, 1 (January 1964), 2–7.

Gibney, Thomas C., and Judith A. Lengel, "Utilizing a Flowchart in Teaching Ninth Grade Mathematics," *School Science and Mathematics*, 68, 4 (April 1968), 292–296.

Harvey, R. B., "Grade Seven and a Computer," *School Science and Mathematics*, 68, 2 (February 1968), 91–94.

Hausner, Melvin, "On an Easy Construction of a Computing Machine," *Mathematics Teacher*, 59, 4 (April 1966), 351–355.

Hoffman, Irwin, and Larry Kanvar, "Polynomial Synthetic Division," *Mathematics Teacher*, 63, 5 (May 1970), 429–431.

Hoffman, Walter, *et al.*, "Computers for School Mathematics," *Mathematics Teacher*, 57, 5 (May 1965), 393–401.

Hughes, Helen S., "Gauss, Computer-Assisted," *Mathematics Teacher*, 64, 2 (February 1971), 155–166.

Indelicato, Brother Arthur, "Evaluation of Polynomials Using a Computer," *School Science and Mathematics*, 65, 9 (December 1965), 768–769.

Kieren, Thomas E., "Quadratic Equations—Computer Style," *Mathematics Teacher*, 62, 4 (April 1969), 305–309.

Krulik, Stephen, "Using Flow Charts with General Mathematics Classes," *Mathematics Teacher*, 64, 4 (April 1971), 311–313.

La Frenz, Dale E., and Thomas E. Kieren, "Computers for All Students: A New Philosophy of Computer Use," *School Science and Mathematics*, 69, 1 (January 1969), 39–41.

McPherson, Ann, and Douglas Cruikshank, "Newton's Computer Program," *School Science and Mathematics*, 69, 3 (March 1969), 191–194.

Schery, Stephen D., "Topics in Numerical Analysis for High School Mathematics," *Mathematics Teacher*, 63, 4 (April 1970), 313–317.

Travers, Kenneth J., "Mathematics Education and the Computer Revolution," *School Science and Mathematics*, 71, 1 (January 1971), 24–34.

Zoet, Charles J., "Computers in Mathematics Education?," in Douglas B. Aichele and Robert E. Reys (eds.), *Readings in Mathematics Education*, Prindle, Weber, and Schmidt, Boston, 1971, pp. 371–376.

Mathematics Department Facilities 26

Teachers can be more effective when they have the setting and the resources that permit a wide variety of activities. For this reason the mathematics department space should include offices, conference rooms, storage space, and laboratories, as well as classrooms—facilities designed to provide the best possible setting for the activities of the students, the teachers, and the department chairman. When the department is planned, facilities should be made adaptable to unforeseen future needs. Thus, for instance, many room separations should be temporary, so that spaces can easily be rearranged for teaching machines, laboratories, special programs for students of different ability levels, team teaching, computers, programmed instruction, flexible scheduling, computer-mediated instruction, small-group and individualized instruction, and teaching of larger groups.

Staff Facilities

The first requirement for a mathematics department is a well-trained, balanced, and dedicated staff. But this is a necessary, not a sufficient, condition for quality teaching. School systems are wasting tax dollars when they fail to provide their teachers with adequate resources to do the job for which they were hired. Since one of the most effective ways to change behavior is to change the environment, the mathematics department facilities can be used to promote new and better pedagogy.

Teachers need quiet, comfortable offices or workrooms where they can think, study, relax, and write. Here the teacher can prepare his lessons. This preparation might include the duplication of worksheets and laboratory guides, the preparation of audio-visual aids, and the building of demonstration models. Other duties outside the classroom include the preparation and grading of examinations, conferences with students, and professional self-improvement. One of the easiest ways to promote superior teaching is to provide the teacher with facilities such as duplicating equipment, typewriter, professional library, construction material, and classroom supplies in a setting that relates directly to mathematics.

A secondary school that has a mathematics department consisting of several teachers should have a department chairman. He should have responsibility for planning the curriculum, supervising instruction, purchasing supplies and textbooks, hiring new teachers, and conducting inservice training. To do this he will need resources such as time, finances, a private office, space for conferences, and secretarial assistance.

At a minimum, a mathematics staff's space allotment should provide a private office for the department chairman, and a larger space for use as office–library–meeting room for the entire mathematics staff. An example of layout is given in Figure 26.1. Next in priority is a resource center that provides library, equipment, kits, games, audio-visual supplies, and other instructional materials and equipment.

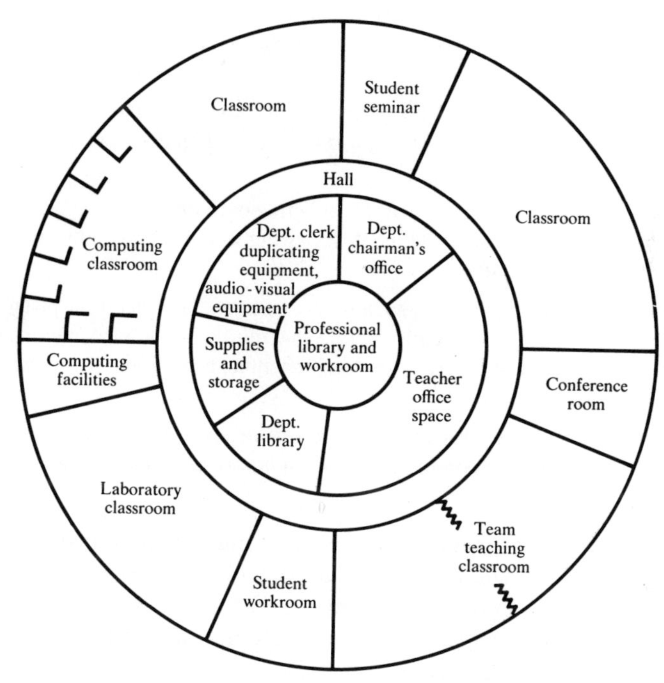

Figure 26.1 ***Plan for a Mathematics Department***

Classroom Facilities

The classroom is the place where the teacher and pupil spend the major portion of their school day, where the tools of learning are available and used, and where the motivation and guidance of learning activities are concentrated. Therefore, the classroom should be planned and equipped so that it is a pleasant, stimulating, comfortable room in which teacher and student will enjoy working.

The classroom must first provide for the physical needs of the pupils: adequate light, proper acoustics, comfortable temperature and ventilation, adequate space, and furniture suited to the activities of the classroom. But this is only part of the story. As in an efficient shop, laboratory, or office, the equipment must be appropriate, adequate, and properly located. The materials and tools needed should be readily available to the teacher, so that he can put them to use with ease and convenience and with a minimum of distraction. Materials are usually used in direct proportion to their accessibility, and perhaps too in direct proportion to their distance from the classroom.

If the mathematics classroom is to meet the needs of modern instructional practice, provision should be made for the following types of activities:

Class learning. Group discussion of ideas, reading, listening, writing on the chalkboard, viewing an audio-visual presentation, correcting papers, preparing an assignment, taking a test.

Laboratory lessons. Performing experiments, collecting data, operating a business enterprise, making measurements, making a model, completing a drawing, making a chart, playing a game.

Small-group activities. Completing a bulletin board display, planning and completing a committee project, reading for enrichment, reviewing lessons, remedial instruction, viewing a film loop, tutoring another student.

Independent study. Library reading and research, completing a programmed unit, preparing a demonstration, constructing a model, having an individual conference, completing a make-up assignment or test, using a teaching machine, completing a project, using a tape cassette or independent study kit.

Computational work. Operating a slide rule or a desk calculator, participating in a practice game, programming a computer, making a nomograph, using a tachistoscope.

To meet the major problems of modern mathematics instruction, a mere rectangular space with blackboards and fixed desks is completely out of date. The modern classroom must have plenty of space that provides for a variety of seating arrangements, adequate instructional materials, and the equipment and facilities necessary for a variety of activities.

The proper size for the mathematics classroom is determined by the number of students per class and the type of activity for which the room is used. Ordinarily 1,000 square feet of space is barely adequate for conventional instruction, but much more space is needed for group work, for library work, and for preparation of displays and models.

An ideal classroom suite includes a large room for group instruction and three small rooms or partially sectioned areas for individual or group work. Folding screens, bookcases, or cabinets on casters can provide a useful and versatile means of forming these areas if it is impossible to have individual rooms. One of these small rooms or sections could be the teacher's work area. It would be the place for the teacher to keep his materials, plan his lessons, and conduct student conferences. Another small room could be used as a workroom, which would include the materials and equipment for building models, making exhibits, drawing charts, preparing projectuals, duplicating material, and completing individual projects. The third small room would be a classroom library, where books, pamphlets, games, films and film loops, slides, kits, periodicals, and clipping files would be available for committee work, remedial instruction, recreational activities, and individual projects. The placement of these rooms between two classrooms would make it possible for two teachers to share them. Windows with draw curtains could allow for privacy or supervision.

Team teaching also places new demands on facilities. In order to function properly in a cooperative program, teachers need classrooms that are easily modifiable to serve very large (up to 150) or very small (a half dozen or fewer) student groups. Such an arrangement as that described in the previous paragraph is suitable for all but the large-group activities. Separate facilities for large groups may be provided on a share basis with another department.

Classroom Equipment

After the classroom space is provided, it should be properly equipped as follows:

Adequate desk, files, and seating for the teacher.

Proper tables (desks) and seating for students.

Equipment for demonstrations, movable demonstration tables, and audio-visual aids.

Facilities for displays, exhibits, and books.

Storage space for class material and students' and teacher's belongings.

Miscellaneous equipment such as pencil sharpener, clock, coat rack, and sink. (Since the clock is for the teacher, it should be placed in the back of the room. Likewise, classroom doors should not be at the front of the room.)

The teacher needs a large desk and a comfortable chair. Some teachers also use a movable demonstration table, which is convenient for transporting books, models, or supplies in and out of the classroom. Most mathematics teachers also use an overhead projector, which should be available at all times.

The movable demonstration table, such as the one shown in Figure 26.2 is an ideal way of providing for accessible teaching materials. Here are the various things a demonstration table may provide:

Space for performing experiments or displaying demonstration models

A projector stand for the overhead projector, film projector, or television set

Storage space for models, texts, chalkboard material, graph paper, games, student supplies, worksheets, tests, laboratory manuals, kits, and the like

Display space for library books and pamphlets

A distribution area for materials for laboratory lessons

Lectern for lesson plan or text

Electrical outlet for projection equipment

Storage for projectuals, filmstrips, films, charts

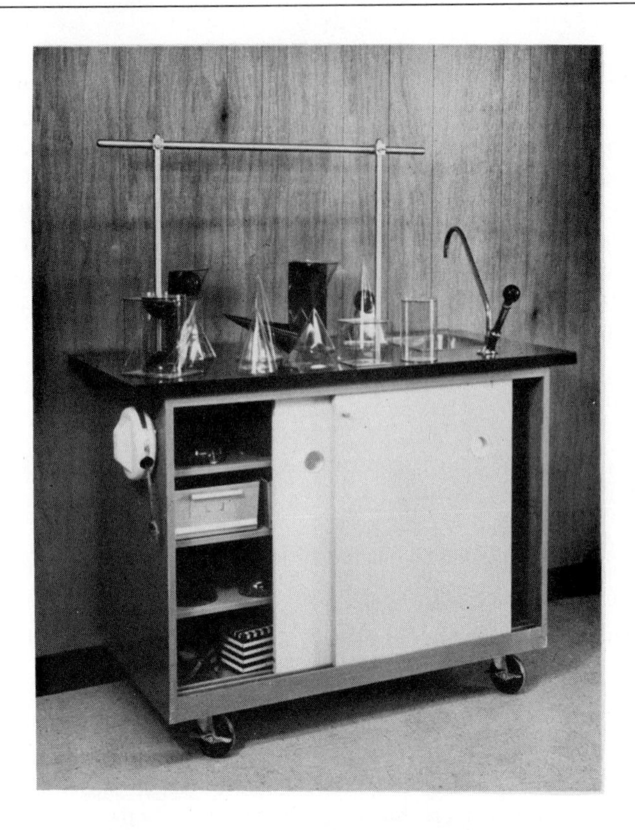

Figure 26.2 Demonstration Table

Student desks in the mathematics classroom should have adequate writing space. Most teachers prefer tables with a formica writing surface to be used as desks. Desks and tables specifically designed for the mathematics classroom have such aids as coordinate graphing grids or tables of trigonometric functions imprinted in the formica. Before selecting these special desks, the mathematics department should decide how they will be used and what they will contribute to the learning of mathematics.

Every mathematics classroom should be equipped with a screen for projections. Overhead projections, films, videotapes, and filmstrips are likely to be as essential as the chalkboard in the near future. Similarly, every mathematics classroom should have walls used for chalkboard, bulletin board, and display equipment. Books, pamphlets, models, and charts should be constantly on display. An exhibit case with a glass-enclosed cabinet that may be locked will permit the exhibit of materials too valuable to be handled by students. This case may be positioned in the wall between the classroom and the hall, so that it is available to both classroom and hall. As much wall space as possible should be utilized for bulletin boards and displays. Efficient use of walls should provide *both* chalkboard and display space; one need not be sacrificed for the other.

Every possible space should be utilized for storage, since the number of student supplies as well as teaching aids will undoubtedly increase in the future. Built-in storage should include spaces of different dimensions for models, charts, books, calculators, construction materials, laboratory equipment, games, and supplies. Some classrooms are designed to provide individual drawers for students to store their books, drawing instruments, and supplies.

It is likely that storage space for department materials will be needed in a separate room. This room should be available for storing material such as the following:

Tools
Construction supplies
Texts
Duplicating equipment
Calculators
Field instruments
Tests
Audio-visual equipment and supplies
Kits
Models and demonstration devices
Games
Chalk and chalkboard instruments
Charts
Measuring instruments
Laboratory equipment
Bulletin board exhibit materials
Drawing equipment

Instructional Materials

Every mathematics department should have the necessary equipment and supplies for the production of instructional materials. This means a typewriter with a keyboard that includes common mathematical symbols, duplicating devices so that written materials for class use can be easily produced, and equipment for the production of projectuals for the overhead projector. In addition, such items as tape recorders, television receivers, and movie projectors—formerly considered luxuries—are fast becoming necessities. Other potential equipment for the near future includes computers or computer terminals, calculating devices, and teaching machines. All this equipment and these supplies must be readily accessible to teachers when they wish to use them. (Specific materials for laboratory work are listed in Chapter 27.)

Special Rooms

Wherever possible, provision should be made for a departmental reading room or library. Sometimes this may be a separate room attached to the school library. This room should contain hundreds of books for enrichment reading and independent study as recommended in Appendix C and reference material for mathematics lessons. This room might be suitable for seminar groups or committee projects. When space is limited, the room could also be used by the mathematics staff for office and meeting space, but it must be accessible to students.

In the future it is likely that mathematics departments will need additional space divided into small cubicles or carrels. In these cubicles, students will be learning through the use of teaching machines. These cubicles might also have terminals connected to a central computer. These terminals may be used to communicate to the computer programs, problems, tests, or lessons. Through these terminals the computer can ask questions, correct answers, work out sample problems, complete a proof, administer tests, and perform other routine activities.

Use of Facilities

In the past, many teachers have been able to excuse inadequate programs because adequate facilities have not been available. Now, however, many new schools are building the needed facilities and providing the money to purchase necessary equipment and supplies. Such facilities are sometimes misused, and therefore several suggestions pertaining to their proper use are appropriate here:

Students and staff should be made aware of all the facilities of the mathematics department. This may be accomplished by means of written notices, but it is better accomplished by a program of visits to special facilities with demonstrations of their use. Even an in-service education course for teachers may be needed to ensure the proper use of the many new instructional aids.

Staff meetings should be devoted to discussion of specific uses of specific equipment. Different teachers may take responsibility for presentations of the use of particular facilities. Opportunity should then be provided for discussion so that all teachers may contribute ideas.

A teacher's ability and willingness to use facilities should be taken into consideration when he is being hired. Sometimes a newcomer may provide special talents to supplement those of the current staff.

School or district audio-visual personnel should be invited to present ideas for using facilities and equipment to the staff. Such presentations should always be followed by discussions focusing on the general uses of such devices for special mathematics problems.

Individual teachers should try techniques that are new to them and report frankly on their success or failure. Such sharing of experiences not only heads off repetitions of poor experiences but also may provide better direction to the experimenter.

Within the framework of cooperative department activity, teachers should be frank about their hesitation to use some of the available supportive material. Such conservative attitudes are helpful in determining the worth of the new material. No facility, no equipment, should be used unless it *improves* the program for students or teachers in some way.

Scheduling

The day of the fixed time schedule is becoming as much a part of the past as the traditional curriculum of the last century. Mathematics teachers must assess their teaching programs in order to determine schedules that make optimum use of facilities and best serve the students.

Many schools have adopted flexible scheduling that provides class periods of varying length: longer periods to allow for lessons that require extended activities, shorter periods to respond to the reduced attention span that is characteristic of other activities. Often the data gathering and analysis of laboratory lessons call for longer time periods, while classes devoted largely to single tasks like review of homework may be scheduled for less time. Breaking down the fixed structure allows a variety of patterns and often provides the opportunity of changing the pattern during the school year on the basis of experience with it.

Related to variable scheduling is the concept of mini-courses, short courses varying in length from a few days to a semester or more. These courses provide an opportunity for teachers to teach content of special interest to them, and for students to elect a wider variety of mathematics courses. Some examples of topics for short courses are: continued fractions, computer programming, transformation geometry, symmetry and design, limits, special curves, game theory and linear programming, problem solving, statistics. In each case, the intent of the course is to introduce students to new content and to involve them in the development and organization of that content—in other words, to give them an opportunity to participate in the activities of a mathematician.

Some schools set aside a month of the school year—often January—for special courses like this in all subject areas. This, of course, restricts mini-courses to a definite time period. More often mini-courses are offered concurrently with the regular program following the pattern long established by English departments. The full year's schedule of courses is announced in September so that students can incorporate desired topics in their program.

Team Teaching

Another aspect of teaching closely related to facilities is team teaching. To form a teaching team, two or more teachers combine their talents and resources to develop a joint instructional program. For example, all algebra students could be assigned to a given time period and the teachers of these students could then work together to teach the entire group. If the facilities are available, the larger group may then be taught in a variety of ways. At one time, all students might meet as a group for a demonstration or an examination. At other times, the larger group could be broken down into smaller groups of a variety of forms.

Some of the advantages of team teaching are:

Cooperative planning and materials development are encouraged.

The grouping freedom allows opportunities to identify and serve top students as well as students needing remedial assistance without freezing them into these categories.

Much teacher time is saved through elimination of duplication of effort. One teacher instead of all can monitor an examination or present a demonstration. This allows more preparation time for other team members.

The interaction between teachers is a beneficial experience and provides an intensive in-service training program for all participants.

Despite these and other strong advantages of team teaching, teachers should embark upon such a program only with careful preparation and intensive pre-planning. Teachers who participate in successful teaching teams say that they must work much harder in this pattern than they did in the self-contained classroom. And there are unsuccessful teaching teams that merely waste teacher talent or student time. There is no justification whatsoever, for example, for a team-teaching program that utilizes only large group instruction.

In planning creatively for the future, we should eliminate preconceived ideas about the shape, size, facilities, and scheduling for the mathematics classroom. We should consider the elimination of stereotyped furniture, the fixed time schedule, and the traditional curriculum. The acceptance of daily classes, constant time for each class and each student, and the same equipment for each classroom is no longer defensible.

Learning Exercises

1. Many schools today employ teacher aides (in educational jargon, paraprofessionals), adults whose role is to assist classroom teachers. Because they are noncertified personnel, these aides are not allowed to perform direct instruction. What are some of the tasks that such a person could do to assist you as a teacher?
2. Most secondary-school teachers today have a "free period" as part of their schedule. This is their only unscheduled time during regular school hours. Indicate three or more activities related to their instructional program that teachers can perform during this period.
3. Here is an opportunity you may never have again. You have unlimited funds. The sky's the limit. Design and stock a classroom for 25 students. Include all the equipment you want, but be sure you leave room for your students.
4. Explain and justify your attitude toward:
   (a) Students working at the chalkboard
   (b) Grouping in the classroom
   (c) Team teaching

Suggestions for Further Reading

Archer, Allene, *How to Use Your Library in Mathematics*, National Council of Teachers of Mathematics, Washington, D. C., 1958.

Bartnick, L. P., *A Design for the Mathematics Classroom*, National Council of Teachers of Mathematics, Washington, D. C., 1957.

Frame, J. Sutherland, "Facilities for Secondary School Mathematics," *Mathematics Teacher*, 57, 6 (October 1964), 379–391.

Johnson, Donovan A., "Why Use Instructional Materials in the Mathematics Classroom?," in Douglas B. Aichele and Robert E. Reys (eds.), *Readings in Secondary School Mathematics*, Prindle, Weber, and Schmidt, Boston, 1971, pp. 349–351.

The Mathematical Laboratory 27

Teaching is an invigorating adventure, especially when our students get excited about learning mathematics. This is why teachers are enthusiastic about laboratory lessons for mathematics students. The informality, the opportunity for independent exploration, and the realism of laboratory experiments appeal to students of all levels of ability. This is the way young children enjoy learning before their curiosity is killed by formal school experiences. We must try to get back to this natural interest in learning about the world around us. It may be that mathematics laboratory lessons hold the key to our students' minds. The inquiry approach helps the student to help himself. It gives him experiences in thinking for himself, in discovering ideas, and in communicating his discoveries. This shifts the emphasis from teaching to learning, from prescribing and proscribing to helping, and from authority to inspiration. It is putting into practice the maxim: "Each student is a light to be turned on rather than a boat to be filled until it sinks."

Instead of expecting students to store and retrieve information, the teacher should direct the students to build and test theories for themselves by the use of open-ended questions and experiments. Students who play it safe by waiting for the teacher to hand out knowledge become dependent learners. Even though a student may be naturally inquisitive, he does not want to run the risk of a rejected wrong response. The spirit of laboratory lessons includes experimentation and guessing, and a student will often learn by his failures. Such a try will create a favorable classroom atmosphere, in which curiosity and independence grow.

What is a Mathematics Laboratory?

The mathematics laboratory means different things to different teachers. To some it is a computer center where students use calculators or computer terminals to solve problems. To others it is a room that is a resource center—a room for special activities such as tutoring students, working on projects, playing games, or writing make-up tests. Some schools have a special room called the mathematics laboratory to which classes are brought for special enrichment lessons, but for some it is the normal classroom in which they teach their classes every day. For this book, the mathematics laboratory means instead a way of teaching mathematics. It is discovery teaching of topics of the regular curriculum. It is essentially a systems-development lesson for any topic. This means it is a lesson that uses the best resources available to provide the student with experiences that he needs in order to learn the lesson at hand. It may involve performing an experiment, viewing a film, playing a game, discussing, reading, programming a computer, building a model, solving a problem, making a survey, drawing a design, making a graph, finding applications, presenting a mathematical skit, completing a test, proving a theorem, or just quiet thinking.

Of course these things can best be done in a room that has the space, equipment, and resources for this variety of activities. In general, laboratory lessons require more space so that groups of students or individuals can work on different activities. It is important to have storage space and the normal classroom facilities described in Chapter 26. Carpeting, accoustical tile, and draperies can help keep noise level down. Sometimes background music is appropriate to keep noises from distracting students. Projection screens, chalkboards, bulletin boards, exhibit cases are needed as in a regular classroom. Study carrels, room dividers, or even dividers on tables improve conditions for independent work. In any case, tables rather than fixed desks are highly desirable.

The Unique Role of Laboratory Lessons

Laboratory lessons give a new approach to learning mathematics and hence are especially successful in many ways.

They provide *success* for the learner who has difficulty learning the formal, abstract treatment of the typical textbook. The laboratory lesson provides a non-threatening, realistic, concrete approach to learning. In this way, laboratory lessons are particularly helpful in dealing with individual differences, in reaching the culturally deprived, and in preventing dropouts.

The possibilities for *individual, independent work* make laboratory lessons especially appealing and profitable for the creative, talented student. Each student has responsibility for his own learning at his own rate, and with his individual learning style. He uses his own ingenuity in planning activities to find the answer to an inquiry.

Similarly, opportunities for working in *small groups* encourage leadership, shared responsibility, and team effort.

As a result of the related independence of laboratory lessons, *positive attitudes* toward mathematics and the mathematics teacher are likely to be a major outcome. The only way to break down hostility toward mathematics is to provide pleasant, enjoyable experiences. These lessons reduce futility and bring joy back to the classroom.

The completion of an experiment results in an individual product or an independent discovery that gives tangible *evidence of progress* to the student. This individual production is much more satisfying than the completion of a textbook assignment. Although his generalizations may lack the precision of textbook statements, they represent his understanding. We accept immature statements until more experiences build greater maturity.

The similarity of laboratory work to that of the science class and to the mathematics embedded in everyday events insures greater *transfer of learning* than the usual classroom procedures. Laboratory lessons often use common materials such as string, rulers, or scales to explore phenomena such as motion, force, weight, location, speed, time, and direction. Thus laboratory lessons have that essential ingredient, relevance.

The laboratory lesson is most effective in getting every student to *participate actively*. Each student must do some thinking as he collects data, plays a game, or conducts an experiment. The student's active involvement is the key to his successful learning.

Kinds of Laboratory Lessons

Mathematics is a study of patterns, a language for expressing relationships, and a tool for dealing with approximate data. Consequently, mathematics is uniquely suited for exploring ideas in a laboratory setting. These laboratory lessons may be of several types.

Make measurements of two variables in an everyday setting to determine the relationship between them. A typical example of this would be the comparison of measures of the circumference and diameter of round objects.

Measure the circumference and diameters of many round objects (wheels, wastebaskets, tin cans, clock dials, lamp shades). Tabulate the data and compute the sum, difference, product, and quotient of each pair of measures, as shown in Table 27.1. From the pattern of the data obtained in this table, it appears that $C \div d = 3.1$. From this consistent result, it is estimated that $C = 3.1\,d$, or $C = \pi\,d$.

Table 27.1

Object	Circumference (C)	Diameter (d)	$C + d$	$C - d$	$C \times d$	$C \div d$
Wheel	7	21	28	14	147	3.0
Basket	46.5	15	61.5	31.5	697.5	3.1
Can	6.2	2	8.2	4.2	12.4	3.1
Clock	10.5	3.4	13.9	7.1	35.7	3.1
Lamp	36.8	11.5	48.3	25.3	423.2	3.2

Note that the student is required to compare C and d by all four operations of arithmetic rather than by division alone. His results lead him to the generalization, but do not focus only on that operation.

Another linear relationship is that between the stretch of a rubber band and the weights attached to it. Attach a rubber band or spring to a hook above the chalkboard. Attach weights. Record the stretch for each weight added. Draw a graph of the data. What equation of the form $y = mx + b$ best fits your graph?

Similar experiments can be performed with levers, bouncing balls, and falling objects.

Find a pattern by drawings or constructions. An example of this type of experimentation is the exploration of the formula for the maximum number of parts into which the interior of a circle can be divided by a given number of lines in the plane. The results are, shown in Table 27.2. Since second differences are constant, the formula is quadratic, ($r = n^2/2 + n/2 + 1$.)

Table 27.2

Number of lines (n)	**Drawing**	**Maximum number of regions (r)**	**Difference in number of regions**
0		1	
			1
1		2	
			2
2		4	
			3
3		7	
			4
4		11	
			5
5		16	

Another experiment can be done by marking points on the circumference of a circle. Collect data for the number of chords determined by these points.

Similar experiments can be done to discover the Pythagorean theorem, relations on networks on Möbius strips, the sum of the angles of a polygon, the polygon formed by connecting the midpoints of the sides of a quadrilateral, and many others.

Discover a scientific principle by performing an experiment with simple equipment. A simple example of this type is the determination of the geometric series that represents the bounce of a "super" ball. Form a cylindrical tube from a large sheet of clear plastic. Place this tube upright on a hard surface. Drop a ball from the top of the tube and record the height of each bounce. Repeat the experiment many times to get the average height of each bounce. The collected data will give an approximation to a geometric series.

Another experiment can explore the results when pulleys are used to lift objects. An interesting formula relates the distance the object is raised as compared to the weight raised and the force required to raise it.

Similar experiments can explore specific gravity, coefficient of friction, and the centrifugal force of a rotating object.

Explore a relationship by manipulating simple objects. A simple experiment to determine the permutations of different numbers of objects can be performed by arranging a series of cards. The results can be related to factorials ($n!$) to obtain a formula ${}_nP_n = n!$

This leads to other experiments in sampling. Draw random samples of marbles or cards. Predict the makeup of the population. Check your results by recording the population or reverse the process—given the population, predict the results of samples. Then draw samples of different size to compare results.

Other common objects like pencils, paper, tin cans, coins, and string can be used to illustrate mathematical ideas.

Collect original data from a survey to determine what variables are related. Data from students may suggest that there is a relationship between grade in mathematics and number of hours of homework, number of hours spent viewing television, or car ownership.

One of the major ways of exploring scientific law is to find cause and effect. Often simultaneous events are confused with cause and effect. Surveys of data on accidents, costs, absences, traffic, or wages can be used to distinguish among these relationships.

Collect data by performing an arithmetic operation. Explore the relationship between triangular numbers and square numbers. Find a shortcut for squaring numbers ending in 5.

Table 27.3

n	Factors	Number of factors	Sum of factors
1	1	1	1
2	1, 2	2	3
3	1, 3	2	4
4	1, 2, 4	3	7
5	1, 5	2	6
6	1, 2, 3, 6	4	12
7	1, 7	2	8
8	1, 2, 4, 8	4	15

A typical example of this type is the pattern of factors of counting numbers (see Table 27.3). To determine the pattern of these data, it is helpful to ask questions such as these:

What numbers have exactly two factors? (Primes.)

What numbers have exactly three factors? (Squares of primes.)

What numbers have exactly four factors? (Cubes of primes and products of prime pairs.)

How does the number of factors of a number compare with the number of its prime factors? (Less than or equal to the sum of the relatively prime factors.)

How does the sum of factors of prime numbers compare with the prime number? (One more than the prime.)

What formula gives the sum of factors of 2^x? 3^x? 2^x3^x?

Perform indirect measurements to determine unknown distances. Figure 27.1 illustrates a simple experiment of this type, the determination of the height of a tree by measuring its shadow and comparing this with the shadow of a known height.

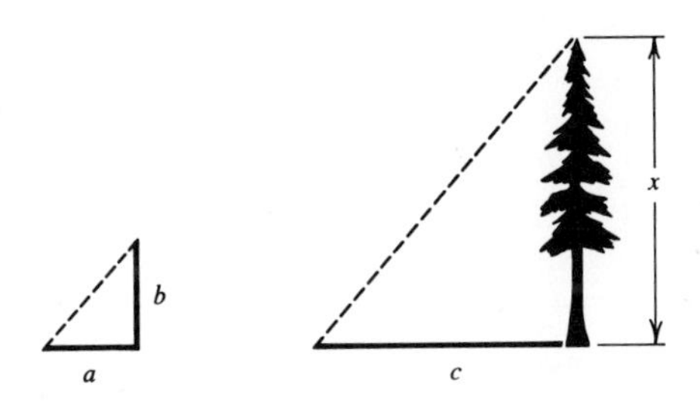

Figure 27.1 $a/b = x/c$

Extend the solution of a specific problem to a generalization. Consider the problem of a drawer filled with single brown and blue socks. If one selects individual socks in the dark, how many selections are needed to guarantee two socks of one color? The easy answer is three. To generalize this problem, we can increase the number of colors and increase the number of feet needing matching socks. The analysis can be done with objects representing socks or by logically analyzing each situation. The results may be recorded as in Table 27.4. The guaranteed selections are $S = C(F - 1) + 1$, where $C(F - 1)$ represents the possible tie results.

Represent a mathematical idea by building a simple model. For example, it is impossible to represent the fourth dimension by a two-dimensional model. Using balsa strips, pins, and glue, students can build a tesseract or superprism in one class period. This model then becomes a basis for dimensional analysis and the meaning of dimensions beyond three. Tabulation of data as in Table 27.5 adds meaning.

Other simple models for laboratory construction include nomographs, slide rules, abaci, linkages, conic sections, proportional dividers, alidade, curve stitching, dial calculators, and Trigtrackers.

Table 27.4

Number of feet (F)	Available colors (C)	Number of selections that may be tie result	Number of selections that guarantee matched socks (S)
1	1	—	1
1	2	—	1
1	3	—	1
2	1	—	2
2	2	2	3
2	3	3	4
3	1	—	3
3	2	4	5
3	3	6	7
4	1	—	4
4	2	6	7
4	3	9	10
5	1	—	5
5	2	8	9
5	3	12	13

Table 27.5

Figure	No. dimensions	No. boundaries	Dimensions of boundaries	Coordinates needed to locate a point on figure	Distance formula
Point	0	0	0	0	
Line segment	1	2	0	1	$d = \sqrt{(x_1 - x_2)^2}$
Region of plane	2	4	1	2	$\sqrt{(x_1 - x_2)^2 + (y_1 - y_2)^2}$
Bounded three-space	3	6	2	3	$\sqrt{\Delta x^2 + \Delta y^2 + \Delta z^2}$
Bounded four-space	4	8	3	4	$\sqrt{\Delta x^2 + \Delta y^2 + \Delta z^2 + \Delta w^2}$

Complete an individual project or report. When time and construction materials are available, students can create original devices to demonstrate mathematical ideas. For example, a student may use balsa strips to build a three-dimensional graph of data consisting of three variables.

Perhaps he would be interested in growing crystals that illustrate polyhedra or building wire frames for soap bubbles that illustrate mathematical surfaces. Logic can be explored by building models such as electric circuits or a logic computer. Other areas of interest include space travel, topology, non-Euclidean geometry, and topics as listed in Appendix D.

Illustrate mathematical ideas by folding paper. For example, flexagons are fascinating devices that mathematicians have investigated as a simple, new branch of mathematics. A hexaflexagon can be folded and glued together in a few minutes. The balance of the period can be devoted to the story of its development, a study of the properties, and the extension to multisided flexagons.

Other paper folds can be used to illustrate all the constructions of Euclidean geometry. Folding bisectors of angles of triangles, the altitudes of triangles, the perpendicular bisectors of the sides of a triangle, or medians to the sides of a triangle show the concurrency of these lines. The envelopes of conics can be illustrated by folding wax paper. For instruction on paper folding read the pamphlet *Paper Folding for the Mathematics Class.*[1]

All these activities have a mathematical aspect. Performing them should have the objective of learning a mathematical idea. For example, prove that the following folds result in a 30-60-90–degree triangle.

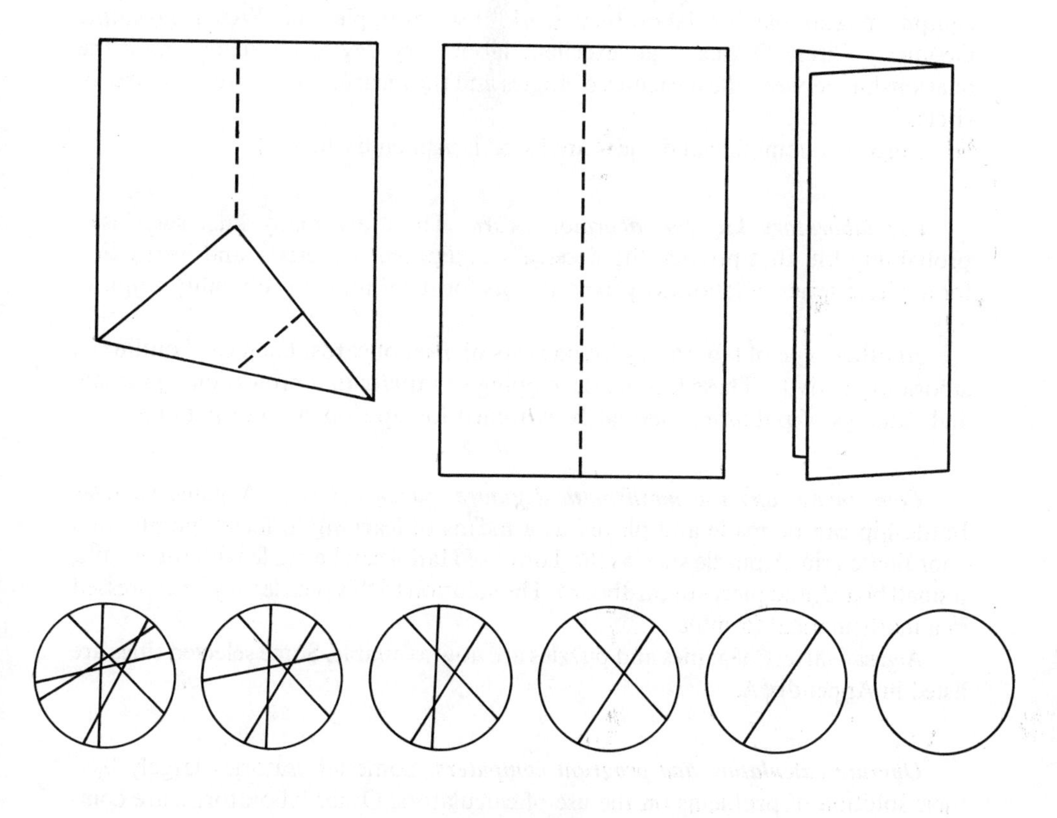

Figure 27.2

[1] National Council of Teachers of Mathematics, Washington, D. C.

Fold a rectangular sheet into two rectangles along a midline.

Fold one corner of the rectangle on this middle line so that the crease passes through the adjacent corner of the rectangle (see Figure 27.2).

The triangle formed is a 30-60–degree right triangle. Why?

Build exhibits, charts, bulletin board displays. An excellent way to build communication skill is the construction of a display. This construction requires concrete and visual representation that should add new meaning to the idea. For example, one group of students may use Tinker-Toy pieces to illustrate the meaning of algebraic expressions such as x, $3x$, x^2, $x + y$, $(x + y)^2$, and x^3.

Examine the publication *Bulletin Board Displays for Mathematics*,[2] for ideas on how to build bulletin board exhibits. Suggested titles are given in Appendix A.

Use commercial laboratory devices to perform experiments. There is a variety of equipment suitable for laboratory work. For example, the Welch Dynamic Geometry Circle Device is an excellent laboratory device for determining the relationship between the measures of angles and the measures of intercepted arcs of circles.

Suggested commercial devices are listed in Appendix A.

Use laboratory kits for laboratory work. There are many kits, such as a probability kit, that provide the necessary equipment materials, and instruction for a whole series of laboratory experiments for determining probability empirically.

Another type of laboratory kit consists of a set of cards. Each card outlines a laboratory activity. These kits are developing so rapidly that professional journals and catalogs of publishers should be searched for up-to-date information.

Plan, make, and use mathematical games, puzzles, stunts. A game such as Battleship can be made and played as a means of learning to locate points on a coordinate grid. A puzzle such as the Tower of Hanoi can be made with three nails, a small board, and pieces of cardboard. The solution to this puzzle may be expressed in a mathematical formula.

A great variety of games and puzzles are now available. Some selected titles are listed in Appendix A.

Operate calculators and program computers. Some laboratories largely base their solution of problems on the use of calculators. Other laboratories are com-

[2] Donovan A. Johnson, and Charles H. Lund, *Bulletin Board Displays for Mathematics*, Dickenson Publishing Co., Inc., Encino, Cal., 1967.

puter oriented, with a desk-top computer or a computer terminal. These computer devices are used to learn programming and to find solutions to programmed problems. See Chapter 25 for further details.

Organize and operate a business enterprise. The establishment of an insurance company that insures students for specific items such as breakage, textbook loss, illness, or accident can be a realistic way of teaching probability and insurance. A stock brokerage firm, bank, credit union, and a school store are other examples of possible business ventures to operate as learning experience.

Use an audio-visual aid as the basis for laboratory work. Students may draw or take photographs for a series of slides which can be used as a means of learning applications. A series of overlays for the overhead projector may be prepared to demonstrate a topic or problem. Other students may take movies and write a script to present a mathematical idea.

Some laboratory lessons will involve a combination of the types outlined above. At other times part of a lesson will be a laboratory session and part a practice session. Not all lessons are suitable for laboratory sessions. But too often laboratory lessons are not used because the teacher lacks the ideas, the time, or the material to present them. With the variety of kits of laboratory activities and activity-oriented texts now available, every teacher can use laboratory lessons to supplement text lessons.

How to Teach Laboratory Lessons

A laboratory lesson, like any other lesson, requires careful planning to be successful. However, laboratory lessons tend to be more difficult than textbook lessons for several reasons. For one thing, ready-made material for these lessons may not be available. Thus, most laboratory-lesson material must be developed by the teacher. Also, laboratory lessons require the use of a variety of materials, which need to be scheduled and prepared well in advance. In addition, the laboratory situation is naturally more permissive and unstructured and thus needs to be well organized to avoid undesirable disruptions. Even so, teachers who use this method find that the results are well worth the effort.

Some teachers prepare for laboratory lessons by making up a laboratory kit. This kit contains the guide sheets for students, the materials needed by the students, and the necessary teacher instructions. Some kits include an inventory of materials, the objectives or content to be developed, and perhaps an evaluation of the lesson as previously experienced. Sometimes kits include a test as a follow-up of the lesson. Consider, for example, a kit on topology. It includes adding machine tape to be used to make Möbius strips. It includes string for a student demonstration, a pattern for a topological puzzle like the boot puzzle, and a guide sheet that gives instructions and questions on the Möbius strip. This guide sheet may also include exercises on networks.

In conducting a laboratory lesson, the teacher acts as a guide and supervisor. He initiates the lesson with instructions and provides students with the material they need. He establishes groups to work together and assigns space for the activities. He coordinates the work of individuals and answers questions that arise. He gives time signals so that housekeeping details are taken care of before the period ends. He assigns responsibilities to students for distributing, collecting, and storing materials so that all is in order for the next session. Even though the atmosphere is permissive, he remains the adult in authority who makes sure that behavior is appropriate and equipment is properly cared for. However, his role is more of a counselor and helper than referee.

Specific Procedures for Laboratory Lessons

Here are some specific procedures to follow in order to have successful laboratory lessons:

Prepare guide sheets for the students so they will know what material they need and what they are to investigate. These guide sheets must not overstructure the lesson and leave no decisions for the student to make. At the same time, laboratory lessons should usually be completed during a regular class session; therefore, considerable efficiency will be required of both student and teacher. As in discovery teaching, laboratory guide sheets should include a series of carefully prepared questions. These questions should suggest possible variations, common properties, or patterns. Also, the questions should direct attention to interrelationships, analogies, or special cases. Replication is encouraged so that no counter example is possible. Note how the following pattern of arithmetic computation might lead one to a false conclusion.

$$\begin{array}{ll} 2 \times 2 = 4 & 2 + 2 = 4 \\ \frac{3}{2} \times 3 = 4\frac{1}{2} & \frac{3}{2} + 3 = 4\frac{1}{2} \\ \frac{4}{3} \times 4 = 5\frac{1}{3} & \frac{4}{3} + 4 = 5\frac{1}{3} \\ \frac{5}{4} \times 5 = 6\frac{1}{4} & \frac{5}{4} + 5 = 6\frac{1}{4} \end{array}$$

The generalization that $a \times b = a + b$ can be disproved by the simple example $3 \times 3 \neq 3 + 3$.

The specific conditions under which this works are shown by these equations.

$$\left(\frac{x}{x-1}\right)x = \frac{x^2}{x-1} \qquad \frac{x}{x-1} + x = \frac{x + x^2 - x}{x-1} = \frac{x^2}{x-1}$$

The sample guide sheets below are suggestive of possible format and content.

LABORATORY EXERCISE: LAW OF THE LEVER

Inquiry:

How are the weights and lengths from the fulcrum related when a lever is balanced?

Objectives:

(a) To collect data on the weights and lengths of a lever that is balanced.
(b) To investigate whether there is a pattern in the data that can be expressed by a simple formula.

Equipment:

Meter stick, fulcrum, weights, thread.

Directions:

(a) Suspend by a thread or support with a fulcrum a meter stick at its midpoint. If it does not quite balance, place a rider made of a bent piece of foil on the lighter side to establish equilibrium.
(b) At some point on the meter stick attach a weight (W) at any distance (d) to the left of the fulcrum.
(c) At some point on the other side of the fulcrum, hang another weight (R) that balances W. Measure the distance (a) to the fulcrum. It is not necessary to wait for the lever to come to rest, for it is in equilibrium when it swings through equal distances on opposite sides of the horizontal position.
(d) Repeat at least four times, using different sets of weights and distances.
(e) Record both the weights (W and R) and their distances (d and a) from the fulcrum when the lever is balanced.

Diagram:

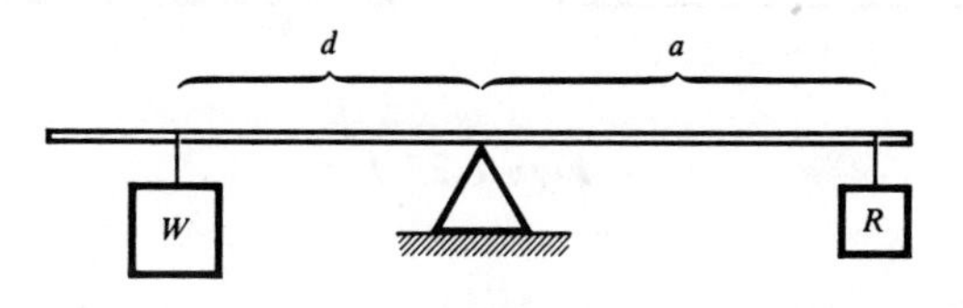

Figure 27.3

Table:

Trial	W	R	WR	W/R	d	a	da	d/a	a/d
1									
2									
3									
4									

Results:

(a) What products or ratios above appear to be related?
(b) What formula expresses this relationship?
(c) What are some applications of this principle?
(d) Was Archimedes correct when he said, "If I had a long enough lever and a fixed point for a fulcrum, I could move the earth"? Explain your answer.

Further exploration:

How can a lever be used to find the mean of a set of data.

LABORATORY EXERCISE: THE AREA OF A TRIANGLE

Inquiry:

What formula expresses the area of a triangle?

Use a rectangle as shown in Figure 27.4a and move the small triangle to the upper base of the rectangle to form a right triangle as shown in Figure 27.4b.

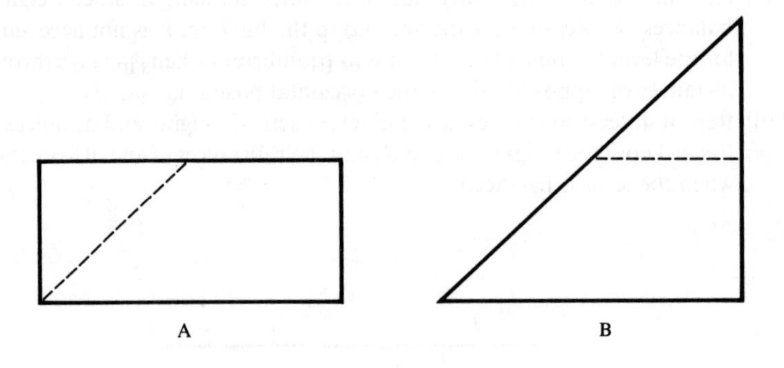

Figure 27.4

What is the area of this triangle? How does the measure of the base of the triangle compare with the length of the rectangle?

How does the measure of the altitude of the triangle compare with the width of the rectangle?

Is the area of the interior of a triangle equal to the product of the base and altitude?

What number sentence expresses the relationship of the area of a triangle to the measures of the base and altitude?

(Questions such as these should lead the pupil to the conclusion: Area equals $\frac{1}{2}$ base times altitude, or base times altitude/2.)

Further exploration:

How can the formula for the area of a trapezoid be determined from this experiment?

LABORATORY EXERCISE

Inquiry:

What is the product of two negative numbers?

Here is an example of how a number pattern can be developed to discover the product of two negative numbers. Study the pattern of products below:

$$\begin{aligned} {}^{+}12 \times {}^{+}5 &= {}^{+}60 \\ {}^{+}12 \times {}^{+}3 &= {}^{+}36 \\ {}^{+}12 \times {}^{+}2 &= {}^{+}24 \\ {}^{+}12 \times {}^{+}1 &= {}^{+}12 \\ {}^{+}12 \times 0 &= 0 \\ {}^{+}12 \times {}^{-}1 &= \; ? \end{aligned}$$

How is the change in the product related to the change in one factor? Why is it logical to expect that $12 \times {}^{-}1$ is a product less than zero?

If $12 \times {}^{-}1 = {}^{-}12$, what is $12 \times {}^{-}2$ equal to?

The results above suggest that the product of a positive number and a negative number is a ________ number.

The laboratory lesson needs adequate materials. For some lessons only paper and pencil are needed. For others, measuring instruments, graph paper, or a laboratory device is needed. These materials must be ready for use when the laboratory lesson begins.

The laboratory lesson needs a classroom with some flexibility and certain special equipment. Student stations should be tables instead of desks. Storage cabinets are needed to keep equipment at hand. Filing cabinets and book shelves are needed to store resource materials. Electrical outlets should be plentiful, and a sink with running water should be provided. At least one table should be a workbench with a vise and other tools for simple woodworking.

Each student should participate in an activity in which he can have some success. Ideally, the mathematics laboratory needs laboratory assistants as in a science laboratory. These assistants may be auxiliary workers such as teacher aides or student assistants. In any case, students in the laboratory need considerable individual attention if each one is to complete the experiments correctly and arrive at generalizations. Be sure to allow ample time for the completion of the experiment.

The students must be properly prepared for the laboratory lesson. They must know their responsibility for independent work and for the care of equipment. Sharing equipment, cleaning up debris, and being orderly must be the accepted responsibility of each student. Provide supplementary activities for the learner who completes his work far in advance of others. Sometimes he may also use this extra time to act as a laboratory assistant. At other times, he extends the generalizations to a new situation as illustrated by the laboratory lesson on levers above.

Specific Laboratory Lessons

The number of laboratory lessons appropriate for learning mathematics is limited only by the ingenuity of the teacher and the facilities available. Most concepts can be discovered by experimentation if time is taken to plan and devise appropriate experiments. The suggested lessons that follow indicate the wide range of possibilities. In each of the following cases, the reader must extend the idea in order to make a complete laboratory lesson. This extension will often involve a search for related literature which will offer further suggestions and a more complete discussion.

Numbers and Calculations

Build nomographs for computations, such as addition, subtraction, multiplication, division, squaring, square root, the Pythagorean formula $a^2 + b^2 = c^2$, and roots of quadratic equations.

Make models of Napier's bones out of cardboard. Make the bones for different bases to illustrate or check computations with numerals in bases other than 10.

Using a sheet of graph paper and a string, make a percentage computer, as described in Chapter 16.

Using two squares of cardboard on nails pounded into a board, make a binary abacus. A string of lights, such as those used as Christmas tree decorations, can also be used to represent the 1 and 0 of the binary system.

Discover the patterns of multiples of numbers by writing the multiples in a vertical column. What is the pattern of the sums of the digits? What is the pattern of the units digits? Of the tens digit?

Write a 10 by 10 table of numbers from 1 to 100, listing them consecutively left to right in rows. Examine the patterns in this table of multiples and primes. Cardboard templates uncovering certain multiples may be cut out to show various patterns.

Use dots to represent the values of counting numbers. What numbers are triangular numbers? Square numbers? How are triangular numbers related to square numbers? What numbers are pentagonal numbers? How do triangular numbers and square numbers combine to form pentagonal numbers?

Start Pascal's triangle. What number series are formed by numbers found in different rows, columns, and diagonals?

Learn to construct magic squares.

Explore counting-number games such as Nim with rules of increasing complexity and generality.

Discover shortcuts and explain why they are appropriate.

Investigate the pattern of Pythagorean triples and derive the formulas that express the relationship discovered.

Investigate the shapes and surfaces of soap bubbles.

Explore the mathematics of codes and ciphers.

Statistical Surveys

Collect data such as the following for graphs, statistical analysis, or the study of cause and effect:

Scoring records or athletes' percentages at athletic contests

School records, such as attendance, tardiness, registrations, illness

School sales in the lunchroom, bookstore, or candy counter

Daily community vital statistics, such as births, deaths, fires, accidents

Student activities, such as playing games, seeing shows, watching television, listening to music, and reading books

Marks and test grades in different courses

Market quotations of stocks, bonds, grain, cattle

Utility bills such as electric, gas, telephone, water

Sizes, weights, heights of students

Tax rates, school costs, public debt, business conditions

Traffic records, such as accidents, amounts of traffic, cost of automobile operation, kinds of automobiles

Probability Experiments

Toss coins to compare the predicted results with experimental results for equally likely events.

Toss thumbtacks, corks, bottle caps to determine the probability of falling in various positions.

Make a spin dial. Check the probability of the arrow's stopping within a given section of the dial.

Roll a hexagonal pencil or toss a die to find the probability of a given side turning up. Roll two pencils or toss two dice to build the sample space of events and find the probability of combinations of events. Shave an edge of the pencil to check its effect on events.

Make duodecagon dice from cardboard. Determine the probability of certain tosses of one or two of these polyhedra.

Draw cards at random from a deck of homemade number cards or playing cards. Determine the probability of drawing certain cards.

Toss different numbers of coins. Compare the events with the numbers in Pascal's triangle.

Draw samples from bags of colored marbles or decks of cards. Predict the composition of the collection. Compare results obtained from different-sized samples.

Make a probability game in which players capture points by spinning a dial or by predicting events.

Collect data on the frequency of letters in a paragraph or the frequency of digits in telephone numbers. Predict the frequency of all paragraphs and check the results.

Compute a value for π by tossing sticks (or toothpicks) on parallel lines.

Collect data and compare results with the normal curve. Experiment with a probability board and compare results with Pascal's triangle and the normal curve.

Keep a record of weather predictions for several days. What is the probability that the weather forecast is correct?

Find the batting average of a certain baseball player, the average yards gained by a certain football player, or the average points scored per game by a certain basketball player. Use these results to predict events of the next game.

Record the kinds of cars either parked in a parking lot or passing over a busy street in a 10-minute time period. Use these results to approximate the number of Fords, Chevrolets, Plymouths, or Volkswagens in the United States. Compare results with data in the *World Almanac*. What are some reasons why the suggested sampling may not be a proper guide?

Glue two dice together. Toss and collect data to illustrate dependent events.

Construct a probability game based on tossing a thumbtack. Each player gets 20 points if he predicts correctly whether the tack will land with its head up or down. To estimate the probability of the tack turning up or down, each player can have as many sample tosses as desired. However, one point is lost for each sample toss. How many sample tosses should a player take before he is willing to predict an event for 20 points? The answer depends on the results of the sample tosses. If the tack lands head down the first six times, he may be willing to predict "down." If the sample tosses give four down and two up, he may want more sample tosses before he predicts for 20 points.

Ratio and Proportion

Determine the ratios of speeds and revolutions for gears and pulley trains. Use a bicycle to make this determination.

Use a meter stick lever to determine the law of the lever.

Draw geometric figures on graph paper. Determine the ratios of areas as dimensions are varied. Compare perimeters and areas.

Compare the volumes of different cylinders, cones, or prisms. Determine the ratios of volumes as dimensions are varied.

Measure the variation in the period of a pendulum as the length of the pendulum is changed, to determine the relationship between these variables.

By measurement determine the ratios between the parts of an automobile in a picture, and the ratio that occurs when the automobile itself is measured.

Draw rectangles that represent the golden section. Investigate its occurrence in daily life and the unusual properties of the related Fibonacci series.

Area and Volume

Use a rectangular sheet of graph paper marked off into square inches. Cut into parts and reassemble to form a parallelogram, triangle, or rhombus. Use the dimensions of the new figure to find the appropriate formula.

Use containers such as cylinders, cones, prisms, and pyramids with equal altitudes and bases. Compare the volumes by pouring sand or salt from one container to another.

Cut a circle into small equal sectors (Figure 27.5a). Reassemble them into a shape that is like a parallelogram (Figure 27.5b). The altitude of this pseudo-parallelogram will be the radius, and the base will be $\frac{1}{2}C$. Then the area of a circle is $\frac{1}{2}C \times r = \frac{1}{2}(2\pi r) \cdot r$ or $A = \pi r^2$.

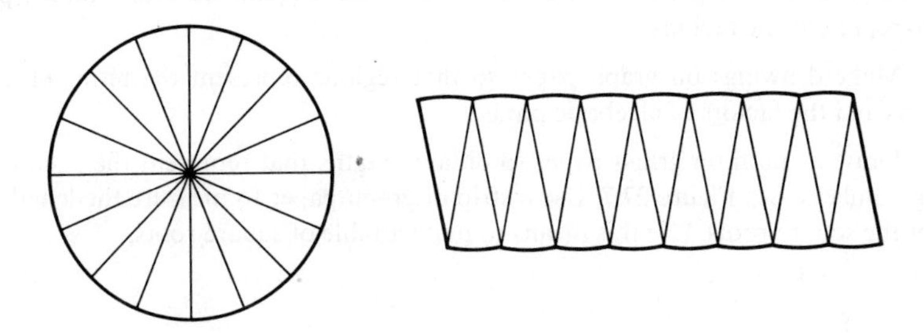

Figure 27.5

Illustrate the Pythagorean theorem by comparing areas as in Figure 27.6.

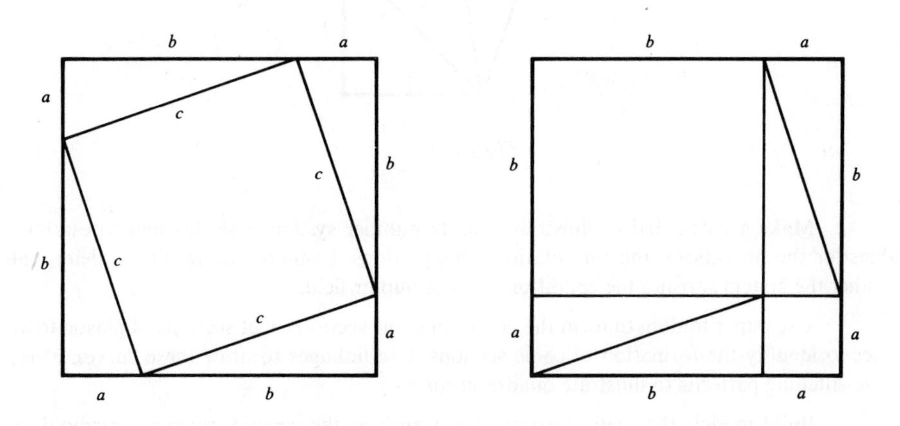

Figure 27.6

Cover a great circle of a sphere or a hemisphere with heavy string by winding around a nail at the center or pole. Use the length of string as a measure of area to compare areas.

Draw right triangles on graph paper. Outline the squares on each leg. Cut a square of graph paper to represent the square of the hypotenuse. Compare the areas of these squares by counting the squares of graph paper.

Show the relationship of area and perimeter by drawing different rectangles on graph paper or pegboard. If the perimeter is held constant, what constraints are put on the area? If the area is held constant, how may the perimeter vary? Extend these relationships to volume and surface area.

Algebra

Use sticks or dowels to form the representation for algebraic phrases such as $a, a + b, 3a, ab, a^2, a^3, (a + b)^2$.

Represent the graphs of linear equations by elastic thread on pegboard, acoustical tile, or on a commercial equation board. Do the same for inequalities, using colored paper or plastic to represent the regions.

Make drawings on graph paper so that regions represent the terms of binomial expansions and the factors of algebraic phrases.

Draw a spiral on graph paper to obtain lengths that represent the square root of counting numbers. See Figure 27.7. Use a strip of graph paper to measure the lengths which represent the square roots. Use this means to build a table of square roots.

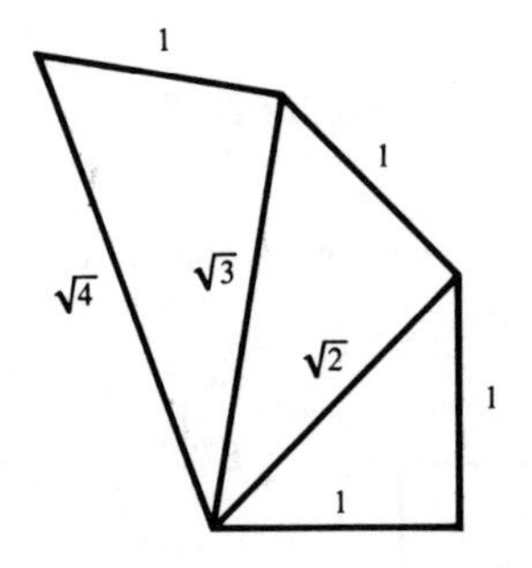

Figure 27.7

Make a clock dial to illustrate a finite number system. Use this dial to establish tables for the operations and for solutions to equations. Examine the results to determine whether the system satisfies the conditions for a group or field.

Use paper folding to form the curves of conic sections. Cut sections of plastic-foam cones to identify the formation of conic sections. Use linkages to draw these curves. Make curve-stitching patterns to illustrate quadratic curves.

Build models that will illustrate curves such as the cycloid, catenary, cardioid, or spiral.

Design and make a game based on mathematical ideas, as those described in *Games for Learning Mathematics.*[3]

Develop the flow chart for an algorithm such as computing a square root. Write the computer program for this flow chart and check it on a computer. Find out how a calculator computes a square root.

Build a three-dimensional coordinate system, using three squares of coarse screen. Represent graphs of equations with elastic thread for lines and plastic sheets for planes.

Compare the extension of a spring to the weight that causes the extension. (a) Attach weights to the bar of a wire coat-hanger. Measure the sag for each weight. (b) Attach a meter stick over the edge of a table. Attach weights to the end of the meter stick and measure the distortion for each weight.

Measure the distance an object falls in space and compare with the time measured by a stop watch. Or time a marble as it rolls down an incline.

Establish area formulas for composite figures such as those in Figure 27.8.

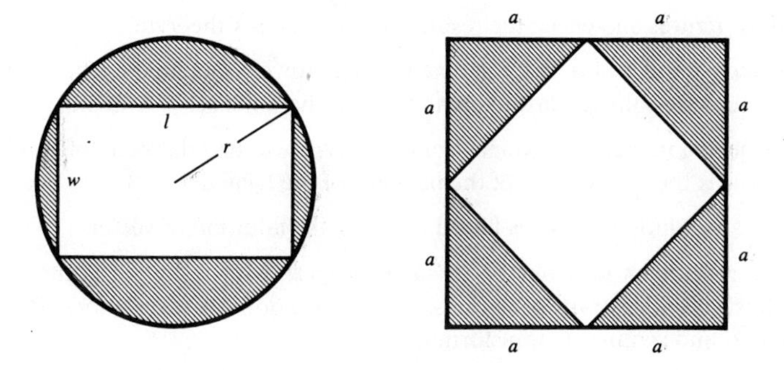

Figure 27.8

Use toys or models of cars, trains, airplanes to build a concrete representation of a verbal problem. Use these models also to illustrate travel on the number line as an introduction to the addition or multiplication of directed numbers.

Find examples or pictures of curves and geometric patterns in nature such as the snowflake, a honeycomb, the spiral shell, crystals, and elliptical orbits.

Geometry

Make simple electric circuits to represent truth tables.

Form regular polygons by folding paper. Perform all the usual constructions of Euclidean geometry by paper folding as described in *Paper Folding for the Mathematics Class.*[4]

---

[3] Donovan A. Johnson, *Games for Learning Mathematics*, J. Weston Walch, Portland, Me., 1960.

[4] Donovan A. Johnson, *Paper Folding for the Mathematics Class*, National Council of Teachers of Mathematics, Washington, D. C., 1957.

Outline regular polygons on graph paper or by elastic thread on pegboard. Measure angles and area to determine formulas.

Count vertices, edges, and regions of polygons and polyhedra to determine Euler's formula. Compare results of concave and convex, simple closed polyhedra, and composite figures.

Measure the angles and sides of parallelograms to determine equalities. Use a linkage to discover these relationships.

Make drawings of triangles with specified conditions. Cut them out and compare by superposition to determine conditions for congruency.

Draw triangles of various shapes and sizes. Determine relationships of angle bisectors, altitudes, medians, and perpendicular bisectors of the sides. This can be done with commercial instruments, by constructions, and by paper folding.

Use drawings on acetate film or a commercial device to find the relationship between the measures of angles between two intersecting lines and the measures of the intercepted arcs of a circle.

Draw figures and check the results for Desargues's theorem.

Make a model of a finite geometry. Use dowels mounted vertically to a base to represent points. Use colored elastic thread to form the finite set of triangles.

Suspend a model of a triangle from each vertex to find its center of gravity. Check to see whether this is the intersection of the medians of the triangle.

Use a resolution-of-forces board to check the addition of vectors.

Use reflections in a mirror to show properties of reflective symmetry. Use wire models to illustrate other transformations, such as rotations or translations. Build and use a kaleidoscope to find symmetric transformations.

Investigate paradoxes such as "every triangle is an isosceles triangle."

Use or make models of pseudospheres and spheres to illustrate non-Euclidean geometry.

Make models for topology and investigate the properties of the Möbius strip, the hexaflexagon, and the four-color map problem on a plane, sphere, and torus.

Make drawings and three-dimensional models of symmetries.

Make a collection of optical illusions. Find the basis for the illusion, as described in *Experiments in Optical Illusion.*[5]

Use a semicircle to illustrate the relations between the arithmetic mean and the geometric mean.

Devise a system of computation based on line segments and geometric constructions.

Examine drawings for perspective. Make drawings based on vanishing points. Make models to demonstrate perspective.

Determine the basis for tesselations, mosaics, stained-glass window design, and other patterns.

---

[5] Nelson F. Beeler and Franklyn M. Branley, *Experiments in Optical Illusion*, Thomas Y. Crowell Company, New York, 1951.

Use pendulums of various types to trace pendulum patterns.

Determine projections of lines and closed curves by shadows.

Make models of geometric solids. Use these as a basis for developing Euler's formula and determining volume and surface-area relationships.

Construct two- and three-dimensional optical illusions. Use a camera to photograph staged illusions based on expected geometric relationships.

Trigonometry

Draw a number of right triangles on graph paper. Use the ratios of the measures of sides to build tables of sines, cosines, and tangents. Use unit lengths for the adjacent side or the opposite side so that the measure of the hypotenuse is the value of the sine or cosine.

Make a model to illustrate the winding or wrapping function. Use it to build a table of sines and cosines.

Demonstrate wave action with a rope, waves on water, alternating current on the oscilloscope.

Make a Trigtracker out of cardboard. Use it to make a table of sine, cosine, and tangent ratios.

Make a three-dimensional rectangular coordinate system with three squares of cardboard or plastic. Use one of the three axes for the scale of imaginary numbers. Draw the graphs of quadratic equations on these planes extending the graph to include the complex roots.[6]

Facilities Needed for a Mathematical Laboratory

The classroom described in Chapter 26 will provide a proper setting for laboratory lessons. In addition, facilities, supplies, and tools are essential. For laboratory work, every mathematics classroom needs the following materials:

cardboard
tackboard
fiberboard
colored corrugated cardboard
plywood
pegboard
balsawood
cork panels
plastic foam
plastic sheets (plain and colored)
felt
glue
rubber cement
adhesive wax
paint
crayons
colored chalk
spray paint
nails
thumbtacks
brads
staples
map tacks
construction paper
wax paper
drawing paper
graph paper (rectangular, polar coordinate, logarithmic)

[6] For complete instructions see *Multi-Sensory Aids in the Teaching of Mathematics*, National Council of Teachers of Mathematics, Washington, D. C., 1946.

aluminum foil
adhesive tape
masking tape
colored plastic tape
string
yarn
elastic thread
colored rubber bands
pegs
golf tees
soda straws
plastic tubes
toothpicks
letter stencils
letters for mounting
title board
beads
marbles
chips
rings
tongue depressors
pill boxes
modeling clay
powdered plaster
fishline weights
balloons
wire
sandpaper
brush pens
nylon point pens
paint brushes

To work with these materials, tools such as the following are helpful:

pliers
hammers
saws
screw drivers
shears
staplers
paper punch and eyelets
soldering iron
hand drill and assorted drills
paper cutter
planer or rasp
knives
bench vise
gluing clamps
carpenter's square
drawing set
drawing board
protractors

For laboratory lessons, the following demonstration models and commercial devices are needed:

Demonstration devices:

abacus
base and place model
counting frame
binary abacus
area boards
area aids
balance
binomial cube
calculator
computer terminals
caliper, vernier
micrometer
circle device
cube (dissectable)
cylinder
cone
conic sections
data graphing board
equation board
flannel board
force table
geometry devices
pegboards
geometric solids
graph board
inverse squares apparatus
lines and planes models
map projections
measuring disk for determining pi
micrometer demonstrator
number base blocks
number lines
orthographic projection models
pattern dial
perimeter measurement device
polyhedra (plastic)
probability board
prisms
pyramids (dissectable)
Pythagorean theorem demonstration set
reflectors (parabolic)
slated globe
slide rules
solids of rotation
sphere (plastic, dissectable)
polyhedra models
string models
transit
trigonometric functions device
vector demonstrator model
vernier demonstration
volume and unit cube demonstration

Drawing devices:

compass
drafting machine
drawing board
drawing board set
drawing instrument kit
ellipse drawing device
pantograph
parallel rules
proportional dividers
protractor
ruler
triangles (plastic)

Kits:

base and place
contour mapping
curve stitching
drawing
geometric construction
geometriclaboratory
mapping
mathematical shapes
measurement
multiplication and division
numerative
probability and statistics
sampling
solid shapes
surveyors
time learning

Measurement instruments:

alidade
altimeter
angle mirror
arrows
caliper
carpenter's rule
carpenter's square
clinometer—hypsometer
compass—magnetic needle
Jacob's staff
level
leveling rod and target
meter sticks
micrometer
odometer
plane table
planimeter
proportional dividers
protractor
spherometer
steel tape
stopwatch
sun dial
transit
tripod

Toys and puzzles: Appendix A.

Commercial games: See Appendix A.

Science apparatus (to be shared with science department):

balance and weights
seconds pendulum
levers
inclined plane
resolution of forces
pulley set
wheel and axle
spring scales

Much of the material listed above can be purchased at hardware or department stores, toy stores, lumber yards, or school-supply stores. The commercial devices are available from the companies listed in Appendix A.

Summary

In laboratory lessons, the emphasis is on independent discovery. The student explores situations and looks for patterns, for applications, for relationships, and for important ideas. As he does this, he records results and writes conclusions. In the process, he may invent ways of shortening labor, he may learn to compute better, and he should find illustrations of mathematical ideas. For many students it is a new experience that gives a new outlook on school life and often gives him new confidence in his ability to learn.

Learning Exercises

1. Cut out a circular piece of paper. (Use a cup base or small plate to draw your circle.) Determine a procedure to fold this paper into the largest possible equilateral triangle.
2. After you have performed exercise 1, analyze the experience as a learning activity. Compare the exercise with standard textbook exercises.
3. Make instruction sheets for laboratory lessons related to two of the following:
   (a) Algebra
   (b) Seventh-grade mathematics
   (c) Geometry
   (d) Probability
4. Prepare a laboratory lesson based on an instrument like a micrometer or a hypsometer.
5. Examine textbooks to find a laboratory lesson at each of the following levels:
   (a) Seventh-grade mathematics
   (b) Algebra
   (c) Geometry
6. Examine a high school science textbook to find a laboratory lesson appropriate to the mathematics classroom.

Suggestions for Further Reading

Auclair, Jerome A., and Thomas P. Hillman, "A Topological Problem for the Ninth-grade Mathematics Laboratory," *Mathematics Teacher*, 61, 5 (May 1968), 505–507.

Bruce, Matthew H., "Using the Cathode Ray Oscilloscope in the High School Trigonometry Class," *School Science and Mathematics*, 60, 8 (November 1960), 593–602.

Cundy, H. M., and A. P. Rollett, *Mathematical Models*, Clarendon Press, Oxford, 1951.

Dunn-Rankin, Peter, and Raymond Sweet, "Enrichment: A Geometry Laboratory," *Mathematics Teacher*, 56, 3 (March 1963), 134–140.

Hillman, Thomas P., "A Current Listing of Mathematics Laboratory Materials," *School Science and Mathematics*, 68, 6 (June 1968), 488–490.

Hochstein, A. E., "Trisection of an Angle by Optical Means," *Mathematical Teacher*, 66, 7 (November 1963), 522–524.

Johnson, Larry K., "The Mathematics Laboratory in Today's Schools," *School Science and Mathematics*, 62, 8 (November 1962), 586–592.

Kidd, Kenneth P., "Measuring the Speed of a Baseball," *School Science and Mathematics*, 66, 4 (April 1966), 360–364.

———, Shirley S. Meyers, and David M. Cilley, *The Laboratory Approach to Mathematics*, Science Research Associates, Inc., Chicago, 1970.

Kluttz, Marguerite, "The Mathematics Laboratory: A Meaningful Approach to Mathematics Instruction," *Mathematics Teacher*, 56, 3 (March 1963), 141–145.

Perham, Father Arnold, "An Exercise for the Mathematics Laboratory," *Mathematics Teacher*, 58, 2 (February 1965), 114–117.

Spencer, Richard V., "Discovery of Basic Inversion Theory by Construction," *Mathematics Teacher*, 57, 5 (May 1964), 303–306.

Spitznagel, Edward L., Jr., "An Experimental Approach in Teaching of Probability," *Mathematics Teacher*, 61, 6 (October 1968), 565–568.

Sweet, Raymond, "Do Math Labs Just Happen?," in Douglas B. Aichele and Robert E. Reys (eds.), *Readings in Secondary School Mathematics*, Prindle, Weber, and Schmidt, Boston, 1971. pp. 352–356.

———, "Organizing a Mathematics Laboratory," *Mathematics Teacher*, 60, 2 (February 1967), 117–120.

Wilkinson, Jack, "Teaching General Mathematics—A Semi-laboratory Approach," *Mathematics Teacher*, 63, 7 (November 1970), 571–577.

York, Howard P., "The Coat-hanger Cannon—Being an Exercise in Making a Projectile Go Where You Want It To," *School Science and Mathematics*, 71, 3 (March 1971), 226–228.

Mathematics Outside the Classroom 28

One of our basic axioms is that learning mathematics should be pleasant. If the classroom is joyless, hostilities grow and teaching becomes futile. To add pleasure and realism to mathematics lessons, mathematics should be taught in its natural setting. We can find much important mathematics on the street, in nature, in recreations, and in the newspaper.

Almost any objects outside the classroom can be sampled, measured, or observed to illustrate or apply a mathematical idea. Consider them in three categories: nature, technology, and recreation.

The World of Nature

Mankind enjoys the out-of-doors—this seems to be his natural habitat. Sunshine and moonlight, flowers and trees, birds and animals, lakes and mountains —these are the things that add beauty and tranquility to life.

At the same time, the world of nature illustrates patterns and variations that are closely related to mathematics. Why then don't we teach the mathematics of the out-of-doors? Perhaps if we did, we could catch the mathematical sense that seems to pervade nature.

Birds, insects, fish, and animals seem to have a natural sense of distance and direction. In building their homes, creatures are architects, engineers, and artists. Bees build their hexagonal cells in a manner that engineers cannot improve. Spiders build webs that are spirals, ideal for catching prey and for informing the spider of the location of the catch. Birds build symmetric nests that sustain the stresses of wind, rain, and enemies. Beavers build homes that use the water behind their dam as a safety factor.

Creatures also seem to have a sense of ratio and proportion. All animals seem to know that a straight line is the shortest distance between two points. Bees perform a dance to tell other bees the direction and distance to a new-found source of honey.

In addition to the mathematical sense of creatures, the world of nature seems to be a mathematical world. Symmetry, design, curves, and mathematical laws abound. Crystals of minerals such as quartz are formed in the shape of polyhedrons. Snowflakes are lacey and each one different and yet each one has a regular six-sided pattern. No wonder it has been said of mathematics that "its greatest creation is the universe itself." In this sense, we can indeed look for mathematics in the world of nature.

The objects to be observed in nature include the following:

Plants. Leaves, flowers, seeds, roots, stems

Animals. Size, temperature, speed, power, food, beat of calls, cost

Birds. Nests, feathers, flight, size, food, calls

Insects. Shape, webs, cocoons, wings, flight, work, cells

Trees. Leaves, branches, trunks, seeds, blossoms

Water. Reflection, ripples, waves, speed, pollution

Sky. Clouds, wind, rain, lightning, shadows, sun, moon, stars, constellations

Soil. Crystals, pebbles, moisture, composition, temperature

Sea urchins and fish. Shape, design, quality

Humans. Speed, power, proportions, sight, reaction time, units of measure

Snakes and worms. Shape, crawl

Fruits and vegetables. Shape, volume, cost, quality

The World of Technology

We are living in a world of exploding technology largely based on mathematics. To find the mathematics in our society, our students should explore problems in the community in which they live. They can make measurements, find information, or collect data such as the following:

Streets and highways. Width, slope, curvature, traffic signals

Buildings. Shape, size, height, volume, floor, wall, and window area

Bridges. Shape, models of different types, length, height, strength

Automobiles. Shape, size, wheels, gear ratio, compression ratio, cylinder volume, speed, pollution

Motorcycles. Gear ratios, gas mileage, tires, power, speed, force

Bicycles. Shape, cycloid, gear ratios, tires, speed

Tools. Shape, mechanical advantage, force, helix of screws

Household articles. Shape, gear ratios, area, volume, power, light

Advertisements. Design, size, hidden assumptions, variation in price

Newspapers. Size, content, data, probability, prices, cartoons, sports, financial reports, vital statistics

Photography. Similarity, symmetry, light, scales

Airplanes. Speed, power, surface, pollution, angle, navigation, cost

Trains. Speed, power, cost, pollution, tracks

Banks. Interest rates, checking accounts, money, change

Stores. Boxes, prices, displays

Restaurants. Prices, quality, wages, calories, permutations of orders

Gas stations. Prices, popularity, computing pumps, air pressure

Factories. Wages, production, taxes, pollution, employment

One way to explore the technology of the community is to take students on a field trip or to plan an excursion in indirect measurement. These activities can be done even with limited time and resources. A field trip takes students to a local business or institution where they see mathematics at work, while a measurement excursion takes them outside the classroom to apply their learning in a realistic setting.

Field Trips

A field trip often results in more vivid learning than other forms of educational experience. We usually recall travel experiences much more accurately and intensively than classroom learning experiences.

A carefully planned field trip is particulary well suited to attain goals such as the following:

Field trips *provide motivation* for the study of a unit. A trip to a factory may not only show applications of topics but also may impress the boys with the importance of competence in mathematics in order to be eligible for desirable jobs. A visit to a computing center with its electronic computers may excite interest in the role of mathematics in the future or in the study of computer programming.

Field trips *enrich mathematical learnings* by relating school work to actual life situations. A visit to an insurance company will demonstrate the use of probability, mortality tables, accident rates, premiums, dividends, and loss ratios.

Field trips *provide specific materials* for use in the classroom. Many useful business forms and flow charts of operations can be obtained from a visit to a bank, a tax-collection office, or an architectural concern.

Field trips *generate realistic situations for group planning.* The class may participate in choosing a date, gathering materials for field work, making contacts to arrange the trip, scheduling a bus, sending a courtesy note. It is especially useful to have students work cooperatively with representatives of the community to make arrangements.

Field trips *present material in its natural setting.* In an airport, the mapping of flights, the measurement of ceiling height, the interpretation of weather reports, and the locus of radio beams are matters of life or death. In a laboratory, precision measurements, analysis of data, and the use of formulas are constant activities.

Field trips *integrate subject matter of different courses.* At the highway office, planning the new toll road involves knowledge of weather conditions, rock and soil formation, accident causes, transportation surveys, as well as indirect measurements. At the airport, aerodynamics, engine performance, weather conditions, and radar patrol are related to work in science courses.

Field trips *provide a means for many students and community citizens to participate in the school program.* At the place visited or in follow-up activities, several people will participate as guides, demonstrators, or speakers.

Places to Go for Field Trips

Local communities will differ greatly in the opportunities for field trips. The following list is suggestive of places where application of mathematics may be seen.

Government agencies. Tax collector's office, social-security office, highway department, county surveyor's office, weather bureau, post office, military recruiting office or base, civil air patrol, forestry service, civil-service office, water- or crop-control office, and land-reclamation office.

These government agencies will be using mathematics in a variety of ways, such as determining tax rates, completing tax forms, checking records, determining benefit payments, computing refunds, making maps, surveying land, estimating, computing areas and volumes, determining operating costs, collecting and analyzing data, determining the competence of applicants for positions.

Community institutions. Museums, art centers, churches, laboratories, observatory, planetarium.

Museums may have a variety of exhibits in which the role of mathematics is important, as in, for example, map projections or ancient measuring devices. Art centers often have displays of abstract or modern art that emphasize symmetry, perspective, and proportion. Churches have stained-glass windows and architectural forms that show applied geometry. Laboratories have precise measuring devices and experimental techniques that illustrate applied mathematics. Observatories and planetariums utilize a variety of mathematics such as gear ratios, elliptical orbits, indirect astronomical measurements, temperature determinations, space navigation, large units of length, the fourth dimension, formulas related to gravitational attraction.

Business enterprises. Insurance companies, banks, stock exchange, grain exchange, brokers' offices, factories, contractors, computing and recording firms, engineering offices, supermarkets, farms, architects' offices.

Most business firms are highly dependent on mathematics. Arithmetical computations such as wages, costs, depreciations, overhead, brokerage fees, market quotations, interest rates, dividends, rents are involved. Most of these enterprises collect data, prepare tables, draw graphs, predict markets, determine probability, and use a variety of statistical tools and computers. Planning buildings, laying out machine work, measuring stresses, estimating reconstruction costs require many indirect measurements. Most of these concerns use a variety of business forms and computing machines for efficient operation.

Transportation centers. Airport, train station, bus depot, freight office, railway express office.

Travel by land, sea, or air involves the use of maps, schedules, rates, profit sharing, traffic control, and time zones. Air and water travel are highly dependent on principles of navigation, aerodynamics, and vectors. Measurements are made by compasses, sextants, astrolobes, and driftmeters. Radar and sonar are used every day. Research in accident control, operating efficiency, and new design is constantly being carried on.

Public utilities. Telephone center, electric company, gas company, water plant.

Measurement of the amount of service used is the daily problem of public utilities. Their measurements may involve varied types of meters and units. Computing charges, bookkeeping, planning extended services, designing new equipment, planning new rates are constant activities involving mathematics at the public-utility plant.

Thus, it is apparent that mathematics is a part of many everyday activities of establishments in your community. The listing above includes topics from many fields of mathematics, from elementary arithmetic through trigonometry. Which field trip would be most appropriate for a particular course or unit must be determined by the teacher and his students.

Planning the Field Trip

The success of a field trip depends on careful planning. Successful planning involves selecting an appropriate trip, making the necessary arrangements, arranging for supervision while on the visit, and providing suitable follow-up activities.

On field trips, it is important that you help students to identify or discover the mathematical aspects of what they see. These aspects should be tied in with what has been learned in the class. As an illustration, a trip was made by a general mathematics class to a bank. The mathematical aspects were outlined in a guide sheet prepared by the teacher. A flow chart was made of processing a check. Attention was called to the use of calculators and computers. Illustrations were given of interest tables and comparisons made of different types of investments. Qualifications were given for establishing credit. Questions about requirements and opportunity for employment were suggested. Without the guide sheet, the mathematics could well have been overlooked among the many processes and situations observed.

The World of Recreation

Almost everyone is interested in sports, games, or recreations. The school grounds, parks and commercial enterprises provide a rich setting for applied mathematics. The following are some suggestions for mathematics projects:

Courts. Size, shape, angles, lines, tracks, lengths, layout

Play equipment. Shape, area, volume, weight, strength of fish lines

Players. Performance records, track records, tests of coordination, pulse, strength

Paths or curves. Parabolas of objects, ellipses of stadia or pools, track layouts, catenaries of soils, cycloid slopes for skiing

Scoring. How to read and record game records, probability of scores

Timing. Travel time for game piece, reaction time, game time, speed of objects

Graphs. Seat locations, designs for cheers, bar graphs, line graphs

Sports data. Team standings, scoring records, costs, attendance, odds.

Measurement Excursions

Excursions in measurement will add reality and interest to the study of measurement, geometry, and trigonometry. These excursions can be made to illustrate the historical development of geometry and trigonometry and to illustrate the measurements frequently made by civil engineers and others in technical work.

Indirect-measurement excursions are appropriate at many different grade levels. At an elementary level, scale drawings and approximations can be made with simple instruments. At a higher level, similar triangles, congruent triangles, and the Pythagorean theorem are used. At the next level, trigonometric functions are used to compute the inaccessible measurement.

Students who are taken on indirect-measurement excursions should have some background in geometry. They will need to be familiar with ideas such as angles and angle measurement, triangles, parallel lines, perpendicular lines, areas, and principles of direct measurement. They will be expected to solve formulas, equations, and proportions. In addition, a familiarity with the role of precision and accuracy in computing with measures is useful for the student on these excursions.

Another basic requirement for excursions outside the classroom is a student's sense of responsibility. Students must be willing and able to conduct independent activities without the immediate presence of the teacher.

Excursions with Simple Materials

There are a great many excursions possible with very simple materials. These are crude approximations, but the mathematical principles involved are sound. The following are some suggested activities:

Measure an inaccessible height by measuring shadows
Measure a distance by timing an echo
Determine an east-west line with shadows or a watch
Measure an inaccessible height by measuring angles of elevation
Measure an inaccessible distance by laying out congruent or similar triangles
Map a small region with an alidade and plane table
Measure the contour of a hill and make a contour map
Plot a traverse by measuring angles and distances
Lay out a treasure hunt by establishing points, directions, and distances.

Materials Needed for Measurement Excursions

To complete measurement excursions successfully, the mathematics department should have the following measuring and drawing instruments:

protractors
compasses
ruler
triangles
T-square
parallel rulers
drawing board and drawing instrument kit
pantograph
proportional dividers
carpenter's square
transit
hypsometer
clinometer
level sextant
angle mirror
stadia tube
alidade
plane table
yard stick
meter stick
measuring tape
ranging pole
arrows
tripod
micrometer
caliper
spherometer
odometer
planimeter
stop watch

Many of these devices can be purchased from local engineer's supply stores or mail order stores. For a more complete list of supplies, write to the following companies:

Yoder Instruments
East Palestine, Ohio

Keufel and Esser Co.
Hoboken, New Jersey

Eugene Dietzen Co.
2425 Sheffield Ave.
Chicago, Illinois

W. M. Welch Scientific Co.
1515 Sedgwich Street
Chicago, Illinois

Where commercial devices are not available, many measuring devices can be made by students. These include equipment such as the following:

hypsometer	angle mirror
proportional dividers	telemeter
sundial	range finder
pantograph	angle trisector

Preparing for an Excursion Outside the Classroom

Before leaving the classroom, the student should know exactly what to do, how to do it, and what materials will be needed. He will need to know the mathematical ideas involved, the data to collect, and how to operate the instruments to be used. He should complete some type of report of his observations.

Excursions in indirect measurement often can be done in your classroom, on the school grounds, or at home. The gymnasium has adequate space for an indoor excursion if the weather is inclement. Most measurement excursions should be made up of groups of two to four persons. Each member of the group will have a different job, such as recording and keeping records, measuring distance, operating instruments, or placing sighting poles. The equipment should be selected in advance so that each student will have everything needed. Complete written instructions should be given to the group so that the group is clear about what the necessary measurements are. The following guide sheet is an example for an out-of-doors lesson:

1. Find a small object that you have never seen before: pebble, leaf, flower, branch, insect.
2. Examine it carefully to find its features and its structure.
3. Make a sketch of it. Label each part.
4. Measure and record its characteristics: size, weight, hardness, color, texture.
5. Describe the object in words. Classify it in a set that would identify it.
6. Indicate what mathematics could apply to it.
7. Find further information about the object in reference books.

There are no limits to the possibilities for mathematics outside your classroom. If you run out of ideas, ask your students to plan their own lesson. You may be surprised by the ingenuity and complexity of the lessons they propose.

Learning Exercises

1. Write a lesson plan for a general mathematics class on one of the following topics:
   (a) Perimeter and area of your school
   (b) Probability experiments
   (c) Speed of cars passing the school building
2. Investigate phyllotaxis and the Fibonacci sequence. How could this topic be utilized in the school program?
3. Make a hypsometer. Write a series of activities related to this instrument. Include one set of instructions for constructing the instrument.
4. PROJECT. Visit a local establishment and plan a field trip to it for your class.
5. PROJECT. Investigate the availability and cost of commercial measuring instruments. Make a selection for a specific course.

Suggestions for Further Reading

Boeckmann, Hermann, "Elementary Field Survey—An Enrichment Course for High School Students," *School Science and Mathematics*, 67, 2 (February 1967), 132–134.

———, "Surveying for High School Students," *School Science and Mathematics*, 64, 5 (May 1964), 347–352.

Hamilton, W. W., "Field Work Modifies Our Work in Arithmetic," *School Science and Mathematics*, 51, 7 (October 1951), 527–531.

Kidd, Kenneth P., "Measuring the Speed of a Baseball," *School Science and Mathematics*, 66, 4 (April 1966), 360–364.

Ransom, William R., "Some Mirror Trigonometry," *School Science and Mathematics*, 55, 8 (November 1955), 599–600.

Shuster, Carl N., and Fred A. Bedford, *Field Work in Mathematics*, Yoder Instruments, East Palestine, Ohio, 1953.

Woodby, Lauren G., "The Angle Mirror Outdoors," *Arithmetic Teacher*, 17, 4 (April 1970), 298–300.

The Future

What Lies Ahead? 29

Our society is a dynamic, continually changing one in which knowledge and resources increase rapidly and unpredictably. Recent technological advances have created numerous devices for communicating mathematics and therefore have wide application for new types of mathematics classroom instruction. Federal support for education provides funds for developing new programs for learners of varied ability. Furthermore, experimental projects such as CSMP (Comprehensive School Mathematics Program) and UICSM (University of Illinois Committee on School Mathematics) continue to prepare revised curriculum. And professional organizations are offering journals, supplementary publications, and conferences to keep teachers informed of innovations.

Trends in Mathematics Teaching

In order to predict the future we need some information for extrapolation. We can provide one location by considering mathematics teaching a generation or more ago. Next, we can fix a second point by describing the present situation. These two points will give us a trend line from which to predict what the classroom a generation from now will be like.

Objectives

Past. In the past, teaching objectives were largely limited to the memorization of facts and development of computation skills. Rote memorization of rules, shortcuts, and mechanical manipulation were considered satisfactory achievements.

Present. Present objectives for students include both computational skill and mastery of ideas with the emphasis on computation with understanding. Also, present objectives include broad concepts, understanding of structures, and the ability to solve problems. On the other hand, we pay lip service to developing positive attitudes toward mathematics and appreciation for mathematical ideas as objectives but rarely devote classroom time or provide material for them.

Future. In the future, it is likely that objectives will broaden and include those difficult to attain: creativity, positive attitudes, learning how to learn, and values. Our scientific society is dependent upon creating new knowledge and new products. Hence, preparation for becoming creative will be an objective in all fields. Since knowledge will continue to expand we will need to become proficient in learning how to learn—the technique basic to continued adult learning. This will involve learning to use computers as auxiliary memories and for information retrieval. Even so, the traditional goals of computational skill and mastery of concepts will remain important, if only as a basis on which higher cognitive skills may be developed. The increasing concern for social and especially ecological values will produce a greater emphasis on application not just of skills but of thinking and decision making techniques as well.

In our technological society, the educated person has so much power that it is necessary for educators to try to develop in the student values that will enable this power to be directed toward worthy ends. By virtue of his instilling these human values and concerns in a machine-oriented age, the teacher will be of increasing importance.

Content

Past. The mathematics sequence of the past was organized into narrow, unsophisticated courses: arithmetic in the elementary school, algebra in ninth grade, plane geometry in tenth, calculus in college, modern algebra in graduate school.

Present. School mathematics today is a precise, integrated mathematics sequence that emphasizes the structure of the number system, logic, functions, and other contemporary topics. In general, topics are treated at an earlier grade than in the past but geometry remains a traditional treatment of deduction at the tenth-grade level.

Future. New topics will be introduced and classical topics will be taught at a lower level. Topics from analytical geometry, linear algebra, and calculus will be integrated into the secondary school sequence. Space concepts normally taught in tenth-grade geometry will be relegated to lower grades and new topics such as transformations, vectors, convexity, combinatorial topology, and projective geometry may be examined in the geometry of tomorrow.

The trend in content has been to lower the grade level at which topics are introduced, keeping most of the traditional topics but presenting the content in more precise, sophisticated language and then adding new topics at the end of the current sequence. The future will see a counteracting trend when the proponents of depth of treatment, problem-solving approaches, and independent study call for a slower pace to allow concentration on these noncontent goals.

The current trend has also been away from local development of the curriculum to the adoption of curriculum prepared by national groups such as SMSG. The future will face the issue of greater centralization and standardization through national curricula.

Materials

Past. The teacher's tools of the past consisted of chalk, red pencil, and textbook.

Present. Today, the materials of the modern secondary school include demonstration models, slide rules, overhead projectors, drawing instruments, graph stencils, measuring instruments, and some enrichment pamphlets and books.

Future. The trend toward increased availability and use of equipment and supplies is strong and clear. Despite some cutback in government support, the number of instructional aids being invented and produced continues to increase rapidly. Films, commercial projectuals, computer devices and computer terminals, manipulative materials, kits, teaching machines, demonstration equipment, laboratory devices, and supplementary books may require the mathematics department of the future to have extensive storage space in every classroom. There will be equipment and supplies for the duplication of material, the production of audio-visual material, and the construction of manipulative material within the classroom or in a resource center.

The audio-visual equipment will include three-dimensional motion picture films, film and videotapes, and film loops. Specific materials to fit specific purposes will be the trend.

As these things become more readily available, competition will serve the teacher by forcing producers to improve the quality of their product.

Evaluation

Past. In the past, achievement was measured largely by tests constructed by the teacher. These paper-and-pencil tests were concerned with evaluating the students' computational skills and their recall of memorized facts.

Present. Today, the additional use of published tests with established norms is common in most schools. In addition, many textbooks include chapter tests and course tests. Some teachers are even giving open-book tests in order to emphasize understanding, but most current tests measure only content goals.

Future. As the objectives of school mathematics change, new tests will be needed to measure creativity, attitudes, and values. New tests such as performance tests, reading tests, and problem-solving tests will be devised. It is likely that in the future these tests will be administered by a computer. The computer will score the test, give an immediate analysis of the performance, and prescribe remedial instruction. The computer will also store information about each student and provide the teacher with a cumulative record at the time grade reports are to be completed. The computer will also give the teacher an item analysis to be used as a basis for improving his instruction as well as revising the test.

One of the issues to be faced in the near future relates to national or state assessment programs. There are advantages in having national or state norms available for comparisons; but there are great dangers of conformity, for one instance, in the misuse of this powerful program. And there is great danger that assessment will be confined to narrow cognitive facts or skills.

A very positive aspect of the future will be the increased stress on normative (course and student improvement) testing rather than summative (grading) examinations. Individualization of programs will also increasingly pit the student against his own best effort rather than those of other students with differing abilities.

Individual Differences

Past. In the past, little attention was given to the needs of individual students. The mathematics courses of grades 9 through 12 were largely elective, and so a selection process resulted in large numbers of dropouts. There was usually a single-track college preparatory curriculum.

Present. Today, the variation in ability and interests at each grade level is great and is usually responded to by a multiple-track curriculum. As a result, our able students are far superior in achievement to those of the previous generation. The poor student or reluctant learner, however, may be further behind and a greater discipline problem than his predecessors. Cultural differences as well as the increasing amount of course content have accentuated the problems of students' varying abilities to the point where present curriculum planning procedures are extremely inadequate.

Future. In the future, we will have more information about each student and hence will be better able to provide for his needs, interests, and ability. Some of this information will be more readily at hand because of greater and more effective use of the computer. Special courses and multiple-track curricula will be available. Student advancement from one level to another or from one track to another will be determined on an individual basis and arranged for by flexible scheduling. The idea that every learner should spend the same number of days in a given grade, or the same number of minutes in a mathematics class every day, or that he should study the same textbooks with the same class will be abandoned. Individualized instruction will also be presented by specialists—teachers specifically trained for remedial or clinical work or advanced courses. For the talented, there will be greater opportunities for enrichment, for independent study, and for advanced placement.

Identification of specific learning disabilities will be much more specific. For example, a teacher will not just know that a student cannot divide. He will know what underlying skills are missing. In this way, he will be able to mount a more effective instructional program.

Method

Past. The method of the past was largely the stereotype lesson: discussion of questions on homework, discussion of the new process or new theorem, assignment.

Present. The method most commonly used in today's classroom is exactly the same as that of a generation ago. There are only a few teachers who give thoughtful presentations utilizing a variety of materials and who guide the student to discover new ideas.

Future. The methods of the future will be widely varied. With added resources, more time, and a better professional background, mathematics teachers will present lessons based on a systems development approach; the best technique and material selected for each presentation. Increased resources and increased pooling of information and techniques means that many more lessons will encourage and exploit student interests and provide students with exciting participation opportunities. Discovery and laboratory learning techniques will be central to this new approach, as will be micro-courses, flexible scheduling, and computer-managed instruction.

The ideal instruction has always been a one-to-one relationship between student and teacher. But it has been impossible to attain this individualized instruction in our classrooms. We lack the time, the material, and the information about each student needed to make this method effective. However, in the not-too-distant future, an inexpensive computer terminal will provide each pupil with an individual electronic tutor. This teaching device will review an individual student's records to determine his specific background and needs and will choose on this basis content to be presented verbally and visually as well as in printed form. The course will be programmed in the memory of a computer. This computer will act like a teacher as it responds to the learner and presents the ideas, the problems, or the tests that are appropriate for the individual learner. It will even answer the questions of the students. As each learner responds to each question, the computer will record the response, any errors, and the time used.

The computer will then give the teacher information on student performance and select the appropriate tutorial material for the next lesson for each student.[1]

Although the computer will do a great deal of the work of the teacher—ask questions, answer questions, work exercises, assign problems, administer tests—it will never replace the teacher. It is more likely to result in the teacher becoming more important than ever, for it is the teacher who will furnish the human element—the values, the affection, the interest, the specific personal attention to problems

[1] This is essentially what is now being done by Professor Patrick Suppes in his experimental project in the elementary schools of Palo Alto, California. In the future, the content will be broadened to include more than computational skills.

beyond even the computer's capacity, which are so essential in the classroom. The computer will release the teacher from clerical, routine tasks so that he will have more time to plan lessons, prepare materials, work with individual students, and develop creative ideas. At the same time, it will demand close monitoring and much more input from the classroom teacher than is currently the case.

Classroom

Past. The typical mathematics classroom was a rectangular space whose four walls were covered by chalkboards. The furniture consisted of a teacher's desk and students' desks in fixed position.

Present. The mathematics classroom of today is not much different from that of the past. Bulletin boards, projection screens, and bookcases may be recent additions. In most classrooms, the students' desks are now movable and a filing case has been added to the room, but little else is changed.

Future. The classrooms of tomorrow will vary greatly in size and appointments according to their use. Some classrooms will include student stations or carrels much like those of the current language laboratories. Each carrel may have a computer terminal, which may be used as a teaching machine. Other classrooms will be learning laboratories for the exploration and discovery of new ideas.

In addition to learning laboratories, the mathematics department will have seminar rooms; a department library; tape, television, and computer center; storerooms; and a department office. These rooms will be air-conditioned, with wall-to-wall carpeting, with adequate light, and comfortably furnished so that the physical conditions will be optimal for learning.

Staff

Past. The mathematics teacher of the past had a weak background in mathematics and often taught courses in several other fields. He generally taught all day and was expected to shoulder extra janitorial and community responsibilities.

Present. Today's mathematics teacher has a bachelor's degree with a major in mathematics and, usually, some summer institute experience. A major of at least 32 semester hours in mathematics is commonly required for certification to teach mathematics in senior high schools. Many noncertified teachers or teachers with credentials from times when requirements were less are still on staffs today. A slight reduction in teaching load has resulted in teachers having about one preparation period per day.

Future. The increased sophistication of mathematics courses will require at least five years of college preparation and a strong major that includes courses not presently in the college curriculum. At the same time, teachers will be required to continue learning and to attend professional conferences. Weekly seminars, in-service courses, and summer study—often with released time and financial support—will be required. Specialization will increase to the point where there will be special teachers for remedial teaching and special teachers of the gifted. Specialization may even extend to various fields so that one teacher will be a specialist in computer science, for example. The number of auxiliary helpers—clerical staff, laboratory assistants, and maintenance workers—will increase greatly.

As objectives broaden, content varies more, and materials increase, the teacher will become increasingly influential. He will need to render professional judgment in his daily decisions, provide the human touch in the classroom, and establish orderly procedures in truly complex situations. To remain up-to-date, he will constantly call on the computer to retrieve information that will suggest consequences for a given decision, but his decisions will be made on ever higher value levels.

Also, staff organization will certainly change. Programs will be much more fluid and students will not expect to see the same teacher at the same hour of each school day. Cooperative teaching arrangements will allow teachers of the same course to develop specific topics for presentation to multiple groups, thus providing each teacher with more preparation time. Teachers will work with groups more varied in size and studying under conditions strikingly different from those of today.

In all of this, the teacher's amount of available time and energy will be acknowledged as a central problem. No longer will a teacher's time for classroom preparation be less than 10 percent of the time he spends in class. Preparation time will rapidly increase to 50 percent or more, because of the additional demands placed on the teacher's instruction time. That means that teachers will be responsible each day for two or three hours of classroom instruction.

Teacher training will begin to reflect the changing requirements of this task. Teachers will be required to learn more mathematics, but will also have to learn about the technology of the new education and how it is to be adapted to classroom instruction. They will also have to learn more about the inner workings of the human mind so that their classroom decisions will be optimum.

Unresolved Issues

The spirit of innovation is an outstanding characteristic of mathematics education today. Revolutionary changes in school mathematics are altering traditional content, practices, classroom organization, and basic views of learning. This rapid change brings with it the danger that innovations may become established as new orthodoxies without anyone asking where these innovations are leading or how they should be instituted. Here are some of the unresolved issues, implicit in the previous discussions in this book, that are facing mathematics educators today:

What are the goals of teaching mathematics? Are we teaching mathematics for vocational needs, for improved citizenship, or for success in advanced courses? Are we teaching mathematics to change our society or to establish values that will maintain our social order? What is the role of mathematics in the life of all pupils of different abilities and cultures? What is the ideal product of our instruction?

How and by whom shall the objectives and the content be determined? What mathematical ideas, skills, attitudes, and habits can be most effectively developed at a given grade level? The new programs have found that we *can* teach complex ideas to very young children. Now the question is what ideas *should* be taught to our students? What new topics should be introduced? What traditional topics should be dropped?

How shall programs be varied to provide for different levels of ability? What enrichment should be provided? How should we accelerate the learning of the talented at all levels? Should the elementary school, secondary school, college sequence be a unified pattern with no separation of special interests? How can instruction be individualized and at the same time provide group interaction?

How do we teach for transfer so that mathematical principles will be used when needed? What specific applications need to be included? Are the social applications, such as installment buying, to be taught by some other department?

What degree of rigor of mathematical precision in definitions and proofs is appropriate at various grade levels? Should mathematical ideas be presented in simple language that, by virtue of its simplicity, is somewhat lacking in precision? What vocabulary and symbolism should be used?

What emphasis should be placed on formal logic and the structure of mathematics? How important is it to stress the basic axioms of our number system such as commutativity, associativity, or distributivity? And if these basic ideas are to be taught, what is the best time to teach them?

One of the greatest dangers of the new programs is that the reorganization may go too far and confront students with concepts whose degree of abstraction exceeds the youngsters' mathematical maturity. Excessive abstraction might result in students' bewilderment and revulsion against mathematics rather than their increased knowledge.

What is the role of intuition and concrete representation of mathematical ideas? How can we best transfer from the physical representation to the symbolic?

What emphasis should be given to computational skill? Can this skill be attained by means other than drill? What level of competence is considered satisfactory at a given level?

What is the role of the computer in the mathematics program? Should mathematics courses teach computer programming? Should the computer be used to solve problems? Are computers and calculators appropriate tools for the low-ability student?

How do we prepare teachers for the new program? How is the effectiveness of a teacher measured? What are appropriate mathematics courses for the teacher? How can we increase the teacher's concern for students and be sensitive to student needs? What preservice and inservice programs will help teachers to carry out the sophisticated programs of the years ahead?

How do we evaluate the effectiveness of a new mathematics program? What behaviors demonstrate the attainment of objectives? What tests can be used to compare two programs each based on different content?

What criteria should be used in selecting instructional aids? What sequence of textbooks is most appropriate? Should each mathematics class have several texts and supplementary books or pamphlets? What is the role of programmed texts?

How shall the achievement of students of different ability be graded? Should the general mathematics class as well as the accelerated class receive the entire range of marks from A to F?

How are students selected for different curriculum tracks? How can provision be made to transfer from one track to another?

The mathematics program of today is largely the textbook treatment of a generation ago. The activities of the mathematics teacher often consist only in answering questions, working sample exercises, reading definitions, giving assignments, and giving tests. This is not enough. The mathematics teacher must use every possible means to help children to be successful in learning and to enjoy learning mathematics. We can no longer afford to ignore the resources now available for improving mathematics teaching. To prepare for anticipated changes in future school mathematics, we must use every possible means for continued improvement in learning how to teach mathematics. It has been the goal of this text to supply ideas and inspiration to meet this challenge. The appendixes that follow suggest additional resources.

Learning Exercises

1. Indicate some specific activities that will prepare you to teach the school mathematics of the decades ahead, including:
   (a) College courses to elect
   (b) Books to read
   (c) Ways to update your knowledge of new technology
2. What is performance contracting? What has been the result in mathematics of specific performance contracts?
3. Contemporary students are concerned with relevance. If you are to be a successful mathematics teacher, you *must* consider mathematics relevant! Justify the relevance of mathematics in terms that you think would convince a student.
4. Many of the so-called romantic critics of education have suggested that modern education is dehumanizing. Indicate some ways that can be used by mathematics teachers to respond to the need to humanize the instructional program.

Suggestions for Further Reading

Begle, Edward G., and James W. Wilson, "Evaluation of Mathematics Programs," in *Mathematics Education*, Yearbook 69, Part I, National Society for the Study of Education, University of Chicago Press, Chicago, 1970, pp. 335–366.

Bell, Max S., "Teaching and Learning Mathematics: 1991," in Douglas B. Aichele and Robert E. Reys (eds.), *Readings in Secondary School Mathematics*, Prindle, Weber, and Schmidt, Boston, 1971, pp. 485–494.

Botts, Truman, "A Changing Mathematics Program" in *Mathematics Education*, Yearbook 69, Part I, NSSE, pp. 449–460.

Fehr, Howard F., "What are the Issues and Trends Associated with Mathematics Education?," in Aichele and Reys (eds.), *Readings in Secondary School Mathematics*, pp. 471–478.

Ferguson, W. Eugene, "The Junior High School Mathematics Program—Past, Present and Future," *Mathematics Teacher*, 63, 5 (May 1970), 383–390.

Johnson, Donovan A., "Next Steps in Mathematics." *Arithmetic Teacher*, 14, 3 (March 1967), 185–189.

Travers, Kenneth J., "Cooperation for Better Mathematics Teacher Education," *Mathematics Teacher*, 64, 4 (April 1971), 373–379.

Willoughby, Stephen S., "Issues in the Teaching of Mathematics," in *Mathematics Education*, Yearbook 69, Part I, NSSE, pp. 260–281.

Appendix A: Instructional Aids

Selected Guides to Audio-visual Materials

Since there is no one source to which you can refer to obtain a comprehensive listing of all the instructional materials available in mathematics, the following publications should be consulted to obtain media information.

* *Index to 8mm Motion Picture Cartridges*, 1969; *Index to Overhead Transparencies*, 1969; *Index to 16mm Educational Films*, 2nd ed., 1969; *Index to 35mm Educational Filmstrips*, 2nd ed., 1969. R. R. Bowker Co., National Information Center for Educational Media (NICEM), 1180 Ave. of the Americas, New York, N. Y. 10036.

The NICEM indexes are organized by subject area. Each of these four volumes reports titles under the categories of general mathematics, arithmetic, calculus, computers, geometry, history, measurement, sets and number systems, slide rule, teaching methods, and trigonometry.

* *Learning Directory*, 7 vols., 1970. Westinghouse Learning Corp., 100 Park Ave., New York, N. Y. 10017.

This comprehensive directory is a compendium of over 200,000 different media items including all of the standard audio-visual materials as well as multimedia kits.

Educational Media Index, 14 vols., 1964. McGraw-Hill Book Company, 330 W. 42 St., New York, N. Y. 10036.

A project of the Educational Media Council. Volume 10, *Mathematics*, covers such areas as arithmetic, algebra, geometry, trigonometry, and higher mathematics.

Educational Product Report (published monthly, October through June). Educational Products Information Exchange (EPIE) Institute, 386 Park Ave. S., New York, N. Y. 10016.

This organization publishes the results of impartial studies of the availability, use, and effectiveness of educational materials.

Educators Purchasing Master, 1970, vol. 2, *Audio-Visual*, ed. Beryl Wellborn. Fisher Publishing, 3 W. Princeton Ave., Englewood, Colo. 80110.

This publication provides a quick index to the commercial media producers' catalogs. It provides the names of producers who have media to offer under such mathematics headings as addition, angles, area, base ten and other bases, circles, computer math, counting, decimals, division, exponents, factors, fractions, geometry, graphing, inequalities, lines, segments, measurement, ratio, percentage, sets, subtraction, and volume.

---

* Items preceded by an asterisk are those we believe to be of greatest interest and value.

8mm Film Directory, 1969, comp. and ed. Grace Ann Kone. Comprehensive Service Corporation, Educational Film Library Association, 250 W. 64th St., New York, N. Y.

This listing provides brief content summaries and audience levels for films under such mathematics topics as math aids (slide rule, computers, scales, etc.), algebra, geometry, and trigonometry.

Guide to Government-Loan Films (16 mm), 1969. Serina Press, 70 Kennedy St., Alexandria, Va. 22305.

Government-sponsored and/or -produced films available for the cost of return postage are listed in this guide under such topics as binary system, probabilities, algebra, geometry, and trigonometry.

Library of Congress Catalog: Motion Pictures and Filmstrips, published with three quarterly cumulations and one for the year; *Educators Guide to Free Films*, 30th ed. 1970, comp. and ed. Mary Foley Horkheimer and John W. Diffor; *Educators Guide to Free Filmstrips*, 22nd ed. 1970, comp. and ed. Mary Foley Horkheimer and John W. Diffor; *Educators Guide to Free Tapes, Scripts, and Transcriptions*, 17th ed., 1970, comp. and ed. Walter A. Wittich and Gertie Hanson Halsted. Educators Progress Service, Randolph, Wis. 53956.

This organization reports free materials dealing with mathematical subjects. For example, the *Films* catalog lists Measurement in Modern Technology as a subject; the *Filmstrip* list offers materials dealing with computers and space navigation; and the *Tapes, Scripts, and Transcriptions* index lists materials under the topics of mathematics and algebra.

Games for Secondary Mathematics Classes

Title	Company	Price
**Krypto*	Creative Publications	$1.50
**Attribute Games*	SEE; McGraw-Hill	8.85
**Make One*	Garrard	1.50
**Prime Drag*	Creative Publications	2.50
**Polyhedra*	Scott, Foresman	1.25
**Quinto*	SEE; 3M	8.95
**TUF*	TUF	5.60
Numble (Number Scrabble)	Creative Publications	4.00
Euclid	Midwest	6.50
Chrominoes	Creative Publications	5.00
Checking (Tic-Tac-Toe in 3 dimensions)	Creative Publications	2.00
Real Numbers	WFF'N PROOF	2.25
Tac-Tickle (Game of Strategy)	WFF'N PROOF	1.25
Tri Nim	WFF'N PROOF	4.50
Come Out Even	Holt	1.92
Orbit the Earth	Scott, Foresman	5.00
Imout	Imout	5.30
Numo	Midwest	8.95
Equations	WFF'N PROOF	3.50
Fraction Dominoes	SEE	2.75

Title	Company	Price
Concentrate	Hope	$1.25
Think a Dot	Childcraft	2.50
Radix Playing Cards	James Lang	1.50
Heads Up	Creative Publications	5.00
Vectors	SEE; Cuisenaire	5.95
Twixt	3M	8.95
Sum-Up	3M	4.50
Comput-A-Tutor	World Wide Computer	6.95
Math Music	Cadaco	6.00
Stocks and Bonds	3M	8.95
Ranko: Arithmetic Drill and Strategy	Midwest	7.95
Acquire	3M	8.00
Bee Line	SEE	4.25
Go Game	Math Media	6.50
Basis	WFF'N PROOF	2.25
Kalah	Creative Publications	3.50
Symmetry Dominoes	SEE	2.75
Haar Hoolim	SEE	3.50
On Sets	WFF'N PROOF	5.50
Scan	Math Media	3.00
I Win	Scott, Foresman	1.25
Prime Factor	Creative Publications Mig-8	4.50
Block It	Creative Publications Mig-119	2.95
Math Match	Creative Publications MGP-35	2.50
One	Creative Publications Mig-5	1.85
Operations Bingo	Creative Publications Mig-3	7.50

Selected Puzzles for Mathematics Classes

Title	Company	Price
**Bali-Buttons*	Creative Publications	$1.00
**Fascinating "15"*	Creative Publications	1.00
Hexogram	Creative Publications	4.50
**Two Piece Pyramid Puzzle*	Worldwide Games	1.45
Instant Insanity	Creative Publications	1.00
Lo-Man	Are-Jay	3.00
Nine Block Puzzle	Worldwide Games	4.95
**Ox Bow Solitaire Game*	Miles Kimball	1.19
Pythagoras	Kohner	1.00
**Soma Puzzle*	Creative Publications	2.25
Tangrams	Creative Publications	1.25
**Tower Puzzle*	Creative Publications	3.50
Pentagram Puzzle	Creative Publications	1.00
Construct-a-Cube	SEE	2.35
Psychepaths	Cuisenaire	4.50

Companies Producing Mathematical Games and Puzzles

Are-Jay Game Company, Inc., 7509 Denison Ave., Cleveland, Ohio 44102
Cadaco, Inc., 310 W. Polk St., Chicago, Ill. 60607
Caddy-Imler Creations, Inc., Box 5097, Inglewood, Cal.
Childcraft Equipment Co., 155 E. 57th St., New York, N. Y. 10010
Creative Playthings, Inc., Princeton, N. J. 08540
Creative Publications, P. O. Box 328, Palo Alto, Cal. 94302
Cuisenaire Company of America, Inc., 12 Church St., New Rochelle, N. Y. 10805
Garrard Press, Champaign, Ill.
Holt, Rinehart & Winston, Inc., Box 2334, Grand Central Station, New York, N. Y. 10017
Hope–Math Games–Turnpike Press, Inc., Box 170, Annandale, Va. 22003
Imout Arithmetic Drill Games, 706 Williamson Bldg., Cleveland, Ohio 44114
Kohner Bros., Inc.—Tryne Game Division, P. O. Box 294, East Paterson, N. J. 07407
James Lang, P. O. Box 224, Mound, Minn. 55364
Math Media, Inc., P. O. Box 345, Danbury, Conn. 06810
McGraw-Hill Book Co., 330 West 42 St., New York, N. Y. 10036
Midwest Publications Co., P. O. Box 307, Birmingham, Mich. 48012
Miles Kimball Co., 41 W. Eighth Ave., Oskosh, Wis. 54901
Milton Bradley Co., 74 Park St., Springfield, Mass.
3M: Minnesota Mining and Manufacturing Co., St. Paul, Minn. 55119
Scott, Foresman and Company, 1900 E. Lake Ave., Glenview, Ill. 60025
*SEE: Selective Educational Equipment, 3 Bridge St., Newton, Mass. 02195
TUF: Avalon Hill Co., 4517 Harford Rd., Baltimore, Md. 21214
WFF 'N PROOF, Box 71-SL, New Haven, Conn. 06501
Worldwide Computer Services Inc., Fortune Bldg., Hartsdale, N. Y. 10530
Worldwide Games, Inc., Box 450, Delaware, Ohio 43015

Toys for the Mathematics Laboratory

Erector Set, A. C. Gilbert Co., New Haven, Conn.
Fiddlestraws, Samuel Gabriel Sons, New Haven, Conn.
Flexagons, Creative Playthings, Herndon, Pa.
Kaleidoscope, Creative Playthings, Herndon, Pa.
Magic Designer, Northern Signal Company, Milwaukee, Wis.
Make-It-Toy, W. R. Benjamin, Granite City, Ill.; Playschool Manufacturing Co., 1750 N. Lawndale Ave., Chicago, Ill.
Mek-N-Ettes, Judy Toy Co., Minneapolis, Minn.
Moby Lynx Construction Set, Kendrey Co., P. O. Box 629, San Mateo, Cal. 94401

Spirograph, Kenner Products Co., Cincinnati, Ohio 45202
Tinker-Toy, A. G. Spaulding and Sons, Evanston, Ill.

Principal Suppliers of Mathematical Models or Classroom Equipment

Berger Scientific, 37 Williams St., Boston, Mass. 02119
Milton Bradley Co., 74 Park St., Springfield, Mass. 01101
Cambosco Scientific Co., 37 Antwerp St., Boston, Mass. 02135
Central Scientific Co., 2600 South Kostner Ave., Chicago, Ill. 60623
*Creative Publications, P. O. Box 328, Palo Alto, Cal. 94302
*Cuisenaire Company of America, 12 Church St., New Rochelle, N. Y. 10805
Denoyer-Geppert Co., 5235 Ravenswood Ave., Chicago, Ill. 60640
Edmund Scientific Co., 100 Edscorp Bldg., Barrington, N. J. 08007
Educational Supply and Specialty Co., 2823 Gaye Ave., Huntington Park, Cal. 90255
Hans K. Freyer, Inc., P. O. Box 245, Westwood, Mass. 02090
Gamco Products, Box 305, Big Spring, Tex. 79720
Houghton Mifflin Co., 2 Park St., Boston, Mass. 02107
Ideal School Supply, 11000 S. Laverghe Ave., Oak Lawn, Ill. 60453
Keuffel and Esser Co., 300 Adams St., Hoboken, N. J. 07030
*Lano Co., 4741 W. Liberty St., Ann Arbor, Mich. 48103
*LaPine Scientific Co., 6001 S. Knox Ave., Chicago, Ill. 60629
Lufkin Rule Co., 1730 Hess St., Saginaw, Mich.
*Math-Masters Labs, Inc., Box 310, Big Spring, Tex. 79720
*Math Media, Inc., P. O. Box 345, Danbury, Conn. 06810
*Math-U-Matic, Inc., 3017 North Stiles, Oklahoma City, Okla. 73105
Pickett, P. O. Box 1515, Santa Barbara, Cal. 93102
Frederick Post Co., 3650 No. Avondale Ave., Chicago, Ill.
Scott, Foresman and Company, 1900 E. Lake Ave., Glenview, Ill. 60025
Scott Scientific Inc., Box 2121, Fort Collins, Colo. 80521
Selective Educational Equipment, 3 Bridge St., Newton, Mass. 02195
Sheldon Equipment Co., Muskegon, Mich. 49443
3 D Magna-Graph Corp., Box 261, Park Ridge, Ill.
Viking Importers, 113 So. Edgemont St., Los Angeles, Cal.
Vis-x-Co., Box 107, Los Angeles, Cal.
Wabash Instruments and Specialties, Box 194, Wabash, Ind. 46992
*W. M. Welch Scientific Co., 7300 N. Linder Ave., Skokie, Ill. 60076
*Yoder Instruments, East Palestine, Ohio 44413

Commercial Projectuals for the Overhead Projector

Visuals or master copies for making projectuals are available for almost any mathematical topic from the following companies. Write to them for catalogs and up-to-date information.

Admaster Prints, Inc., 425 Park Ave., New York, N. Y. 10016
Charles Beseler Co., 219 S. 18 St., East Orange, N. J. 07018
Channing L. Bete Co., Inc., 45 Federal St., Greenfield, Mass. 01301
*John Colburn Associates, Inc., 265 Alice Drive, Wheeling, Ill. 60090
Creative Publications, P. O. Box 328, Palo Alto, Cal. 94302
Creative Visuals, Inc., Box 1911, Big Spring, Tex. 79720
DCA Educational Products Inc., 4865 Stenton Ave., Philadelphia, Pa. 19144
Donnelly's Teaching Aids Inc., 7000 Marlboro Pike S.E., Forestville, Md. 20028
Educational Audio-Visual, 29 Marble Ave., Pleasantville, N. Y. 10570
Encyclopaedia Britannica Films, 425 N. Michigan Ave., Chicago, Ill. 60611
Gamco Products, Box 305, Big Spring, Tex. 79720
C. S. Hammond and Co., 515 Valley St., Maplewood, N. J. 07040
Houghton Mifflin Co., 110 Tremont St., Boston, Mass. 02107
Hubbard Scientific Co., P. O. Box 150, Northbrook, Ill. 60062
Instructo Products Co., 1635 N. 55th St., Philadelphia, Pa. 19131
Keuffel and Esser Co., 20 Whippany Rd., Morristown, N. J. 07960
LaPine Scientific Company, 6001 South Knox Ave., Chicago, Ill. 60629
*Math-Masters Lab, Box 310, Big Spring, Tex. 79720
Math-U-Matic, 607 W. Sheridan, Oklahoma City, Okla.
McGraw-Hill Book Company, 330 W. 42 St., New York, N. Y. 10036
Milliken Publishing Co., 611 Olive St., St. Louis, Mo. 63101
*Minnesota Mining and Manufacturing Co., Visual Products Div., St. Paul, Minn. 55119
Nystrom, 3333 Elston Ave., Chicago, Ill. 60618
Photo and Sound Co., 515 Sunset Blvd., Los Angeles, Cal. 90028
RCA Records, Educ. Dept., 1133 Ave. of Americas, New York, N. Y. 10036
Scott, Foresman and Company, 1900 E. Lake Ave., Glenview, Ill. 60025
Study Scope Co., P. O. Box 689, Tyler, Tex. 75701
*Technifax, Scott Educ. Div., Box 391, Holyoke, Mass. 01040
*Tweedy Transparencies, 308 Hollywood Ave., E. Orange, N. J. 07018
*United Transparencies, Box 688, Binghamton, N. Y. 13902
Visual Materials, Inc., 2549 Middlefield Rd., Redwood City, Cal. 94063
Western Publishing Educational Services, 1220 Mound Ave., Racine, Wis. 53404

Companies Supplying 16 mm Films

Acme Film Labs, Inc., 1161 N. Highland Ave., Hollywood, Cal. 90038

Association Films, 347 Madison Ave., New York, N. Y. 10017

American Telephone & Telegraph Co., Information Dept., 195 Broadway, New York, N. Y. 10007

Audio-Visual Sound Studios, National Education Association, 1201 16th St. N.W., Washington, D. C. 10016

Bailey Films, Inc., 6509 De Longpre Ave., Los Angeles, Cal. 90028

Bank of America, 300 Montgomery, San Francisco, Cal. 94120

Brandon Films, 221 W. 57 St., New York, N. Y. 10019

Calvin Productions, Inc., 1105 Truman Rd., Kansas City, Mo. 64106

Carpenter Center for the Visual Arts, Harvard University, 19 Prescott St., Cambridge, Mass. 02138

Cenco Educational Films, 2600 S. Kostner Ave., Chicago, Ill. 60623

John Colburn Assoc., Inc., P. O. Box 236, Wilmette, Ill. 60091

Contemporary Films, McGraw-Hill Book Company, 330 W. 42 St., New York, N. Y. 10036

Coronet Films, 65 E. South Water St., Coronet Bldg., Chicago, Ill. 60601

Davidson Films, 1757 Union St., San Francisco, Cal. 94123

Walt Disney Production, Educational Film Div., 350 S. Buena Vista Ave., Burbank, Cal. 91503

Charles and Ray Eames, 901 Washington Blvd., Venice, Cal. 90291

Educational Service Inc., 47 Galen St., Watertown, Mass. 02100

Encyclopaedia Britannica Educational Corp., 425 N. Michigan Ave., Chicago, Ill. 60611

*Film Association of California, 11559 Santa Monica Blvd., Los Angeles, Cal. 90025

Gateway Productions, Inc., 1859 Powell St., San Francisco, Cal. 94111

General Electric, Educational Films, 60 Washington Ave., Schenectady, N. Y. 12305

General Motors Corp., 3044 W. Grand Blvd., Detroit, Mich. 48202

Holt, Rinehart & Winston, 383 Madison Ave., New York, N. Y. 10017

International Business Machines, Armonk, N. Y. 10504

*International Film Bureau, 332 S. Michigan Ave., Chicago, Ill. 60604

*Knowledge Builders, 31 Union Square W., New York, N. Y. 10003

Library Film, Inc., 723 Seventh Ave., New York, N. Y. 10019

Martin Moyer Productions, 900 Federal Ave., Seattle, Wash. 98102

McGraw-Hill Text Films, 330 W. 42 St., New York, N. Y. 10036

Minnesota Mining and Manufacturing Co., Medical Film Library, 2501 Hudson Rd., St. Paul, Minn. 55119

Modern Film Rentals, 2323 New Hyde Park Rd., New Hyde Park, N. Y. 11040

*Modern Learning Aids, 16 Spear St., San Francisco, Cal. 94105

Moody Institute of Science, 1200 E. Washington Blvd., Whittier, Cal. 90606

National Council of Teachers of Mathematics, 1200 Sixteenth St. N.W., Washington, D. C. 20036

David Nulsen Enterprises, 3211 Pico Blvd., Santa Monica, Cal. 90405

Penn State University, Psych. Cinema Register, University Park, Pa. 16802
Science Research Association, 259 E. Erie St., Chicago, Ill. 60611
Silver-Burdett, Park and Colombia Rd., Morristown, N. J. 07960
L. S. Starrett Co., 121 Crescent St., Athol, Mass. 01331
Sturgis-Grant Productions, 328 E. 44 St., New York, N. Y. 10017
Universal Education & Visual Arts, Div. of Universal City Studies, Inc., 221 Park Ave. S., New York, N. Y. 10003
University of California, Ext. Medical Center, Film Dist., 2223 Fulton St., Berkeley, Cal. 94720
Webster College, Webster Groves, Mo. 63119

Companies Distributing 35 mm Mathematics Filmstrips

American Book Co., 55 Fifth Ave., New York, N. Y. 10016
Gilbert Altschull Productions, 909 W. Diversey Pkwy, Chicago, Ill. 60614
Bailey Film Association, 11559 Santa Monica Blvd., Los Angeles, Cal. 90025
Stanley Bowmar Co., 4 Broadway, Valhalla, N. Y. 10595
Herbert E. Budek, P. O. Box 307, Santa Barbara, Cal. 93102
Cenco Educational Films, 2600 S. Kostner Ave., Chicago, Ill. 60623
John Colburn Assoc. Inc., 1215 Washington Ave., Wilmette, Ill. 60091
*Colonial Films, 752 Spring St. S.W., Atlanta, G. 30308
Coronet Films, Coronet Bldg., 65 E. South Water St., Chicago, Ill. 60601
Cuisenaire Co. of America Inc., 235 E. 50 St., New York, N. Y. 10022
*Curriculum Materials Corp., 1319 Vine St., Philadelphia, Pa. 19107
Educational Filmstrips, Box 1031, Huntsville, Tex. 77340
Herbert M. Elkins Co., 10031 Commerce Ave., Tujunga, Cal. 91042
Encyclopaedia Britannica Educational Corp., 425 N. Michigan Ave., Chicago, Ill. 60611
*Eye Gate House Inc., 146–01 Archer Ave., Jamica, N. Y. 11435
*Filmstrip House, 432 Park Ave. S., New York, N. Y. 10016
Jam Handy Organization, 2781 E. Grand Blvd., Detroit, Mich. 48211
International Film Bureau, 322 S. Michigan Ave., Chicago, Ill. 60604
Math-U-Matic, 607 W. Sheridan, Oklahoma City, Okla.
*McGraw-Hill Text Films, 330 W. 42 St., New York, N. Y. 10036
Photo and Sound Co., 5515 Sunset Blvd., Los Angeles, Cal. 90028
Popular Science Publishing Co., 239 Fairview Blvd., Inglewood, Cal. 90302
Sigma Educational Films, P. O. Box 1235, Studio City, Calif. 91604
*Society for Visual Education, Inc., Div. Gen. Precision Equip. Corp., 1345 Diversey Pkwy., Chicago, Ill. 60614
Sterling Educational Films, 241 E. 34 St., New York, N. Y. 10016
Stipes Publishing Co., 10 Chester St., Champaign, Ill. 61820
Visual Education Consultants, Inc., 2066 Helena St., Box 52, Madison, Wis. 53701
Visual Sciences, Box 599, Suffern, N. Y. 10901

Companies Distributing 8 mm Film Loops

American Film Productions, Inc., 1540 Broadway, New York, N. Y. 10036
Avis Films, 2408 W. Olive Ave., Burbank, Cal. 91506
Bailey Film Association, 11559 Santa Monica Blvd., Los Angeles, Cal. 90025
BFA Educational Media, 2211 Michigan Ave., Santa Monica, Cal. 90404
Ealing Corp., 2225 Massachusetts Ave., Cambridge, Mass. 02140
Encyclopaedia Britannica Educational Corp., 425 N. Michigan Ave., Chicago, Ill. 60611
Gateway Educational Films, Ltd., 470–472 Green Lanes, Palmers Green, London N.13, Eng.
Halas and Batchelor, 317 Kean St., London N.13, Eng.
Hester and Associates, P. O. Box 20812, Dallas, Tex. 75220
International Communications Films, 1371 Reynolds Ave., Santa Ana, Cal. 92705
Macmillan Company, 866 Third Ave., New York, N. Y. 10022
McGraw-Hill, Text Films, 330 W. 42 St., New York, N. Y. 10036
National Film Board of Canada, Toronto, Ontario, Canada.
Arthur Rank Organization, 11 Cumberland Ave., London, Eng.
Selective Educational Equipment, Inc., 3 Bridge St., Newton, Mass. 02195
Sterling Educational Films, P. O. Box 8497, Universal City, Los Angeles, Cal. 91608
Thorne Films, Inc., 1229 University Ave., Boulder, Colo. 80302
Universal Education and Visual Aids, 221 Park Ave. S., New York, N. Y. 10003

Commerical Models for a Mathematics Laboratory

Abacus: bases 10, 5, 2
Area boards
Attribute blocks
Base and place demonstrator
Binary counter
Binomial squares, rectangles, strips, cubes, prisms
Calculator, computer terminal, Napiers bones
Circle demonstration board
Circle area demonstrator
Dissected cone
Flannel board
Fraction and percent demonstration devices
Fraction and percent slide rules, nomographs
Geometric figures: plane and solid
Geometric demonstration devices: triangles, quadrilaterals, circles
Geo-strips

Graph stencil, pegboard graph, magnetic graph board
Indirect measurement devices such as clinometer, transit, angle mirror
Map projection model
Map reading model
Measurement devices for length, area, angles, volume, weight
Measuring wheel for determining pi
Micrometer demonstrator, vernier calipers
Mirror cards
Number line
Orthographic projection models
Pantograph
Pegboard, geoboard
Plastic models of polyhedra, pyramids, cones, cylinders, sphere
Probability board, probability kit
Projection models
Pythagorean theorem demonstrator
Scale balance, sampling kit
Spherometer
Slated globe
Slide rule
Sticks for polygons, polyhedrons, binomial expansions
String models
Trigonometric functions device
Volume measurement set

Computer Companies

Clary Corporation, 408 Juniper St., San Gabriel, Cal. 91776

Computer Control Company, Old Connecticut Path, Framingham, Mass. 13017

Control Data Corporation, 8100 South 34 Ave., Minneapolis, Minn. 55420

Digital Equipment Corporation, 146 Main St., Maynard, Mass. 01754

Fabri-Tek, 1261 S. Boyle Ave., Los Angeles, Cal. 90054

Frieden Inc., 2350 Washington Ave., San Leandro, Cal. 94577

Hewlett Packard, Palo Alto, Cal. 94304

Honeywell, 2701 So. 4 Ave., Minneapolis, Minn. 55408

IBM, Data Processing Division, 112 East Post Road, White Plains, N. Y. 10601

Mathatronics Inc., 257 Crescent St., Waltham, Mass. 02154

National Cash Register, Main and K Street, Dayton, Ohio 45409

Olivetti Corp. of America, 500 Park Ave., New York, N. Y. 10022

UNIVAC Division, Sperry Rand Corporation, 1200 Ave. of the Americas, New York, N. Y. 10036

Wang Laboratories, 836 North St., Tewksbury, Mass. 01876

Westinghouse Electric Corp., Westinghouse Bldg., Gateway Center, Pittsburgh, Pa. 15222

Bulletin Boards

Suggested Bulletin Board Displays

What's My Name? Biographies of mathematicians for identification.
Curves for Swinging in Space. Spiral, cardioid, ellipsis, cycloid parabola, catenary.
The Magicians' Geometry. Topology.
What's the Trick about Magic Squares?
How to Confuse Your Friends. Paradoxes.
Why Take Chances? Probability.
There's a Formula in Your Future.
How Far from Here to There? Map projections.
Measuring Distances in Space. Indirect measurement.
How to Lose a Square. Fallacy, such as 64 = 65.
Number Shapes. Triangular, square, and pentagonal numbers.
How Smart Are You? Puzzles.
Mathematics Predicts. Predictions of weather, eclipse, location of Pluto.
Mathematical Art. Curves, symmetrics, op art, moive patterns.
These Things Never End. Infinities, sequences.
Where Is the Fallacy? Proof that every triangle is isosceles.
One Way To Solve Problems. Flow charts and computer programs.
This Series Is Golden. Fibonacci Series.
Mysteries in Mathematics. Unsolved problems.
Say It with Formulas. Applications of mathematics in science.
Mathemagic. Tricks with numbers.
What's New in Mathematics? Linear programming.
The House that Math Built. A finite mathematical system.
Math Made Easy. Shortcuts.
Where Did the Bees Learn Mathematics? The mathematics of the honeycomb.

Bulletin Board Booklets

Caroline, Sister Mary, *Bulletin Boards for the New Math.* F. A. Owen Publishing Co., New York, 1965.

Johnson, Donovan A., and Charles Lund, *Bulletin Board Displays for Mathematics*, Dickenson Publishing Co. Inc., Belmont, Cal., 1967.

Johnson, Donovan A., *How to Use Your Bulletin Board*, National Council of Teachers of Mathematics, Washington, D. C., 1954.

Koskey, T. A., *Baited Bulletin Boards*, Fearon Publishers, Belmont, Cal., 1957.

Appendix B: Professional Periodicals, Books, and Pamphlets

Professional Journals

American Mathematical Monthly. Mathematical Association of America, 1225 Connecticut Ave. N.W., Washington, D. C. 20036.

Journal for Research in Mathematics Education. National Council of Teachers of Mathematics, 1201 Sixteenth Street N.W., Washington, D. C. 20036.

Mathematics Magazine. Mathematical Association of America, 1225 Connecticut Avenue N.W., Washington, D. C. 20036.

Mathematics Teaching. Association of Teachers of Mathematics, Market Street Chambers, Nelson, Lancashire, Eng.

The Pentagon. Kappa Mu Epsilon, Central Michigan University, Mount Pleasant, Mich. 48858.

The Science Teacher. 1201 Sixteenth Street N.W., Washington, D. C. 20036.

Scientific American. 415 Madison Ave., New York, N. Y. 10017.

Scripta Mathematica. Yeshiva University, Amsterdam Ave. and West 186th St., New York, N. Y. 10033.

See also list of reference tools before Chapter 1.

Mathematics Students Journals

The Mathematical Log. Mu Alpha Theta, Box 504, University of Oklahoma, Norman, Okla.

Mathematical Pie. 9 Naseby Road, Solihull, Warwick, Eng..

The Mathematics Student Journal. National Council of Teachers of Mathematics, 1201 Sixteenth Street N.W., Washington, D. C. 20036.

O. U. Mathematics Letter. O. U. Mathematics Service Committee, University of Oklahoma, Norman, Okla. 73069.

Pythagoras. Fanfare Educational Publishing Co., Fanfare House, 174 Chingford Mount Rd., London E4, Eng.

Newsletters

Association Happenings, School Science and Mathematics Association, P. O. Box 246, Bloomington, Ind. 47401.

Bulletin for Leaders. National Council of Teachers of Mathematics, 1201 Sixteenth Street N.W., Washington, D. C. 20036.

CBMS Newsletter. Conference Board of the Mathematical Sciences, 2100 Pennsylvania Avenue N.W., Suite 834, Washington, D. C. 20037.

NCTM Newsletter. National Council of Teachers of Mathematics, 1201 Sixteenth Street N.W., Washington, D. C. 20036.

Science Education News. American Association for the Advancement of Science, 1515 Massachusetts Avenue N.W., Washington, D. C. 20005.

NCTM Publications

Yearbooks and pamphlets are available from the National Council of Teachers of Mathematics, 1201 Sixteenth St., N.W., Washington, D. C. 20036. Copies of out-of-print yearbooks are available from AMS Press, Inc., 56 East 13th St., New York, N. Y. 10003.

Yearbooks

1. *A General Survey of Progress in the Last Twenty-five Years*, 1926.
2. *Curriculum Problems in Teaching Mathematics*, 1927.
3. *Selected Topics in the Teaching of Mathematics*, 1928.
4. *Significant Changes and Trends in the Teaching of Mathematics Throughout the World Since 1910*, 1929.
5. *The Teaching of Geometry*, 1930.
6. *Mathematics in Modern Life*, 1931.
7. *The Teaching of Algebra*, 1932.
8. *The Teaching of Mathematics in the Secondary School*, 1933.
9. *Relational and Functional Thinking in Mathematics*, 1934.
10. *The Teaching of Arithmetic*, 1935.
11. *The Place of Mathematics in Modern Education*, 1936.
12. *Approximate Computation*, 1937.
13. *The Nature of Proof*, 1938.
14. *The Training of Mathematics Teachers*, 1939.
15. *The Place of Mathematics in Secondary Education*, 1940.
16. *Arithmetic in General Education*, 1941.
17. *A Source Book of Mathematical Applications*, 1942.
18. *Multi-Sensory Aids in the Teaching of Mathematics*, 1945.
19. *Surveying Instruments: Their History and Classroom Use*, 1947.
20. *The Metric System of Weights and Measures*, 1948.
21. *The Learning of Mathematics: Its Theory and Practice*, 1953.

22. *Emerging Practices in Mathematics Education*, 1954.
23. *Insights into Modern Mathematics*, 1957.
24. *The Growth of Mathematical Ideas, Grades K–12*, 1959.
25. *Instruction in Arithmetic*, 1960.
26. *Evaluation in Mathematics*, 1961.
27. *Enrichment Mathematics for the Grades*, 1962.
28. *Enrichment Mathematics for High School*, 1963.
29. *Topics in Mathematics for Elementary School Teachers*, 1964.
30. *More Topics in Mathematics for Elementary School Teachers*, 1968.
31. *Historical Topics for the Mathematics Classroom*, 1969.
32. *A History of Mathematics Education in the United States and Canada*, 1970.
33. *The Teaching of Secondary School Mathematics*, 1971.
34. *Instructional Aids in Mathematics*, 1972

Pamphlets

Administrative Responsibility for School Mathematics
Bibliography of Recreational Mathematics, vol. 1
Bibliography of Recreational Mathematics, vol. 2
Boxes, Squares, and Other Things
Chips from the Mathematical Log
Computer-assisted Instruction
Computer Facilities for Mathematics Instruction
Computer Oriented Mathematics
Continued Fractions
Continuing Revolution in Mathematics
Cumulative Index: The Mathematics Teacher, 1908–1965
Elementary School Mathematics
High School Mathematics Library
How to Study Mathematics
How to Use the Overhead Projector
How to Use Your Bulletin Board
In-Service Education
Introduction to an Algorithmic Language (BASIC)
Mathematical Challenges
Mathematics for Elementary School Teachers (paperback)
Mathematics Library—Elementary and Junior High School
Mathematics Teaching as a Career
Mathematics Tests Available in the U.S.
Math Teaching with Special Reference to Epistemological Problems

Numbers and Numerals
Number Stories of Long Ago (paperback)
Number Story, The
Paper Folding for the Mathematics Class
Piagetian Cognitive-Developmental Research and Mathematical Education
Polyhedron Models for the Classroom
Portrait of 2
Puzzles and Graphs
Pythagorean Proposition
Readings in Geometry from the Arithmetic Teacher
Readings in the History of Math Education
Research and Development in Education: Mathematics
Research in Mathematics Education
Revolution in School Mathematics
School Mathematics Contests
Secret Codes
Some Ideas about Number Theory
Soviet Secondary Schools for the Mathematical Talented
20th Century Algebra
Vectors in Three Dimensional Geometry

Sources of Free or Inexpensive Materials

The following companies have provided pamphlets and charts for mathematics teachers. Write to them for information regarding materials that are currently available and for possible costs.

American Automobile Association, 1712 G St. N.W., Washington, D. C. 20006
American Bankers Association, Public Relations Dept., 90 Park Ave., New York, N. Y. 10018
American Telephone & Telegraph Co., Long Lines Dept., 32 Ave. of the Americas, New York, N. Y. 10013
Automobile Manufacturers Association, Educational Services, 320 New Center Bldg., Detroit, Mich. 48202
Bausch & Lomb, Inc., Rochester, N. Y. 14602
Bell Telephone Laboratories, Mountain Ave., Murray Hill, N. J. 07974
Boy Scouts of America, National Council, New Brunswick, N. J. 08903
Chase Manhattan Bank Museum of Moneys of the World, 1254 Ave. of the Americas, New York, N. Y. 10020
Dunn and Bradstreet, Inc., 99 Church St., New York, N. Y. 10007

The Duodecimal Society of America, 20 Carlton Place, Staten Island, N. Y. 10304
Francis I. Dupont Co., 1 Wall St., New York, N. Y. 10005
Federal Reserve Bank, Minneapolis, Minn. 55440
Ford Motor Company, Educational Affairs Dept., The American Rd., Dearborn, Mich. 48120
General Electric Co., Educational Publications, One River Road, Schenectady, N. Y. 12305
General Motors Corp., Public Relations Staff, Detroit, Mich., 48202
Household Finance Corp., Money Management Institute, Prudential Plaza, Chicago, Ill. 60601
IBM Editorial Promotions, Armonk, N. Y. 10504
Institute of Life Insurance, Educational Div., 277 Park Ave., New York, N. Y. 10017
Internal Revenue Service, Training Div., Washington, D. C. 20224
Keuffel and Esser Company, 20 Whippany Rd., Morristown, N. J. 07960
Litton Industries, Inc., Beverly Hills, Cal. 90213
Lufkin Rule Co., P. O. Box 728, Apex, N. C. 27502
Merrill, Lynch, Pierce, Fenner, & Smith, 70 Pine St. New York, N. Y. 10033
Metric Association, Inc., 2004 Ash St., Waukegan, Ill. 60085
Monroe International, 550 Central Ave., Orange, N. J. 07051
National Aerospace Education Council, 616 Shoreham Building, Washington, D. C. 20055
National Better Business Bureau, 230 Park Ave., New York, N. Y. 10017
National Consumer Finance Assoc., 1000 Sixteenth St. N.W., Washington, D. C. 20036
National Industrial Conference Board, Road Map Educational Program, 845 Third Ave., New York, N. Y. 10022
New York Stock Exchange, School and College Relations, 11 Wall St., New York, N. Y. 10005

Publishers of Mathematics Tests

Bureau of Educational Research and Service, State University of Iowa, Iowa City, Iowa 52240
California Test Bureau, Del Monte Research Park, Monterey, Cal.
Educational Test Bureau, Publishers Building, Circle Pines, Minn. 55014
Educational Testing Service, 20 Nassau St., Princeton, N. J. 08540
Harcourt Brace Jovanovich, 757 Third Ave., New York, N. Y. 10017
Houghton Mifflin Company, 2 Park St., Boston, Mass. 02107
Psychological Corporation, 304 E. 45 St., New York, N. Y. 10017
Science Research Associates, 57 W. Grand Ave., Chicago, Ill. 60610

Books for the Professional Library of a Mathematics Teacher

See also list of reference materials before Chapter 1.

Association of Assistant Masters in Secondary Schools, *The Teaching of Mathematics*, Cambridge University Press, New York, 1957.

Association of Teachers of Mathematics, *Some Lessons in Mathematics*, Cambridge University Press, Cambridge, 1964.

Brown, Claude H., *The Teaching of Secondary Mathematics*, Harper & Row, Publishers, Inc., New York, 1953.

Davis, D. R., *The Teaching of Mathematics*, Addison-Wesley Publishing Co., Inc., Reading, Mass., 1951.

Dienes, Z. P., *Building Up Mathematics*, Hutchinson Educational, Ltd., London, 1960.

Hadamard, Jacques, *The Psychology of Invention in the Mathematical Field*, Dover Publications, Inc., New York, 1954.

Hedges, William D., *Testing and Evaluation for the Sciences*, Wadsworth Publishing Co., Inc., Belmont, Cal., 1966.

Johnson, Donovan, and Robert Rahtz, *The New Mathematics in Our Schools*, The Macmillan Company, New York, 1966.

Kenna, L. A., *Understanding Mathematics with Visual Aids*, Littlefield Adams & Company, Totowa, N. J., 1962.

Kinney, Lucien B., and Richard C. Purdy, *Teaching Mathematics in the Secondary School*, Holt, Rinehart & Winston, Inc., New York, 1952.

Kinsella, John J., *Secondary School Mathematics*, Center for Applied Research in Education, New York, 1965.

Land, F. W., *New Approaches to Mathematics in Teaching*, The Macmillan Company, New York, 1963.

Lovell, K., *The Growth of Basic Mathematical and Scientific Concepts in Children*, Philosophical Library, Inc., New York, 1961.

Mager, Robert F., *Developing Attitudes toward Learning*, Fearon Publishers, Belmont, Cal., 1968.

Midonick, Henrietta O. (ed.), *The Treasury of Mathematics* (Anthology of Historical Items), Philosophical Library, Inc., New York, 1965.

Piaget, Jean, *The Child's Concept of Number*, Routledge and Kegan Paul, Ltd., London, 1961.

Progressive Educational Association, *Mathematics in General Education*, Appleton-Century-Crofts, New York, 1941.

Saaty, T. L., and F. J. Weyl, *The Spirit and Uses of the Mathematical Sciences*, McGraw-Hill Book Company, New York, 1969.

Sawyer, W. W., *The Search for Pattern*, Penguin Books, Inc., Baltimore, 1970.

———, *Vision in Elementary Mathematics*, Penguin Books, Inc., Baltimore, 1964.

Books and Pamphlets Containing Laboratory Lessons

Anderson, James T., *Space Concepts through Aestheometry*, Aestheometry Inc., 1903 Coyton Ave., Artesia, N. M., 1968.

Biggs, Edith, and J. R. MacLean, *Freedom to Learn*, Addison-Wesley (Canada) Ltd., Don Mills, Ont., 1969.

Buckeye, D. A., W. A. Ewbank, J. L. Ginther, *A Cloudburst of Math Lab Experiments*, vols. 1, 2, and 3, Midwest Publications, Birmingham, Mich., 1971.

Cohn, Donald, *Inquiry in Mathematics Via the Geo-Board*, Walker & Company, New York, 1967.

Cundy, H. M., and A. P. Rollet, *Mathematical Models*, Oxford University Press, New York, 1951.

Del Grande, John J., *Geoboards and Motion Geometry*, Scott, Foresman Co., Chicago, 1972.

Experiences in Mathematical Ideas, vols. 1 and 2, National Council of Teachers of Mathematics, Washington, D. C., 1970.

Jacobs, H. R., *Mathematics: A Human Endeavor*, W. H. Freeman and Co. Publishers, San Francisco, 1970.

Johnson, D. C., L. L. Hatfield, P. W. Katzman, T. E. Kieren, D. E. LaFrenz, J. W. Walther, *Computer Assisted Mathematics Program*, Scott, Foresman and Company, Glenview, Ill., 1968.

Johnson, Donovan, Viggo Hansen, Wayne Peterson, Jesse Rudnik, Ray Cleveland, Carey Bolster, *Activities in Mathematics, First Course*; *Activities in Mathematics, Second Course*; *Applications in Mathematics, Course A*; *Applications in Mathematics, Course B*; Scott Foresman and Company, Glenview, Ill., 1972.

Johnson, Donovan, *Paper Folding for the Mathematics Class*, National Council of Teachers of Mathematics, Washington, D. C., 1957.

Jones, Madeline, *The Mysterious Flexagons*, Crown Publishers, Inc., New York, 1966.

Kadesch, Robert R., *Math Menagerie*, Harper & Row, Publishers, New York, 1970.

Kapur, J. N., *Suggested Experiments in School Mathematics*, vols. 1 and 2, Arya Book Depot, New Delhi-5, India, 1968.

Kidd, Kenneth, S. S. Myers, D. W. Cilley, *The Laboratory Approach to Mathematics*, Science Research Associates, Chicago, 1970.

Krulik, Stephen, *A Handbook of Aids for Teaching Junior–Senior High School Mathematics*, W. B. Saunders Company, Philadelphia, 1971.

———, *A Mathematics Laboratory Handbook for Secondary Schools*, W. B. Saunders Company, Philadelphia, 1972.

National Council of Teachers of Mathematics, *Experiences in Mathematical Ideas*, vol. 112, Washington, D. C., 1970.

Nelson, L. D., and W. W. Sawyer, *Mathex*, Encyclopaedia Britannica, Toronto, 1970.

Pearcy, J. F. F., and K. Lewis, *Experiments in Mathematics*, Stages 1, 2, and 3, Houghton Mifflin Company, Boston, 1967.

School Mathematics Project, Books 1, 2, 3, and 4, Cambridge University Press, New York, 1965. (Available from Cuisenaire Company of America, Inc., 12 Church St., New Rochelle, N. Y. 10805.)

Seymour, Dale, *Aftermath*, vols. 1, 2, 3, and 4, Creative Publications, Palo Alto, Cal., 1970.

Shuster, C. N., and Fred L. Bedford, *Field Work in Mathematics*, American Book Company, New York, 1935. (Available only from Yoder Instrument Co., East Palestine, Ohio.)

Steinhaus, H., *Mathematical Snapshots*, Oxford University Press, New York, 1950.

Swartz, Clifford E., and Roy A. Gallant, *Measure and Find Out*, Books 1, 2, and 3, Scott, Foresman and Company, Glenview, Ill., 1969.

Wenninger, M. J., *Polyhedron Models for the Classroom*, National Council of Teachers of Mathematics, Washington, D. C., 1966.

Principal Publishers of Mathematics Textbooks

Academic Press, Inc., 111 Fifth Ave., New York, N. Y. 10003
Addison-Wesley Publishing Co., Inc., 508 South St., Reading, Mass. 01867
Allyn & Bacon, Inc., 470 Atlantic Ave., Boston, Mass. 02210
American Book Company, 450 W. 33 St., New York, N. Y. 10001
Appleton-Century-Crofts, 440 Park Ave. S., New York, N. Y. 10016
Barnes & Noble, Inc., 105 Fifth Ave., New York, N. Y. 10003
Blaisdell Publishing Co., 275 Wyman St., Waltham, Mass. 02154
The Bobbs-Merrill Co., Inc., 4300 W. 62 St., Indianapolis, Ind. 46268
Brooks/Cole Publishing Co., Belmont, Cal. 94002
Bruce Publishing Co., 400 N. Broadway, Milwaukee, Wis. 53201
Byrne Publishing, 105 Fifth Ave., New York, N. Y. 10003
Cambridge University Press, 32 E. 57 St., New York, N. Y. 10022
Chelsea Publishing Co., Inc., 159 E. Tremont Ave., Bronx, N. Y. 10453
Creative Publications, P. O. Box 328, Palo Alto, Cal. 94304
Thomas Y. Crowell Company, 201 Park Ave. S., New York, N. Y. 10003
Cuisenaire Company of America, Inc., 12 Church St., New Rochelle, N. Y. 10805
The John Day Company, Inc., 257 Park Ave. S., New York, N. Y. 10010
Dickenson Publishing Co., Inc., Encino, Cal. 91316
Doubleday & Company, Inc., 277 Park Ave., New York, N. Y. 10017
Dover Publications, Inc., 180 Varick St., New York, N. Y. 10014
Emerson Books, Inc., 251 W. 19 St., New York, N. Y. 10011
Encyclopaedia Britannica, Inc., 425 N. Michigan Ave., Chicago, Ill. 60611
Fearon Publishers, 6 Davis Drive, Belmont, Cal. 94002
Franklin Teaching Aids, 847 N. East St., San Bernadino, Cal. 92410
W. H. Freeman & Co., Publishers, 660 Market St., San Francisco, Cal. 94104
Ginn and Company, Statler Bldg., Back Bay, P. O. Box 191, Boston, Mass. 02117
Glencoe Press, 8701 Wiltshire Blvd., Beverly Hills, Cal. 90211
Harcourt Brace Jovanovich, Inc., 757 Third Ave., New York, N. Y. 10017
Harper & Row, Publishers, Inc., Elhi Division, 2500 Crawford Ave., Evanston, Ill. 60201
D. C. Heath & Company, 125 Spring St., Lexington, Mass. 02173
Holbrook Press, Inc., 470 Atlantic Ave., Boston, Mass. 02210
Holden-Day, Inc., 500 Sansome St., San Francisco, Cal. 94111
Holt, Rinehart & Winston, Inc., 383 Madison Ave., New York, N. Y. 10017
Houghton Mifflin Co., 2 Park St., Boston, Mass. 02107

Laidlaw Brothers, Thatcher and Madison Sts., River Forest, Ill. 60305
Lyons & Carnahan, 407 E. 25 St., Chicago, Ill. 60616
The Macmillan Company, 866 Third Ave., New York, N. Y. 10022
McCromick-Mathers Publishing Company, Inc., 450 W. 33 St., New York, N. Y. 10001
McGraw-Hill Book Company, 330 W. 42 St., New York, N. Y. 10036
Charles E. Merrill Publishing Co., 1300 Alum Creek Drive, Columbus, Ohio 43216
Oxford University Press, Inc., 200 Madison Ave., New York, N. Y. 10016
Pergamon Press, Inc., Maxwell House, Fairview Park, Elmsford, N. Y. 10523
Prentice-Hall, Inc., Educational Books Div., Englewood Cliffs, N. J. 07632
Prindle, Weber & Schmidt, Inc., 53 State St., Boston, Mass. 02109
Rand McNally & Co., Box 7600, Chicago, Ill. 60680
Random House/Singer School Div., 201 E. 50 St., New York, N. Y. 10022
W. B. Saunders Company, W. Washington Sq., Philadelphia, Pa. 19105
Science Research Associates, Inc., 259 E. Erie St., Chicago, Ill. 60611
Scott, Foresman and Company, 1900 E. Lake Ave., Glenview, Ill. 60025
Charles Scribner's Sons, 597 Fifth Ave., New York, N. Y. 10017
Silver Burdett Company, 250 James St., Morristown, N. J. 07960
L. W. Singer Co., Inc., 249–259 W. Erie Blvd., Syracuse, N. Y. 13202
Van Nostrand Reinhold Company, 450 W. 33 St., New York, N. Y. 10001
A. C. Vroman, Inc., 2085 E. Foothill Blvd., Pasadena, Cal. 91109
Wadsworth Publishing Co., Inc., Belmont, Cal. 94002
Webster Division, McGraw-Hill Book Company, Manchester Rd., Manchester, Mo. 63011
J. Weston Walch, Box 1075, Portland, Me. 04104
John Wiley & Sons, Inc., 605 Third Ave., New York, N. Y. 10016

Appendix C: Enrichment Materials

Mathematics Books for the School Library

The following publications of the National Council for Teachers of Mathematics give extensive bibliographies of books for the school library:

Mathematics Library—Elementary and Junior High School
The High School Mathematics Library

Junior High School

Abbott, E. A., *Flatland*, Dover Publications, Inc., New York, 1932.

Adler, I., *The Giant Golden Book of Mathematics*, Golden Press, New York, 1960.

Bakst, A., *Mathematics, Its Magic and Mastery*, Van Nostrand Reinhold Company, New York, 1952.

Barr, S., *Experiments in Topology*, Thomas Y. Crowell Company, New York, 1964.

Beiler, A. H., *Recreations in the Theory of Numbers*, Dover Publications, Inc., New York, 1964.

Bergamini, D., *Mathematics*, Time Inc., New York, 1963.

Bowers, H., and J. E. Bowers, *Arithmetical Excursions*, Dover Publications, Inc., New York, 1961.

Burger, Dionys, *Sphereland: A Fantasy about Curved Surfaces and an Expanding Universe*, Thomas Y. Crowell Company, New York, 1965.

Diggins, J., *String, Straightedge and Shadow*, The Viking Press, Inc., New York, 1965.

Fadiman, C. *Fantasia Mathematica*, Simon and Schuster, Inc., New York, 1958.

Frolichstein, J., *Mathematical Fun, Games, and Puzzles*, Dover Publications, Inc., New York, 1962.

Gardner, M., *Mathematical Puzzles and Diversions*, Simon & Schuster, Inc., New York, 1959.

———, *Mathematics Magic and Mystery*, Simon & Schuster, Inc., New York, 1956.

———, *New Mathematical Diversions from Scientific American*, Simon & Schuster, Inc., New York, 1966.

Greenblatt, M. H., *Mathematical Entertainments*, Thomas Y. Crowell Company, New York, 1965.

Heath, R. V., *Mathemagic, Magic Puzzles and Games with Numbers*, Dover Publications, Inc., New York, 1933.

Hogben, L., *Wonderful World of Mathematics*, Random House, Inc., New York, 1955.

Hunter, J. A. H., and J. S. Madachy, *Mathematical Diversions*, Van Nostrand Reinhold Company, New York, 1963.

Jacoby, Oswald, *Mathematics for Pleasure*, McGraw-Hill Book Company, New York, 1962.

Kasner, E., and J. Newman, *Mathematics and the Imagination*, Simon & Schuster, Inc., New York, 1940.

Kendall, P. M. H., and G. M. Thomas, *Mathematical Puzzles for the Connoisseur*, Thomas Y. Crowell Company, New York, 1964.

Lukas, C., and E. Tarjan, *Mathematical Games*, Walker & Company, New York, 1968.

Madachy, J. S., *Mathematics on Vacation*, Charles Scribner's Sons, New York, 1966.

Newman, J. R., *The World of Mathematics*, Simon & Schuster, Inc., New York, 1956.

Northrup, E., *Riddles in Mathematics*, Van Nostrand Reinhold Company, New York, 1944.

Rappaport, S., and H. Wright, *Mathematics*, New York University Press, New York, 1964.

Ravielli, A., *An Adventure in Geometry*, The Viking Press, Inc., New York, 1957.

Simon, W., *Mathematical Magic*, Charles Scribner's Sons, New York, 1964.

Steinhaus, H., *Mathematical Snapshots*, Oxford University Press, New York, 1969.

Valens, E. F., *The Number of Things: Pythagoras, Geometry and Humming Strings*, E. P. Dutton & Co., Inc., New York, 1964.

Vergara, W. C., *Mathematics in Everyday Things*, Harper & Row, Publishers, Inc., New York, 1959.

Senior High School

All books on the junior high school list are also recommended for senior high school. The list below includes content that is somewhat more sophisticated.

Adler, I., *A New Look at Geometries*, The John Day Company, Inc., New York, 1966.

———, *Probability and Statistics for Everyman*, The John Day Company, Inc., New York, 1963.

Ahrendt, M. H., *The Mathematics of Space Exploration*, Holt, Rinehart & Winston, Inc., New York, 1965.

Bakst, A., *Mathematical Puzzles and Pastimes*, Van Nostrand Reinhold Company, New York, 1954.

Barnard, D. St. Paul, *Adventures in Mathematics*, Hawthorn Books, Inc., New York, 1965.

Bell, E. T., *Men of Mathematics*, Simon & Schuster, Inc., New York, 1937.

———, *The Last Problem*, Simon & Schuster, Inc., New York, 1961.

Boyer, C. B., *A History of Mathematics*, John Wiley & Sons, Inc., New York, 1968.

Courant, R., and H. Robbins, *What Is Mathematics?* Oxford University Press, Inc., New York, 1941.

Court, N. A., *Mathematics in Fun and in Earnest*, The Dial Press, New York, 1958.

Coxeter, H. S. M., *Introduction to Geometry*, John Wiley & Sons, Inc., New York, 1969.

Dunn, A., *Mathematical Bafflers*, McGraw-Hill Book Company, New York, 1964.

Eves, H. W., *An Introduction to the History of Mathematics*, Holt, Rinehart & Winston, Inc., New York, 1968.

———, *Fundamentals of Geometry*, Allyn & Bacon, Inc., Boston, 1969.

———, *In Mathematical Cirles*, vol. 1 and 2, Prindle, Weber & Schmidt, Inc., Boston, 1969.

Fadiman, C., *The Mathematical Magpie*, Simon & Schuster, Inc., New York, 1962.

Gardner, M., *Relativity for the Million*, The Macmillan Company, New York, 1962.

Golomb, S. W., *Polyominoes*, Charles Scribner's Sons, New York, 1965.

Hilbert, D., and S. Cohn-Vossen, *Geometry and the Imagination*, Chelsea Publishing Co., Inc., Bronx, N. Y., 1952.

Hogben, L., *Mathematics in the Making*, Doubleday & Company, Inc., New York, 1960.

Huff, D., *How to Take a Chance*, W. W. Norton & Company, Inc., New York, 1959.

——— and I. Geis, *How to Lie with Statistics*, W. W. Norton & Company, Inc., New York, 1934.

Infeld, L., *Whom the Gods Love: The Story of Evariste Galois*, McGraw-Hill Book Company, New York, 1948.

James, G., and R. C. James, *Mathematics Dictionary*, Van Nostrand Reinhold Company, New York, 1959.

Kline, M., *Mathematics: A Cultural Approach*, Addison-Wesley Publishing Co., Inc., Reading, Mass., 1962.

——— (ed.), *Mathematics in the Modern World*, W. H. Freeman & Co., Publishers, San Francisco, 1968.

Kramer, Edna E., *The Main Stream of Mathematics*, Oxford University Press, Inc., New York, 1951.

———, *The Nature and Growth of Modern Mathematics*, Hawthorn Books, Inc., New York, 1970.

Langman, Harry, *Play Mathematics*, W. W. Norton & Company, Inc., New York, 1954.

Lieber, L., *The Education of T. C. Mits*, W. W. Norton & Company, Inc., New York, 1951.

———, *Human Values and Science, Art and Mathematics*, W. W. Norton & Company, Inc., New York, 1961.

———, *Mits, Wits, and Logic*, Institute Press, New York, 1954.

Longley-Cook, L. H., *New Math Puzzle Book*, Van Nostrand Reinhold Company, New York, 1970.

Loomis, E. S., *The Pythagorean Proposition*, National Council of Teachers of Mathematics, Washington, D. C., 1968.

Maxwell, E. A., *Fallacies in Mathematics*, Cambridge University Press, New York, 1959.

Messick, D. M., *Mathematical Thinking in Behavioral Sciences*, W. H. Freeman & Co., Publishers, San Francisco, 1968.

Muir, J., *Of Men and Numbers*, Dodd, Mead & Co., New York, 1961.

Ogilvy, C. S., *Through the Mathescope*, Oxford University Press, Inc., New York, 1956.

———, *Tomorrow's Math: Unsolved Problems for the Amateur*, Oxford University Press, Inc., New York, 1962.

———, *Excursions in Geometry*, Oxford University Press, Inc., New York, 1969.

Pedoe, D., *The Gentle Art of Mathematics*, The Macmillan Company, New York, 1958.

Peter, R., *Playing with Infinity*, Simon & Schuster, Inc., New York, 1962.

Rademacher, H., and O. Toeplitz, *The Enjoyment of Mathematics*, Princeton University Press, Princeton, N. J., 1957.

Reid, C., *A Long Way from Euclid*, Thomas Y. Crowell Company, New York, 1963.

———, *From Zero to Infinity*, Thomas Y. Crowell Company, New York, 1955.

———, *Hilbert*, Springer-Verlag New York, Inc., New York, 1970.

Saaty, T. L., and F. J. Weyl, *The Spirit and Uses of the Mathematical Sciences*, McGraw-Hill Book Company, New York, 1969.

Shklarsky, D. O., *et al.*, *The USSR Olympiad Problem Book*, W. H. Freeman & Co., Publishers, San Francisco, 1962.

Stein, S. K., *Mathematics; The Man-Made Universe*, W. H. Freeman & Co., Publishers, San Francisco, 1969.

Sullivan, J. W. N., *Isaac Newton: 1642–1727*, The Macmillan Company, New York, 1938.

Trumbull, H., *The Great Mathematicians*, New York University Press, New York, 1961.

Wiener, N., *I Am a Mathematician*, Doubleday & Company, Inc., New York, 1956.

Williams, J. D., *The Compleat Strategyst*, McGraw-Hill Book Company, New York, 1954.

Wolf, P., *Breakthroughs in Mathematics*, Signet Books, New York, 1963.

Enrichment Pamphlets and Monographs

Pamphlets available from J. Weston Walch, Box 658, Portland, Me. 04104. Send to publisher for prices.

Colorful Teaching of Mathematics.
Games for Learning Mathematics.
Geometric Models and Demonstrations.
Geometry Teaching Aids You Can Make.
Dramatizing Mathematics.
Successful Devices in Teaching Geometry.
Yes, Math Can Be Fun.
The Math Wizard.
Tessellation and Dissection.
Mathematical Bingo.
A Collection of Cross Number Puzzles.
Introduction to Optical Illusions.
Graphing Pictures.
Math in Nature.

"New Mathematics Library." Random House, Inc., 201 E. 50 St., New York, N. Y. 10022, or The L. W. Singer Co., 249–259 W. Erie Blvd., Syracuse, N. Y. 13202.

Numbers: Rational and Irrational.
What Is Calculus About?
Introduction to Inequalities.
Geometric Inequalities.
The Contest Problem Book.
The Lore of Large Numbers.
Uses of Infinity.
Geometric Transformations.
Continued Fractions.
Graphs and Their Uses.
Hungarian Problem Book I.
Hungarian Problem Book II.
Episodes from the Early History of Mathematics.
Groups and Their Graphs.
Mathematics of Choice.
From Pythagoras to Einstein.
The Contest Problem Book II.
First Concepts of Topology.
Geometry Revisited.

"Exploring Mathematics on Your Own." McGraw-Hill Book Company, Webster Division, Manchester Rd., Manchester, Mo. 63011. American edition now out of print. Available from John Murray, London, 50 Albemarle St., W1X 4BD.

Sets.
Pythagorean Theorem.
Topology.
Numeration System.
Fun with Mathematics.
Invitation to Mathematics.
World of Statistics.
Shortcuts in Computing.
Computing Devices.
World of Measurement.
Adventures in Graphing.
Finite Math Systems.
Logic and Reasoning.
Basic Concepts of Vectors.
Probability and Chance.
Geometric Constructions.
Curves in Space.

"Thinking With Mathematics Series." D. C. Heath & Company, 125 Spring St., Lexington, Mass. 02173.

Mathematics Projects Handbook.
The Concept of a Function.
Graphing Relations and Functions.
An Introduction to Linear Programming.
The Natural Numbers.
The Integers.
The Rational Numbers.
The Real Numbers.
The Complex Numbers.
Finite Mathematical Structures.
An Introduction to Transfinite Mathematics.
Congruence and Motion in Geometry.
An Introduction to Sets and the Structure of Algebra.

"Topics in Modern Mathematics." Ginn & Company, Statler Bldg., Back Bay, P.O. 191, Boston, Mass. 02117.

Limits and Limit Concepts.
Random Numbers—Mathematical Induction—Geometric Numbers.
Digital Computers and Related Mathematics.
The Nature of the Regular Polyhedra—Infinity and Beyond—Introduction to Groups.
Pythagorean Numbers—Congruences, A Finite Arithmetic—Geometry in the Number.
An Introduction to Sets and the Structure of Algebra.
Introduction to Logic and Sets.
Principles and Patterns of Numeration Systems.

"Popular Lectures in Mathematics Series." Blaisdell Publishing Co., 275 Wyman St., Waltham, Mass. 02154.

The Method of Mathematical Induction.
Fibonacci Numbers.
Some Applications of Mechanics to Mathematics.
Geometrical Constructions Using Compasses Only.
The Ruler in Geometrical Constructions.
Inequalities.

"Topics in Mathematics" (translations from the Russian). D. C. Heath & Company, 125 Spring St., Lexington, Mass. 02173.

Configuration Theorems.
What Is Linear Programming?
Equivalent and Equidecomposable Figures.
Mistakes in Geometric Proofs.
Proof in Geometry.
Induction in Geometry.
Computation of Areas of Oriented Figures.
Areas and Logarithms.
Summation of Infinitely Small Quantities.
Hyberbolic Functions.
How to Construct Graphs and Simplest Maxima and Minima Problems.
The Method of Mathematical Induction.
Algorithms and Automatic Computation Machines.
An Introduction to the Theory of Games.
The Fibonacci Numbers.
Convex Figures and Polyhedra.
Eight Lectures in Mathematical Analysis.
Geometric Constructions in the Plane.
Geometry of the Straightedge and Geometry of the Compass.
Infinite Sets.
Isoperimetry.
Multicolor Problems.
Probability and Information.
Problems in the Theory of Numbers.
Random Walks.

"Mathematics Enrichment Series." Houghton Mifflin Co., 2 Park St., Boston, Mass. 02107.

Legislative Apportionment.
Topics From Inversive Geometry.
Sequences.
Four by Four.
Induction in Mathematics.
Mosaics.
Stereograms.

"Topics from Mathematics." Cambridge University Press, 32 E. 57th St., New York, N. Y. 10022.

Computers.
Cubes.
Statistics.
Circles.
Solid Models.
Tessellations.

"Programmed Junior High Enrichment Units." The Macmillan Company, 866 Third Ave., New York, N. Y. 10022.

Sets, Operations and Circuits.
Bases and Numerals.
What Are the Chances?
Modular Systems.
Clear Thinking.
Points, Lines, and Planes.
Points, Lines, and Space.
Number Sentences.
Factors and Primes.
From the Naturals to the Reals.

"Contemporary School Mathematics Series." Houghton Mifflin Co., 2 Park St., Boston, Mass. 02107.

Matrices 1.
Sets and Logic 1.
Computers 1.
Shape, Size and Place.
An Introduction to Probability and Statistics.
Matrices 2.
Sets and Logic 2.
Computers 2.

"Supplementary and Enrichment Series." A. C. Vroman, Inc., 2085 E. Foothill Blvd., Pasadena, Cal. 91109.

Plane Coordinate Geometry.
Inequalities.
Numeration.
Algebraic Structures.
Factors and Primes.
Mathematical Systems.
Systems of First Degree Equations in Three Variables.
Radioactive Decay.
Absolute Value.
Mathematical Theory of the Struggle for Life 1 + 1 = ?
Order and The Real Numbers: A Guided Tour.
The Mathematics of Trees and Other Graphs.

"SMSG Reprint Series." A. C. Vroman, Inc., 2085 E. Foothill Blvd., Pasadena, Cal. 91109.

The Structure of Algebra.
Prime Numbers and Perfect Numbers.
What is Contemporary Mathematics?
Mascheroni Constructions.
Space, Intuition and Geometry.
Nature and History of Pi.
Computation of Pi.
Mathematics and Music.
The Golden Measure.
Geometric Constructions.
Memorable Personalities in Mathematics: Nineteenth Century.
Memorable Personalities in Mathematics: Twentieth Century.
Finite Geometry.
Infinity.
Geometry, Measurement and Experience.
Functions.
Circular Functions.
The Complex Number Systems.
The System of Vectors.
Non-Metric Geometry.

"Mathematics in the Making." Houghton Mifflin Co., 2 Park St., Boston, Mass. 02107.

Pattern, Area and Perimeter.
Binary and Other Number Systems.
Looking at Solids.

Rotation and Angles.
Curves.
Scale Drawing and Surveying.
Transformations and Symmetry.
Networks.
All Sorts of Numbers.
Sets and Relations Graphs.
Statistics.
Introduction to Computer Programming.
Flowcharting.

"Franklin Mathematics Series." Lyons & Carnahan, 407 E. 25 St., Chicago, Ill. 60616.

Making and Using Graphs and Nomographs.
Mathematics around the Clock.
Patterns and Puzzles in Mathematics.
From Fingers to Computers.
Mathematics: Man's Key to Progress Book A.
Mathematics: Man's Key to Progress Book B.
Probability: The Science of Chance.

"Exploring Mathematics." Selective Educational Equipment, Inc., 3 Bridge St., Newton, Mass. 02195.

Number Systems Old and New.
Mathematical Puzzles.
Introducing Algebra.
Simple Line Graphs.
Introducing Polyhedra.
Great Mathematicians.
Calculating Devices.
Measurement.
Number Patterns.
Introducing Sets.
Histograms.
Introducing Geometry.
Introducing Topology.
Statistics.
Cyphers.
Curves.
Decimals.

Pamphlets on Careers in Mathematics

Careers in Mathematics. National Council of Teachers of Mathematics, 1201 Sixteenth St. N.W., Washington, D. C. 20036, 1961. 32 pp. Free.

Mathematics Teaching as a Career. National Council of Teachers of Mathematics, 1201 Sixteenth St. N.W., Washington, D. C. 20036, 1964. 8 pp. Free.

Turner, Nura, *Mathematics and My Career*, National Council of Teachers of Mathematics, Washington, D. C., 1971. $1.30.

You Will Need Math. Mathematics Association of America, 1225 Connecticut Ave. N.W., Washington, D. C., 20036, 1967. Free.

Appendix D: Projects, Exhibits, and Displays

The topics listed below are suitable for reports, exhibits, or projects. Exhibits are usually judged on the following criteria: originality, completeness, clarity, interest value, craftsmanship, and mathematical thought. In order to organize an exhibit that is a success the following suggestions may be helpful:

Select a topic that has interest potential. The topics listed below suggest some of the many possibilities.

Find as much information about the topic as possible. Check bibliographies, such as those by Schaaf in *Recreational Mathematics* and in *The Mathematics Teacher*. Check *Reader's Guide* and *Education Index* as well as encyclopedias. The four-volume *The World of Mathematics*, edited by James R. Newman, published by Simon and Schuster, is a gold mine of ideas and material. Other important source books are *Mathematics and the Imagination*; *Mathematics in Western Culture*; *Mathematics, Its Magic and Mastery*; and *Fundamentals of Mathematics*. From these sources collect information, ideas, materials.

Prepare and organize your material into a concise, interesting report. Include drawings in color, pictures, applications, examples that will get the readers' attention and add meaning to your exhibit. Build models, mock-ups, or devices that add interest and understanding of the topics.

Build an exhibit that will tell the story of your topic. Use models, applications, charts that lend variety. If possible, prepare materials that viewers can manipulate. Give your exhibit a catchy, descriptive title. Label everything with brief captions or legends so that viewers will understand the principles involved. Make the display simple, but also attractive and dramatic. Use color for emphasis. Write captions in a unique way, such as with rope, pipe cleaners, plastic tubing, or yarn. Show craftsmanship, creativeness, diligence in arranging the exhibit. Have a brief summary of the basic ideas, plans, and references for your topic available for distribution.

Be able to demonstrate the topic of your exhibit. Speak clearly and correctly. Be well informed so that you can answer questions.

Additional references for these topics will be found in W. L. Schaaf, *A Bibliography of Recreational Mathematics*, vols. 1 and 2, published by the National Council of Teachers of Mathematics.

Sidelights on Mathematics

Ancient Computing Methods

How do you multiply with Roman numerals? What is the scratch system, the doubling system, the lattice method of computations? What changes in our methods of long multiplication and long division are being suggested in recent arithmetic textbooks? How is the abacus used for computation? How are Napier's bones used for multiplication? How did the old computing machines work? How were logarithms invented? Who invented the slide rule?

Exhibit suggestions. Charts of sample computations by ancient methods. Ancient number representations such as pebbles, tally sticks, tally marks in sand, Roman numeral computations, abaci, Napier's bones, old computing devices.

References:

Hogben, L., *Wonderful World of Mathematics* (Random House, 1955).

Kokomoor, F. W. *Mathematics and Human Affairs* (Prentice-Hall, 1942).

Larsen, Harold D., and H. Glenn Ludlow, *Arithmetic for Colleges* (Macmillan Company, 1958).

Cryptography, Codes, and Ciphers

How are codes made? What is the difference between codes and ciphers? How are machines used to make codes and to decipher codes?

Exhibit suggestions. Devices or charts for writing and deciphering codes. Code messages. Illustrations of famous codes of the past.

References:

Andree, R. V., "Cryptography as a Branch of Mathematics" (*Mathematics Teacher*, November 1952).

Bakst, A., *Mathematics, Its Magic and Mastery* (Van Nostrand Reinhold, 1952), Chap. 6.

Laffin, J., *Codes and Cyphers* (Abelard-Schuman, 1964).

Peck, L., *Secret Codes* (NCTM, 1961).

Schaaf, W. L., *Recreational Mathematics* (NCTM, 1955), pp. 107–112.

Smith, L. D., *Cryptography* (Dover, 1943).

Curves and Curve Drawing

What are conic sections, cycloids, spirals, catenaries, cardioids? What are the equations of these curves? How are these curves drawn? Where do these curves occur? How are these curves represented in polar coordinates? What are curves of constant breadth?

Exhibit suggestions. Devices for drawing curves. Models representing these curves. Applications of curves. Curves formed by curve stitching. Graphs and curves in rectangular coordinates, logarithmic coordinates, or polar coordinates. Models of three-dimensional curved surfaces. Designs graphed with sets of equations.

References:

Bakst, A., *Mathematics, Its Magic and Mastery* (Van Nostrand Reinhold, 1952), Chap. 24.

Baravalle, H. von, "Geometric Drawing," in *Multi-Sensory Aids in the Teaching of Mathematics*, Yearbook 18 (NCTM, 1945), pp. 64–81.

Johnson, D. A., *Curves in Space* (Webster Publishing Co., 1963).

Kline, M., *Mathematics in Western Culture* (Oxford, 1953), Chap. 12.

Kramer, E. E., *The Main Stream of Mathematics* (Oxford, 1951), Chap. 7.

Lockwood, E. H., *A Book of Curves* (Cambridge, 1961).

Ogilvy, C. S., *Through the Mathescope* (Oxford, 1956).

Curve Stitching

How can curves be formed by straight lines? What is the envelope of a curve? What is a pencil of lines? What curves can be formed by stitching? How can mathematical curves be combined to form beautiful models?

Exhibit suggestions. Stitch a variety of curves with brightly colored yarn or string in two or three dimensions.

References:

Anderson, J. T., *Space Concepts through Aestheometry* (Artesia, N. M.: Aestheometry, Inc.).

Cundy, H. M., and A. P. Rollett, *Mathematical Models* (Oxford, 1951), p. 38.

Johnson, D. A., *Curves in Space* (Webster Publishing Co., 1963).

McCamman, C. V., "Curve Stitching in Geometry," in *Multi-Sensory Aids in Teaching of Mathematics*, Yearbook 18 (NCTM, 1945), pp. 82–85.

Seymour, D., and J. Snider, *Line Designs* (Creative Publications, 1969).

Graphing

What are the different kinds of graphs and graph paper? How do you make three-dimensional graphs? How is graph paper used in designing, enlarging drawings, planning cheering section designs, solving verbal problems, representing numbers, making slide rules? What is a nomograph? Stereograph? How can graphs be used in finding slopes, lengths, areas, perimeters? How do we graph inequalities?

Exhibit suggestions. Types of graphs, types of graph paper, graphs of equations. Graphing on acoustical tile or pegboard with elastic thread for graph lines. Models of stereographs or object graphs. Designs based on graphs of equations or the use of graph paper. Graphs of the common solutions of linear and quadratic graphs.

References:

Bristol, J. D., *Graphing Relations and Functions* (D. C. Heath, 1963).

Huff, D., and I. Geis, *How to Lie with Statistics* (W. W. Norton, 1934).

Reynolds, J. A. C., *Shape, Size and Place* (Houghton Mifflin, 1964).

Shelov, G. E., *How to Construct Graphs* (D. C. Heath, 1963).

Magic Squares

What is a magic square? Where and how did magic squares originate? How are magic squares made? How can you make magic squares with algebraic expressions? What games can be played with the principle of the magic square? What are the different kinds of magic squares? How can you make a magic cube?

Exhibit suggestions. Sample magic squares of varied numbers of cells. Illustrations of magic square construction with variations.

References:

Kraitchik, M., *Mathematical Recreations* (Dover, 1942), pp. 142–192.

Meyer, J. S., *Fun with Mathematics* (Dover, 1952).

Mathematical Forms in Nature

What forms in nature represent the graphs of formulas? What shapes in nature are geometric and/or symmetric? What activities in nature illustrate mathematical functions?

Exhibit suggestions. Samples of spiral shells, crystals, ellipsoidal stones or eggs, spiraling sunflower seed pods, branch distributions illustrating Fibonacci series, snowflake patterns, body bones acting as levers, ratio of food consumed to size.

References:

Boys, C., *Soap Bubbles* (Dover, 1959).

Holde, A., and P. Singer, *Crystals and Crystal Growing* (Doubleday, 1960).

Newman, J. R., *World of Mathematics* (Simon & Schuster, 1956), "Crystals and the Future of Physics," pp. 871–881; "The Soap Bubble," pp. 891–900; "On Being the Right Size," pp. 925–927; "On Magnitude," pp. 1001–1046.

Thompson, D. Arcy, *On Growth and Forms* (Cambridge, 1952).

Wenninger, M. J., *Polyhedron Models for the Classroom* (NCTM, 1966).

Measurement and Approximations

How are units of measure invented? What is the difference between accuracy and precision? What special rules must be followed when computing with measurements? Why is progress in science so very dependent on precise measurement?

Exhibit suggestions. Samples of a variety of units of measure, for example, square yard, cubic foot, cubit, furlong, Canadian gallon, grain. Examples of daily measurements, such as shoe size, hats, nails, stockings, calories, board foot. Models of odd measuring devices, such as pedometer, Geiger counter.

References:

Bakst, A., *Mathematics, Its Magic and Mastery* (Van Nostrand Reinhold, 1952), Chap. 7.

Johnson, D. A., *The World of Measurement* (Webster Publishing Co., 1961).
Kokomoor, F. W., *Mathematics and Human Affairs* (Prentice-Hall, 1942), Chap. 2.
Shuster, C. N., and Fred L. Bedford, *Field Work in Mathematics* (American Book, 1935; available from Yoder Instrument Company, East Palestine, Ohio).

Number Curiosities

What are prime numbers? Perfect numbers? Amicable numbers? What are some unusual number relations? What are imaginary numbers? Irrational numbers? Complex numbers? Invent a number system that needs three coordinates to express its value. How have numbers been related to superstitions and magic? How have numbers been related to religion? Are there any references in the Bible to the use of mathematical ideas?

Exhibit suggestions. Charts of number relations. Proof that 2 is a prime number. Graphs of complex numbers. Illustrate the size of billion. Illustrate how rapidly a quantity grows by doubling.

References:

Bakst, A., *Mathematics, Its Magic and Mastery* (Van Nostrand Reinhold, 1952).
Beiler, A. H., *Recreation in the Theory of Numbers* (Dover, 1964).
Gamow, G., *One, Two, Three, Infinity* (Viking, 1947).
Johnson, D. A., and W. H. Glenn, *Number Patterns* (Webster Publishing Co., 1960).
Kasner, E., and J. Newman, *Mathematics and the Imagination* (Simon & Schuster, 1940), Chaps. 1 and 2.

Optical Illusions

What are some optical illusions? Why do these patterns cause illusions? Are these illusions still apparent with three-dimensional representations? What tricks depend on these illusions? How are these illusions used in advertising, dress designing, and architecture?

Exhibit suggestions. Charts, models, ads, pictures of illusions.

References:

Bakst, A., *Mathematics, Its Magic and Mastery* (Van Nostrand Reinhold, 1952), pp. 469–478.
Beeler, N. F., and F. M. Branley, *Experiments in Optical Illusions* (Crowell, 1951).

Paper Folding

How can all the constructions of Euclidean geometry be done by folding paper? What assumptions are made when paper folding is used to construct geometric figures? What polygons and polyhedrons can be formed by folding paper? What is a hexaflexagon? How can conic sections be formed by folding paper? What puzzles and tricks are based on paper folding?

Exhibit suggestions. Models, geometric constructions, polyhedrons, conic sections, puzzles formed by paper folding.

References:

Cundy, H. M., and A. P. Rollett, *Mathematical Models* (Oxford, 1951).

Johnson, D., *Paper Folding for the Mathematics Class* (NCTM, 1957).

Row, S., *Geometrical Exercises in Paper Folding.*

Wenninger, M. J., *Polyhedron Models for the Classroom* (NCTM, 1966).

Paradoxes and Fallacies

What fallacies result when an expression is divided by zero? What fallacies occur from incorrect constructions in geometry? What fallacies can be explained by limits? What fallacies depend on false probabilities?

Exhibit suggestions. Charts and illustrations of famous paradoxes. Experiments with coins or dice to illustrate false probabilities.

References:

Bakst, A., *Mathematics, Its Magic and Mastery* (Van Nostrand Reinhold, 1952).

Dunn, A., *Mathematical Bafflers* (McGraw-Hill, 1964).

Kasner, E., and J. Newman, *Mathematics and the Imagination* (Simon & Schuster, 1940), Chap. 6.

Kendall, P. M. H., and G. M. Thomas, *Mathematical Puzzles for the Connoisseur* (Thomas Y. Crowell, 1964).

Kline, M., *Mathematics in Western Culture* (Oxford, 1953), Chap. 20.

Northrup, E., *Riddles in Mathematics* (Van Nostrand Reinhold, 1944).

Pi, i, e

What are several ways of obtaining the value of pi—measurement, limit of series, empirical probability, polygon perimeter? What are applications of i? What are the sources of the constant e? What formula expresses the relationship of π, i, and e? What series are used in obtaining values for π and e?

Exhibit suggestions. Experiments for the determination of π. Graphical representation of i. Development of the formula connecting π, i, e. Applications of π, i, e.

References:

Bakst, A., "Mathematical Recreations" (*Mathematics Teacher*, October 1952).

———, *Mathematical Puzzles and Pastimes* (Van Nostrand Reinhold, 1954), pp. 110–116.

Baravalle, H. von, "The Number π" (*Mathematics Teacher*, February 1953).

Kasner, E., and J. Newman, *Mathematics and the Imagination* (Simon & Schuster, 1940), Chap. 3.

Ogilvy, C. S., *Through the Mathescope* (Oxford, 1956), Chap. 10.

Contemporary Mathematics

Numeration Systems

What were the numeration systems of ancient man? Where did our number symbols come from? What are the different ways of building a numeration system, for example, using words, letters, repetition of symbols, or place value? What are different ways of representing numbers, such as tally marks, rope knots, or standard notations? What are different number bases that have been used? Why are different bases used and what are the advantages of each base? What puzzles or tricks can be performed with different number bases? Can you invent a new numeration system with new symbols? Can you build a computing device for a number base other than our decimal system?

Exhibit suggestions. Charts or number symbols, numeration systems, and computations in each system. Computing devices for various number bases, such as abacus, Napier's bones, electric binary abacus, binary computer. Pamphlets and books on numerous systems.

References:

Bakst, A., *Mathematics, Its Magic and Mastery* (Van Nostrand Reinhold, 1952).
Hogben, L., *Wonderful World of Mathematics* (Random House, 1955).
Johnson, D. A., and W. H. Glenn, *Exploring Mathematics on Your Own* (Doubleday, 1961).
Kramer, E. E., *The Main Stream of Mathematics* (Oxford, 1951).
Newman, J. R., *The World of Mathematics* (Simon & Schuster, 1956), vol. I, part 3.
Reid, C., *From Zero to Infinity* (Thomas Y. Crowell, 1955).

Mathematics, What Is It?

Mathematics is a science, a language, a system of logic. The fields of mathematics are many and are constantly increasing. Mathematics is not an old, dead subject; rather it is very much alive, with new discoveries and new inventions occurring constantly.

References:

Courant, R., and H. Robbins, *What Is Mathematics?* (Oxford, 1941).
Johnson, D. A., and W. H. Glenn, *Invitation to Mathematics* (McGraw-Hill, 1960).
Kasner, E., and J. Newman, *Mathematics and the Imagination* (Simon & Schuster, 1940), pp. 357–362.
Kline, M., *Mathematics in Western Culture* (Oxford, 1953), Chaps. 1–3.
Newman, J. R., *World of Mathematics* (Simon & Schuster, 1956), vol. III, parts 11 and 12.
Rappaport, S., and H. Wright, *Mathematics* (New York University, 1964).
Stein, S. K., *Mathematics, the Man-Made Universe* (W. H. Freeman, 1969).

Topology

Why is topology called rubber-sheet geometry? How can a strip of paper or a bottle have only one surface? What is the famous Königsberg bridge problem? What is the four-color problem in mapping? How has topology used networks to build a mathematical structure?

What are some new applications of topology? What are some stunts, tricks, and puzzles based on topology?

Exhibit suggestions. Königsberg bridge problem, four-color mapping problem, puzzles, Möbius strips. Klein bottle, network analysis, proofs.

References:

Barr, S., *Experiments in Topology* (Thomas Y. Crowell, 1964).

Courant, R., and H. Robbins, *What Is Mathematics?* (Oxford, 1941), pp. 235–271.

Fadiman, C., *Fantasia Mathematica* (Simon & Schuster, 1958).

Gamow, G., *One, Two, Three, Infinity* (Viking, 1947), chap. III.

Gardner, M., *Mathematical Puzzles and Diversions* (Simon & Schuster, 1959).

Johnson, D. A., and W. H. Glenn, *Topology, the Rubber Sheet Geometry* (McGraw-Hill, 1960).

Kasner, E., and J. Newman, *Mathematics and the Imagination* (Simon & Schuster, 1940), pp. 265–298.

Newman, J. R., *World of Mathematics* (Simon & Schuster, 1956), vol. I, pp. 573–599.

Sets and the Logic of Algebra

How are sets used in logic, electric networks, and probability? What is the union and intersection of sets? What are the commutative, associative, distributive laws for sets?

Exhibit suggestions. Analysis of statements with Venn diagrams. Electric circuits representing the relationships of sets.

References:

Allendoerfer, C. B., and C. O. Oakley, *Principles of Mathematics* (McGraw-Hill, 1955), Chap. 5.

Johnson, D. A., and W. H. Glenn, *Sets, Sentences and Operations* (Webster Publishing Co., 1960).

Kemeny, J. G., J. L. Snell, and G. L. Thompson, *Introduction to Finite Mathematics* (Prentice-Hall, 1957), Chaps. 2 and 3.

Montague, H. F., and M. D. Montgomery, *The Significance of Mathematics* (Charles E. Merrill, 1963).

Swain, R. L., and E. D. Nichols, *Understanding Arithmetic* (Holt, Rinehart & Winston, 1957), Chap. 2.

The Pythagorean Theorem

What are some unusual proofs of the Pythagorean theorem? What are some of the usual relationships that exist between Pythagorean numbers? What models can be made to visualize and prove the relationship? How can this relationship be used in indirect measurement?

Exhibit suggestions. Models using cardboard, marbles, BB shot, plastic to show the Pythagorean relationship. Designs in tile, cloth, wallpaper, which are based on this relationship.

Charts of unusual proofs. Historical sidelights. Mock-ups of applications in indirect measurement.

References:

Berger, E. J., "A Model for Visualizing the Pythagorean Theorem" (*Mathematics Teacher*, April 1953).

Glenn, W. H., and D. A. Johnson, *The Pythagorean Theorem* (McGraw-Hill, 1960).

Hart, P. J., "Pythagorean Numbers" (*Mathematics Teacher*, January 1954).

Loomis, E. S., *The Pythagorean Proposition* (NCTM, 1968).

Schaaf, W. L., "The Theorem of Pythagoras" (*Mathematics Teacher*, December 1951).

Valens, E. G., *The Number of Things: Pythagoras, Geometry and Humming Strings* (Dutton, 1964).

Statistics

How can statistics be summarized? What are the measures of central tendency? What are measures of variation? How is the relationship between sets of measures determined? What is a normal curve? What is quality control? How is the accuracy of a sample measured? How are statistics used to draw conclusions?

Exhibit suggestions. Sample distributions of original data, with analysis and graphs. Examples of statistics found in advertisements and newspapers. Examples of misuses of statistics. Models of random sampling devices and Gauss probability board.

References:

Adler, I., *Probability and Statistics for Everyman* (John Day, 1963).

Huff, D., *How to Take a Chance* (W. W. Norton, 1959).

———, and I. Geis, *How to Lie with Statistics* (W. W. Norton, 1934).

Johnson, D. A., and W. H. Glenn, *The World of Statistics* (Webster Publishing Co., 1961).

Kline, M., *Mathematics in Western Culture* (Oxford, 1953), Chap. 12.

Mosteller, F., *et al.*, *Probability and Statistics* (Addison-Wesley, 1961).

Newman, J. R., *World of Mathematics* (Simon & Schuster, 1956), vol. III, part 8.

Wallis, W. A., and H. V. Roberts, *The Nature of Statistics* (Macmillan, 1962).

Probability and Chance

What is probability? What is the risk involved in driving a car? What are the chances you will die this year? What are the odds of winning a lottery? What is the mathematical expectancy in a carnival game? Why doesn't gambling pay? How is the law of disorder related to nuclear fission? How is probability used to derive new facts?

Exhibit suggestions. Illustrate with toys or models the probability of accidents, crop damage from storms, winning bridge hands, slot machine odds, coin tosses, dice totals, contest winnings, lock combinations, multiple births.

References:

Adler, I., *Probability and Statistics for Everyman* (John Day, 1963).

Huff, D., *How to Take a Chance* (W. W. Norton, 1959).

Kemeny, J. G., J. L. Snell, and G. L. Thompson, *Introduction to Finite Mathematics* (Prentice-Hall, 1957), Chap. 4.

Kline, M., *Mathematics in Western Culture* (Oxford, 1953), Chap. 23.

Newman, J. R., *The World of Mathematics* (Simon & Schuster, 1956), vol. II, part 7.

Fourth Dimension and Relativity

What is the fourth dimension? How is the fourth dimension related to time? How is the speed of light related to relativity? Could you live a thousand years if you traveled at the speed of light? What formulas are used in relativity? How can you use mathematics to represent any dimension—fourth, fifth, or n-th?

Exhibit suggestions. Illustrate the meanings of the fourth dimension with tesseracts or superprisms. Models to illustrate the relativity of motion.

References:

Abbott, E. A., *Flatland* (Dover, 1932).

Bakst, A., *Mathematics, Its Magic and Mastery* (Van Nostrand Reinhold, 1952), Chaps. 22–23.

Gamow, G., *One, Two, Three, Infinity* (Viking, 1947), Chap. 4.

Kasner, E., and J. Newman, *Mathematics and the Imagination* (Simon & Schuster, 1940), Chap. 4.

Kline, M., *Mathematics in Western Culture* (Oxford, 1953), Chap. 27.

Kramer, E. E., *Main Stream of Mathematics* (Oxford, 1951), Chap. 12.

Lieber, L., *The Education of T. C. Mits* (W. W. Norton, 1954), pp. 187–211.

Newman, J. R., *The World of Mathematics* (Simon & Schuster, 1956), vol. II, pp. 1107–1145.

Zippin, L., *Uses of Infinity* (Random House, 1962).

Game Theory

What kind of games are the basis for game theory? What definitions, assumptions, axioms are used in game theory? How can game theory be applied to recreational games? How is game theory related to linear programming for electronic computers?

Exhibit suggestions. Charts outlining basic principles of game theory. Simple analysis of two-by-two games and matrix games. Simple models or circuits for programming games.

References:

Barson, A. S., *What Is Linear Programming?* (D. C. Heath, 1964).

Bristol, J. D., *An Introduction to Linear Programming* (D. C. Heath, 1963).

Newman, J. R., *World of Mathematics* (Simon & Schuster, 1956), pp. 1285–1293.

Richardson, M., *Fundamentals of Mathematics* (Macmillan, 1958), pp. 390–401.

Williams, J. D., *The Compleat Strategyst* (McGraw-Hill, 1954).

Groups and Fields

What are groups and fields? How do groups and fields illustrate a mathematical structure? How are theorems and problems solved involving groups or fields? Examples and theorems.

References:

Allendoerfer, C. B., and C. O. Oakley, *Principles of Mathematics* (McGraw-Hill, 1955), Chaps. 3 and 4.

Richardson, M., *Fundamentals of Mathematics* (Macmillan, 1958), Chap. 17.

Sawyer, W. W., *Prelude to Mathematics* (Penguin, 1955), Chap. 7.

Stabler, E. R., *Introduction to Mathematical Thought* (Addison-Wesley, 1953), Chaps. 4, 8, and 9.

Finite Geometry

What kind of figures do you get when the number of points and lines in space is restricted? What relationships or theorems can be found to be true? What kind of constructions are possible?

References:

Lieber, L., *The Education of T. C. Mits* (W. W. Norton, 1954).

Norton, M. S., *Finite Mathematical Systems* (Webster Publishing Co., 1963).

Sawyer, W. W., *Prelude to Mathematics* (Penguin, 1955), Chap. 13.

Stabler, E. R., *An Introduction to Mathematical Thought* (Addison-Wesley, 1953), Chap. 7.

Yarnelle, J. E., *Finite Mathematical Structure* (D. C. Heath, 1963).

Infinity

What is infinity? Is one infinity more than another? Can you compute with infinite quantities?

Exhibit suggestions. Illustrations and comparison of infinite amounts. Paradoxes of the infinite.

References:

Gamow, G., *One, Two, Three, Infinity* (Viking, 1947), Chap. 1.

Gardner, M., *Relativity for the Million* (Macmillan, 1962).

Kasner, E., and J. Newman, *Mathematics and the Imagination* (Simon & Schuster, 1940), Chap. 7.

Kline, M., *Mathematics in Western Culture* (Oxford, 1953), Chap. 25.

Lieber, L. R., *Infinity* (Holt, Rinehart & Winston, 1953).

Newman, J. R., *World of Mathematics* (Simon & Schuster, 1956), vol. III, part 10.

Northrop, E., *Riddles in Mathematics* (Van Nostrand Reinhold, 1944), Chap. 7.

Peter, R., *Playing with Infinity* (Simon & Schuster, 1962).
Thomas, G. B., *Limits* (Addison-Wesley, 1963).
Zippin, L., *Uses of Infinity* (Random House, 1962).

Logarithms

What are logarithms? How were they discovered? How are the tables of logarithms computed? What is most useful about logarithms? How are logarithms used to build a slide rule?

Exhibit suggestions. How to compute with logarithms. Graphs of exponential equations on logarithmic graph paper. Slide rule made from logarithmic graph paper. Problems solved with logarithms.

References:

Bakst, A., *Mathematics, Its Magic and Mastery* (Van Nostrand Reinhold, 1952), Chap. 18.
Kokomoor, F. W., *Mathematics in Human Affairs* (Prentice-Hall, 1942), Chap. 10.
Ransom, W. R., "Elementary Calculation of Logarithms" (*Mathematics Teacher*, February 1954).
Shuster, C. N. "The Calculation of Logarithms in High School" (*Mathematics Teacher*, May 1955).

Logic

What are the laws of logic? What symbols are used in logical analysis? What are truth tables? What is a syllogism? What are different methods of proof? What are Venn diagrams? What is Boolean algebra? How are electric circuits related to the analysis of statements? How do logic machines work?

Exhibit suggestions. Logic machine which analyzes syllogisms. Unusual problems, puzzles, or proofs using logic. Truth analysis of a statement. Electric circuits which give truth table analyses.

References:

Adler, I., *Logic for Beginners* (John Day, 1964).
Allendoerfer, C. B., and C. O. Oakley, *Principles of Mathematics* (McGraw-Hill, 1955), Chap. 1.
Gardner, M., *Logic Machines and Diagrams* (McGraw-Hill, 1958).
Jones, B. W., *Elementary Concepts of Mathematics* (Macmillan, 1963), Chap. 1.
Kemeny, J. G., J. L. Snell, and G. L. Thompson, *Introduction to Finite Mathematics* (Prentice-Hall, 1957), Chap. 1.
Kline, M., *Mathematics in Western Culture* (Oxford, 1953).
Lieber, L., *Mits, Wits, and Logic* (Institute Press, 1954), Chaps. 17–24.
Newman, J. R., *World of Mathematics* (Simon & Schuster, 1956), vol. III, part 13.

Pfeiffer, J., "Symbolic Logic" (*Scientific American*, December 1950).

Richardson, M., *Fundamentals of Mathematics* (Macmillan, 1958), Chaps. 2 and 17.

Non-Euclidean Geometry

What is the basis for non-Euclidean geometry? What is a pseudosphere? How do the geometries of Lobachevski and Riemann differ? What are the proofs of some non-Euclidean theorems?

Exhibit suggestions. Models of sphere and pseudosphere, with geometric figures drawn on the surfaces. Charts comparing theorems in different geometries.

References:

Encyclopedia Titles: Nikolai Lobachevski; Johann Bolyai; George Riemann; Non-euclidean geometry.

Kasner, E., and J. Newman, *Mathematics and the Imagination* (Simon & Schuster, 1940), pp. 135–153.

Kline, M., *Mathematics in Western Culture* (Oxford, 1953), Chap. 26.

Lieber, L., *The Education of T. C. Mits* (W. W. Norton, 1951), pp. 138–153.

———, *Mits, Wits, and Logic* (Institute Press, 1954).

Newman, J. R., *The World of Mathematics* (Simon & Schuster, 1956), vol. III.

Richardson, M., *Fundamentals of Mathematics* (Macmillan, 1958), Chap. 16.

Sawyer, W. W., *Prelude to Mathematics* (Penguin, 1955), Chap. 6.

Wolfe, H. E., *Introduction to Non-euclidean Geometry* (Holt, Rinehart & Winston, 1945).

Number Theory

What is a number? What definition, axioms, and propositions are used to build a mathematical structure about numbers? What are different kinds of numbers, such as complex numbers? What is a modulus number system?

Exhibit suggestions. Number tree. Proofs of theorems. Device to show how to obtain the value of irrational numbers. Computing device in a modulus system.

References:

Allendoerfer, C. B. and C. O. Oakley, *Principles of Mathematics* (McGraw-Hill, 1955), Chap. 2.

Courant, R., and H. Robbins. *What Is Mathematics?* (Oxford, 1941), Chaps. 1 and 2.

Dantzig, T., *Number: The Language of Science* (Macmillan, 1954).

Jones, B. W., *Elementary Concepts of Mathematics* (Macmillan, 1963).

Ogilvy, C. S., *Through the Mathescope* (Oxford, 1956), Chap. 2.

Richardson, M., *Fundamentals of Mathematics* (Macmillan, 1958), Chaps. 3, 4, and 15.

Permutations and Combinations

How can you compute the number of ways in which objects can be arranged? How can you find out the number of different ways dice may fall or coins turn up heads and tails? How many different hands of bridge can be dealt? How many different committees can be formed from a certain number of boys and girls? How many different four-letter code words can be made from the letters of your name? How many different basketball teams can be formed from ten players?

Exhibit suggestions. Contrast permutations and combinations with letters, digits, or models. Show the analysis of a variety of conditions where permutations or combinations are involved with charts and objects. Show how permutations are used by detectives, politicians, statisticians, engineers, scientists.

References:

Bakst, A., *Mathematics, Its Magic and Mastery* (Van Nostrand Reinhold, 1952), Chap. 21.

Johnson, D. A., *Probability and Chance* (Webster Publishing Co., 1963).

Jones, B. W., *Elementary Concepts of Mathematics* (Macmillan, 1963).

Perspective, Projective Geometry, and Transformations

How are three-dimensional objects represented on two dimensions? What are different ways of projecting lines or surfaces on a plane?

Exhibit suggestions. Models of perspective. Analysis of pictures showing perspective.

References:

Courant, R., and H. Robbins, *What Is Mathematics?* (Oxford, 1941).

Ivins, W. M., *Art and Geometry* (Dover, 1964).

Kline, M., *Mathematics in Western Culture* (Oxford, 1953), Chaps. 10 and 11.

———, "Projective Geometry" (*Scientific American*, January 1955).

Yaglom, I. M., *Geometric Transformations* (L. W. Singer Co., Random House, 1964).

Progressions

What is an infinite series? How can we find the sum of an infinite series? What are Fibonacci numbers? What are some applications of progressions in nature, industry, science?

Exhibit suggestions. Illustrate progressions with models, such as a super ball bouncing in a plastic tube, pyramid box display, branches on a plant, an ancestral tree, a ball rolling down an incline. Unusual problems, such as doubling wages or compound interest.

References:

Bakst, A., *Mathematics, Its Magic and Mastery* (Van Nostrand Reinhold, 1952), Chap. 17.

Kokomoor, F. W., *Mathematics and Human Affairs* (Prentice-Hall, 1942), Chap. 12.

Kramer, E. E., *Main Stream of Mathematics* (Oxford, 1951), Chap. 9.

Valens, E. G., *The Number of Things: Pythagoras, Geometry and Humming Strings* (Dutton, 1964).

Nomographs

How are nomographs made? What kind of problems can be solved with nomographs? How are nomographs related to computing devices?

Exhibit suggestions. Examples of a variety of nomographs including the simple one for the addition of directed numbers. Charts showing how nomographs are constructed and used.

References:

Adams, D. P., "The Preparation and Use of Nomographic Charts in High School Mathematics" in *Multi-Sensory Aids for Teachers of Mathematics*, Yearbook 18 (NCTM, 1945), pp. 164–181.

Meyer, J. S., *Fun with Mathematics* (Dover, 1952).

———, *More Fun with Mathematics* (Fawcett, 1963).

Vectors and Matrices

What are vectors? How do we compute with vectors? How are vectors related to the solution of equations?

Exhibit suggestions. Charts and devices to demonstrate applications and solutions.

References:

Kemeny, J. G., J. L. Snell, and G. L. Thompson, *Introduction to Finite Mathematics* (Prentice-Hall, 1957), Chap. 5.

Matthews, G., *Matrices I and II* (Houghton Mifflin, 1964).

Norton, M. S., *Basic Concept of Vectors* (Webster Publishing Co., 1963).

Sawyer, W. W., *Prelude to Mathematics* (Penguin, 1955), Chap. 8.

Schuster, S., *Elementary Vector Geometry* (Wiley, 1962).

Applications of Mathematics

Space Travel and Ballistics

What is the trajectory of a bullet? How is the orbit of a satellite determined? How can man navigate in space where directions are no longer east-north-south-west? What formulas, equations, fields of mathematics are used in planning space travel?

Exhibit suggestions. Models, pictures, charts of space missiles, with analysis of the trajectories or orbits. Formulas, graphs, computations involved in space travel. Automatic pilots in space. Problems of gravitation, pressure, friction, radio communication.

References:

Ahrendt, M. H., *The Mathematics of Space* (Holt, Rinehart & Winston, 1960).

Bakst, A., *Mathematics, Its Magic and Mastery* (Van Nostrand Reinhold, 1952), Chap. 36.

Kline, M., *Mathematics, A Cultural Approach* (Addison-Wesley, 1962).

Newman, J. R., *World of Mathematics* (Simon & Schuster, 1956), vol. I, part 4.

Art and Mathematics

Perspective, symmetry, balance are all mathematical ideas connected with painting. Reflections, rotations, constructions, similarity, and proportionality are also involved in painting.

References:

Kline, M., *Mathematics, A Cultural Approach* (Addison-Wesley, 1962).

———, *Mathematics in Western Culture* (Oxford, 1953), Chaps. 9 and 11.

Newman, J. R., *World of Mathematics* (Simon & Schuster, 1956), "Mathematics of Aesthetics," pp. 2182–2197; "Mathematics as an Art," pp. 2012–2023.

Valens, E. G., *The Number of Things: Pythagoras, Geometry and Humming Strings* (Dutton, 1964).

Ethics, Philosophy, Religion, and Mathematics

The logic of mathematics has frequently been the basis for philosophical thought.

References:

Kline, M., *Mathematics in Western Culture* (Oxford, 1953).

Lieber, L., *Human Values and Science, Art and Mathematics* (W. W. Norton, 1961).

Newman, J. R., *World of Mathematics* (Simon & Schuster, 1956), "A Mathematical Approach to Ethics," pp. 2198–2290; "Mathematics and Metaphysicians," pp. 1576–1590; "Meaning of Numbers," pp. 2312–2347; "The Locus of Mathematical Reality: An Anthropological Footnote," pp. 2348–2365.

Schaaf, W. L., "Science, Mathematics, and Religion" (*Mathematics Teacher*, January 1954).

Literature with Mathematical Overtones

Mathematics is frequently the basis for essays, fiction, poems, and plays. The proof of guilt in a trial may follow a logical pattern. The intrigue of a mystery may be analyzed like a mathematical problem. The critical aspect of mathematics may lend beauty to a poem.

References:

Fadiman, C., *Fantasia Mathematica* (Simon & Schuster, 1958).

Gardner, M. (ed.), *The Annotated Alice* (Clarkson N. Potter, 1960).

Newman, J. R., *World of Mathematics* (Simon & Schuster, 1956), "Cycloid Pudding," pp. 2214–2220; "Young Archimedes," pp. 2221–2249; "Geometry in the South Pacific," pp. 2250–2260; "Inflexible Logic," pp. 2261–2267; "The Law," pp. 2268–2273; "Mathematics for Golfers," pp. 2456–2459; "Common Sense and the Universe," pp. 2460–2470.

Plotz, H., *Imagination's Other Place* (Thomas Y. Crowell, 1955).

Music and Mathematics

Even Pythagoras set up ratios for a musical scale. Someone has said music is rhythmic counting. Music's symbolism and pleasing qualities have many other relationships to mathematics. A composer must learn the mathematics of music.

References:

Kline, M., *Mathematics in Western Culture* (Oxford, 1953), Chap. 19.

Newman, J. R., *World of Mathematics* (Simon & Schuster, 1956), "Mathematics of Music," pp. 2274–2311.

Ridout, T. C., "Sebastian and the 'Wolf' " (*Mathematics Teacher*, February 1955).

Science and Mathematics

Astronomers and physicists have always been dependent on mathematics for leadership in discoveries. A dramatic illustration of this is Einstein's mathematical discovery in 1905 of the formula for releasing energy by nuclear fission. Forty years later, August 1945, the terrible atomic bomb over Hiroshima announced the application of his formula. The discovery of Pluto, after its existence was proclaimed by mathematics, is another illustration of the use of mathematics in prediction. More recently, the law of disorder and the tools of statistics have aided research in many fields of science.

References:

Gamow, G., *One, Two, Three, Infinity* (Viking, 1947).

Hooke, R., and D. Shafer, *Math and Aftermath* (Walker & Company, 1964).

Kline, M., *Mathematics, A Cultural Approach* (Addison-Wesley, 1962).

———, *Mathematics in Western Culture* (Oxford, 1953), Chaps. 13 and 14.

———, *Mathematics and the Physical World* (Thomas Y. Crowell, 1959).

Newman, J. R., *World of Mathematics* (Simon & Schuster, 1956), vol. I, pp. 546–569; vol. II, Part 5.

Saaty, T. L., and F. J. Weyl, *The Spirit and Uses of the Mathematical Sciences* (McGraw-Hill, 1969).

Map Projections and Cartography

What are ways of locating points on a plane, a sphere, in space? How can a spherical surface be projected to a plane surface? How can the distortions due to projection be reduced?

Exhibit suggestions. Spherical projection device to project maps on plastic cylinder, cone, plane. Map samples. Analysis of projections and distortions.

References:

Greenhood, D., *Mapping* (University of Chicago Press, 1964).

Schaaf, W. L., "Map Projections and Cartography" (*Mathematics Teacher*, October 1953).

The Role of Mathematics in Western Civilization

What are significant events in the development of mathematics? Who are some of the famous persons contributing to mathematical knowledge? What are some of the amusing anecdotes from the lives of famous mathematicians? How has mathematics been involved in science, politics, military campaigns, architecture, transportation, philosophy, art, music, literature?

References:

Bell, E. T., *Men of Mathematics* (Simon & Schuster, 1937).

Bergamine, D., *Mathematics* (Time Inc., 1963).

Eves, H. W., *In Mathematical Circles*, Vols. 1 and 2 (Boston: Prindle, Weber & Schmidt, 1969).

Hogben, L., *Wonderful World of Mathematics* (Random House, 1955).

Kline, M., *Mathematics in Western Culture* (Oxford, 1953).

Kramer, E. E., *Main Stream of Mathematics* (Oxford, 1951).

Newman, J. R., *World of Mathematics* (Simon & Schuster, 1956), vol. I.

Ogilvy, C. S., *Tomorrow's Math* (Oxford, 1962).

Trumball, H., *The Great Mathematicians* (New York University Press, 1961).

Index